CHEMISTRY OF SUPERCONDUCTOR MATERIALS

CHEMISTRY OF SUPERCONDUCTOR MATERIALS

Preparation, Chemistry, Characterization and Theory

Edited by

Terrell A. Vanderah

Naval Weapons Center
China Lake, California

NOYES PUBLICATIONS
Park Ridge, New Jersey, U.S.A.

Copyright © 1992 by Noyes Publications
No part of this book may be reproduced or utilized in any form or by any means, electronic or mechanical, including photocopying, recording or by any information storage and retrieval system, without permission in writing from the Publisher.
Library of Congress Catalog Card Number: 90-27624
ISBN: 0-8155-1279-1
Printed in the United States

Published in the United States of America by
Noyes Publications
Mill Road, Park Ridge, New Jersey 07656

10 9 8 7 6 5 4 3 2 1

Library of Congress Cataloging-in-Publication Data

Chemistry of superconductor materials : preparation, chemistry, characterization and theory / edited by Terrell A. Vanderah.
       p.   cm.
   Includes bibliographical references and index.
   ISBN 0-8155-1279-1 :
    1. Superconductors--Chemistry. I. Vanderah, Terrell A.
QC611.97.C54C48   1991
537.6'23--dc20   90-27624
                                        CIP

For Jumbo Wells

MATERIALS SCIENCE AND PROCESS TECHNOLOGY SERIES

Editors

Rointan F. Bunshah, University of California, Los Angeles *(Series Editor)*
Gary E. McGuire, Microelectronics Center of North Carolina *(Series Editor)*
Stephen M. Rossnagel, IBM Thomas J. Watson Research Center *(Consulting Editor)*

Electronic Materials and Process Technology

DEPOSITION TECHNOLOGIES FOR FILMS AND COATINGS: by Rointan F. Bunshah et al

CHEMICAL VAPOR DEPOSITION FOR MICROELECTRONICS: by Arthur Sherman

SEMICONDUCTOR MATERIALS AND PROCESS TECHNOLOGY HANDBOOK: edited by Gary E. McGuire

HYBRID MICROCIRCUIT TECHNOLOGY HANDBOOK: by James J. Licari and Leonard R. Enlow

HANDBOOK OF THIN FILM DEPOSITION PROCESSES AND TECHNIQUES: edited by Klaus K. Schuegraf

IONIZED-CLUSTER BEAM DEPOSITION AND EPITAXY: by Toshinori Takagi

DIFFUSION PHENOMENA IN THIN FILMS AND MICROELECTRONIC MATERIALS: edited by Devendra Gupta and Paul S. Ho

HANDBOOK OF CONTAMINATION CONTROL IN MICROELECTRONICS: edited by Donald L. Tolliver

HANDBOOK OF ION BEAM PROCESSING TECHNOLOGY: edited by Jerome J. Cuomo, Stephen M. Rossnagel, and Harold R. Kaufman

CHARACTERIZATION OF SEMICONDUCTOR MATERIALS—Volume 1: edited by Gary E. McGuire

HANDBOOK OF PLASMA PROCESSING TECHNOLOGY: edited by Stephen M. Rossnagel, Jerome J. Cuomo, and William D. Westwood

HANDBOOK OF SEMICONDUCTOR SILICON TECHNOLOGY: edited by William C. O'Mara, Robert B. Herring, and Lee P. Hunt

HANDBOOK OF POLYMER COATINGS FOR ELECTRONICS: by James J. Licari and Laura A. Hughes

HANDBOOK OF SPUTTER DEPOSITION TECHNOLOGY: by Kiyotaka Wasa and Shigeru Hayakawa

HANDBOOK OF VLSI MICROLITHOGRAPHY: edited by William B. Glendinning and John N. Helbert

CHEMISTRY OF SUPERCONDUCTOR MATERIALS: edited by Terrell A. Vanderah

CHEMICAL VAPOR DEPOSITION OF TUNGSTEN AND TUNGSTEN SILICIDES: by John E.J. Schmitz

(continued)

Ceramic and Other Materials—Processing and Technology

SOL-GEL TECHNOLOGY FOR THIN FILMS, FIBERS, PREFORMS, ELECTRONICS AND SPECIALTY SHAPES: edited by Lisa C. Klein

FIBER REINFORCED CERAMIC COMPOSITES: by K.S. Mazdiyasni

ADVANCED CERAMIC PROCESSING AND TECHNOLOGY—Volume 1: edited by Jon G.P. Binner

FRICTION AND WEAR TRANSITIONS OF MATERIALS: by Peter J. Blau

SHOCK WAVES FOR INDUSTRIAL APPLICATIONS: edited by Lawrence E. Murr

SPECIAL MELTING AND PROCESSING TECHNOLOGIES: edited by G.K. Bhat

CORROSION OF GLASS, CERAMICS AND CERAMIC SUPERCONDUCTORS: edited by David E. Clark and Bruce K. Zoitos

Related Titles

ADHESIVES TECHNOLOGY HANDBOOK: by Arthur H. Landrock

HANDBOOK OF THERMOSET PLASTICS: edited by Sidney H. Goodman

SURFACE PREPARATION TECHNIQUES FOR ADHESIVE BONDING: by Raymond F. Wegman

FORMULATING PLASTICS AND ELASTOMERS BY COMPUTER: by Ralph D. Hermansen

Foreword

The word "Chemistry" in the title of this book just radiates a host of associations, questions, and relations. What role is there for the science of molecules in the study of a phenomenon that is basically physical? Why were these remarkable compounds not made before? Do we need to understand a physical property to make advances with it?

You cannot investigate the properties of a molecule before making it. Oh, some theoretical colleagues of mine would argue otherwise, and they do have a case for some small molecules in the interstellar medium, and a few other isolated successes as well. But for molecules of reasonable size one needs the beast in hand before one can tame it. And who are the experienced designers of molecules? Chemists, of course. Not all of them, mind you. Physical chemists have a lot of trouble making molecules, and physicists—well, they're just not supposed to be good at this archetypical chemical enterprise, the transformation of recalcitrant matter from one bonding arrangement to another.

So here is the first surprise of the new developments of the eighties in superconductivity. Complex materials, compounds of as many as seven elements, were (and are) made by people who weren't supposed to be good at making them, physicists in particular. To some chemists this came as a shock. And to say that it was easy, that this solid state synthesis, the "shake-and-bake" methodology, is trivial, only allows the question to form in our minds, "If so, why did chemists not discover them earlier?"

Perhaps they did, or could have, come close, except that they didn't possess the capability to make the crucial measurements indicating superconductivity. You might think "never again," that given what we have learned of the high T_c materials, no chemist could possibly avoid searching for superconductivity in newly synthesized conducting materials. But that is not so. About 90% of the interesting materials that I, as a theoretician, have picked up from the literature of the last three years, remain untested. Only their synthesis and structure are reported, perhaps the briefest indication of electric and magnetic properties. Obviously the community has been sensitized, but lags behind in exploring the richness it itself wrought. Either the instruments are not available, or people do not have friends cooperative enough to do the measurements, compartmentalized as our specialized disciplines often are. Or they, the synthesizers, remain content in the paradigmatic exercise of their molecular trade, unwilling to be bothered by Brillouin zones or the art of attaching leads. And, just to set the balance right, many of their more progressive friends, the fortunate ones to whom those measurements come easy, intellectually and instrumentally, would rather spend their time making a tiny stoichiometric wrinkle of a perturbation than look at a phase that bears no resemblance to known successes.

Did we really understand superconductivity before 1985? And, does one need to understand the physics and chemistry of the phenomenon today, before making new, better materials? I think the answer to *both* these questions is "no and yes." And the questions and answers addressed in this new chemistry of superconductivity, in this book, are revealing, probing what we mean by "understanding."

The scientific community at large has bought into reductionism as the ideology of understanding. A phenomenon in chemistry is said to be understood when it is reduced to the underlying physics, when we know the physical mechanism(s) contributing to it, how they vary with macroscopic or microscopic perturbations. In that sense the BCS theory *did* provide a detailed understanding of the physics of superconductivity. But in another sense we should have known all along that we did not really understand the phenomenon. For we could never predict whether a given material would be superconducting or not, not to speak of its T_c. "Oh," we would say–"if only we could calculate the electron-phonon coupling exactly," and some day, in that millennium of the clever theoretical chemist coupled to whatever will come beyond the supercomputer, we would indeed be able to do so. But until then, well, we had to make do with the intuition of people such as the late Bernd Mattias.

I think we didn't understand superconductivity, in the sense of being able to use the physics to build a molecule with predictable properties. Our understanding was deconstructive or analytical rather than constructive or synthetic. I think there is a difference between these modes of understanding.

Do we need to understand superconductivity to make better superconductors? Well, I think we do need to have (a) the systematic experimental experience requisite for intuition to develop; (b) a set of theoretical frameworks, none necessarily entirely logical or consistent, to provide us with the psychological incentive to do the next experiment; (c) a free exchange of results; (d) people wishing to prove others wrong as well as themselves right; (e) the material resources to investigate what needs investigating. Together these criteria make up the working *system* of science, not the catechism of how science should work. Note that understanding, in the reductionist sense, enters only in (b), and even there not in a controlling way. The frameworks of understanding that will move us forward can be simplistic chemical ones, just as much as fancy couplings in some weirdly truncated Hamiltonian. Let's take a look, ten years from now, at what principles will have guided the discoverers, chemists and physicists, of the materials that may make this book obsolete.

Cornell University
Ithaca, New York
August, 1990

Roald Hoffmann

Preface

The post-1986 explosion in research on complex cuprates found to be superconducting above the boiling point of liquid nitrogen has sharply focussed our growing need to attack scientific frontiers using multi- and interdisciplinary approaches. Nearly four years of intensive research have produced several new families of compounds with T_c values as high as 125 K, yet both physicists and chemists remain largely baffled by these amazing oxides. Our lack of theoretical understanding of high-temperature superconductivity is underscored by a continuing and profound inability to predict its appearance; our lack of chemical understanding, by the notorious difficulties encountered in the preparation of samples with reproducible chemical compositions, structures, and superconductive properties. This inherently difficult situation, compounded by a breakneck pace of research unleashed by the historical importance of Bednorz and Mueller's breakthrough, quite predictably resulted in a veritable flood of detailed studies on the physical properties of impure, ill-characterized materials. Not since the semiconductor revolution has there existed such a well-defined need for chemists and physicists to collaborate in the chemical tailoring and fundamental understanding of electronically important solids—clearly, the tantalizing payoffs loom large for the advancement of technology as well as fundamental knowledge.

Concomitant with a more reasonable pace of research has come a general acceptance that progress in our understanding and control of high-temperature superconductivity hinges on systematic and meaningful investigations of the intrinsic properties of high-quality, well-characterized samples. However, the chemical-structural nature of these unusual compounds; e.g., thermodynamic instability, nonstoichiometry, and, for the cuprates, two-dimensional structures notoriously prone to defects and intergrowths, has presented the synthesis with a formidable challenge to develop easily reproducible protocols that yield pure materials with optimal superconducting properties. It was quickly learned that subtle differences in reaction conditions and sample history could, and often did, drastically affect superconducting properties. Yet, the kinetic factors involved in solid state reactions are not well understood, measured, or controlled. Furthermore, the relationship of defects and other "micro-scale" variations in chemistry and structure to the superconducting properties is not yet clear. A striking example is the peculiar chemistry of the "n-type" superconductor $Nd_{1.85}Ce_{0.15}CuO_4$: neither polycrystalline samples nor single crystals (usually) superconduct after initial formation of the structure in air. Upon a high-temperature anneal in an inert gas, however, superconductivity above 20 K can (frequently) be observed. The chemical and structural changes induced by the inert-gas anneal are so subtle that, despite intense research since the discovery of the compound almost two years ago, a clear picture of the changes that we can associate with the "turning on" of superconductivity still eludes us.

The purpose of this book is to address this difficult starting point for every experimentalist—the sample. In this regard, we have endeavored to produce a useful reference text for those interested in the preparation, crystal chemistry, structural and chemical characterization, and chemistry-structure-property relations of these fascinating materials. This book has in large part been written by members of that subspecies of inorganic chemistry called solid state chemistry, a specialization that evolved during the heyday of the semiconductor device, and which now seems uniquely suited for the challenging chemistry of high-T_c compounds. Hence, the approach taken to the material is predominantly that of the solid state chemist. We are grateful, however, that two of our physicist colleagues consented to keep company with us and contribute chapters in their own specialties concerning the proper measurement and interpretation of transport and magnetic properties.

The book is extensively referenced and will hopefully serve the experimentalist as a day-to-day source of information (e.g., synthetic procedures, crystal structures, and tables of X-ray diffraction data) as well as a review of the literature into 1990. The introductory chapter describes the detailed chemical history of superconductivity, beginning with the first observation of the phenomenon in the element Hg in 1911. Thereafter, the chapters are partitioned into three sections emphasizing synthesis and crystal chemistry, sample characterization, and chemistry-structure-property relations. The section on Structural and Preparative Chemistry contains ten chapters, three of which, using different approaches, describe the detailed crystal chemistry of the various cuprate systems. Three chapters discuss synthesis exclusively and include a basic review of solid state synthetic methods as well as detailed procedures for the preparation of single-crystal and polycrystalline samples of the various superconducting systems. Four other chapters in this section include both synthetic and structural discussions of $YBa_2Cu_3O_7$-related compounds, the non-cuprate Ba-K-Bi-O and Ba-Pb-Bi-O systems, and the "n-type" superconductors derived from Nd_2CuO_4. The section entitled Sample Characterization contains six chapters. A basic review of phase identification by X-ray diffraction is followed by a compilation of single-crystal structural data and calculated powder patterns for nearly all of the superconducting cuprates. Two chapters cover structural and chemical characterization by electron microscopy, a technique particularly important for these typically defective, nonstoichiometric compounds. The remaining three chapters cover wet chemical methods and the measurement of transport and magnetic properties. In the last section of the book, two chapters discuss chemistry-structure-property relationships from the point of view of the chemist. Finally, we have included two appendices that will hopefully aid the reader in finding synthetic procedures for desired phases as well as review articles and texts in superconductivity and solid state chemistry.

I am most grateful to the contributors of this book, every one of whom took the time to write a fine chapter in a timely, enthusiastic fashion. As editor, it has been my privilege to work with internationally recognized leaders in the field of solid state chemistry and superconductivity. Above all, I thank the chapter-writers for all that I have learned from them during the detailed reading of each manuscript. The tireless efforts of George Narita were responsible for both the genesis and realization of this project, and I will always cherish the unwavering support and encouragement given to me by Charlotte Lowe-Ma and David Vanderah.

China Lake, California
October 9, 1990

Terrell A. Vanderah

Contributors

Brian G. Bagley
Bellcore
Red Bank, New Jersey

Phillipe Barboux
Universite Pierre et
  Marie Curie
Paris, France

Jeremy K. Burdett
The James Franck Institute
The University of Chicago
Chicago, Illinois

Robert J. Cava
AT&T Bell Laboratories
Murray Hill, New Jersey

Bertrand L. Chamberland
University of Connecticut
Storrs, Connecticut

Anthony K. Cheetham
University of Oxford
Oxford, England

Ann M. Chippindale
University of Oxford
Oxford, England

Pratibha L. Gai
E.I. du Pont de Nemours
  and Co., Inc.
Wilmington, Delaware

Daniel Groult
Centre Des Materiaux
  Supraconducteurs
ISMRA
Caen Cedex, France

Daniel C. Harris
Naval Weapons Center
China Lake, California

Maryvonne Hervieu
Centre Des Materiaux
  Supraconducteurs
ISMRA
Caen Cedex, France

Anthony C.W.P. James
AT&T Bell Laboratories
Murray Hill, New Jersey

Donald H. Liebenberg
Office of Naval Research
Arlington, Virginia

Claude Michel
Centre Des Materiaux
  Supraconducteurs
ISMRA
Caen Cedex, France

Donald W. Murphy
AT&T Bell Laboratories
Murray Hill, New Jersey

Michael L. Norton
University of Georgia
Athens, Georgia

Bernard Raveau
Centre Des Materiaux
  Supraconducteurs
ISMRA
Caen Cedex, France

Anthony Santoro
National Institute of
 Standards and Technology
Gaithersburg, Maryland

Lynn F. Schneemeyer
AT&T Bell Laboratories
Murray Hill, New Jersey

Arthur W. Sleight
Oregon State University
Corvallis, Oregon

Hugo Steinfink
The University of Texas
 at Austin
Austin, Texas

Stephen A. Sunshine
AT&T Bell Laboratories
Murray Hill, New Jersey

J. Steven Swinnea
The University of Texas
 at Austin
Austin, Texas

Jean Marie Tarascon
Bellcore
Red Bank, New Jersey

Charles C. Torardi
E.I. du Pont de Nemours
 and Co., Inc.
Wilmington, Delaware

Terrell A. Vanderah
Naval Weapons Center
China Lake, California

Eugene L. Venturini
Sandia National Laboratories
Albuquerque, New Mexico

NOTICE

To the best of the Publisher's knowledge the information contained in this publication is accurate; however, the Publisher assumes no responsibility nor liability for errors or any consequences arising from the use of the information contained herein. Final determination of the suitability of any information, procedure, or product for use contemplated by any user, and the manner of that use, is the sole responsibility of the user.

The book is intended for informational purposes only. The reader is warned that caution must always be exercised when dealing with chemicals, products, or procedures which might be considered hazardous. (Particular attention should be given to the **caution notes in chapters 6 and 16.**) Expert advice should be obtained at all times when implementation is being considered.

Mention of trade names or commercial products does not constitute endorsement or recommendation for use by the Publisher.

Contents

PART I
INTRODUCTION

1. **HISTORICAL INTRODUCTION AND CRYSTAL CHEMISTRY OF OXIDE SUPERCONDUCTORS** 2
   Bertrand L. Chamberland
   1. **Introduction** 2
      1.1 Discovery of Superconductivity 4
      1.2 Superconductivity–A Brief Survey 4
      1.3 Search Within the Chemical Elements–
          Pure Metals and the Elements 10
      1.4 An Overview of the Superconducting Binary
          Alloy Systems 11
      1.5 Ventures Into Ceramic Materials–Binary
          Borides, Carbides, and Nitrides 15
      1.6 Ventures Into Ceramic Oxides–Simple Binary
          and Ternary Systems 17
      1.7 Major Milestones in Oxide Superconductivity
          Research 21
      1.8 Ternary Chemical Compounds–Complex Borides
          and Sulfides 23
      1.9 Non-Transition Metal Systems–$(SN)_x$ and
          Others 25
      1.10 Organic Superconductors 28
   2. **Studies on Superconducting Oxides Prior to 1985** 30
      2.1 Studies of Superconducting Oxides with the
          Sodium Chloride Structure 30

xviii Contents

    2.2 Studies of Superconducting Oxides with the Perovskite-Type Structure 34
    2.3 Studies of Superconducting Oxides with the Spinel Structure 49
    2.4 Post-1985 Entry of Copper Oxide Superconductors 52
3. **Structural Features and Chemical Principles in Copper Oxides** 52
    3.1 The Fascinating Chemistry of Binary and Ternary Copper Oxides 52
    3.2 Copper to Oxygen Bond Distances–Ionic Radii .. 55
4. **Physical Property Determination on Ternary Copper Oxides–Studies on Copper Oxide Systems Prior to 1985** 61
    4.1 Studies on La_2CuO_4 and Its Derivatives 61
    4.2 Startling Discovery by Müller and Bednorz 70
    4.3 Corroboration of the Discovery and Further Developments 76
    4.4 Major Copper Oxide Superconductors Presently Being Investigated 84
5. **Chemical Substitutions–Crystal Chemistry** 84
    5.1 Chemical Substitutions in the La_2CuO_4 Structure 84
    5.2 Chemical Substitutions in the Perovskite Structure 84
6. **References** 93

PART II
STRUCTURAL AND PREPARATIVE CHEMISTRY

2. **THE COMPLEX CHEMISTRY OF SUPERCONDUCTIVE LAYERED CUPRATES** 106
    Bernard Raveau, Claude Michel, Maryvonne Hervieu, Daniel Groult
    1. **The Structural Principles** 106
    2. **Oxygen Non-Stoichiometry and Methods of Synthesis** 114
    3. **Extended Defects** 124
        3.1 Defects in $YBa_2Cu_3O_7$ 124
        3.2 Intergrowth Defects in Thallium Cuprates 129
    4. **Incommensurate Structures and Lone Pair Cations** .. 133
    5. **Concluding Remarks** 139
    6. **References** 141

Contents xix

3. **DEFECTIVE STRUCTURES OF $Ba_2YCu_3O_x$ AND $Ba_2YCu_{3-y}M_yO_z$ (M = Fe, Co, Al, Ga, ...)** 146
   Anthony Santoro
   1. Introduction 146
   2. Discussion of the Structure of $Ba_2YCu_3O_{7.0}$ 147
      - 2.1 Structural Changes as a Function of Oxygen Stoichiometry 150
      - 2.2 Twinning, Twin Boundaries and Model of the Structure of $Ba_2YCu_3O_{7.0}$ 154
      - 2.3 Oxygen Vacancy Ordering in $Ba_2YCu_3O_x$ 161
      - 2.4 Mechanisms of Oxygen Elimination From the Structure of $Ba_2YCu_3O_x$ 169
      - 2.5 Metal Substitutions 174
   3. References 185

4. **CRYSTAL CHEMISTRY OF SUPERCONDUCTORS AND RELATED COMPOUNDS** 190
   Anthony Santoro
   1. Introduction 190
   2. Description of Layered Structures 191
      - 2.1 Structural Types of Superconductors 200
      - 2.2 Compounds with the Perovskite Structure 201
      - 2.3 Compounds with Crystallographic Shear 205
      - 2.4 Compounds with the Rocksalt-Perovskite Structure 213
   3. References 220

5. **CRYSTAL GROWTH AND SOLID STATE SYNTHESIS OF OXIDE SUPERCONDUCTORS** 224
   Lynn F. Schneemeyer
   1. Overview 224
   2. Solid State Synthesis 224
      - 2.1 Oxide Synthesis 225
      - 2.2 Preparation of Samples Containing Volatile Constituents 227
      - 2.3 Ceramic Characterization 228
   3. Bulk Crystal Growth 229
      - 3.1 Introduction to the Growth of Single Crystals .. 229
      - 3.2 Flux Growth 232
      - 3.3 Growth of Superconducting Oxides 236
      - 3.4 Characterizing Single Crystals 247
   4. References 250

6. **PREPARATION OF BISMUTH- AND THALLIUM-**

BASED CUPRATE SUPERCONDUCTORS 257
Stephen A. Sunshine and Terrell A. Vanderah
   1. Introduction 257
   2. Synthetic Methods 263
   3. Synthesis of Bi-Based Cuprate Superconductors 265
      3.1 Single Cu-O Layer Phase [2201], T_c = 10 K ... 265
      3.2 Cu-O Double Layers [2212], T_c = 80 K 266
      3.3 Triple Cu-O Layers [2223], T_c = 110 K 270
   4. Synthesis of Thallium-Based Cuprate
      Superconductors 273
      4.1 Preparations in Air or Lidded Containers 273
      4.2 Preparations in Hermetically Sealed
          Containers 275
      4.3 Preparations Under Flowing Oxygen 279
   5. Conclusion 280
   6. References 281

7. **SYNTHESIS OF SUPERCONDUCTORS THROUGH SOLUTION TECHNIQUES** 287
Phillipe Barboux
   1. Introduction 287
      1.1 Ceramic Processing 287
      1.2 Interest in Solution Techniques 288
      1.3 The Different Solution Techniques 289
   2. Synthesis Procedures 289
      2.1 General Principles of Synthesis 290
      2.2 The Different Solution Processes 292
      2.3 The Different Precursors 293
      2.4 Thermal Processing 298
      2.5 Carbon-Free Precursors 302
   3. Conclusion 305
   4. References 306

8. **CATIONIC SUBSTITUTIONS IN THE HIGH T_c SUPERCONDUCTORS** 310
Jean Marie Tarascon and Brian G. Bagley
   1. Introduction 310
   2. Materials Synthesis 313
   3. Results 314
      3.1 $La_{2-x}Sr_xCuO_4$ 314
      3.2 $YBa_2Cu_3O_7$ 322
      3.3 $Bi_2Sr_2Ca_{n-1}Cu_nO_y$ 328
   4. Discussion 335
   5. References 342

9. **THE CHEMISTRY OF HIGH T_c IN THE BISMUTH BASED OXIDE SUPERCONDUCTORS $BaPb_{1-x}Bi_xO_3$ AND $Ba_{1-x}K_xBiO_3$** 347
   Michael L. Norton
   1. Introduction 347
   2. Theoretical Underpinnings of Bismuthate Research 348
   3. Crystallography 354
   4. Materials Preparation 355
      4.1 Bulk Growth 355
      4.2 Crystal Growth 356
      4.3 Thin Film Preparation 358
   5. Physical Properties 359
      5.1 Electrical Transport Properties 359
      5.2 Magnetic Properties 361
      5.3 Optical and Infrared Properties 361
      5.4 Specific Heat 362
      5.5 Pressure Effects 363
      5.6 Isotope Effects 364
   6. Theoretical Basis for Future Bismuthate Research 365
   7. Applications 367
   8. Conclusion 369
   9. References 369

10. **CRYSTAL CHEMISTRY OF SUPERCONDUCTING BISMUTH AND LEAD OXIDE BASED PEROVSKITES** 380
    Robert J. Cava
    1. Introduction 380
    2. $BaBiO_3$ 382
    3. $BaPbO_3$ 392
    4. $BaPb_{1-x}Bi_xO_3$ 396
    5. $BaPb_{1-x}Sb_xO_3$ 406
    6. $Ba_{1-x}K_xBiO_3$ 410
    7. Conclusion 419
    8. References 422

11. **STRUCTURE AND CHEMISTRY OF THE ELECTRON-DOPED SUPERCONDUCTORS** 427
    Anthony C.W.P. James and Donald W. Murphy
    1. Introduction 427
    2. The T'-Nd_2CuO_4 Structure 428
    3. Systematics of Electron Doping 431
    4. Chemical Synthesis and Analysis 437

5. Single Crystals and Thin Films 442
 6. Summary 444
 7. References 445

PART III
SAMPLE CHARACTERIZATION

12. X-RAY IDENTIFICATION AND CHARACTERIZATION OF COMPONENTS IN PHASE DIAGRAM STUDIES 450
J. Steven Swinnea and Hugo Steinfink
 1. The Gibbs Phase Rule 451
 2. Phase Diagrams 454
 3. Phase Diagram Studies 464
 4. X-Ray Diffraction 465
 5. Tying It All Together 476
 6. References 484

13. STRUCTURAL DETAILS OF THE HIGH T_c COPPER-BASED SUPERCONDUCTORS 485
Charles C. Torardi
 1. Introduction 485
 2. Structures of the Perovskite-Related $YBa_2Cu_3O_7$, $YBa_2Cu_4O_8$, and $Y_2Ba_4Cu_7O_{15}$ Superconductors 488
   2.1 123 Superconductor 488
   2.2 124 and 247 Superconductors 490
 3. Structures of the Perovskite/Rock Salt Superconductors 490
   3.1 Lanthanum-Containing Superconductors 490
   3.2 Bismuth-Containing Superconductors 491
   3.3 Thallium-Containing Superconductors 493
   3.4 Thallium-Lead Containing Superconductors ... 495
 4. Structures of Pb-Containing Copper-Based Superconductors 495
 5. Distortions in the Rock Salt Layers and Their Effect on Electronic Properties 496
 6. Correlations of T_c with In-Plane Cu-O Bond Length .. 500
 7. Tables of Crystallographic Information 501
 8. References 541

14. CHEMICAL CHARACTERIZATION OF OXIDE SUPERCONDUCTORS BY ANALYTICAL ELECTRON MICROSCOPY 545
Anthony K. Cheetham and Ann M. Chippindale

    1. Introduction 545
    2. Analytical Electron Microscopy: A Brief Survey 547
    3. Experimental Method 548
    4. Data Collection and Analysis of Standards 551
    5. Analysis of the "2212" Compound (36) 554
    6. Discussion 555
    7. References 558

15. ELECTRON MICROSCOPY OF HIGH TEMPERATURE
    SUPERCONDUCTING OXIDES 561
    Pratibha L. Gai
    1. Introduction 561
    2. Techniques of High Resolution Transmission
       Electron Microscopy (HREM), Transmission EM
       (TEM) Diffraction Contrast, and Analytical
       EM (AEM) 562
       2.1 HREM 563
       2.2 TEM Diffraction Contrast 564
       2.3 High Spatial Resolution Analytical EM 564
    3. Microstructural and Stoichiometric Variations 566
       3.1 Substitutional Effects in La-Based Super-
           conductors 566
       3.2 Y-Based Superconductors 570
       3.3 Substitution of Ca in Tetragonal $YBa_2Cu_3O_6$ ... 576
       3.4 Bi-Based Superconductors 578
    4. Tl-Based Superconductors 589
       4.1 Tl-Ba-Ca-Cu-O Superconductors 591
       4.2 (Tl,Pb)-Sr-Ca-Cu-O Superconductors:
           $(Tl,Pb)Sr_2CaCu_2O_7$ (n=2) and
           $(Tl,Pb)Sr_2Ca_2Cu_3O_9$ (n=3) 597
    5. References 604

16. OXIDATION STATE CHEMICAL ANALYSIS 609
    Daniel C. Harris
    1. Superconductors Exist in Variable Oxidation
       States 609
       ...And We Don't Know What Is Oxidized 610
    2. Analysis of Superconductor Oxidation State
       by Redox Titration 611
       2.1 Iodometric Titration Procedure 614
       2.2 Citrate-Complexed Copper Titration
           Procedure 616
    3. Reductive Thermogravimetric Analysis 616
    4. Oxygen Evolution in Acid 619

5. Electrochemical Investigation of Super-
   Conductor Oxidation State 621
6. Assessment of Analytical Procedures 624
7. References 624

17. TRANSPORT PHENOMENA IN HIGH TEMPERATURE SUPERCONDUCTORS 627
Donald H. Liebenberg
1. Introduction 627
2. Resistivity Measurement 627
   2.1 Survey of Results of Resistivity Measurements .. 632
   2.2 Theoretical Notes 637
3. Critical Current Density Measurements 639
   3.1 Measurement of Critical Current 639
   3.2 Results of Critical Current Measurements 645
4. Dissipation in the Intermediate State 652
5. Thermal Conductivity 656
6. Thermopower 657
7. Hall Effect 658
8. Tunneling Transport 660
   8.1 Josephson Effect 662
   8.2 Tunneling Results 663
9. References 667

18. STATIC MAGNETIC PROPERTIES OF HIGH-TEMPERATURE SUPERCONDUCTORS 675
Eugene L. Venturini
1. Introduction 675
2. Low-Field Measurements 677
   2.1 Normal State Response 677
   2.2 Diamagnetic Shielding by a Superconductor ... 681
   2.3 Magnetic Flux Exclusion and Expulsion 687
3. High-Field Measurements: Hysteresis Loops
   and Critical Current Density 691
4. Magnetization Relaxation or Giant Flux Creep 696
5. Problems with Porous and Weak-Linked Ceramics ... 700
6. Concluding Remarks 705
7. References 706

PART IV
STRUCTURE-PROPERTY CONSIDERATIONS

19. ELECTRONIC STRUCTURE AND VALENCY IN OXIDE SUPERCONDUCTORS 714

Arthur W. Sleight
   1. Introduction 714
   2. Mixed Valency and the Partially Filled Band 714
   3. Valent States vs Real Charges 718
   4. Stabilization of O^{-II} and High Oxidation States 720
   5. Polarizibility 721
   6. Defects and Inhomogeneities 723
   7. Stability 726
   8. T_c Correlations 729
   9. Mechanism for High T_c 731
   10. References 733

20. ELECTRON-ELECTRON INTERACTIONS AND THE ELECTRONIC STRUCTURE OF COPPER OXIDE-BASED SUPERCONDUCTORS 735
    Jeremy K. Burdett
    1. Introduction 735
    2. One- and Two-Electron Terms in the Energy 736
    3. Energy Bands of Solids 748
    4. The Peierls Distortion 753
    5. Antiferromagnetic Insulators 756
    6. Electronic Structure of Copper Oxide Superconductors 759
    7. The Orthorhombic-Tetragonal Transition in 2-1-4 766
    8. Superconductivity and Sudden Electron Transfer 770
    9. Some Conclusions 773
    10. References 774

APPENDIX A: GUIDE TO SYNTHETIC PROCEDURES 776

APPENDIX B: FURTHER READING IN SUPERCONDUCTIVITY AND SOLID STATE CHEMISTRY 784

FORMULA INDEX 790

SUBJECT INDEX 802

Part I
Introduction

1

Historical Introduction and Crystal Chemistry of Oxide Superconductors

Bertrand L. Chamberland

1.0 INTRODUCTION

The objectives of this chapter are to introduce the reader to the interesting and important phenomenon of superconductivity in various materials. We begin with the original discovery of superconductivity and proceed to the investigation of the chemical elements, then metallic alloys, and finally to ceramic and other inorganic materials which also exhibit this unusual phenomenon. After the brief overview of discovery and the observation of superconductivity in a number of substances, we will focus on research studies conducted on oxide materials. Several oxide systems were investigated during the early years, but due to experimental temperature limitations, very few of these materials were found to superconduct above 2 K. This led to frustration and disappointment for those scientists seeking superconductivity in the most promising candidates.

The research effort on oxides began in 1933 and our focus will be on the published results up to 1985, just prior to the major breakthrough in superconductivity which occurred in the fall of 1986 with the discovery of the Cu-O superconductors. The critical parameters for a superconductor are: its critical transition temperature (here, given on the Kelvin scale), its critical magnetic field (given in Tesla or Gauss units), and its critical current (given in units of Amperes). Our presentation is aimed principally on the single property of critical transition temperature (T_c), since it is controlled primarily by the chemical composition and the structural arrangement of atoms. These two factors are of primary concern to the chemist and material scientist. The transition temperature can be determined in a number of ways; by magnetic, electrical, or calorimetric experiments; and the temperature can be given as the onset of transition, the mid-point of the transition, or at the terminus of transformation (for example, when the resistance in the material can no longer be measured). The T_c data presented in this chapter were obtained by any of those methods mentioned above, and are often those temperatures reported in the literature. More recently, the T_c values are those obtained at the onset region since these represent the highest temperatures which can readily be reproduced in both the magnetic and electrical experiments. The original term, high T_c superconductivity, was used to designate a material with a superconducting transition temperature greater than 18 K, the highest value obtained for the intermetallic compound Nb_3Sn (T_c = 18.07 K) in the late 1950's. This temperature approaches the boiling point of liquid hydrogen (20 K), which could be used as the cryogen instead of liquid helium. Today the term high T_c has taken on a new meaning, where the onset temperature for superconductivity is greater than 30 K. The overall chronological development of oxide superconductors will be presented in a summary fashion. The research on non-oxide systems is presented in a highly abbreviated form, and only a sampling of different materials will be given to acquaint the reader with some specific examples of these materials and their transition temperatures, for comparison with the oxide compounds.

The chemistry of copper, especially in oxide compounds, is presented from a crysto-chemical viewpoint. The known binary and ternary compounds are reviewed and the important features of geometry, covalency, ionic radii, and bond length are discussed in great depth. This section concludes with a brief description of the Bednorz and Müller discovery, followed by Chu and co-workers' important discovery of superconductivity at 95 K in the so-called "1-2-3" copper-oxide system.

This chapter will exclude all detailed descriptions of physical properties and experimental results presented in several of the Physics journals. The theoretical aspects of superconductivity will also be omitted, as well as the practical application and engineering aspects of these new materials.

4   Chemistry of Superconductor Materials

1.1 Discovery of Superconductivity

At very low temperatures, the movement of atoms, ions, and molecules is greatly reduced. As the temperature of a substance is decreased, the states of matter can change from gas to liquid, and then to a solid or the condensed state. A technical challenge in the early 1900's was the liquefaction of all known room-temperature gases to their liquid state. Research at achieving very low temperatures led the Danish physicist Heike Kamerlingh Onnes in 1908 to first obtain liquid helium at 2.18 K. Liquid helium displays many remarkable properties, one of which was given the name "super-fluidity" by Kamerlingh Onnes. Superfluidity is a phenomenon by which the liquid exhibits a completely frictionless flow, which allows it to pass easily through small holes (less than 10^{-6} cm in diameter), and to flow up the walls of any container! The superfluid can rise up the inside walls, pass over the top edge, and escape by flowing down the outside walls of the container. The phenomenon of superconductivity was discovered (1) by Kamerlingh Onnes at the University of Leiden in 1911, three years after he had achieved the first liquefaction of helium. In the critical experiment, his students were measuring the (relative) resistance of solid mercury which was 125 $\mu\Omega$ at 4.5 K and, on cooling to lower temperatures, a sharp discontinuity in resistance occurred at ~4.3 K; below this temperature, the resistance in mercury was less than 3 $\mu\Omega$ (see Figure 1).

Superconductivity is the sudden and complete disappearance of electrical resistance in a substance when it is cooled below a certain temperature, called the critical transition temperature, T_c.

1.2 Superconductivity— A Brief Survey

The phenomenon of superconductivity, a new state of matter, has intrigued physicists, metallurgists, electrical engineers, and material scientists ever since its discovery. In addition to the total loss of electrical resistance to the passage of a direct current in these materials, there also occurs an unusual magnetic behavior: that tendency to expel a magnetic field from its interior below the T_c, this results in the formation of a perfect diamagnetic material. This effect allows superconductors to repel magnets and act as magnetic shields with high efficiency.

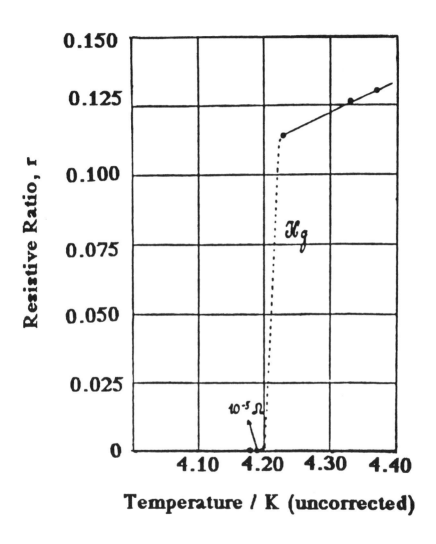

Figure 1: The resistive ratio of solid mercury *versus* absolute temperature (uncorrected scale) as actually reported by H. Kamerlingh Onnes (1). This observation marked the discovery of superconductivity.

This perfect diamagnetic behavior is called the Meissner effect, after the physicist who first observed (2) it in superconductors. The Meissner effect is responsible for the levitation phenomenon in superconductors (see Figure 2). Because certain superconductors can carry high electrical currents, coils with several turns of superconducting wire can generate very strong magnetic fields— some strong enough to levitate an entire train for high-speed travel on a smooth magnetic cushion.

These two properties, <u>zero resistance</u> and <u>perfect diamagnetism</u>, are critical parameters for superconductivity and these two effects are used as the criteria for establishing superconductivity in materials.[1] The superconductive state can be removed, not only by heating above T_c, but also by applying a strong magnetic field (above a certain threshold value), or by applying too high an electrical current, H_c or J_c, respectively.

The superconducting transition temperatures for selected chemical elements and certain metallic alloys are presented in Table 1. These data and those presented throughout this Chapter have been taken, for the most part, from the excellent compilation by B. W. Roberts (3).

In Table 2, a brief chronology of certain important discoveries and breakthroughs in the field of superconductivity is outlined.

Superconductivity has not only been beneficial to science and technology but also has been highly rewarding to its scientists. Thus far, Nobel Prizes in Physics have been awarded on four occasions to scientists working in this area. The first of these was for the discovery of superconductivity by Kamerlingh Onnes, awarded in 1913. In 1972 the prize went to John Bardeen, Leon Cooper, and Robert Schrieffer for the BCS theory. The following year (1973), the Prize was awarded to Brian Josephson, L. Esaki and I. Giaever for the

---

[1] Because of the recent claims of room-temperature superconductivity in materials by several research groups, two additional criteria are further required to certify that a material is superconducting. These are: long-term stability, and reproducibility in the preparation and physical property determination.

Figure 2: The Meissner Effect, or the levitation of a strong magnet by the internal diamagnetic field of a high T_c superconductor.

TABLE 1: Superconducting Transition Temperatures for Selected Elements and Alloys.

Element	T_c (K)	Alloy	T_c (K)
Tc	11.2	Nb_3Ge	23.2
Nb	9.46	Nb_3Sn	18.05
Pb	7.18	Nb_3Al	17.5
Ta	4.48	Mo_3Ir	8.8
Hg	4.15	NiBi	4.25
In	3.41	AuBe	2.64
Al	1.19	$PdSb_2$	1.25
Cd	0.56	TiCo	0.71

TABLE 2: Chronology of Important Discoveries in Superconductivity.

Year	Discovery	Discoverer(s)
1911	Superconductivity in Hg	H. Kamerlingh Onnes
1933	Magnetic flux exclusion effects	W. Meissner/R. Ochsenfeld
1934	"Two-fluid model", a thermodynamic relationship	Gorter and Casimir
1935	Phenomenological (London) equations	Fritz and Heinz London
1950	Isotope effect, $T_c \sim M^{-1/2}$	Maxwell/Reynolds, et al.
1950	Phenomenological theory of superconductivity.	V. Ginzburg and L. Landau
1953	Coherence length	A. B. Pippard
1953	Experimental evidence for energy gap	Goodman, et al.
1957	BCS theory	Bardeen/Cooper/Schrieffer
1959	Foreign magnetic impurity destroying electron pairing	H. Suhl and B. Matthias
1960	Tunnel effect	Giaever and Esaki
1961	"Hard" or "High-field" superconductors	Kunzler
1962	Josephson effect, supercurrent flow through a tunnel barrier	B. Josephson, et al.
1964	Prediction of high-T_c in organic compds.	W. A. Little
1973-4	Critical temperature of 23 K is reached in Nb_3Ge, surpassing the liq. H_2 barrier	J. Gavaler/Westinghouse, and also Bell Labs.
1986	Evidence of high-T_c in La-Ba-Cu-oxides.	Müller and Bednorz
1987	Observation of 95 K T_c in Y-Ba-Cu-O.	Chu and Wu

10   Chemistry of Superconductor Materials

tunnelling effect. Quite recently (in 1987) it was presented to K. Alex Müller and J. Georg Bednorz for their discovery of superconductivity in copper oxides.

In terms of the electronic age, which includes the invention of the radio, television, calculator, and computer, it has been claimed that the discovery of high T_c superconductors has resulted in a "third electronics revolution"; preceded by the transistor (1947), and the vacuum tube (1904). It now appears that we have shifted from "silicon valley" to a "copper-oxide valley" with the new discoveries in the field of high T_c superconductivity.

1.3   Search within the Chemical Elements— Pure Metals and the Elements

Following the discovery of superconductivity in mercury in 1911, the low-temperature physics community immediately began a systematic search for this unusual phenomenon in all the other metallic elements of the Periodic Table. A thorough study, first at ambient pressure and then at higher applied pressures, yielded only a handful of superconducting chemical elements. Additional research on the discovery of superconductivity in the metallic elements continued and now there are presently 26 superconducting elements at ambient pressure, and an additional 4 formed under high-pressure conditions. The ferromagnetic metals and our best metallic conductors (Cu, Ag, Pt, and Au) were not found to be superconducting to the lowest temperatures measured. Worthy of note are Pb and Nb, both readily available metals, with superconducting transition temperatures of 7.18 and 9.46 K, respectively, at ambient pressure (see Table 1). For the transition metals, a plot of transition temperature *versus* the total number of valence electrons shows two maxima. A large number of transition metals exhibit a maximum at about 5 electrons per atom and a secondary peak appears at about 7 electrons per atom.

Non-metals, such as silicon, can also become superconducting when pressure is applied. At 120-130 kbar pressure, silicon exhibits a T_c of 6.7 to 7.1 K. Sulfur has also recently been converted into the superconducting state at 200 kbar with a transition temperature of 5.7 K. In 1989, hydrogen was obtained in the condensed state and under 2.5 megabars pressure, it becomes opaque. This observation indicates that the element is possibly transforming into a metal. Several scientists have predicted that metallic hydrogen could become a superconductor under sufficiently high pressures. The transition temperature for superconductivity in metallic hydrogen has been theoretically predicted to be very high, possibly near room temperature.

1.4 An Overview of the Superconducting Binary Alloy Systems

Realizing that metallic behavior at room temperature was possibly necessary for a superconductor candidate, the search for superconductors shifted to the metallic-conducting alloys, especially those containing metals having the highest T_c's; i.e., Pb and Nb acting as one component of the solid solution. Metallurgists joined the searching party in this endeavor. During the period 1940-1970 many such alloys (intermetallics) were found to be superconductors, but by 1970 the critical temperature had reached a peak value of only 20 K. Figure 3 shows the progress in increasing the transition temperature with new materials as plotted against the year of discovery. The slope of the line corresponds to a 3 degree increase in transition temperature per decade of research. In this chapter we will present only an overview of the results obtained on alloy systems through 30 years of research by several groups and many scientists.

The results of this research also indicated a relationship between transition temperature and the average number of electrons for the intermetallics. Peaks in the graphs of these two parameters showed maxima at 4.7 and ~6.5 electrons per atom for many binary and ternary alloy systems (Figure 4). A structural relationship was also observed in these superconductors. Some good alloy candidates were found to possess certain crystalline structures, whereas other crystal types produced few, if any, interesting or important superconductors. The primary crystal structures which yielded the most promising superconductors were: Cr_3Si (or β-W), α-Mn, and $MgZn_2$-type (Laves) phases. The importance of structure and number of available conducting electrons became key factors in the search for superconducting alloys.

In Table 3, some data for a family of the most promising A-15 type materials are presented with their critical magnetic fields, and the derivative of critical field with respect to transition temperature (all about 2.5 in value). These data were reported by Hulm and

12 Chemistry of Superconductor Materials

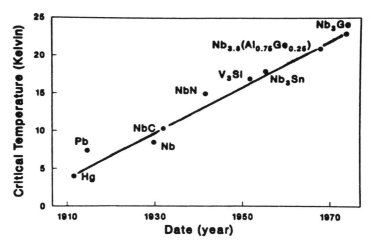

Figure 3: The increase in superconducting transition temperature (T_c / K) as a function of year, from the discovery of superconductivity in 1911 to 1973.

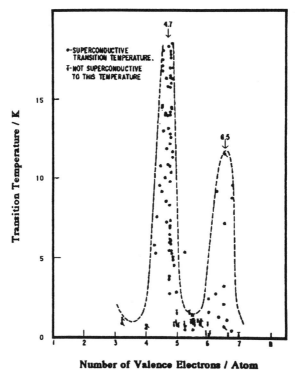

Figure 4: A plot of critical temperature (T_c/K) *versus* the number of valence electrons per atom in the intermetallic A-15 materials.

TABLE 3: High T_c Systems based on the A-15 (β-W) Structure.

Composition	T_c (K)	H_{c2} (Tesla)	$-(dH_{c2}/dT_c)$ (K^{-1})	Remarks
Nb_3Ge	23.2	37.1	2.4	Sputtered film
Nb_3Ge	21.7	-	-	Chemical vapor deposition
$Nb_3Ge_{0.9}Si_{0.1}$	20.3	-	-	Film
Nb_3Ga	20.3	34.1	2.4	Quenched sample
Nb_3Sn	18.3	29.6	2.4	
Nb_3Al	18.7	32.7	2.5	
V_3Si	17.0	34.0	2.9	
V_3Ga	14.8	34.9	3.4	

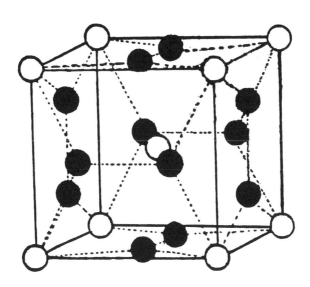

Figure 5: The β-W or A-15 (Cr_3Si) structure, as re-drawn from Reference 5. Dark circles = W (or Cr) atoms; white circles = O (or Si) atoms.

Matthias (4) for the series of practical materials with technological importance at the time.

An interesting sidelight to the A-15 story is the fact that one of its structural prototypes is the β-tungsten structure. This structural form of tungsten appears to be stabilized by small amounts of oxygen, and some have suggested that its chemical composition is actually W_3O, or possibly W_3O_{1-x}. Therefore, it is a structural prototype for this large and important family of intermetallic superconductors, some having the highest T_c values known. The A-15 binary alloys were developed during the sixties and seventies, and can be considered as derived from this "oxide structure" (Figure 5).

The structural relationship between the A-15 structure, or Cr_3Si, and that of W_3O will now be given. In the W_3O structure, the metal atoms (as a dumbbell of metal atoms on each cubic face) occupy the 6c sites of Space Group Pm3n (# 223), and the oxygen atoms are at the origin (the corners) and also at the center of the cubic unit cell. In this structure the metal atoms are tetrahedrally coordinated to the oxygen atoms with two additional near-metal neighbors. The oxygen atoms are XII-coordinated to the metals having a triangulated dodecahedral geometry. The metal atoms in this structure are compressed in chains parallel to the three cubic axes. The adoption of this structure may be caused by strong metal-metal interactions, leading to intermetallic distances which are not reconcilable with the packing of spherical metal atoms (5). See Section 1.6 for further insight into the oxides with the beta-tungsten structure investigated during this interval.

Several myths were introduced in the search for new superconductors. It was believed that the intermetallic candidates could only contain metallic elements which were themselves superconducting, and this would exclude those elements such as Cu, Ag, and Pt. Research, however, showed this to be false and now there are several alloy systems which contain the non-superconducting metals. Another misconception was that the new alloy systems could not contain any of the ferromagnetic elements. Once again, alloys such as: $CoZr_2$ (T_c = 5 to 7 K), Ti_2Co (T_c = 3.44 K), and $Fe_{0.5}Mn_{0.5}U_6$ (T_c = 2.8 K) were prepared and found to be superconducting. Also it was believed that ferromagnetism and superconductivity were mutually exclusive properties in a material. However, gadolinium-doped $InLa_{2.93-3.0}$ alloys show a superconducting transition temperature of ~9.0 K, and

Historical Introduction and Crystal Chemistry of Oxide Superconductors 15

at lower temperatures they become ferromagnetic with a Curie temperature (T_C) of 3 K. Gadolinium is a ferromagnetic element with an ordering temperature of 292 K. In later studies on ternary ceramic phases, most of these myths were completely abandoned when exceptions were found to all the "guidelines" generated during these intervening years. Other misconceptions, however, were about to surface as the search for new superconducting materials continued.

1.5 Ventures into Ceramic Materials— Binary Borides, Carbides, and Nitrides

The quest for higher transition temperatures in superconductors took a strange turn when ceramic materials, possessing good, room-temperature metallic conductivity, were investigated. A study of simple binary compounds such as: ZrN (T_c = 10.7 K), NbC (T_c = 11.1 K), MoC (T_c = 14.3 K), NbN (T_c = 16.0 K), $LaC_{1.3-1.6}$ (T_c = 11.1 K), and the mixed phases: $NbC_{0.35}N_{0.65}$ (T_c = 17.8 K), and $Y_{0.7}Th_{0.3}C_{1.55}$ (T_c = 17.0 K), generated much interest. Some selected binary carbides, nitrides, and borides are presented in Table 4 with their superconducting transition temperatures.

Here again certain trends were observed, and the most influential factor was the crystal structure which the superconducting material adopted. The most fruitful system was the NaCl-type structure (also referred to as the B1 structure by metallurgists). Many of the important superconductors in this ceramic class are based on this common structure, or one derived from it. Other crystal structures of importance for these ceramic materials include the Pu_2C_3 and MoB_2 (or $ThSi_2$) prototypes. A plot of transition temperature *versus* the number of valence electrons for binary and ternary carbides shows a broad maximum at ≈5 electrons per atom, with a T_c maximum at ~13 K.

The superconductivity in transition metal sulfides, selenides, and phosphides possessing the NaCl structure is presented in the excellent review (7).

As the cubic system was often found to be an important structural class for good superconductors, another myth was generated that suggested one should focus on compounds having a cubic-type crystalline structure, or a structure possessing high symmetry. This myth was also abandoned when lower symmetry systems were found

TABLE 4: Selected Binary Carbides, Nitrides, and Borides with Transition Temperature.

Carbide	T_c (K)	Nitride	T_c (K)	Boride	T_c (K)
MoC	14.3	MoN	12.0	Mo_2B	5.85
NbC	11.1	NbN	16.0	NbB	8.25
La_2C_3	11.0	VN	8.5	TaB	4.0
Nb_2C	9.1	Nb_2N	8.6	$NbB_{2.5}$	6.4
TaC	10.35	ZrN	10.7	ZrB_{12}	6.02
$ThC_{1.45}$	8.2	ThN	3.3	HfB	3.1
$YC_{1.45}$	11.5	TiN	5.49	Re_2B	4.6

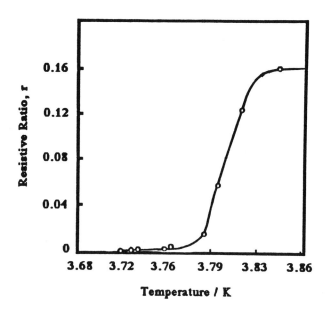

Figure 6: The original plot of relative resistance *versus* temperature (T / K) for "SnO", as published in Reference 6.

to generate interesting superconductors in much the same way as materials adopting a cubic crystal structure.

1.6 Ventures into Ceramic Oxides— Simple Binary and Ternary Systems

If carbides, nitrides, and borides could generate such intriguing T_c values, what promise did the ceramic, metallic-conducting oxides hold?

Since this chapter is aimed at presenting much of the information leading to the discovery of the new high T_c oxide superconductors, we shall present here an overview of the oxide systems which have been studied for superconducting properties prior to 1985. In the next section of this chapter, we will give a more detailed and descriptive narration of the work performed on oxide systems, presented in terms of the crystal classes which have yielded the most important oxide superconductors.

Meissner, Franz, and Westerhoff (6) were probably the first group of physicists to study oxides as superconductor candidates. In their 1933 publication, they presented results of resistivity measurements on NbO, SnO, Pb_2O, PbO_2, Tl_2O_3, MoO_2, a mixture of SnO + SnO_2, and several "tungsten bronzes". Their experiments were conducted from room temperature to 1.3 K, the lowest temperature they could then achieve by pulling a vacuum on liquid helium. They determined that, of these oxides, only NbO and possibly SnO were superconducting.

The ratio of the measured resistance normalized to the room-temperature resistance, $(r = R/R_o)$, was found to decrease steadily for NbO until it reached an exceedingly low value at 1.54 K. From these data, they concluded that NbO became superconducting near that temperature. Later authors disagreed, but the most recent results on NbO are in support of that conclusion.

In their "SnO sample" they observed an abrupt change in the resistive ratio as a function of temperature (see Figure 6).

Their data suggested a superconducting onset near 3.83 K, and zero resistance at 3.7 K. By increasing the applied current at the lowest temperature, however, they noticed that the resistance increased and that superconductivity in the sample could not be maintained. They concluded that bulk superconductivity was not achieved in "SnO", but they were at a loss at explaining their results. It should be noted that they prepared their "SnO" sample by the decomposition of $Sn(OH)_2$ in vacuum at 400°C. Most probably their "Sn_2O_3" sample, or a mixture of "SnO" and SnO_2, was also similarly contaminated with superconducting Sn metal. In our research studies on pure SnO, we have found that this compound spontaneously disproportionates at relatively low temperatures (near 400°C) to yield Sn (a known superconductor, with a T_c of 3.72 K) and inert SnO_2 according to the following chemical equation:

$$2\ SnO \xrightarrow{\text{heat} > 400°C} Sn + SnO_2$$

It is highly probable that the resistive data in Figure 6 simply represents the presence of superconducting Sn admixed with SnO_2. X-Ray data on residues of our SnO samples that were heated to 450°C in gettered argon indicate that Sn is a definite by-product of this thermal treatment. It is interesting to note that a highly useful Josephson device has been realized (8) by coupling the superconductivity behavior of Sn with the non-conducting oxide barrier of SnO. This Sn-SnO-Sn "sandwich", a Josephson device, shows a maximum current at zero magnetic field and an oscillating signal with applied (positive or negative) magnetic field. Meissner and co-workers did not observe superconductivity in any of the other oxide compounds they investigated. Their oxide list also included certain "tungsten bronzes", namely, Li_xWO_3, K_xWO_3, and Rb_xWO_3, which exhibited low resistive ratios but showed no superconductivity to 1.32 K.

Research studies on other binary and ternary oxide systems were quite limited during the next 20 years. In a 1952 comprehensive review, entitled "A Search for New Superconducting Compounds", B. T. Matthias and J. K. Hulm (9) stated:

> "Oxides and hydrides will be omitted from the present (review) since hardly any progress has been made (to date) in studying superconductivity in these compounds."

This bold statement by two leading research authorities possibly set back oxide superconducting research for several decades.

In another research summary (10), this same point was driven home when the authors further indicated that NbO, TiO, and VO, the three most promising superconducting oxide candidates, were not observed to superconduct down to 1.20 K. By 1965, after much research, scientists were greatly astonished when the following results were finally obtained: NbO, T_c = 1.38 K; and TiO, T_c = 1.28 K.

Such low transition temperatures! What future would there be for oxide superconductors?

Oxides Investigated But Not Found to Be Superconducting: As previously mentioned, several binary oxides were investigated at low temperatures (to appoximately 1.28 K), but never transformed into the superconducting state. These oxides can be listed into two broad categories; the metallic-conducting oxides, and the insulating class (which shall include the semiconductors). The first grouping of good metallic conductors that never became superconducting includes: MoO_2, PbO_2, VO, V_2O_3, Tl_2O_3, Ag_3O, Ti_2O_3, ReO_2, ReO_3, and WO_2. The insulating (narrow-band and broad-band semiconductors) oxide compositions that were not observed to superconduct are: Ag_2O, CdO, CoO, CuO, Cu_2O, Mn_2O_3, Mo_2O_5, NiO, Rh_2O_3, SnO + SnO_2, UO_2, WO_3, $LaNiO_3$, Pb_3O_4, and V_2O_5.

Gloom for Oxide Superconductors: Dismayed at the progress through the years, even with the most promising room-temperature metallic, binary oxides, many scientists abandoned the search for new high temperature oxide superconductors. Also, it should be mentioned that a deep-rooted prejudice had developed which claimed that the BCS theory had imposed a maximum transition temperature limit of 25 K for all superconducting materials, and that this temperature had already been achieved in certain alloys of niobium. Some scientists, however, were steadfast in their determination to break this barrier, optimistic in their outlook, and they continued their search for this unusual phenomenon in other metallic oxide systems.

Of the good anion-formers from Group VI of the Periodic Table, (the chalcogens), it has been claimed (11) that oxygen has yielded the least number of superconducting compounds. Large families of superconductors had been reported for the other Group VI congeners, namely, S, Se, and Te, in contrast to oxygen which, by the end of 1973, had generated only a handful.

In Roberts' report (3), a compilation of data to 1975 for over 3000 substances investigated as superconductors to different low temperatures, only 42 different oxide compositions are listed. These oxides can be classified into separate groups: binary oxides = 12; ternary oxides = 22; and multinary or complex oxides = 8. Of these, only 25 were found to superconduct and the highest superconducting transition temperature achieved in oxides was 2.3 K for nearly stoichiometric TiO. In 1978, Roberts presented (11) an addendum to his 1975 data set with the addition (or revision) of 8 new (mostly superconducting) oxides.

Metal-rich binary systems (also referred to as metallic sub-oxides), however, exhibited slightly higher transition temperatures than the stoichiometric oxide compounds, but for these interstitials the T_c was never any higher than that of the superconducting metal. For example, $Nb_{0.94}O_{0.06}$ was observed to have a T_c of only 9.02 K. Other interstitial niobium oxides had transition temperatures to 9.23 K, and similar tantalum oxides showed a maximum T_c value of 4.18 K.

As mentioned above in the intermetallic Section, beta-tungsten, which is chemically W_3O, is the prototype of the A-15 structure. The interest in W_3O, or W_3O_{1-x}, is not only structural but is also based on the fact that this material is superconducting! Further surprising is that, for the first time, this oxide superconductor has a higher transition temperature than that of the metal itself. Pure tungsten metal has a T_c of 15.4 mK, whereas the oxide W_3O has a reported T_c of 3.35 K. Other oxide compounds such as Cr_3O and "Mo_3O", which are isostructural with W_3O, do not superconduct above 1.02 K.

Thin films of an amorphous-composite In/InO, which contain a metal-dielectric interface, have been studied (12)(13) and found to superconduct when the electron carrier density reaches a value of approximately $10^{20} cm^{-3}$. This system, having a T_c of 2.5 to 3.2 K, has been described as exhibiting "interface-dominated superconductivity".

By going to mixed-metal systems and incorporating small amounts of oxygen, certain high transition temperatures could be maintained. These metal-rich phases, once again were found to have "interesting" T_c's. These stoichiometric, ternary-metal oxide systems include: ReTiO (T_c = 5.74 K); Zr_3V_3O and Zr_6Rh_3O (both cubic phases, with T_c's of 7.5 and 11.8 K, respectively). Certain superconducting films also showed enhanced transition temperatures when small amounts of oxygen were introduced. In a study by Kirschenbaum (14), thin films of niobium, which had been briefly exposed to an oxygen atmosphere, were observed to superconduct in the region 9.22 to 9.34 K. These films were found to contain only slight traces of oxygen, 0.001 to 0.015 mol%.

High T_c "Oxide" Films Possessing the A-15 Structure

Composition	T_c (K)	Remarks
$Nb_3Ge_{1-x}O_x$	21.5	Film (resistivity ratio = 1.1 to 2.6)
Nb_3GeO_x	19.5-20.5	Film deposited at 950-1050°C.

The causes for the enhancement of the T_c's in these "oxide" films have never been specified.

What can be concluded from these studies is the fact that oxygen was not responsible for the destruction of the superconducting state in "oxygen-containing materials".

There are only three broad structural categories into which most of the reported oxide superconductors can be classified; i.e., sodium chloride, perovskite, and spinel. It is interesting to note that these three structures possess cubic symmetry in their most idealized state. A detailed discussion of the research performed on oxide compounds derived from these three structures will be presented in Section 2.0 below. But before we continue with the general study of superconductivity in other materials, an overview of the oxide work is given in chronological fashion (to 1975) in the following Section.

1.7 Major Milestones in Oxide Superconductivity Research

At this point, it may be informative to present a chronological listing of the different discoveries in oxide superconductors reported prior to 1975. In this listing, Table 5, we present the year that the oxide compound was first reported, then the year in which superconductivity was first observed in the system and the group credited for the discovery. Of particular interest is the compound $Ba(Pb_{1-x}Bi_x)O_3$ discovered by Sleight at du Pont in 1975. This oxide material adopts the perovskite-type structure and <u>contains no transition metals</u>.

TABLE 5: Milestones in Binary and Multinary Superconducting Oxides to 1975.

Compound/System	Synthesis Year	Group	Superconductivity Year	$T_c(K)$	Reference to Supercon. report
NbO	1932	Noddack/Ger.	1933	~1.5	6. Meissner,
$SrTiO_{3-x}$	1927	Goldschmidt	1964	~0.3	15. Schooley,
A_xWO_3	1823	F. Wöhler	1964	3.0	16. Raub,
TiO	1939	P.Ehrlich/Ger.	1965	~1.0	17. Hulm,
A_xMoO_3	1966	du Pont Co.	1966	4.0	18. Sleight,
$Ag_7O_8{}^+NO_3{}^-$	1935	Braekken/Swe.	1966	1.04	19. Bell Labs.
A_xReO_3	1966	du Pont Co.	1969	4.0	18. Sleight,
$Li_{1+x}Ti_{2-x}O_4$	1953	Grenoble,Fr.	1973	13.7	20. Johnston,
$Ba(Pb_{1-x}Bi_x)O_3$	1975	du Pont Co.	1975	13.2	21. Sleight,

1.8 Ternary Chemical Compounds — Complex Borides and Sulfides

After the initial attempt to prepare alloy and interstitial superconductors, several ceramists, chemists, and materials scientists joined the group of physicists and metallurgists in search of other superconducting materials. These scientists turned to ternary compounds and to more complex systems. From the mid-60's to the mid-70's, several new "inorganic materials" were found to exhibit the superconducting phenomenon.

B. T. Matthias (22) predicted in 1977 that **ternary compounds** held the promise of a fresh approach to new, improved superconducting materials with higher T_c's. The search took these scientists to complex borides where the unusual phenomena of superconductivity and magnetism co-existed. This re-entrant magnetism (below the superconducting transition temperature) generated much interest in 1977. This particular co-existence was first observed (23) in $ErRh_4B_4$ by B.T. Matthias and co-workers. The structural, electronic, and crystallographic properties of many new ternary borides and sulfides have been the subject of two books, edited (24)(25) by Fischer and Maple, and these editions were dedicated to Matthias who passed away in 1980. In addition to the vast number of ternary borides and silicides described in these two volumes, several ternary and multinary sulfides are also reported. These sulfides, known as the **Chevrel phases** after their discoverer, Roger Chevrel (in collaboration with M. Sergent and J. Prigent), have the general chemical formula: $M_xMo_6S_8$. These unusual sulfides have an octahedral metal cluster within a cubical cage of sulfur atoms. Their structures are slightly distorted from perfect cubic symmetry, but can be considered "cubic" in their idealized state. More recent studies on these metal cluster compounds have revealed that condensation by face-sharing of metal clusters can lead to many more complex systems having superconducting properties. A selection of the superconducting "Chevrel phases" would include: $PbMo_6S_8$ (T_c = 15.2 K), $SnMo_6S_8$ (T_c = 14 K), $Cu_xMo_6S_8$ (T_c = 10.8 K), and $LaMo_6S_8$ (T_c = 11.4 K). As can be expected, sulfur can be replaced by other non-metallic elements, namely Se, Te, I, Br, and Cl. Another unusual property of these "Chevrel phases" is their very high critical fields (H_{c2}). These compounds generate magnetic fields of 500 to 600 kGauss at very low temperatures in their superconducting state. They exhibit the highest critical (magnetic) fields known for superconducting materials.

Some Superconducting Borides and Sulfides

$LnRh_4B_4$ (Ln = lanthanide element). Tetragonal. Discrete Rh tetrahedra and B_2 atomic pairs. Superconducting and ferromagnetic. $T_c = 8.7$ K and $T_c = 0.9$ K for Ln = Er. $T_c = 11.7$ K for Ln = Lu

$LnRhB_2$ (Ln = lanthanide element). Orthorhombic. $T_c = 10.0$ K, for Ln = Lu.

$TiB_{1.6-2.0}$ Superconducting at 22°C (295 K) claimed by Fred W. Vahldiek, but not yet verified or published in literature.

$M_xMo_6S_8$ (M = Pb or Ln; x = 1) Chevrel phases, distorted cubic structure. Isolated Mo_6S_8 clusters. ($T_c = 15$ K for M = Pb)

Pd_3Ag_2S Contains no superconducting elements. $T_c = 1.13$ K.

M_xMoS_2 (M = Sr; x = 0.2) Intercalation-type compound, layer structure. $T_c = 5.6$ K for the composition above.

CdS   "Irreproducible Superconductor" as a thin film. Anomalously large diamagnetic signal with a "T_c" above liquid nitrogen temperature.

$CuRh_2S_4$ Spinel structure. $T_c = 4.8$ K.

$CuRh_2Se_4$ Spinel structure. $T_c = 3.50$ K.

1.9 Non-Transition Metal Systems— $(SN)_x$ and Others

In the course of the broad research effort on preparing radically new superconducting materials, other inorganic systems were found to exhibit this phenomenon; a sampling of these are presented below. The unusual feature of these materials is the fact that few contain "the best" superconducting elements in their chemical formulations.

$Hg_4(AsF_6)_2$ — Linear chains of Hg atoms arranged in a criss-cross fashion. (T_c = 4.2 K)

Ag_2F (Silver subfluoride) — Anti-CdI_2 structure. III coordinated Ag atoms in contact. (T_c = 66 mK)

$(SN)_x$ (with doping) — Helical inorganic chain polymer which contains no metallic atoms. (T_c = 0.2 to 6.3 K)

$K_2Pt(CN)_4X_{0.3}\cdot(H_2O)_3$ (X = Cl, Br) — Highly conducting 1-D system. Undergoes a Peierls transition at low temperature. Nearly superconducting. Stack of superpositioned, square-planar $Pt(CN)_4$ groups.

NbPS — Orthorhombic, high-pressure phase. T_c = 12.5 K.

ZrRuP — Hexagonal phase. T_c = 13.0 K.

The strange case of CuCl: During 1978 a Russian group (26) reported the observation of a strong diamagnetic signal from CuCl that had been subjected to high pressure. This finding caused great excitement because the transition was reported to occur at approximately 150 K. The strong Meissner signal strongly suggested superconductivity, but difficulties were encountered when further measurements were carried out on the sample. Attempts to reproduce the magnetic results on this metastable, high-pressure product were unsuccessful. The Moscow researchers could not obtain complete diamagnetism, and they were unable to observe low electrical resistivity in the sample.

The report of such a diamagnetic anomaly in CuCl created a stir of activity in the USA. Two groups immediately attempted to confirm these results, but only observed a fleeting glimpse of some unusual behavior. Chu and co-workers (27), working at the University of Houston, found only a 7% Meissner effect in CuCl. Joint researchers at Stanford University also observed (27) anomalies in the CuCl system but could not confirm the existence of bulk superconductivity. The magnetic data observed by the Russian group was most exciting, and today the published magnetic data for this CuCl sample appears strikingly similar to that observed and reported for the 1-2-3 copper oxide phases. The magnetic data published by the Moscow group is reproduced as Figure 7 below.

The magnetic anomaly in CuCl can be explained by postulating a probable disproportionation reaction occurring at high pressure with the formation of Cu and $CuCl_2$. The metal-semiconductor interface behaves as a diamagnetic superconductor through the generation of eddy-current shielding effects. The experimental data were not reproducible and the resistivity *versus* pressure plots could not be recycled. This high-pressure CuCl phase and other similar compounds, such as CdS and NbSi, which show anomalous diamagnetic behavior, have been classified (28) as "irreproducible superconductors".

Graphite-Intercalation Compounds: The high conductivity of graphite within the planes attracted some researchers to this material as a prototype for other new superconductors. Prior research studies had been conducted on the intercalation of alkali metals and other inorganic compounds in graphite. These studies yielded interesting metallic conducting compounds possessing a variety of colors. This graphitic system appeared to be quite similar in physical properties to the "tungsten bronzes", but their chemical properties and structures are rather different. Some of the graphite/alkali-metal intercalation compounds that have been investigated (29)(30) are listed below with their superconducting transition temperatures.

Both $CuCl_2$ and $Cu(NO_3)_2$ have been introduced (31) between the layers of graphite (stage-five intercalation) by E. Strumpp in W.

Historical Introduction and Crystal Chemistry of Oxide Superconductors 27

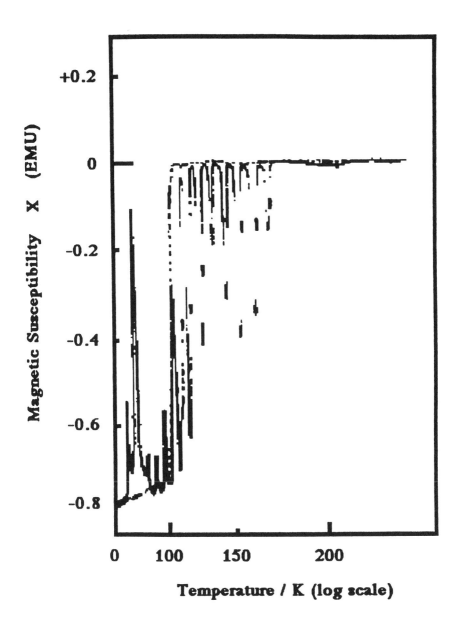

Figure 7: The original plot of magnetic susceptibility *versus* log absolute temperature (T / K) for CuCl under ~5 kbar pressure. Reference 26.

Germany, but the physical properties of these intercalates have not been reported. Japanese workers have focussed their research on other alkali-metal derivatives, and have incorporated some heavy metals between the layers. Compositions such as C_4KHg, C_8KHg, $C_4KTl_{1.5}$, and $C_8KH_{-0.10}$ have been isolated.

Graphite Intercalation Compounds with Their T_c Values (29)

Intercalation Compound	T_c (mK)
CsC_8 (golden)	20 to 135
CsC_{16} (blue)	< 11
KC_8 (golden)	550, 390
KC_{16} (blue)	< 11
RbC_8 (golden)	23 to 151
RbC_{16} (blue)	< 11

1.10 Organic Superconductors

In 1964, W.A. Little (32)(33) predicted that higher superconducting transition temperatures would be theoretically possible if the mode for electron coupling occurred through electron polarization. He suggested that a conjugated polyene, a long-chained organic molecule, might become superconducting (with a $T_c > 300$ K) if it contained some appropriately polarizable side-groups. This announcement set off an army of organic chemists in search of "room-temperature" organic superconductors. The success rate was very poor, yielding at best only a few metallic organic conductors possessing low resistivity at 1 K. The first such organic metal was TTF.TCNQ, or chemically the tetrathiafulvalene.tetracyanoquinodimethane salt. It was then realized that an electron donor (generating a cation) could be coupled with an electron acceptor (forming an anion) to yield a metallic conducting salt or compound. In most cases the donor, and sometimes the acceptor molecule, was a large planar organic molecule. These charge-transfer products were tested for superconductivity but little success was achieved in obtaining high transition temperatures. A listing of good electron donors and acceptors is presented in Table 6.

However, when pressure was applied to these metallic organic systems, the phenomenon of superconductivity was detected. The

TABLE 6

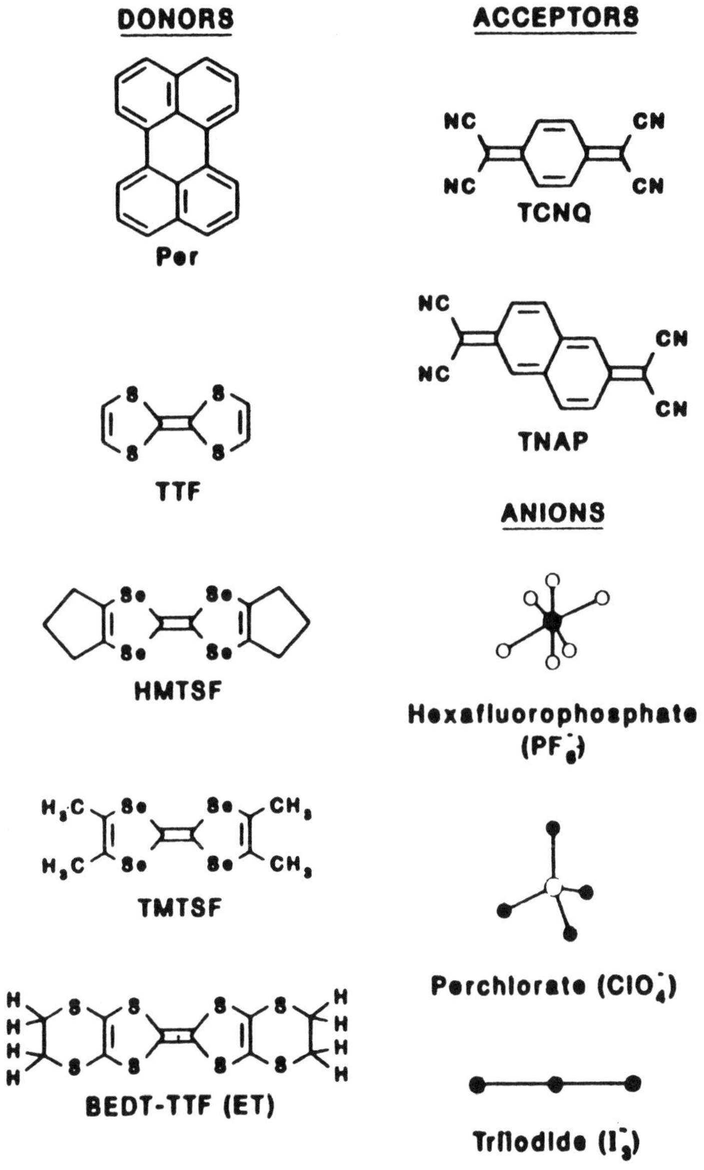

DONORS

Per

TTF

HMTSF

TMTSF

BEDT-TTF (ET)

ACCEPTORS

TCNQ

TNAP

ANIONS

Hexafluorophosphate (PF_6^-)

Perchlorate (ClO_4^-)

Triiodide (I_3^-)

transition temperature for superconductivity was always very low, for example, $(TMTSF)_2PF_6$ showed a T_c of 1.3 K at 6 kbar pressure. We shall not comment on this class of superconducting materials any further, except to mention that the present record holder for highest T_c in this category of organic superconductors is κ-$(ET)_2Cu(SCN)_2$, having an onset of superconductivity of 10 K at ambient pressure. The transition temperature in certain organic superconductors rapidly decreases as pressure is applied.

Metallic conduction has recently been observed in specially-prepared organic compounds, such as polyacetylene, polypyrrole, and polyaniline, having conductivities of the order 10^{-9} (ohm-cm)$^{-1}$; but by proper doping these conductivities can be increased to 10^2 (ohm-cm)$^{-1}$. Some of the organic metallic systems have also been converted into the superconducting state by proper doping, but in all cases the T_c remains at very low temperature.

Several chapters, reviews, and a recent book have been published on these materials, and the area is rapidly growing, becoming known as the "Organic Solid State". This supplementary reading is presented as references 34-42.

2.0 STUDIES ON SUPERCONDUCTING OXIDES PRIOR TO 1985

Few oxide superconductors were known prior to 1985 and we shall now return to these so that we can discuss these materials in reference to their crystal structure classes. There are only three broad structural categories in which most of the oxide superconductors occur. The important structural types include: sodium chloride (rocksalt, or B1-type), perovskite ($E2_1$), and spinel ($H1_1$).

2.1 Studies of Superconducting Oxides with the Sodium Chloride Structure

The sodium chloride structure, AX systems. Cubic Fm3m (Space Group #225): The sodium chloride or rock salt, NaCl, structure has a simple face-centered cubic unit cell (Figure 8) with alternating cations-anions along the three cubic axes.

Historical Introduction and Crystal Chemistry of Oxide Superconductors 31

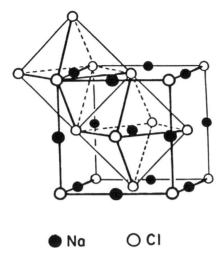

Figure 8: The NaCl (rock salt) structure, as re-drawn from Reference 5. Dark circles = Na, or metal atoms; white circles = Cl, or non-metal atoms.

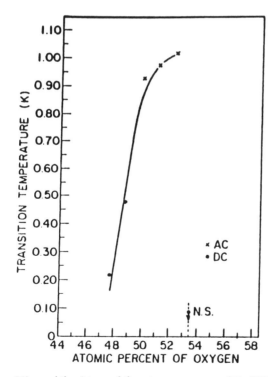

Figure 9: The critical transition temperature (T_c/K) plotted against oxygen content in TiO_{1-x}, as published in Reference 17.

This atomic arrangement leads to octahedral coordination about each cation and anion. In oxide systems, metallic-conducting materials occur when the cations are low valent transition metals. This is the case for TiO, VO, and NbO; all are room temperature metallic conductors crystallizing with the NaCl structure, or one closely derived from it, as in the case of NbO which has ordered metal and oxygen vacancies. We have already mentioned (6) the report of "near-zero" resistivity in NbO by Meissner in 1933. Further studies on this system in recent years have shown that superconductivity does exist in this compound, but that the transition is sluggish. The highest reported superconducting transition temperature for NbO is 1.55 K. The vanadium analog, VO, has been studied to 63 mK and has not been found to exhibit superconducting behavior. Similar negative results have been reported for non-stoichiometric VO samples.

The case for TiO is very different. Superconductivity in TiO was sought during the early sixties but was not observed due to cryogenic temperature limitations. At that time, the low temperature of 1.2 K could only be attained by pulling a vacuum on liquid helium. In 1965, a temperature of 0.02 K could be achieved by using the demagnetization method. This equipment was necessary for the study of most of the binary and ternary oxide compounds. It was during this time that superconductivity was finally observed (17) in TiO by Hulm and co-workers. In their mutual inductance experiments they noted that the transition temperature was dependent on the exact composition of the TiO samples. The dependence of T_c on the oxygen content of TiO_x is reproduced in Figure 9.

It was further noted that a complete loss of superconductivity occurred in oxygen-rich compositions, as in the $TiO_{1.10}$ sample, even though it also crystallized in the same structure and had similar unit cell dimensions to those of the superconducting phases. Several other studies on TiO_x compositions were carried out and these results are presented in Table 7. In 1968 Doyle et al. reported (43) further on the strong dependence of T_c on the composition (oxygen-content). High-pressure experiments on TiO_x samples were carried out in an attempt to remove the defects which are generated in materials prepared at high temperatures. In certain experiments, a monoclinic form of TiO_x was obtained (44). A maximum T_c of 2.0 K was observed (45) on a sample with nominal composition $TiO_{1.24}$, annealed at 60 kbar and 1300°C. This treatment resulted in a 15% decrease in

TABLE 7: Table of Superconducting NaCl-type Oxides.

Composition	T_c (K)	Reference
TiO	<1.20	46
TiO	<1.20	10
TiO	0.58	17
TiO	0.60	47
$TiO_{0.96 - 1.17}$	<0.5	48
$TiO_{0.85 - 1.25}$	<1.3 to 2.0	45
$TiO_{0.86 - 0.91}$	<1.3	45
$TiO_{0.95}$	0.65	48
$TiO_{1.06}$	0.94	48
$Ti_{1-x}O_{1-x}$	0.6 to 2.3	43
$TiO_{1.24}$	2.0	45
NbO	<1.20	46
NbO	<1.20	10
NbO	1.20	50
NbO	1.25 br.	49
$NbO_{1.02}$	1.38	48
NbO	~ 1.4	6
$NbO_{0.96-1.02}$	1.37, 1.55	48
$NbO_{1.00}$	1.61	48
VO	<1.20	46
VO	<1.20	10
VO	<0.063	49

the number of vacancies within the sample. The maximum value of T_c for the TiO_x system is 2.3 K for a stoichiometric composition with unit cell parameter, a = 4.207 Å.

It should be mentioned that, of the other first-row transition metal oxides crystallizing with the NaCl structure, none has been found to superconduct down to 2.5 K. Some of these oxides undergo magnetic ordering at low temperature and most behave as semiconductors at all temperatures. These would include MnO, FeO, CoO, and NiO. Studies performed on CuO, which has a different crystalline structure, showed only semiconducting behavior to very low temperatures (1.9 K).

2.2 Studies of Superconducting Oxides with the Perovskite-type Structure

The Perovskite Structure, ABX_3 Systems. Cubic Pm3m (Space Group #221): A cubic structure was assigned to the mineral perovskite, $CaTiO_3$, but this particular compound was later found to actually possess orthorhombic symmetry. Today, however, we refer to the perovskite structure in its idealized form as having cubic symmetry and it is normally represented by a simple unit cell (Figure 10).

The ideal cubic perovskite structure contains a large A cation in a cuboctahedral XII-coordination site, a smaller transition metal ion with octahedral coordination, and anions that are VI-coordinated to the metal atoms. The transition metal octahedral groups are vertex-connected along the three crystallographic axes to generate the structure (see Figure 10, above). The general formula ABX_3 is capable of representing several hundred different chemical compositions. With slight distortion of the structure, or the lowering of symmetry, this structure type comprises a myriad of inorganic compounds. Vacancies can also be accommodated in all the different sites; viz. $A_{1-x}BO_3$, $AB_{1-x}O_3$, and ABO_{3-x}. The well-known "tungsten bronze" family is an example of the $A_{1-x}BO_3$ class of defect compounds. This family will be discussed later. Another important class of defect perovskites is the anion-deficient oxide series, from which several superconductors have been isolated. The perovskite structure is relatively simple, and can accommodate a number of large mono-, di-, or tri-valent cations in the cuboctahedral sites of the unit cell. Similarly, a multitude of low or high valent transition metal ions can

Historical Introduction and Crystal Chemistry of Oxide Superconductors 35

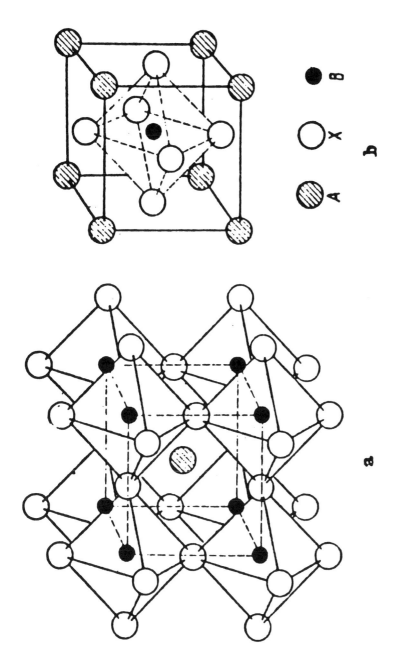

Figure 10: The Perovskite (CaTiO$_3$) structure. A polyhedral representation for an unit cell having different origins. From Reference 169.

adopt the octahedral site. We shall limit our presentation once again to oxide compositions since this is the primary focus of this chapter.

The search for superconductivity in the perovskite structure evolved from <u>two</u> different research groups, simultaneously. Cohen was investigating the possible existence of superconductivity in certain semiconductors, such as Si, Ge, and in the solid solution of Si in Ge. During 1964, he theoretically predicted (51)(52) that superconductivity could exist in certain semiconductive materials if the current density could be increased to relatively high values. Cohen also believed (53) that the T_c would be in the 0.1 K range, strongly dependent on the carrier concentration, and the systems would be Type II, bulk superconductors. Only one such semiconducting oxide, $SrTiO_{3-x}$, has thus far been observed to superconduct from Cohen's effort. The second approach, by Matthias and Raub (16), evolved from a study of ferroelectric conductors. Raub suggested an investigation of the "tungsten bronzes", unaware of the previous studies by Meissner and co-workers (6). In 1964, with the use of the new ^{3}He refrigerators, Matthias and Raub observed a low-temperature transformation in this class of materials. The experimental results on $SrTiO_{3-x}$ and the "tungsten bronze" were published (15)(16) in the same journal, within months of each other, in 1964.

Oxygen-deficient perovskites, ABO_{3-x}: Since the semiconductive-superconductive transformation is limited to only one oxide example, we shall present this case first. Strontium titanate, $SrTiO_3$, is a colorless, wide-band semiconductor (insulator) that also exhibits pseudo-ferroelectric properties when properly doped. It crystallizes with the cubic perovskite-type structure, a = 3.905 Å. This compound can be reduced under thermal treatment in a high vacuum system or under a reducing gas, such as hydrogen. The resulting product is a red crystalline solid, maintaining the same structure but possessing a slightly smaller unit cell. A superconducting transition temperature at ~0.30 K was observed in $SrTiO_{3-x}$ (see Figure 11).
Further studies (54) suggested that the T_c was dependent on the carrier concentration. A plot of the transition temperature *versus* n_c, the charge carrier concentration, is presented in Figure 12.

The maximum transition temperature of ~0.45 K was observed on a Nb-doped phase containing approximately $10^{20} cm^{-3}$

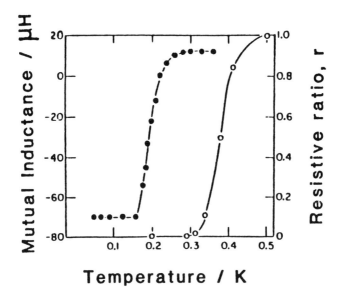

Figure 11: The resistive ratio *versus* absolute temperature (T/K) for $SrTiO_{3-x}$, as published in Reference 54.

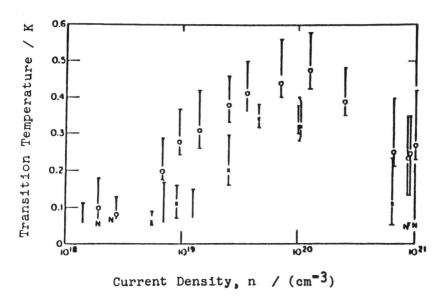

Figure 12: The critical transition temperature (T_c/K) plotted against charge-carrier concentration (n_c) for $SrTiO_{3-x}$, as published in Reference 54.

carriers. Both resistive and magnetic experiments were performed on these $SrTiO_{3-x}$ phases and the results indicated similar onset transition temperatures for superconducting behavior.

The mechanism for superconductivity (Type II) in these degenerate (nonstoichiometric, in this case) semiconductors is believed to occur through a strong coupling between electrons and the optical phonons. Chemically, electrons are produced in the reduction step that gives rise to Ti^{3+} ions, which have a $3d^1$ electronic configuration. These electrons are probably in a metallic conducting band that is strongly affected by the oxygen vacancies.

Many semiconducting oxides were investigated but few of these systems became superconducting to the lowest temperatures achievable (50 mK). Slight doping into the Sr or Ti sites yielded superconducting phases, but in all cases the T_c's for these analogs were near that of the maximum value obtained in the less complex, oxygen-deficient, $SrTiO_{3-x}$ precursor.

Cation-deficient Perovskites, $A_{1-x}BO_3$ Systems: *Space Group: Tetragonal P4/mbm (#127) and Hexagonal P6/mcm (#193):* Earlier work on "tungsten bronzes" by Meissner and co-workers(6). indicated that superconductivity was not observed in Li_xWO_3, K_xWO_3, or Rb_xWO_3 to 1.32 K, the lowest temperature that they could attain in 1933. The family of "tungsten bronzes" comprises several ternary transition metal oxides that possess room-temperature, metallic-conducting properties, and they normally exist as non-stoichiometric phases with vacancies in the large (XII coordinated) cationic sites. As the number of vacancies increases, the crystal symmetry decreases, and the new structures formed contain large tunnels instead of empty cuboctahedral sites. For the chemical formulation $A_{1-x}BO_3$, as x increases the symmetry changes from cubic to two different tetragonal phases, to hexagonal, and then finally to much lower symmetry at small values of x. The metallic-conducting properties are also lost in the orthorhombic, monoclinic, and triclinic phases. Superconductivity is normally observed only in tetragonal and hexagonal "tungsten bronze" phases. Tunnels also appear to be necessary for the existence of superconductivity in these compounds (Figure 13).

In Raub and Matthias' experiment (16)(55), a bluish-black needle of Na_xWO_3, produced by the electrolysis of a sodium tungstate melt in an iron crucible, exhibited a transition at 0.57 K. It was first

Historical Introduction and Crystal Chemistry of Oxide Superconductors 39

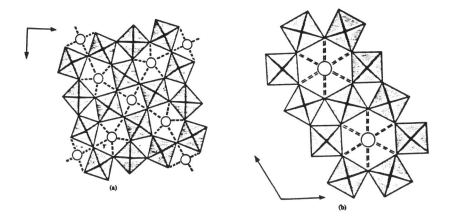

Figure 13: Projections of (a) tetragonal(II), and (b) hexagonal "tungsten bronze" structure. A polyhedral representation showing the large pentagonal and hexagonal tunnels, respectively. From Reference 55.

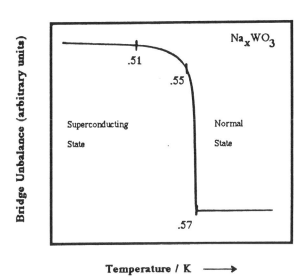

Figure 14: The original plot of bridge unbalance *versus* temperature (T/K) for Na_xWO_3, as published in Reference 16.

believed (56) that ferromagnetism (Fe from the crucible) had produced the strong signal, but on further examination, it was concluded that the abrupt change originated from the superconducting oxide (see Figure 14).

A broad search for superconductivity in the "tungsten bronze" family was initiated. In many such systems, a superconducting transition near 1 K was found. The most fruitful members of the "tungsten bronze" family were those possessing a symmetry lower than that of the cubic, and highly non-stoichiometric materials. Additional defect sites were later introduced in these oxide compounds by replacing some of the oxygen atoms with fluorine. The following Table 8 lists the various "tungsten bronze" compounds that have been studied as superconductor candidates, with their respective onset T_c's.

The highest transition temperature for the "tungsten bronze" family was 7.7 K for an acid-etched (71) sample of composition $Rb_{\sim 0.33}WO_3$. Certain researchers (62), after completing studies on cubic and tetragonal-II (semiconducting) bronzes, made the statement: "It appears as though the (cubic) **perovskite lattice** is not favorable for superconductivity." This statement was made in 1965, prior to the major advances in copper oxides that are considered to have a related-perovskite structure.

Ordered Perovskite-type Compounds, $A_2(BB')O_6$ Systems: Cubic Fm3m: A feature of the perovskite structure is that, with the proper substitutions, many types of ordered structures can readily be formed. This can be accomplished by the substitution of two suitable metal ions (with different oxidation states) in the octahedral sites of the structure. In this case the unit cell is doubled along the three cubic axes to generate an ~ 0.8 Å unit cell (Figure 15). Partial substitution of different transition metal ions in the octahedral sites is also possible; the general formulation for these compounds would be $A_2(B_{2-x}B'_x)O_6$. The parentheses in this formulation enclose atoms occupying the octahedral sites in the structure.

The syntheses of several simple ABO_3 compounds by Art Sleight led him to the first example of oxide superconductors possessing the ordered perovskite structure. It was already known that $BaPbO_3$, a distorted perovskite with pseudo-orthorhombic symmetry, exhibited metallic conducting properties to very low temperatures, but

TABLE 8: Table of "Tungsten Bronze" Superconductors with T_c data.

"Tungsten Bronze" Systems.

Composition	T_c (K)	Reference
Lithium compounds		
Li_xWO_3	<1.3	6
$Li_{0.30}WO_3$ hexag.	2.2	57
Li_xWO_3	~2.9	58
Sodium compounds		
Na_xWO_3 cubic (x=.3, .4, .8)	<0.3	62
$Na_{0.10}WO_3$ tetrag. II	<0.040	59
$Na_{0.20}WO_3$ tetrag.	0.55	59, 60
$Na_{0.28-0.35}WO_3$ tetrag.	0.57	16, 60
$Na_{0.2-0.4}WO_3$ tetrag. I	0.7-3.05	61
$Na_{0.23}WO_3$	~2.2	58
Na_xWO_3 cubic, a = 3.8 Å	<0.011	59, 62
Na_xWO_3 cubic + dist. cubic	<0.04	59
$Na_{0.96}WO_3$ cubic	<1.3	57
Na_xWO_3 hexag.	5.4	57
Potassium compounds		
K_xWO_3	<1.3	6
K_xWO_3	~2.45	58
$K_{0.27-0.31}WO_3$ hexag.	0.5	59
$K_{0.40-0.57}WO_3$ tetrag.	1.5	59
K_xWO_3 powder, hexag.	1.0 - 2.52	60
K_xWO_3 etched, hexag.	3.31 - 5.70	60

(continued)

42   Chemistry of Superconductor Materials

(continued)

Rubidium compounds		
Rb_xWO_3	<1.3	6
$Rb_{0.20-0.33}WO_3$ hexag.	2.15 - 2.9	68
$Rb_{0.20-0.33}WO_3$ hexag.	1.2 - 4.35	68
$Rb_{0.27-0.29}WO_3$ hexag.	1.98	59
$Rb_{0.27}WO_3$	~1.9	58
$Rb_{0.32}WO_3$ hexag.	1.9 - 6.6	57
$Rb_{0.33}WO_3$ acid etched	2.36 - 2.84	60
$Rb_{0.33}WO_3$	~1.75	58
Rb_xWO_3 hexag.	1.88 - 1.97	60
Rb_xWO_3	6.14 - 6.40	60
Rb_xWO_3	3.31 - 4.80	60
Rb_xWO_3	5.35 - 6.25	60
Rb_xWO_3	5.51 - 6.15	60
Rb_xWO_3	5.45 - 6.55	60
Rb_xWO_3	3.90 - 5.15	60
Rb_xWO_3	4.41 - 5.70	60
Rb_xWO_3	6.55 - 5.45	60
Rb_xWO_3	2.74 - 2.36	60
Rb_xWO_3	3.26 - 3.73	60
$Rb_{0.33}WO_3$	4.75	60
$Rb_{0.33}WO_3$	7.7	71
Cesium compounds		
Cs_xWO_3	1.39 - 1.77	60
Cs_xWO_3	2.26 - 4.76	60

(continued)

(continued)

$Cs_{0.30}WO_3$	hexag.	1.1 – 4.8	57
$Cs_{0.32}WO_3$	hexag.	1.12	59, 60
$Cs_{0.20}WO_3$	hexag.	6.7	69
$Cs_{0.30}WO_3$	hexag.	2.0	69
Ammonium compound			
$(NH_4)_{0.33}WO_3$	hexag.	1.4 – 3.2	57
Alkaline-earth compounds			
$Ca_{0.10}WO_3$	hexag.	3.4	63
$Sr_{0.08}WO_3$	hexag.	4.0	63
$Ba_{\sim 0.13}WO_3$	tetrag. I	1.9	62
$Ba_{0.14}WO_3$	hexag.	2.2	63
Misc. compounds			
$Cu_{0.32}WO_3$	hexag.	1.12	70
$In_{0.11}WO_3$	hexag.	<1.25 – 2.8	60
$In_{0.11}WO_3$	etched	~3.8	64
$In_{0.11}WO_3$	hexag.	~2.8	63
$Tl_{0.30}WO_3$	hexag.	1.58	65
$Tl_{0.30}WO_3$	hexag.	2.14	63
$Sn_{0.19}WO_3$	tetrag.	<1.3	57
$Sn_{0.21}WO_3$	hexag.	<1.3	57
$Sn_{0.24}WO_3$	hexag.	<1.3	57
"Bronze" oxyfluorides.			
$K_xWO_{3-x}F_x$		1.9 – 2.1 – 0.8	66

(continued)

(continued)

$K_{0.1}Li_{.02}WO_{2.88}F_{0.12}$	1.1	66
$K_{0.08-.3}WO_{2.7-2.92}F_{0.08-.30}$	1.9 - 2.1	66
$K_{0.08-.3}WO_{2.7-2.92}F_{0.08-.30}$	0.8	66
$Rb_{0.1}Li_{.02-.10}WO_{2.8-2.9}F_{0.12-.20}$	2.1 - 4.0	66
$Rb_{0.08}WO_{2.7-2.92}F_{0.08-.30}$	0.9 - 3.7	66
$Cs_xWO_{3-x}F_x$	1.4 - 4.5	66
$Cs_xWO_{3-x}F_x$ (filament & plate)	5.2	69
$Cs_xLi_yWO_{3-x}F_{x+y}$	2.0 - 3.4	66

"Molybdenum Bronze" Systems.

Sodium compound

Na_xMoO_3 cubic	<1.3	18

Potassium compounds

$K_{\sim 0.5}MoO_3$ tetrag.	4.2	18
$K_{\sim 0.9}MoO_3$	<1.3	18
K_xMoO_3 cubic	<1.3	18

"Rhenium Bronze" Systems.

Sodium compound

Na_xReO_3 cubic	<1.3	18

Potassium compounds

$K_{\sim 0.3}ReO_3$ hexag.	3.6	18
$K_{\sim 0.9}ReO_3$ tetrag.	<1.3	18
K_xReO_3 cubic	<1.3	18

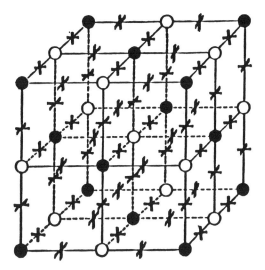

Figure 15: The ordered perovskite structure showing the positions of the B cation positions (white and dark circles). The oxygen positions are represented by the crosses. The eight A cations are omitted from the diagram for purposes of clarity.

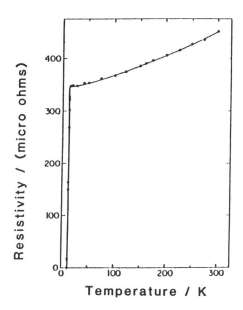

Figure 16: The electrical resistance *versus* absolute temperature (T/K) for $Ba(Bi_{0.20}Pb_{0.80})O_3$, as reported in Reference 21. ($T_c = 11$ K.)

was not a superconductor down to 4.2 K. $BaBiO_3$ had also been previously reported, but its structure and properties were unknown. The bismuth system was further complicated by the fact that at different temperatures of preparation, various oxygen-deficient phases were produced, viz. $BaBiO_{3-x}$, having different crystalline structures (72)(73), all related in some way to the cubic perovskite-type structure.

Controversy arose as to the assignment of the oxidation state for bismuth in $BaBiO_3$. The simple formulation could represent a 4+ oxidation state for bismuth, but an ordered perovskite would indicate a mixture of the more common 3+ and 5+ states. Sleight prepared (21) a $BaBiO_3$ phase at low temperature using the hydrothermal method and isolated a golden, near-stoichiometric product that exhibited insulating properties. Powder neutron diffraction data (74-77) on this compound indicated monoclinic symmetry and a perovskite-related structure with two different octahedral sites, most probably containing ordered arrangements of Bi^{3+} and Bi^{5+}. Other studies on the order-disorder of bismuth ions in $BaBiO_3$ have been carried out by several groups (78-80). In all these studies it has been concluded that, for samples prepared at low temperature, the idealized composition can be written as: $Ba_2(Bi^{3+}Bi^{5+})O_6$, rather than $BaBi^{4+}O_3$. The oxygen content, however, can vary depending upon the synthesis temperature and oxygen-annealing conditions. The structure also changes on heating from monoclinic to rhombohedral (R3) at about 405 K. Three compositional regions are observed between 850 to 1225 K under different oxygen partial pressures. The three $BaBiO_{3-x}$ phases possess different perovskite-related structures.

Sleight then went on to study the solid solution between $BaPbO_3$ and "$BaBiO_3$". These results (21) appeared in 1975 and documented the discovery of a perovskite-type **superconducting oxide** having an interesting T_c (Figure 16). The publication was entitled: "High-Temperature Superconductivity in the $BaPb_{1-x}Bi_xO_3$."

A transition temperature of 13 K had been observed for the composition $Ba(Pb_{0.7}Bi_{0.3})O_3$, an inorganic oxide compound that contained no transition metal ions.

The compositional range for superconductivity in $BaPb_{1-x}Bi_xO_3$ is $0.05 < x < 0.3$ with T_c's varying between 9 and 13 K. Outside this compositional range no superconductivity was observed. Further work has been carried out on this system in the USA, the USSR (81), and in Japan; the highest transition temperature observed thus far is 13.4 K for the composition $BaPb_{0.75}Bi_{0.25}O_3$. A summary report (82) on the various studies of this solid solution was published in 1986. This report contains over 200 references to the published literature on the $BaPb_{1-x}Bi_xO_3$ system.

High-pressure experiments on this oxide superconductor suggested (83) that a re-entrant superconducting transition occurred between the grains of the superconducting particles. The authors further concluded that the mechanism of superconductivity in this oxide system remained unknown. Other resistance measurements were also carried out (84) using pressures of ~125 kbar indicating a possible onset of metallic behavior in this material at room temperature.

With the incorporation of other ions in the oxide superconductor, the transition temperature can be sharpened. The substitution of K^+ for some Ba^{2+} yielded (85) a superconductor with a sharper T_c. The pressure studies on the potassium-doped $Ba(Pb_{1-x}Bi_x)O_3$ compounds showed a decrease in T_c as the hydrostatic pressure was increased to 15 kbar.

Post-1985 Developments in this Structural Class of Compounds: In a recent discovery by Cava (86), the composition $BaPb_{0.75}Sb_{0.25}O_3$, was found to undergo a superconducting transition at only 3.5 K. The substitution of Sb for Bi appears to drive out superconductivity in a manner which was not expected, nor as yet been explained.

In 1988, Cava and co-workers also prepared (88a) a quaternary oxide, Ba/K/Bi/O, and observed superconductivity at ~28 K. This compound was the first "non-transition metal" oxide with a T_c above the legendary "alloy record" of 23 K. Further studies indicated (88a) that the optimum composition for "high temperature" superconductivity in this system was $Ba_{0.6}K_{0.4}BiO_{3-x}$, having a T_c of 30.5 K (Figure 17). The samples were multiphase, and the superconducting fraction varied from 3 to 25%. Superconductivity for the rubidium-substituted compound was observed at ~28.6 K.

Several other research groups (89-91) picked up on this discovery and obtained Ba/K/Bi/O materials with an onset transition temperature of 34 K. The structure of this oxide superconductor is cubic, perovskite-type, having unit cell dimensions, 4.288-4.293 Å. This system appears to be a three-dimensional superconductor and has

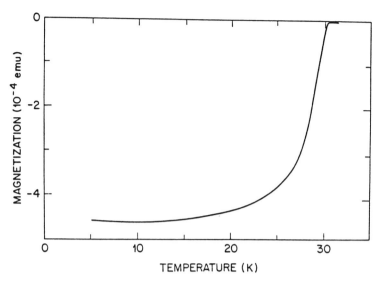

Figure 17: The magnetization *versus* absolute temperature (T/K) for crystals of $Ba_{0.60}K_{0.40}BiO_3$, as reported in Reference 88b. (T_c = 30.5 K.)

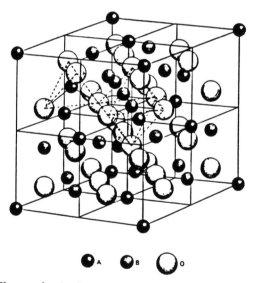

Figure 18: The spinel ($MgAl_2O_4$) structure. Tetrahedral and octahedral polyhedra are depicted by the dotted lines. Not all the atoms are shown for reasons of clarity. Note the oxygen planes along the 111 direction of the unit cell.

recently generated much interest, since the absence of transition-metal ions suggests that the magnetic mechanism for superconductivity may not be an essential requirement for high temperature superconductivity. A controversy also exists over the existence of an isotope effect in this material when ^{18}O is substituted for ^{16}O. Some scientists claim a large isotope effect, while others observe only a slight effect on the T_c of the material by the isotopic oxygen substitution. The effect of pressure on potassium-doped $BaBiO_3$ indicated (92) a slight increase in T_c, of the order 0.05 to 0.10 (dT_c/dP) K/kbar with applied pressure. These investigations were limited to 9 kbar pressure of oxygen.

2.3 Studies of Superconducting Oxides with the Spinel Structure

The Spinel Structure, AB_2O_4. Cubic Fd3m (Space Group #227): The mineral spinel has the chemical composition $MgAl_2O_4$ and is rather scarce in nature. The more common mineral magnetite, Fe_3O_4, also adopts the spinel structure and has been known through the centuries because of its important room-temperature ferrimagnetic properties (as lodestone for use in a compass). The spinel structure contains magnesium ions in tetrahedral sites and aluminum ions in octahedral sites. The coordinations for the different ions can be expressed by the formula: $^{IV}Mg\,^{VI}Al_2\,^{IV}O_4$, which has overall face-centered cubic symmetry in the space group, Fd3m. The structural details are slightly complex and a portion of the unit cell will be presented to better describe the arrangements of cations in the close-packed structure (Figure 18).

Further adding to the complexity of the spinel structure are three possible arrangements of the metal ions in the cubic close-packed anions. The ordering of divalent metal ions (such as Mg^{2+}) on the proper tetrahedral sites and all the trivalent ions (as Al^{3+}) in the correct octahedral sites, will give rise to the **normal** spinel structure. If the divalent ions occupy some of the octahedral sites and half of the trivalent ions move to the tetrahedral sites, the structure is then referred to as the **inverse** spinel structure. The last case exists when the tetrahedral sites and the octahedral sites are occupied by a mixture of di- and tri-valent ions. This type is known to generate the **random** spinel structure, and the exact composition and populations in the respective sites are determined by factors such as final heating temperatures, cooling rates, and the site-preference energies of the different ions for the two polyhedral sites.

Superconductivity has been observed in only one spinel-type oxide thus far, namely $LiTi_2O_4$. The problem of site occupancy and defect concentration has generated many uncertainties as to the exact structural type and chemical composition of this particular compound. D. C. Johnston, while investigating the superconducting properties in the sulfide phase Li_xTiS_2, observed that air-exposed samples showed signs of superconductivity at ~13 K. Johnston believed that the superconductivity was arising from an oxide rather than the intercalated sulfide compound. He immediately initiated a study on the Li/Ti/O system by preparing several ternary oxide phases. In his attempts to isolate a pure spinel-type compound, he noted (20) a large increase in the Meissner fraction for a compound with a nearly-stoichiometric $LiTi_2O_4$ composition (Figure 19). Further studies (93) led him to more detailed structural and compositional information.

His studies indicated that the spinel-type phase could display the variable composition, $Li_{1+x}Ti_{2-x}O_{4-y}$, a **random** spinel structure; i.e., a mixed distribution of cations on the tetrahedral and octahedral sites. The superconducting oxide with the maximum onset T_c of 13.7 K had a cubic unit cell parameter of 8.40_4 Å. This product, however, also contained 20% of the starting reactant, $Li_2Ti_2O_5$, as an impurity phase. High-pressure studies have been carried out (94) on $LiTi_2O_4$ to 16.5 kbars; the onset temperature for superconductivity increases as a function of increasing pressure, according to the expression $(dT_c/dP) = +(12\pm1) \times 10^{-6}$ K bar^{-1}.

The $Li_{1+x}Ti_{2-x}O_4$ system has recently been re-investigated (95) to answer several questions concerning the superconducting behavior in this oxide material. The results are summarized as follows: (1) the $Li_{1+x}Ti_{2-x}O_4$ system transforms from semiconducting to metallic behavior near $x \approx 0.1$ and superconductivity is found in the region $x = 0$ to 0.10, (2) the T_c remains constant at ~12 K and is independent of the value of x in the superconducting region, (3) the Meissner fraction decreases linearly with increasing x, or the random nature, (4) the optimum composition for superconductivity occurs at $x = $ ~0.15, and, (5) the transport properties do not result from the co-existence of stoichiometric $LiTi_2O_4$ and a second phase formed by spinodal decomposition. The system is chemically homogeneous and

Historical Introduction and Crystal Chemistry of Oxide Superconductors 51

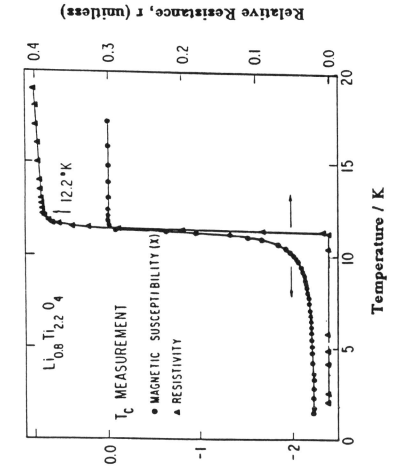

Figure 19: A plot of the magnetic susceptibility (black circles) and the relative resistance (black triangles) as a function of absolute temperature (T/K) for $Li_{0.8}Ti_{2.2}O_4$. From Reference 20.

the additional Li ions are distributed on the empty octahedral sites to generate a random spinel-type product.

The data for several Li/Ti/O spinel compositions are presented in Table 9 with their reported transition temperatures. Unfortunately many of these samples were not homogeneous.

2.4 Post-1985 Entry of Copper Oxide Superconductors

Thus far we have dealt primarily with superconducting oxide systems having relatively common structures and those studied prior to 1985. This was soon to change, for late in 1986 the discovery of superconductivity in copper oxides with perovskite-related structures revolutionized all the work to be carried out in the future on new superconductors. The importance of crystal chemistry in the cuprate systems, superconducting and non-superconducting, has become of prime interest to all scientists in the field. In the following Section, we will present a thorough crysto-chemical description of the element copper and its oxide compounds in an effort to prepare the reader for this startling and exciting discovery.

3.0 STRUCTURAL FEATURES AND CHEMICAL PRINCIPLES IN COPPER OXIDES

General: Of the first-row transition metals, the element copper (Symbol = Cu from the Latin, cuprum) offers an unique and interesting chemistry based on its oxidation states and preferred geometries. Copper can be considered as the "chameleon" of the 3d transition metals in that it can behave as a normal transition-metal species in its formal oxidation state of 2+, and as a post-transition, highly-covalent species when displaying its other common oxidation state of 1+. In its higher valence states of 2+ (common) and 3+ (rare), copper contains partly-filled $\underline{d}$ orbitals. In contrast, the Cu^+ ion, a $3d^{10}$ system, can not be considered a transition metal ion; it shows little complexing power and displays no magnetic or colored species.

3.1 The Fascinating Chemistry of Binary and Ternary Copper Oxides.

TABLE 9: Reported Superconducting Transition Temperatures in the $LiTi_2O_4$ System.

Composition	T_c (K)	Reference
$Li_{2.6}Ti_{1.5-2.7}O_4$	10.9 - 11.4	96
$Li_{2.6}Ti_{1.9}O_4$	11.2	96
$Li_{2.6}Ti_{1.7}O_4$	11.4	96
$Li_{2.6}Ti_{1.3}O_4$	<4.2	96
$Li_{2.6}Ti_{4.0}O_4$	<4.2	96
$LiTi_{2.0}O_{3.95}$	11.2	97
$LiTi_{2.0}O_{3.95}$	11.7	97
$Li_{1.1}Ti_{1.9}O_{3.95}$	9.6	97
$Li_{1.05}Ti_{1.95}O_4$	12.0	97
$LiTi_2O_{3.9}$	11.26	98
$Li_2Ti_{4.0}O_8$	≈ 11.4	99
$Li_{0.8}Ti_{2.2}O_4$	11.20	98
$LiTi_{(2)}O_4$	12.5	93
$LiTi_2O_4$	12.40	100
$LiTi_2O_4$	12.6	100
$LiTi_2O_4$	12.4	100
$Li_{1.05}Ti_{1.95}O_4$	<1.5	93
$Li_{1.33}Ti_{1.67}O_4$	<1.5	93
$Li_{1.33-0.8}Ti_{1.67-2.2}O_4$	1.5-13.7	20

The Cuprous, Cu^+ or Cu(I) State, $3d^{10}$: The chemistry and geometry of the cuprous ion is simple and straightforward. It forms diamagnetic and colorless complexes. In aqueous solution the cuprous ion disproportionates into elemental copper and the more stable cupric ion. As a $3d^{10}$ ion, it utilizes only its low-energy and available s and p orbitals to form "essentially covalent" compounds and complex ions. The most common of these contain a linear two-fold bonding arrangement about the copper atom. In this geometry, "sp" hybrids are presumably formed which generate the required 180° bond angle in the solid state. The binary oxide Cu_2O, known as the mineral cuprite, contains two-coordinate Cu atoms having linear bonding. Cu_2O and Ag_2O are isostructural, and the relationship between Cu^{+1} and Ag^{+1}, is best exemplified by the structures of their oxides and sulfides. In many of the ternary oxides such as NaCuO, KCuO, RbCuO, $CuFeO_2$, $CuCrO_2$, and $LaCuO_2$, Cu^+ maintains its two co-linear oxygen neighbors. Other hybrids, "sp^2" (trigonal planar) and "sp^3" (tetrahedral) geometries, can be formed, especially with the larger, more covalent sulfur ligands.

The Cupric, Cu^{2+} or Cu(II) State, $3d^9$: The most important and stable oxidation state for copper is divalent. There is a well-defined aqueous chemistry of the Cu^{2+} ion, which generates the familiar blue solution when complexed with water. A large number of copper coordination compounds exist and these have been studied extensively. A strong Jahn-Teller distortion is associated with the $3d^9$ electronic configuration of this ion. This implies that a regular tetrahedron or octahedron about the Cu^{2+} ion is never observed, except in the rare occurrence of a dynamic Jahn-Teller effect. The tetragonal distortion about an octahedron can lead to a square-planar coordination which is often observed in Cu(II) oxides.

CuO, the mineral tenorite, and AgO are well known but their structures are quite different. More importantly the valence states in these two compounds are quite different. In CuO, the copper is formally in the divalent state, whereas in AgO, there exist two types of silver atoms, one in formal oxidation state 1+, the other in 3+. These two silver ions also possess strong covalent character. PdO and CuO, however, have similar crystal structures based on chains of opposite edged-shared, square-planar MO_4 groups.

Historical Introduction and Crystal Chemistry of Oxide Superconductors 55

The best indicator of covalent-ionic character in the Cu-O bond is the observed Cu-O bond length (see 3.2 below).

The Cu^{2+} ion, when surrounded by oxygen ligands, tends to adopt an octahedral environment, but because of the Jahn-Teller distortion this octahedron is always distorted (as in La_2CuO_4). Square-planar coordinations about Cu^{2+} are also found (101), for example the ternary oxides $SrCuO_2$, $BaCuO_2$, Ca_2CuO_3, $CaCu_2O_3$, Sr_2CuO_3, and the family of Ln_2CuO_4 (Ln = large lanthanide element). Another family of compounds has been prepared with composition A_2BaCuO_5, where A = Y, Sm, Eu, Gd, Dy, Ho, Er, and Yb.

The Cu^{3+} or Cu(III) state, $3d^8$: Copper(III) is isoelectronic with Ni(II) but can not be isolated from aqueous solution. However, several ternary oxides of Cu(III) have been prepared using strongly oxidizing conditions and this oxidation state has been stabilized using a highly basic cation.

$SrLaCuO_4$ has been prepared (102) under high-pressure oxygen and the product has been found to contain the Cu^{3+} ion in a distorted octahedral environment, the Cu-O bond distances in the compound are: 4 @ 1.88 and 2 @ 2.23 Å.

A Cu-O bond distance of 1.943 Å has been reported (103,104) for the rhombohedral perovskite $LaCuO_3$ which was prepared under very high oxygen pressure. This black metallic conductor is believed to contain Cu^{3+} ions, in a distorted octahedral environment. Other known examples of ternary oxides containing Cu^{3+} ions include $NaCuO_2$, $KCuO_2$, $RbCuO_2$, Li_3CuO_3, and Na_3CuO_3. These compounds are generally stable as dry solids in the absence of air, but decompose rapidly in moist air. The coordination about Cu^{3+} is normally square planar with two additional oxygen neighbors at a longer distance. This same coordination is also found for several Ni^{2+} analogs. Very little is known concerning the physical properties of these ternary Cu(III) oxides.

3.2 Copper to Oxygen Bond Distances— Ionic Radii

The copper to oxygen distance will vary according to the oxidation state of copper, the degree of covalency in the bond, and the geometry about the copper atom. The data given below are the averages of reported (and theoretical "ionic") radii obtained from several different Cu^{2+} oxides and minerals.

Cu-O Bond Distances in Copper(II) Oxide Compounds

Geometry*	IV Sq. planar	V Sq. pyram.	IV "Octahedral"
average, obs.	1.95	2.05	2.12Å
range, obs.	1.85-2.05	2.00-2.15	2.05-2.16Å
ionic, calc.	2.02	---	2.13Å

* all geometries are distorted from the ideal

According to A. W. Sleight, the Cu-O bond length is of the order 1.98 Å for an essentially ionic case, whereas, for covalent species the Cu-O bond length is closer to 1.89 Å. The shorter the Cu-O bond length in "divalent Cu oxide compounds", the greater the degree of covalency in the bond. The presence of Cu^{3+} in the compound will also give rise to a shorter Cu-O bond. From the Cu-O bond distances observed in the Copper(III) compounds listed above, the calculated ionic radius for Cu^{3+} in "VI-coordination" is 0.50 ± .02 Å, while that for Cu^{2+} is much larger, 0.73 Å.

A statistical report of the oxygen coordination environments about divalent metal ions in minerals and other inorganic (oxidic) compounds was published (105) in 1984. The coordinations about the Cu^{2+} ion ($^{VI}r_{Cu}$ = 0.73 Å) shows a pronounced square planar tendency (IV sq. pl.). Quite often, however, one or two additional oxygen atoms are located perpendicular to the square plane, yielding a distorted "V"-coordination (IV + I square pyramid) or a highly distorted octahedron, "VI"-coordination (IV + II), see Figure 20.

This latter distortion is associated with the strong Jahn-Teller effect of the Cu^{2+} ion. The Jahn-Teller distortion can result in the formation of an elongated octahedron (4 short + 2 long); a square pyramid (4 short + 1 long); or quite commonly a square plane (4 short).

The tetrahedral CuO_4 group is rarely observed and no strong evidence has ever been presented for this highly symmetrical environment about the divalent copper ion. Coordinations higher than VI (or IV + II) are also very rare. Statistical data for Cu^{2+} coordination environments in 234 oxysalts and 75 minerals will be summarized.

Historical Introduction and Crystal Chemistry of Oxide Superconductors 57

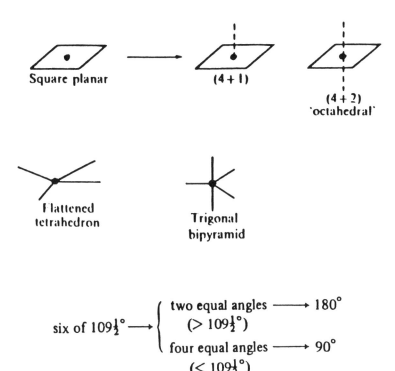

Figure 20: The stereochemistry of copper(II). Different geometric arrangements. The lower region shows the angles involved in converting a regular tetrahedron into a flattened tetrahedron, and then into square-planar geometry. From Reference 5.

These data and more recent results on several multinary copper oxides are presented below.

Statistical Data on Copper Oxide Polyhedra in Minerals and Compounds

Coordination Number	Polyhedron-type	Percentage	Summary of Coord. Frequency
IV	flattened tetrahedron	1%	19%
IV	square plane	18%	
V	trigonal bipyramid	3%	20%
V	square pyramid	17%	
VI	distorted octahedron	60%	60%
>VI	other	1%	1%

A thorough discussion of the square-planar copper oxide compounds has been presented by Müller-Buschbaum (101). This excellent review presents structural data for some 25 ternary copper oxides.

In CuO, the geometry of the 4 oxygen atoms is essentially square planar (4 @ 1.96 Å, with the next two nearest neighbors at 2.78 Å) (106). In the Y_2BaCuO_5 structure (107), the central Cu^{2+} ion has a square-pyramidal geometry. The Cu-O bond distances for this pyramid are: 2 @ 2.01, 2 @ 2.18, and 1 @ 2.29 Å. In La_2CuO_4, the Cu^{2+} ions are in an elongated octahedral environment(108). with four short and two long Cu-O bond distances, (4 @ 1.907 and 2 @ 2.459 Å). The octahedra are also tilted within the ab plane of the orthorhombic cell.

The structures of some multinary copper oxides and copper minerals are further complicated by the fact that several different coordination polyhedra might co-exist within the same structure. In crystallographic terms, Cu can be found in several non-equivalent sites, often with different coordination polyhedra, within the same structure. This complication is exemplified with the compound $Cu_{11}(CrO_4)_4(OH)_4$, which contains six non-equivalent copper atoms (109). The mineral fingerite, $Cu_{11}(VO_4)_6O_2$, is also known (110) to contain copper in six different crystallographic sites, the majority of these having a highly distorted octahedral environment. In $Cu_5V_2O_{10}$, all five copper ions are in different crystallographic sites, once again the majority in distorted octahedral environments, but two Cu ions occupy trigonal bipyramidal sites (111).

One interesting perovskite-type compound that has recently been described (112) is $La_{8-x}Sr_xCu_8O_{20-\epsilon}$ which contains distorted octahedral, square-pyramidal, and square-planar Cu-O polyhedra. This mixed-valent copper oxide forms when $\underline{x}$, the strontium concentration, reaches a value of 1.28 to 1.92. Epsilon can vary depending upon the preparative conditions, but takes on a rather small value, on the order of 0.08 to 0.32. This tetragonal compound crystallizes with the probable space group P4/mbm. The octahedral (CuO_6) group has the following Cu-O distances: 4 @ 2.038 and 2 @ 1.932 Å. The square-planar polygon has two Cu-O @ 1.906 and another pair @ 1.932 Å. Finally, the square pyramid (CuO_5) has Cu-O bond distances of 2 @ 1.860, 2 @ 1.932 and one long apical bond at 2.395 Å. The structure of the compound is related to an oxygen-deficient perovskite. In terms of the number of distinct types of polyhedra in various divalent copper oxides and copper minerals, the following statistics have been presented (105):

Unique type of coordination in the structure	63 %
Two different types of coordination groups in structure	27 %
Three (or more) different types of polyhedra	10 %

The copper ion is not unique in the first-row transition metal series for exhibiting this diversity of polyhedral environments within a single structure, since Mn^{2+} (not a Jahn-Teller ion, but one which has no strong crystal-field stabilization energy for highly symmetrical sites) shows a 20 % preference for occupying multiple crystallographic sites.

It should also be mentioned that unusually large distortions may occur in the octahedral site. In certain CuO_6 octahedra, the longest Cu-O distance can sometimes be 1.0 Å longer than the shortest bond distance, namely 2.85 $\underline{vs.}$ 1.90 Å for Cu(1) in $Cu_5V_2O_{10}$, and

2.96 vs. 1.90 Å for the Cu(4) atom in $Cu_{11}(VO_4)_6O_2$ or fingerite. The octahedral distortions are observed in a large number of Cu(II) compounds; 6 % having a difference in bond length (longest - shortest) of >1.00 Å. This high percentage is the highest value for any element in the first-row transition metal series. The total percentage of known examples increases to 23 % in different Cu^{2+} compounds if the limits of bond length differences are increased to $0.80 < \Delta > 1.00$ Å.

The cubic compound $CaCu_3Mn_4O_{12}$ may be considered related to the perovskite structure; the copper environments (in XII coordination) are given (113) as three sets of four coplanar Cu-O distances, namely 1.942(3), 2.707(3), and 3.181(3) Å. These three rectangular polygons all have mirror symmetry and generate a polyhedron composed of three sets of intersecting rectangles at 90° angles (the oxygen atoms occupying the the corner positions of the rectangles), giving rise to the XII coordination (see diagram below).

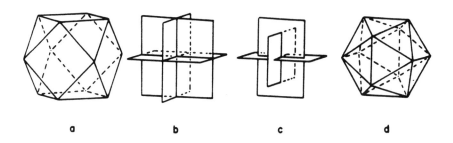

a   b   c   d

Diagram of XII-coordinated site for Cu.

In the diagram above; a represents the regular cuboctahedral polyhedron found about the large A cation in the perovskite structure, b represents the ideal arrangement of three intersecting squares denoting the positions of the vertices (oxygen atoms) for the cuboctahedral polyhedron, c represents the present situation, three sets of rectangles, and finally d represents the high symmetry polyhedron (icosahedron) which also describes XII coordination. The XII coordination site for copper in $CaCu_3Mn_4O_{12}$ as described in the literature is deceptive for the next-nearest neighbors (Cu-O distances) are 39% further apart, and the second nearest-neighbor bond distances are 64% longer than the first. It should also be mentioned that there are 8 nearer Cu-Mn neighbors at 3.135 Å. A similar polyhedral arrangement exists (114) in the manganese analog which contains three Mn^{3+} ions, another Jahn-Teller ion, in this central XII-coordinated site. This particular analog, $Na\{Mn_3\}Mn_4O_{12}$, was the precursor to the copper compound.

For certain Cu^{2+} compounds, ideal or near-ideal octahedral environments can be observed about the copper ion. This is particularly true of salts such as $K_2Pb(Cu(NO_2)_6)$, in which Cu^{2+} has six equally distant N neighbors at 2.12 Å. This symmetrical environment has been explained on the basis of a dynamic Jahn-Teller distortion which comes about by a rapid interchange of long and short Cu-N bonds with a result that all bonds, on the average, become equal. This dynamic Jahn-Teller effect has not yet been observed in oxide systems but one example is known (5) where six equidistant Cu-O bonds exist. In $Cu(Lig.)_3(ClO_4)_2$, where Lig.= $OP_2O_2\{(N(CH_3)_2)\}_4$— a tridentate ligand, there are six equidistant Cu-O bonds at 2.07 Å. The bulky ligand in this case may be responsible for the formation of a highly-symmetrical environment about the copper atom through steric factors. In a few hydrated salts of Cu^{2+}, a small orthorhombic distortion has been observed. In two cases (5) the $Cu-OH_2$ bonds have only a 10 % difference in bond distances; one of these is the common compound, Tutton's salt. The Cu-O bond distances for two examples are presented below:

Cu-O bonds:	2 @	2 @	2 @
Cu salts:			
$(NH_4)_2Cu(SO_4)_2 \cdot 6H_2O$	1.96	2.10	2.22 Å
$Cu(ClO_4)_2 \cdot 6 H_2O$	2.09	2.16	2.28 Å

4.0 PHYSICAL PROPERTY DETERMINATION ON TERNARY COPPER OXIDES— STUDIES ON COPPER OXIDE SYSTEMS PRIOR TO 1985

4.1 Studies on La_2CuO_4 and Its Derivatives

Preparation of La_2CuO_4: In 1960 Foëx, Mancheron, and Line studied (115) the high-temperature reaction of La_2O_3 with NiO using a solar furnace and reported the formation of single crystals of La_2NiO_4, which they believed possessed the tetragonal K_2NiF_4 structure. La_2CoO_4 and La_2CuO_4 were also prepared in the same manner. They noted that these two systems crystallized in a similar, but more complex, structure. La_2CuO_4 was found to decompose above 1200°C to release oxygen and produce La_2O_3 and Cu_2O. A structural study of La_2CuO_4 was performed by Longo and Raccah (116) using a polycrystalline sample prepared at 1000 to 1200°C in air. These high-temperature conditions possibly generated an oxygen-deficient product of general composition La_2CuO_{4-x}, which would contain some Cu(I) ions. These authors concluded that the "unusual" structure for La_2CuO_4 (orthorhombic, space group Fmmm *versus* the tetragonal I4/mmm space group observed for K_2NiF_4) resulted from the Jahn-Teller distortion about the Cu^{2+} ion. This localized distortion generated an elongated octahedron which led to the unusually long c-axis and the high c/a ratio of 3.45 (normalized to the tetragonal K_2NiF_4 structure, which has a c/a of 3.27). The compound underwent a crystallographic transition near 260°C (orthorhombic to tetragonal structural change), with only a minor change in unit cell volume. The magnetic susceptibility could not be determined on the material because the signal was too weak to be measured on their vibrating sample magnetometer at 17 kGauss in the temperature range 4.2 to 300 K. From their studies, however, they concluded that the La_2CuO_4 system was antiferromagnetic, not exhibiting an anomaly at its Néel temperature.[2]

A second-order structural transformation in La_2CuO_4 at 233 ± 5°C was reported (118) in a thermal expansion study. The La_2CuO_4 product was prepared at high temperature (1100°C) by the solid-state reaction of the corresponding binary oxides. The material was found to decompose above 1200°C with the loss of oxygen. Samples of La_2CuO_4, prepared at 1200°C, then maintained at 750°C in vacuum, yielded products having the general composition La_2CuO_{4-x}, or

---

[2] Goodenough (117) stated in a seminar at GTE Labs. in August 1987, "at the time we did not ask the right questions."

$La_2^{3+}(Cu_{2x}^{+}Cu_{1-2x}^{2+})O_{4-x}$. The crystallographic transition temperature for this oxygen-deficient material now occurred at a higher temperature, 263°C.

In companion papers (119)(120), M. Arjomand and D.J. Machin presented a comprehensive study on ternary Ni and Cu oxide compounds. In this survey they outlined the preparation and characterization of several ternary oxides containing Cu and Ni ions in their normal, and higher oxidation states. In particular, their data on orthorhombic La_2CuO_4 suggested antiferromagnetic interactions (they also observed only a low, temperature-independent, magnetic moment in the 80-300 K region).

Structure of La_2CuO_4: A complete three dimensional structural determination was carried out by Grande, Müller-Buschbaum, and Schweizer (108) in 1977 on single crystals of La_2CuO_4. The structural results indicated an orthorhombic distortion (see Figure 21a) of the K_2NiF_4 structure, but was quite different, if not unique, from the structures of all the other lanthanide compounds with the chemical composition Ln_2CuO_4. The structural details on La_2CuO_4 indicated puckered layers of the square planar Cu(II) ions with two additional oxygen neighbors, above and below the plane. This La_2CuO_4 stucture contained Cu(II) in distorted octahedral sites which Longo and Raccah (116) had previously indicated, but the puckering of the layers required a lowering of symmetry and a different space group.

Some of the reported space groups and unit cell parameters (pre-1985) for La_2CuO_4 are presented below.

Space Group	Unit Cell Parameters (Å)			Normalized "c/a"	Ref.
	a	b	c		
Abma	5.406	5.370	13.15	3.452	108
Fmmm	5.363(5)	5.409(5)	13.17(1)	3.458	116
-	5.35_4	5.40_2	13.16	3.462	118
-	5.36	5.41	13.25	3.479	120
Fmmm	5.354(2)	5.400(6)	13.130(6)	3.453	121
-	5.357	5.405	13.15	3.456	122
Abma	5.366(2)	5.402(2)	13.149(4)	3.454	123
Abma	5.342	5.434	13.16	3.454	124
Abma	5.363(3)	5.409(3)	13.17(1)	3.458	125
(I4/mmm)	3.808	3.808	13.20	3.466	126
"I4/mmm"	3.807	3.807	13.17	3.459	127

A value of 3.414 has been calculated as the ideal "c/a" ratio for the K_2NiF_4 structure, if the coordination polyhedra were ideal. From the data presented above, it is evident that the unit cell of La_2CuO_4 has expanded along the c axis, caused by the Jahn-Teller distortion about the Cu^{2+} ion and the tilting of the elongated octahedral groups on the ab plane (see Figure 21b). The orthorhombic crystal class can be represented in a number of ways. By convention, the long axis is normally assigned the b axis, however, the close structural relationship between the orthorhombic La_2CuO_4 structure and the tetragonal K_2NiF_4 structure (possessing the long c axis), has resulted in labeling the long axis in the La_2CuO_4 cell as the non-conventional c axis. For this reason various authors have assigned different space groups, but structurally equivalent structures, to the orthorhombic cell. The space group representation Bmab is based on the unit cell with parameters a < b < c.

Electrical Properties on La_2CuO_4 and Substituted Derivatives: Several Ln_2CuO_4 and Ln_2NiO_4 compounds were prepared and studied (128) in India with an interest in determining their electrical transport properties. Most of the Ln_2CuO_4 compounds were found to be small band-gap semiconductors, but La_2CuO_4 and several Ln_2NiO_4 compounds were observed to show near-metallic behavior. The electrical resistivity and Seebeck coefficients of these systems are reported in the temperature range 125 to 1000 K, and 400 to 850 K, respectively. The physical property data were interpreted by Goodenough (129), utilizing an energy-band model, to suggest a strongly correlated d-type electron within narrow metal bands.

In 1974, the electrical conductivity of La_2CuO_4 was re-determined (130) on a powdered sample prepared at 900-950°C. A 4-probe d.c. cell was used and the electrical data obtained between 100 and 900°C indicated this material to be a metallic conductor. A slight anomaly in the conductivity was noted at ~310°C.

The electrical conducting properties of Ln_2CuO_4 and alkaline-earth substituted $La_{2-x}A_xCuO_4$ were studied (126)(131) by a Japanese group in 1973 and in 1977. Figure 22 illustrates some of their results. These authors also report the electrical resistivity data for several related compounds in the temperature range 300 to 1000°C.

Most of the alkaline-earth substituted compounds are small band-gap semiconductors. However, a metal-semiconductor transition

La₂CuO₄

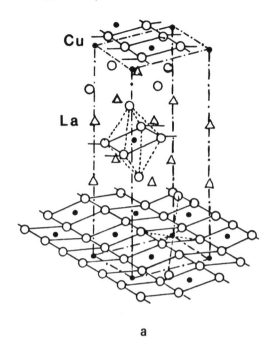

a

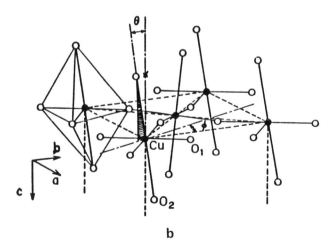

b

Figure 21: The La_2CuO_4 structure. a) Structure showing distorted octahedron and puckered CuO_2 planes. b) Illustration of the tilted octahedra ($\theta = \sim 4.5°$), and rotation about the $\underline{b}$ axis ($\phi = \sim 1°$). From Reference 108.

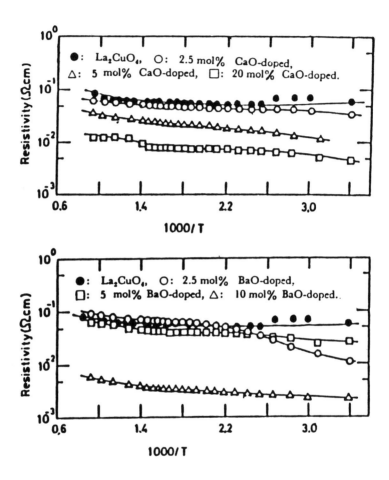

Figure 22: Log resistivity data *versus* the reciprocal of absolute temperature (T/K) for the families $La_{2-x}Ca_xCuO_4$ and $La_{2-x}Ba_xCuO_4$. From Reference 131.

was observed in those compounds having a shortened $\underline{a}$ crystallographic axis. The authors attributed this shortened axis to the additional π-bonding in the Cu-O linkages. Divalent Ca- and Ba-substitution for the La^{3+} ions in La_2CuO_4 was noted to decrease the electrical resistance to much lower values.

The resistivity behavior of La_2CuO_4, and several other substituted La_2CuO_4 derivatives, were studied (121) in the USSR. In many samples, the alkaline-earth ions were substituted for the La^{3+} ion. Their results indicated metallic behavior for La_2CuO_4 and all the measured alkaline-earth derivatives, but semiconductive behavior for the other Ln_2CuO_4 parent compounds and their alkaline-earth derivatives. Selected resistivity data for alkaline-earth doped La_2CuO_4 derivatives are presented in Figure 23.

In addition, the resistivity data for the mixed phases $(La,Pr)_2CuO_4$, and $(La,Tb)_2CuO_4$ were also investigated (122). The compounds Ln_2CuO_4, where Ln = La, Pr, Nd, Sm, Eu, and Gd, were also prepared and their electrical properties were determined. La_2CuO_4 was observed to be a near-metal, whereas all the other rare-earth compounds were found to be n-type semiconductors in the temperature region 78 to 670 K. Mixed compositions $Ln_{2-x}A_xCuO_4$, where A = Ca, Sr, and Ba, were also prepared and their electrical properties are reported in this same temperature region. Metallic-type behavior was once again noted only when the lanthanide ion was lanthanum. In one of their original drawings (see Figure 24 below), a transition region is shown in plot of the resistivity $\underline{vs}$. c/a axial ratio. This region outlines the compositional range where a semiconductor to metal transition is expected.

On the theoretical side, these authors suggested that the unpaired Cu^{2+} electron occupies a lower energy $4p_z$ metal orbital which then forms a π-bond with the $2p_z$ orbital of oxygen, thus mediating the formation of Cu-O conduction bands with covalent character.

Other Ternary Copper Oxides: Two other ternary oxides of copper(II) are worthy of mention. These are $BaCuO_2$ and the Ln_2CuO_4 series in which Ln = Nd, Sm, Eu, and Gd. The former compound, $BaCuO_2$, was first reported in 1975 by Arjomand and Machin (120), and they briefly reported on its powder pattern and magnetic properties. Further studies on $BaCuO_2$ were carried out the

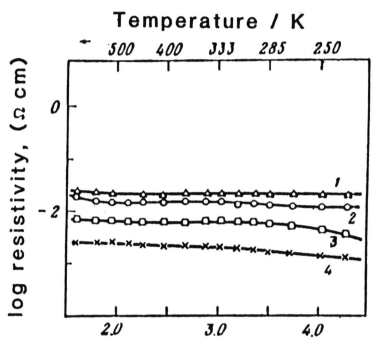

Figure 23: Log resistivity data *versus* the reciprocal of absolute temperature (T/K) for members of the solid solutions $Ln_{2-x}M_xCuO_4$

$1 = La_{1.8}Ca_{0.2}CuO_4$
$2 = La_{1.8}Sr_{0.2}CuO_4$
$3 = La_{1.6}Sr_{0.4}CuO_4$
$4 = La_{1.8}Ba_{0.2}CuO_4$ From Reference 121.

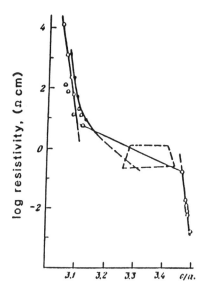

Figure 24: Plot of the specific resistance from a series of $Ln_{2-x}M_x CuO_4$ compounds as plotted against the "c/a" axial ratio. The broken trapezoid (dotted lines) outlines the hypothetical region for materials having a semiconductor to metal transition. From Reference 121.

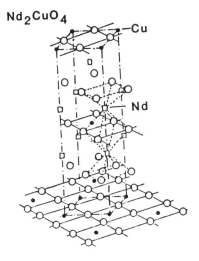

Figure 25: The Nd_2CuO_4 structure. Note the square-planar CuO_2 planes and the cubical arrangement of oxygen atoms about the Nd atoms. From Reference 135.

next year by a French group (132), who prepared the compound by the direct reaction of BaO_2 and CuO in air. A vacuum-treated sample was found to have a cubic structure. The magnetic properties indicated a ferromagnetic component below 45 K, and a magnetic moment of 1.87 μ_B. The crystal structure of $BaCuO_2$ was reported (133) in 1977. The compound has a most unusual structure composed of square-planar CuO_4 groups arranged in polyhedral clusters around the Ba^{2+} cations. The space group for this unusual structure is Im3m and the unit cell parameter is quite large, 18.27 Å. This cell contains some 360 atoms. The unique structure for $BaCuO_2$ has recently been confirmed (134) using powder neutron-diffraction techniques.

The physical properties of the series Ln_2CuO_4, where Ln = Nd, Sm, Eu, and Gd, were extensively studied during the early to mid-1970's. The structure of Nd_2CuO_4 (and Gd_2CuO_4) was found to be quite different from that of La_2CuO_4. Single crystal data for Nd_2CuO_4 were published (135) in 1975. The structure is shown in Figure 25.

Previous authors had noted the shortened "c/a" axis in these compounds, but they had concluded that the Jahn-Teller effect was either not operative, or that the distortion generated a compressed rather than elongated octahedron about the Cu^{2+} ion, i.e., four long and two short Cu-O bonds. Single crystal sudies showed a different structure in which the Cu atoms were surrounded by only four oxygen atoms in a square-planar arrangement. The symmetry was tetragonal, space group I4/mmm, with a = 3.945 and c = 12.171 Å. This structure is also quite different from that of K_2NiF_4, which has the same space group and crystal symmetry. The "c/a" ratio for this Ln_2CuO_4 series was anomalously lower than that for the La_2CuO_4-type compounds or the ideal octahedral K_2NiF_4 structure. The structure of Gd_2CuO_4 was also determined (108) and found to be isostructural with the neodymium prototype. Proper doping of the Nd_2CuO_4-type compounds with cerium or fluorine ions has since produced n-type superconductors having T_c's in the vicinity of 25 K. (See Section 4.4 below.)

4.2 Startling Discovery by Müller and Bednorz

As pointed out above in Section 4.1, the parent compound La_2CuO_4 was studied (115)(117)(126)(136) extensively between 1965 and 1980. Most of these investigations, however, were directed toward the synthesis and property determination of this unique K_2NiF_4-type compound. La_2CuO_4 is an example of a stable, high-temperature, semi-metal and much of the research was performed to examine its unique metallic conducting properties from room temperature to higher temperatures.

Alkaline-earth substitution in La_2CuO_4 was also investigated (121)(128)(131) during the early 1970's in the U.S.S.R., India, and Japan, but unfortunately, these materials were never investigated at sufficiently low temperatures for detection of their superconducting properties. This oversight resulted in a 16 year delay in the discovery of high temperature superconductivity in copper-oxide compounds.

Prior to the events of 1986-87, a substantial effort in the solid state chemistry of simple and complex copper oxides had been established in France. As early as 1980, structural chemists at Caen, under the direction of Michel and Raveau, studied the synthesis and structure of several ternary, quaternary, and multinary Cu-O compounds.

Raveau, et al. reported (137) a series of compounds with general composition: $La_{2-x}A_{1+x}Cu_2O_{6-(x/2)}$, where $\underline{x}$ varied between 0 and 0.14, and A = Sr or Ca. In 1981, the Caen group published (138) the synthesis, X-ray, and electron diffraction data for a series of orthorhombic and tetragonal phases in which Sr^{2+} was substituted for La^{3+} in perovskite-related structures. The values of $\underline{x}$ in the composition $La_{2-x}Sr_xCuO_{4-x/2+\delta}$ varied between the limits $0 < x < 1.34$. The low strontium substitution generated an orthorhombic phase, similar in structure to the La_2CuO_4-type phases, whereas at higher Sr-concentrations, i.e., $0.25 < x < 1.34$, they isolated compounds with the tetragonal K_2NiF_4 structure. Oxygen vacancies were generated at higher values of δ and the copper atoms adopted a square-planar coordination.

A comprehensive report which focussed on the $La_{2-x}Sr_xCuO_{4-x/2+\delta}$ series was published (139) in 1983 by this research group. In this broad review they reported the magnetic and electrical transport properties of these mixed-valent copper oxides in the temperature range 120-650 K. They concluded that the original semiconducting behavior in La_2CuO_4 transformed to semi-metallic behavior as the Cu^{3+} content increased with Sr-substitution. No experiments were conducted below 50 K, and therefore superconductivity was not observed. Three series of compounds, with $0.00 < x < 1.20$ were studied: 1) quenched, 2) annealed in oxygen, and 3) annealed in vacuum. Each series gave quite different experimental results indicating the critical importance of oxygen vacancies on the physical properties of the various copper-oxide products. Both orthorhombic ($0 < x < 0.16$) and tetragonal phases ($x > 0.16$) were prepared and investigated. The following year, another comprehensive review of mixed-valent copper oxides possessing perovskite-related structures was co-authored by Michel and Raveau (140). In this review they clearly state that these compounds can be considered as oxygen intercalation phases. Oxygen vacancies can easily be introduced and these vacancies can be re-populated by an oxygen anneal. These authors[3] also proposed three different classes of compounds (where, A = Ca, Sr, or Ba):

$Ba_3La_3Cu_6O_{14+\delta}$ oxygen-defect perovskite structure
$La_{2-x}A_{1+x}Cu_2O_{6-x/2+\delta}$ oxygen-defect intergrowths, Ruddlesden-Popper phases
$La_{2-x}A_xCuO_{4-x/2+\delta}$ oxygen-defect K_2NiF_4-type

During this same period of time, Michel and Raveau reported (141) the synthesis of $La_3Ba_3Cu_6O_{14+y}$, a new compound having some close relationship to the perovskite structure. This compound was the precursor to a variety of different copper-oxide derivatives; a more complete paper on its structure was published (142) in 1987, and the recent physical property measurements indicated (143) no superconductivity in either the quenched or oxygen-annealed phases.

---

[3] In passing, it should be noted that Michel and Raveau had carried out electrical measurements on several alkaline-earth substituted La_2CuO_4 compounds and presented (140) their experimental results in a comprehensive review during 1984. Their work, however, was primarily structural with a focus on electrical and magnetic properties above 50 K; for this reason they missed observing the superconducting transition at 30 K. It is this author's belief that these Cu-O phases should be called the "Raveau-Michel" phases in much the same way as the molybdenum sulfide superconductors are called the "Chevrel" phases.

The Discovery: Bernd T. Matthias, in 1971, made the statement (144):

"Although superconductivity at room temperature will always remain a pipedream, temperatures (T_c) as high as 25-30 K are a realistic possibility and will trigger a technological revolution."

A determined search for superconductivity in metallic oxides was initiated in mid-summer of 1983 at the IBM, Zurich Research Laboratories in Rüschliken, Switzerland. This research effort was an extension of previous work (145) on oxides, namely, $Sr_{1-x}Ca_xTiO_3$, which exhibited some unusual structural and ferro-electric transitions (see Section 2.2a). During the summer of 1985, the superconductivity research was focussed on copper-oxide compounds. Müller had projected the need for mixed Cu^{2+}/Cu^{3+} valence states, Jahn-Teller interactions (associated with Cu^{2+} ions), and the presence of room temperature metallic conductivity to generate good superconductor candidates. These researchers then became aware of the publication by Michel, Er-Rakho, and Raveau (146) entitled:

The Oxygen Defect Perovskite $BaLa_4Cu_5O_{13.4}$:
A METALLIC CONDUCTOR

The original 1985 paper on $BaLa_4Cu_5O_{13.4}$ discussed the preparation and presented limited electrical properties for the perovskite-type compound that contained a mixture of Cu^{2+}/Cu^{3+} ions. This compound exhibited metallic conductivity from 200 to 600 K; its thermoelectric power was reported for the same temperature region. Pauli paramagnetism was indicated, but no temperature range was given. This compound met all of Müller's criteria for superconductivity in a copper oxide material. Bednorz and Müller set out to prepare this particular compound, and also by varying the Ba-ion concentration, attempted the preparation of several members of a series of copper compounds having the generalized formula: $Ba_xLa_{5-x}Cu_5O_{5(3-y)}$. A thorough study of the electrical properties of the metallic products as a function of temperature was planned. The co-precipitation method was chosen in the synthesis approach, with the formation of a Ba-La-Cu-Oxalate precursor. Thermal decomposition of this mixed oxalate at 900°C for 5 hrs led to a product which contained three distinct phases. Undaunted by this experimental result, the researchers prepared a disk which was sintered in air at 900°C. On 27 January 1986, they found what they were seeking (147). Electrical resistivity data obtained on this disk by the conventional four-probe method indicated bulk superconductivity at **35 K** (see Figure 26).

These data were interpreted as the existence of some superconducting phase with an unusually high T_c. The results of these studies were published (148) in the seminal paper entitled:

POSSIBLE HIGH T_c SUPERCONDUCTORS IN THE Ba-La-Cu-O SYSTEM

X-Ray analysis of the product prepared by Bednorz and Müller suggested a mixture of the following three substances: CuO, a $Ba_xCu_{5-x}O_{5(3-y)}$ phase, and, a K_2NiF_4-type compound. This last phase was believed responsible for the superconductivity since the first two compounds were known to be semiconductors. The K_2NiF_4 system was also known, from previous studies, to generate highly interesting two-dimensional magnetic materials. Müller and Bednorz also realized that this two-dimensional K_2NiF_4 structure was an ideal system for superconductivity. The method used in the synthesis of mixed phases differed from that used by Michel et al. (141) in that the precursor route had been used, and that the thermal treatment of this oxalate precursor occurred at a much lower temperature. The authors believed that the lower reaction temperature was necessary for the formation of the layer-type superconducting phase. Superconductivity is lost if the annealing or sintering is carried out above 1000°C. Further studies in this system indicated that the layer-like oxide was definitely responsible for the observed superconducting properties. To establish the true existence of superconductivity in their Ba-La-Cu-O samples, magnetic susceptibility experiments were planned during the summer and fall of 1986. The results (149) of these magnetic measurements are shown in Figure 27.

In one sample, the crossover from the metallic (Pauli paramagnetic) region to the diamagnetic state occurred at 32 K. The diamagnetism measured was rather weak, on the order of 1% Meissner fraction, as compared to a pure superconductor ($-1/4\pi$, the full Meissner effect).

Historical Introduction and Crystal Chemistry of Oxide Superconductors 75

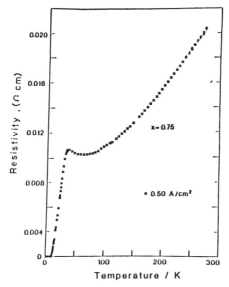

Figure 26: The resistivity of $Ba_{0.75}La_{4.25}Cu_5O_{15-5y}$ *versus* absolute temperature (T/K), as published by Bednorz and Müller in Reference 148.

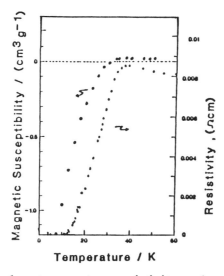

Figure 27: The low temperature resistivity and susceptibility of a (La,Ba)-Cu-O sample where a high-T_c was first observed. Re-drawn from Reference 147.

4.3 Corroboration of the Discovery and Further Developments

The first group to reproduce and confirm the startling discovery in the Ba-La-Cu-O system was that of Tanaka (150-154) at the University of Tokyo in October 1986. This group had studied oxide semiconductors for many years and in the 1970's they shifted their research interests to compound superconductors. Several binary chalcogenides were prepared and screened for superconductivity, and with the report of superconductivity in oxides by Sleight in 1975, they began an extensive study on the $BaPb_{1-x}Bi_xO_3$ phase. This oxide system was prepared in bulk powder form, as single crystals, and finally, as thin films. In November 1986, Tanaka (155) and co-workers became aware of Bednorz and Müller's results on superconductivity at 30 K in the mixed Ba-La-Cu-O phase. They repeated the preparation during the week of 6 to 13 November 1986 and observed a large diamagnetic effect in their sample; their corroborative results were submitted for publication (150) on 22 November. In a second report, submitted on 8 December 1986, they disclosed (151) the chemical composition of the superconducting phase to be $La_{2-x}Ba_xCuO_{4-y}$, possessing the K_2NiF_4-type structure. The onset temperature for superconductivity in their single-phase material exceeded 30 K, based on the magnetic measurements. The samples had been prepared using the ceramic method.

A series of three solid solutions $La_{2-x}(Ba,Sr,Ca)_xCuO_{4-\delta}$ was then prepared (152); the superconducting transition temperatures of the resulting products were determined from a.c. susceptibility measurements. The three-component phase diagram for this chemical system was plotted as a function of T_c. The maximum transition temperature of 37.0 K was observed for the composition, $La_{1.8}Sr_{0.2}CuO_{4-\delta}$, an end-member in the phase diagram.

With the Japanese confirmation, the world "exploded" into research activity on these high T_c superconductors.

Extension of the Discovery: Intensive research programs on

Historical Introduction and Crystal Chemistry of Oxide Superconductors 77

these high T_c (now defined as > 30 K) "ceramic",[4] superconductors began in November 1986 on at least four separate fronts. The effort in Japan, primarily under the direction of Prof. Tanaka (U. of Tokyo), has been previously described in this Section. In the USA, several industrial laboratories immediately reacted to the news of the startling discovery of superconductivity at 30-40 K. These included I.B.M. (Yorktown and Almaden), the Du Pont Co., and A.T.& T. Co. (Bell Labs and Bellcore). Several academic centers also initiated large research efforts on the La-Ba-Cu-O materials; including U. Houston, U. Alabama, Stanford U., UC-San Diego, and UC-Berkeley. Most of the U.S. Government Laboratories also joined the research effort by shifting existing resources to include research on the new ceramic superconductors. These included Argonne, NIST (formerly NBS), NRL, Los Alamos, Sandia, and Brookhaven Labs.

"Paul" Chu, and others at University of Houston, also reproduced Zurich's I.B.M. research results (156). Bell Lab's confirmation of Bednorz and Müller's discovery of high T_c superconductivity in copper oxide compounds was published (157) in the Jan. 1987 issue of Physical Review Letters. The electrical resistivity data from their work showing an onset of superconductivity at 36.5 K for the composition $La_{1.8}Sr_{0.2}CuO_4$ is plotted as Figure 28. This product also showed a 60-70% Meissner effect.

The reported Meissner fraction was an important property to follow during the early days of superconductivity research. The magnitude of this diamagnetic effect, when properly measured, gives a good indication of the homogeneous nature of the materials being prepared in different laboratories. The first reported values were low, 2 to 15%, but as the composition, structure, and synthetic conditions became better known, the values increased to 70-80%— indicative of bulk superconductivity in an essentially pure product.

---

[4] The word **ceramic** is a misnomer since these oxide compounds are all thermally unstable and tend to release oxygen or decompose on melting. Most ceramics are stable at high temperatures and normally do not melt. However, the so-called "ceramic method" is used in the preparation of these cuprate compounds.

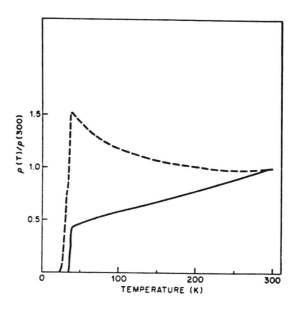

Figure 28: The relative resistivity of $La_{1.8}Sr_{0.2}CuO_4$ as a function of absolute temperature (T/K). The dotted line represents the data obtained from a sample annealed in air, whereas the solid line was obtained from an oxygen-annealed sample. From Reference 157.

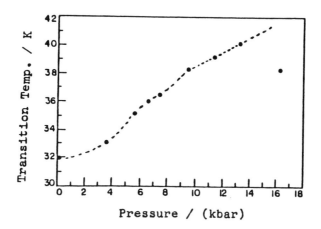

Figure 29: The onset transition temperature (T_c) *versus* pressure (kbar) for a La-Ba-Cu-O sample. The sample was damaged at 16.3 kbar pressure. The dotted line is only a guide to the eye. The data were presented in Reference 156.

Other countries also focussed their research activities on these new oxide materials, including, but not limited to, China, the USSR, Germany, France, India, Israel, and Australia.

Discovery of the 90+ K Superconductor: "Paul" Chu and coworkers at the University of Houston (during October 1986) carried out the synthesis of $(La_{1-x}Ba_x)CuO_{3-y}$ (Type I) and $(La_{1-x}Ba_x)_2CuO_{4-y}$ (Type II) compounds and isolated superconducting phases exhibiting a sharp decrease in resistivity at ~32 K. The best materials, however, showed only a 2% Meissner fraction. By applying pressure to one such product, their **forté** in superconductor research, they observed an increase in transition temperature of 8 degrees at ~14 kbar pressure (see Figure 29). Chu, et al., submitted (156) these results to Physical Review Letters on 15 December 1986, and the publication appeared in the January 26, 1987 issue.
The increase in T_c with increasing pressure immediately suggested to them, and to many other researchers, that the chemical substitution of a smaller ion in the La-site might yield a higher T_c superconductor (through an increase in internal "crystallographic" pressure). In February 1987, Chu and his colleagues discovered (158) a related class of superconducting compounds based on the Y-Ba-Cu-O system which had a superconducting transition temperature of **94 K!** — a critical transition temperature almost three times that of Bednorz and Müller's original result. Figure 30 reproduces their exciting experimental results and their amazing discovery.

Superconductivity was now a phenomenon that could be observed above the boiling point of liquid nitrogen, an inexpensive and readily available cryogen. Research in superconductivity, once the realm of low-temperature physics, now became every scientist's domain.

The crysto-chemical approach to this new copper oxide compound can be described quite simply using the following ionic radii considerations:

La^{3+} : 1.20 Å in X coord.
Y^{3+} : 1.10 Å in IX coord.
and : 1.015 Å in VIII coord.

From the ionic radii above, it is reasonable to expect that the

80 Chemistry of Superconductor Materials

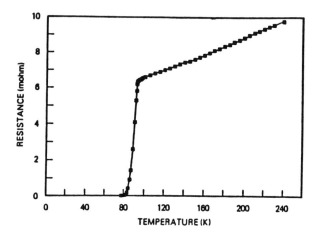

Figure 30: The original plot of resistance *versus* temperature (T/K) (taken in a simple liquid-nitrogen Dewar) for a sample of Ba-Y-Cu-O. This observation marked the beginning of superconductivity above liquid nitrogen temperature. Data from Reference 172.

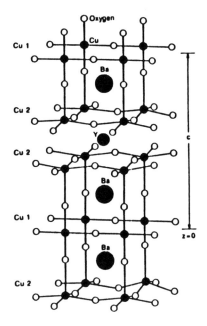

Figure 31: A schematic representation of the 1-2-3 structure, or the 90 K superconducting compound, $Ba_2YCu_3O_{6+x}$. Note the Cu(1) square-planar chains and the Cu(2) puckered planes in the structure.

Historical Introduction and Crystal Chemistry of Oxide Superconductors 81

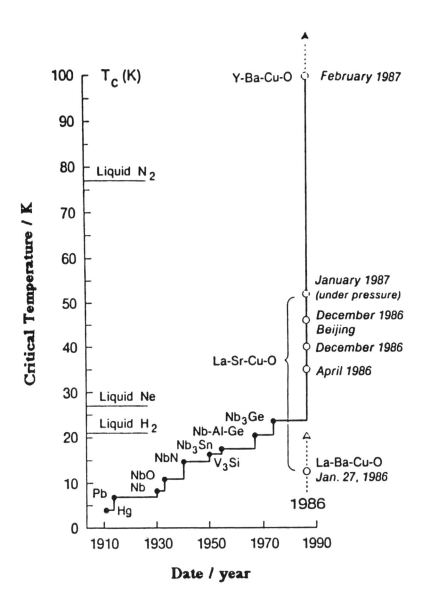

Figure 32: The increase in superconducting transition temperature (T_c / K) as a function of year, from the discovery of superconductivity in 1911 to 1987. The new Tl-Ba-Ca-Cu-O superconductors now hold the record at 110 to 125 K.

substitution of Y^{3+} for La^{3+} would be quite favorable. However, the chemical composition and structure of the new superconducting phase was not immediately known. Chu's product was also multiphase, appearing initially as a mostly-green product. This mixed product was immediately investigated in order to isolate the superconducting phase and determine its crystalline structure. The structure was determined (159-168) on a small black micro-crystalline sliver found in the green mass. In fact, a new structure had been formed; one more closely related to the cubic perovskite (an ordered, tripled cell) than the two-dimensional K_2NiF_4 structure. Once the structure and approximate composition was known, several groups began to prepare the 90 K superconductor in vast amounts. The composition was believed to be $Ba_2YCu_3O_{9-x}$, but the exact oxygen content and positions of these oxygens in the unit cell were not known. The oxygen content was also quite critical to the superconducting properties in this new compound. The latest structural version of this new copper oxide superconductor is shown in Figure 31.

A plot of the increasing superconducting transition temperature (T_c), *versus* year is now revised and this new version appears as Figure 32.

The "Explosion" in Research Activity on Cuprate Superconductors: The research effort in the area of copper oxide superconductors was gradual initially, the disclosure of high T_c's in these materials slowly disseminating into the scientific community. The seminal publication by Bednorz and Müller was slow to appear since it was submitted to a not-too-common journal - Zeitschrift für Physik B, Condensed Matter. The article was received by the journal on April 17, 1986 but was not published until September of that year. The arrival of the journal in the U.S. occurred in November 1986. Confirmation of these results began in Japan soon after the discovery because of the strong interaction between Japanese scientists and the Swiss I.B.M. group in Rüschlikon. M. Takashige of the University of Tokyo worked at the I.B.M. Zurich Laboratory during the summer and fall of 1986, and was a co-author (149) of the second article to be released by this Laboratory in late 1986. The magnetic confirmation of the Meissner effect on the La-Ba-Cu-O sample was submitted to Europhysics Letters (a newly-established European journal of Physics) in late fall and was published in early 1987. Preprints of this article reached a number of American scientists in November of that year.

At the University of Tokyo, Profs. Tanaka and Fueki rapidly formed an interdisciplinary research group directed at the synthesis and structural characterization of these new oxide superconductors. The Japanese academic and industrial centers had never abandoned superconductivity and always maintained a strong research effort in this area over the previous twenty-five years. One reason for the explosion of research activity in this area was that many of those who had "given up" research on superconductors were still "in the wings" waiting for this signal. Many of the physicists, metallurgists, and ceramists were "out there" but inactive. As Bednorz and Müller stated (147), "we expected that confirmation and acceptance of the Rüschlikon discovery could take as much as 2 to 3 years", but the many research groups with expertise in oxide superconductivity were "still in place" and poised for action.

The first oral announcement of Müller and Bednorz's discovery in the U.S. was presented (in an unscheduled talk) by Prof. K. Kitazawa (Tokyo Univ.) at the Materials Research Society (Symposium S on Superconductivity) in Boston, MA on Dec. 4, 1986. Kitazawa, who was working with Prof. Tanaka, had corroborated the original findings on La-Ba-Cu-O and he disclosed the exact composition and confirmed the high T_c. Although the meeting room was sparsely attended, the word on the high T_c materials, then only 30 to 35 K, soon spread quickly throughout the Physics community in the U.S. by telephone and "fax" machines.

In early 1987, the composition and structure of the La-Ba-Cu-O superconductor was still unknown to the general public in the United States. By March of that year certain facts became known from Japanese publications. But at this point in time, a newer, higher T_c (> 90 K) material was announced. This new copper oxide superconductor was quite easy to prepare and, in addition to interested physicists, these new materials could be synthesized by ceramists, chemists, metallurgists, material scientists, or anyone with a knowledge of a chemical approach to solid-state materials. Even high school students developed simple methods for the synthesis of these compounds. The "high" transition temperature and the possible use of liquid nitrogen made research in superconductivity accessible to most scientists and laboratories. The media also capitalized on this worthy news report and published it in newspapers and also presented it on television as a news item.

4.4 Major Copper Oxide Superconductors Presently Being Investigated

With the discovery and disclosure of these events in the area of "High T_c Superconductivity", hundreds, if not thousands, of scientists actively became involved in research on these new materials. Newer materials and higher T_c's soon followed. The competition was fierce and the progress through 1987 and 1988 was moving at a rapid pace with numerous important discoveries. To date, the highest T_c is in the range of 110-125 K, some five times that obtained in 1973 on the revolutionary (A-15) intermetallic materials. These new copper-oxide systems, many of which will be described in detail by other contributors to this book, are presented in Table 10.

5.0 CHEMICAL SUBSTITUTIONS— CRYSTAL CHEMISTRY

5.1 Chemical Substitutions in the La_2CuO_4 Structure

In a previous section, it was shown how the substitution of Y^{3+} for La^{3+} was an important step towards the formation of new compounds for superconductivity research. This substitution attempt led to the discovery of the 1-2-3 (or 90 K) superconducting compound. Substitutions similar to that proposed by Müller and Bednorz are now outlined. The type of chemical substitutions, based on atomic radii, which were proposed for La_2CuO_4, are presented in the following Table 11.

Many of these substitutions were carried out in Japan and in the U.S. immediately after the disclosure of high T_c superconductivity in the barium-doped La_2CuO_4 samples.

5.2 Chemical Substitutions in the Perovskite Structure

TABLE 10: "High T_c" Copper Oxide Superconductors.

Generation	Chemical Formula	T_c (K) range	Discovery date
First	$La_{2-x}M^{2+}_xCuO_{4-x}$ (M = Ca, Sr, Ba)	25-40	Apr. 1986
Second	$Ba_2Ln^{3+}Cu_3O_{6+x}$ (Ln = lanthanide)	90-95	Feb. 1987
Third	$(Tl,Bi)_m(Ba,Sr)_2Ca_{n-1}Cu_nO_{m+2n+2}$	20-120	Feb. 1988
	m = 1, 2 n = 1, 2, 3	"	Mar. 1988
Fourth	$Pb_2Sr_2Ln^{3+}_{0.5}Ca_{0.5}Cu_3O_8$	~ 70	Nov. 1988
Fifth	Ln-Tl-Ca-Sr-Cu-O	~ 90	Dec. 1988
Sixth	$Ln^{3+}_{2-x}Ce^{4+}_xCuO_{4-y}$ (Ln = Pr, Nd, Sm)	15-25	Jan. 1989
	(n-type superconductor)		

TABLE 11: Possible Substitution Studies in La_2CuO_4 According to the Ionic Sizes of Ions in Different Coordination Sites.

Coordinations:	IX	VI	VI
Formula:	La_2	Cu	O_4
Radii: (in Å units)	La^{3+} 1.20	Cu^{2+} 0.73	O^{2-} 1.40
Substitutions:	Ca^{2+} 1.18	Ni^{2+} 0.69	
	Sr^{2+} 1.28	Fe^{2+} 0.65	F^- 1.33
	Ba^{2+} 1.47		
	Other Ln^{3+} 1.06 - 0.85		

The arrangement of ions in the different sites of the perovskite structure (Figure 10) has been described above (Section 2.2). The number of different chemical species which crystallize with this structure (or in a distorted form) is legion. These compositions are listed in two major compilations (Refs. 169 and 170). The large A cation normally carries a low formal oxidation state, such as mono-, di-, or tri-valent. This would suggest that, for oxide systems, the B cation could be chosen from several of the transition metals, because they are of proper size and oxidation state to be accommodated in the octahedral sites. The various formal oxidation states for these metal ions would be 3+ to 5+, as demonstrated below:

$$A^+B^{5+}O_3 \qquad A^{2+}B^{4+}O_3 \qquad A^{3+}B^{3+}O_3$$

The simplest quaternary derivative with the perovskite structure would be one in which two different transition metals might occupy the B-site position. This can be formulated as $A(B'_{1/2}B_{1/2})O_3$, or preferably $A_2(B\ B')O_6$. These compounds can then crystallize with a doubled unit cell, if ordering occurs on the octahedral metal sites. Further compositional and structural adaptions could be obtained, as shown below, all possessing an overall 1:1:3 ratio of A:B:O atoms. In all the following examples and formulations, the proper stoichiometry will be maintained, and oxygen will be the principal anionic species.

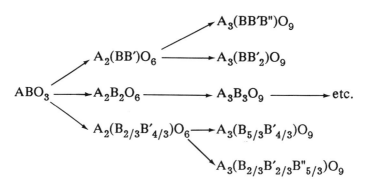

Additionally, the B metal ions can adopt different oxidation states, but maintaining the proper average oxidation state to effect charge balance. That could increase the total number of possibilities for new compositions to very large numbers, even when limited to the oxide anion.

The incorporation of Cu ions in the perovskite structure is known for only a few examples since this particular structure is normally stabilized by or requires a B atom in a high formal oxidation state such as Ti^{4+} in $BaTiO_3$, or Rh^{3+} in $LaRhO_3$. Further, since Cu can not be readily stabilized in its Cu(III) state, and is unknown in the tetravalent state, the simple formation of ternary compounds such as $LaCuO_3$ or $BaCuO_3$ is not expected. Even in the K_2NiF_4 structure, the stabilization of Cu^{4+} as in Ba_2CuO_4 is not expected, but the formation of a stable Cu(II) state is a distinct possibility, as in La_2CuO_4. Copper(II), however, has been introduced in the doubled- or tripled-perovskite structure. Examples of these, which include structural distortions from cubic symmetry, are listed:

Ca_2CuWO_6	not perovskite	Sr_2CuTeO_6	tetragonal
Sr_2CuWO_6	tetragonal	Ba_2CuUO_6	"
Ba_2CuWO_6	"		
$SrLaCuSbO_6$	tetragonal		
$Sr_3CuNb_2O_9$	tetragonal	$Sr_3CuTa_2O_9$	tetragonal
$Ba_3CuNb_2O_9$	"	$Ba_3CuTa_2O_9$	"
$Sr_3CuSb_2O_9$	"		
$Ba_3CuSb_2O_9$	hexagonal		

In addition to these perovskite-related compounds, Raveau and Michel have prepared a series of Cu-containing phases which are structurally related to the perovskite structure. Certain of these derivatives have been discussed in a previous section.

A Tripled-Perovskite — Relationship to the New High T_c Oxide Superconductor: A tripled cell for the perovskite structure is shown in Figure 33. This structure is closely related to that of the new 90 K superconductor in the following way. The structure of the superconductor can be derived from a highly anion-defect, perovskite-type structure. The composition can be obtained as follows.

$$3 \times ABO_3 = A_3B_3O_9 = A_2A'B_3O_9$$

Substituting the proper cations in the latter formula, the composition

$Ba_2YCu_3O_9$ can be formulated. Removing oxygen from CuO_6 octahedra to form square-planar groups around Cu, and incorporating the correct number of anion vacancies, the following composition $Ba_2YCu_3O_7$ is generated. The composition of the original 1-2-3 superconductor was first believed to be $Ba_2YCu_3O_{9-x}$, where $\underline{x}$ was approximately 2 but this varied considerably according to the particular thermal and oxygen-anneal treatment. Figure 31 also shows the structure determined for this high T_c superconductor. In this final structure we observe the formation of puckered CuO_2 sheets or layers and parallel CuO_3 chains which run perpendicular to these sheets. It is for these reasons that the superconductor has been referred to as a layered-perovskite type material. The structure, however, can have additional oxygen vacancies which leads to the variable composition $Ba_2YCu_3O_{7-x}$, or preferably $Ba_2YCu_3O_{6+x}$, since this latter formulaton is the thermally-stable, high-temperature phase.

These vertex-shared, square-planar groups form puckered sheets or layers about the yttrium atom thus generating an unusual eight (cubical) coordination site. This geometry appears to be quite important in forming this unique structure. Another closely related structure and composition is also found in this system, $Ba_2YCu_3O_6$. This stable compound, presumably formed at high temperature during the preliminary synthesis of the superconductor, is a semiconductor at room temperature and does not become superconducting above 3 K, its structure is shown in Figure 34. This compound, based on formal oxidation states for the copper atom, contains mixed Cu^{2+} and Cu^+ ions.

This oxygen variation and the Cu oxidation states play a very important role in the superconducting behavior of this compound. For example, the oxygen content (or x vacancy), the copper oxidation states, and the onset temperatures for superconductivity are listed below for different compositions.

Composition	x value	$Cu^{3+}/Cu^{2+}/Cu^+$ ratio	$T_c (K)$
$Ba_2YCu_3O_7$	0	1/2/0	~ 95
$Ba_2YCu_3O_{6.75}$	.25	0.5/2.5/0	~ 60
$Ba_2YCu_3O_{6.5}$	.5	0/3/0	~ 25
$Ba_2YCu_3O_{6.0}$	1.0	0/2/1	not superconducting

Historical Introduction and Crystal Chemistry of Oxide Superconductors 89

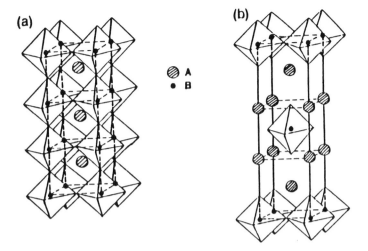

Figure 33: The polyhedral representation of the tripled perovskite-type structure as compared to the K_2NiF_4 structure.

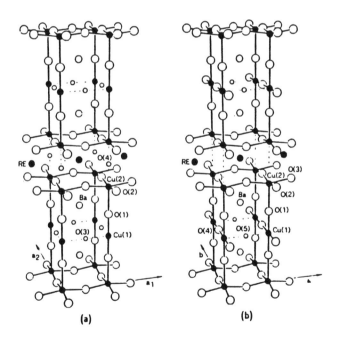

Figure 34: A comparision of Ba_2YCuO_7 and Ba_2YCuO_6 structures. From Reference 168.

The amount and positions (atomic locations) of oxygen atoms in the superconductors are highly critical and determine the properties of the superconductor. The oxygen vacancies (or deficiency) can be ordered in these materials. Neutron-diffraction experiments were required to determine the population parameters and the atomic positions of oxygen in these structures. The superconducting transition temperature in these "ceramic" oxides is a critical balance between the oxygen content and a proper mix of Cu^{2+} and Cu^{3+} ions generated in the anneal or post-heat treatment.

An outline of possible ion substitutions in the new 1-2-3 superconductor (90 K material) based on ionic radii is presented in Table 12.

Various substitution studies (171-173) were conducted in the early stages of research on these new oxide superconductors. One most dramatic result was the facile substitution of other (magnetic) lanthanide ions for yttrium in the VIII-coordinated site of the structure. The incorporation of these magnetic ions had no effect on the superconductivity nor the T_c of the material— quite astounding, since the presence of magnetic ions in superconductors was previously believed to destroy the phenomenon entirely! **Table 13** presents several examples of such substituted compounds.

Substitution of other transition metal ions for Cu, however, was observed (174) to be highly deleterious to superconducting behavior. **Table 14** shows the results of 10% metal-ion substitution in the Cu sites.

TABLE 14: Transition Temperatures for Metal-Substituted $Ba_2YCu_3O_{6+x}$.

$Ba_2YCu_2(Cu_{0.9}M_{0.1})O_{6+x}$	~T_c (K)
M = Cr	90
Ti	75
Ni	60
Fe	45
Co	30
Zn	10

The exact location of these substituted metal ions in the structure has not yet been elucidated in most cases.

Historical Introduction and Crystal Chemistry of Oxide Superconductors 91

TABLE 12: Possible Substitution Studies in $Ba_2YCu_3O_{6+x}$. According to the Ionic Sizes of Ions in Different Coordination Sites.

Coordination:	X	VIII	IV sq.	VI (or V)
Formula:	Ba_2	Y	Cu_3	O_{6+x}
Radii of ions:	Ba^{2+} 1.60	Y^{3+} 0.89	Cu^{2+} 0.62	O^{2-} 1.40
Substituted Ions (in Å)	Sr^{2+} 1.40	Ln^{3+} 1.06 to 0.85 (in VI)	Ni^{2+} 0.60	F^- 1.33
	Ca^{2+} 1.34		Fe^{2+} 0.63	S^{2-} 1.84
	Cd^{2+} 1.31		Ag^{3+} 0.65	N^{3-} 1.50
	Pb^{2+} 1.49	Tl^{3+} 1.00	Au^{3+} 0.70	
		Bi^{3+} 1.11	Pd^{2+} 0.64	Vary $\underline{x}$
		In^{3+} 0.923	Pt^{2+} 0.60	
	K^+ 1.60			
	Na^+ 1.32			
	Rb^+ 1.73			
	La^{3+} 1.32			
	Tl^{3+} 1.00			

TABLE 13: Transition Temperatures (in K) of $Ba_2MCu_3O_{6+x}$ and Substituted Analogs.

Compound	Crystal class	T_c^{onset}	T_c^{mid}	ΔT_c
Ba_2M series				
M = Y	Orthorhombic	98	94	2
Nd	Orthorhombic	91	70	~20
Sm	Orthorhombic	90	85	8
Eu	Orthorhombic	98	92	1
Gd	Orthorhombic	92	86	8
Dy	Orthorhombic	95	91	2
Ho	Orthorhombic	96	92	1
Er	Orthorhombic	94	87	5
Yb	Orthorhombic	93	90	2
Lu	Orthorhombic	91	85	~10
La	Cubic	77	60	29
BaMY series				
M = Ca	?	87	82	1
Sr	?	89	85	2
Ba	Orthorhombic	98	94	2
BaMYb series				
M = Ca	?	85	81	2
Sr	Orthorhombic	85	81	2
Other compositions				
BaCaLa	Cubic	83	80	6
$Ba_2Y_{.75}Sc_{.25}$	Ortho.	92	91	5
$Ba_2Y_{.5}Sc_{.5}$	Ortho.	94	90	4
$Ba_2Eu_{.75}Sc_{.25}$	Ortho.	96	93	5
$Ba_2Eu_{.9}Pr_{.1}$	Ortho.	85	82	5
$Ba_2Eu_{.9}Y_{.1}$	Ortho.	96	94.5	2
$Ba_2Eu_{.75}Y_{.25}$	Ortho.	96	95	2
$Ba_2La_{.5}Y_{.5}$	Ortho.	92	87	10
$Ba_2Lu_{.5}Y_{.5}$	Ortho.	96	92	1

6.0 REFERENCES

1. H. Kamerlingh Onnes. Akad. van Wetenschappen, Proceedings from the Section of Sciences.(Amsterdam), 14:113-115, and 818-821 (1911).

2. W. Meissner and R. Ochsenfeld, *Naturwissensch.* 21:787-788, (1933).

3. B.W. Roberts, "*Survey of Superconductive Materials and Critical Evaluation of Selected Properties.*" Physical and Chemical Reference Data, 5:581-821 (1976), contains nearly 2000 references.

4. J.K. Hulm and B.T. Matthias, *Science*, 208:881-887 (1980).

5. A.F. Wells, "*Structural Inorganic Chemistry*", Fifth Edition, Oxford University Press, Oxford & New York, (1984).

6. W. Meissner, H. Franz, and H. Westerhoff, *Ann. Physik*, 17:593-619 (1933).

7. A.R. Moodenbaugh, D.C. Johnston, R. Viswanathan, R.N. Shelton, L.E.DeLong, and W.A.Fertig, *J.Low Temp.Physics*, 33:175-203 (1978).

8. J.E. Mercereau, in "*Superconductivity in Science and Technology*", M.H.Cohen, Ed., Univ.of Chicago Press, Chicago, pp. 63-76 (1968).

9. B.T. Matthias and J.K. Hulm, *Phys.Rev.*, 87:799-806 (1952).

10. B.T. Matthias, T.H. Geballe, and V.B. Compton, *Rev.Modern Phys.*, 35:1-22 (1963).

11. B.W. Roberts, Properties of Selected Superconductive Materials, 1978 Supplement, National Bureau of Standards, Technical Note 983, October (1978).

12. A.F. Hebard and M.A. Paalanen, *Phys.Rev.*, B 30:4063-4066 (1984).

13. P.L. Gammel, A.F. Hebard, and D.J. Bishop, in "*Novel Superconductivity*", S.A. Wolf, and V.Z. Kresin, Eds., Plenum Press, New York, pp.85-93 (and also p.19) (1987).

14. J. Kirschenbaum, *Phys.Rev.*, B 12:3690-6 (1975).

15. J.F. Schooley, W.R. Hosler, and M.L. Cohen, *Phys.Rev.Lett.*, 12:474-5 (1964).

16. Ch.J. Raub, A.R. Sweedler, M.A. Jensen, S. Broadston, and B.T. Matthias, *Phys.Rev.Lett.*, 13:746-7 (1964).

17. J.K. Hulm, C.K. Jones, R. Mazelsky, R.C. Miller, R.A. Hein, and J.W. Gibson, *Proc.IXth Int.Conf.Low Temp.Phys.*, Plenum Press, New York, (1965). Part A., p.600-603.

18. A.W. Sleight, T.A. Bither, and P.E. Bierstedt, *Solid State Commun.*, 7:299-300 (1969).

19. M.B. Robin, K. Andres, T.H. Geballe, N.A. Kuebler, and D.B. McWhan, *Phys. Rev. Lett.*, 17:917-919 (1966).

20. D.C. Johnston, H. Prakash, W.H. Zachariasen, and R. Viswanathan, *Mater. Res. Bull.*, 8:777-784 (1973).

21. A.W. Sleight, J.L. Gillson, and P.E. Bierstedt, *Solid State Commun.*, 17:27-28 (1975).

22. B.T. Matthias, *Science*, 196:966-968 (1977).

23. W.A. Fertig, D.C. Johnston, L.E. DeLong, R.W. McCallum, M.B. Maple, and B.T. Matthias, *Phys. Rev. Lett.*, 38:987-990 (1977).

24. O. Fischer and M.B. Maple, Eds., *"Superconductivity in Ternary Compounds I, Structural, Electronic and Lattice Properties"*, Springer-Verlag, New York (1982).

25. M.B. Maple and O. Fischer, Eds., *"Superconductivity in Ternary Compounds II, Superconductivity and Magnetism"*, Springer-Verlag, New York (1982).

26. N.B. Brandt, S.V. Kuvshinnikov, A.P. Rusakov, and V.M. Semenov, *Sov. Phys. JETP Lett.*, 27:33-38 (1978).

27. C.W. Chu, A.P. Rusakov, S. Huang, S. Early, T.H. Geballe, and C.Y. Huang, *Phys. Rev.B*, 18:2116-2123 (1978).

28. M.R. Beasley and T.H. Geballe, *Physics Today*, 60-68, October (1984).

29. N.B. Hannay, T.H. Geballe, B.T. Matthias, K. Andres, P. Schmidt, and D. MacNair, *Phys. Rev. Lett.*, 14:225-226 (1965).

30. H. Kamimura, *Physics Today*, 64-71, December (1987).

31. *Chem. & Engin. News*, 25-27, October 4 (1982).

32. W.A. Little, *Phys. Rev.* A, 134:1416-1424 (1964).

33. W.A. Little, *Sci. Amer.*, 212:21-27 February (1965).

34. S. Kagoshima, H. Nagasawa, and T. Sambongi, "*One-Dimensional Conductors*", Springer-Verlag, New York (1982)

35. D.O. Cowan and F.M. Wiygul, "The Organic Solid State", pp. 28-45., *Chem. & Engin. News*, July 21 (1986).

36. R.L. Greene and P.M. Chaikin, "Organic Superconductors", *Physica* B, 126:431-440 (1984).

37. D. Jerome and H.J. Schulz, "Organic Conductors and Superconductors", *Advan. Phys.*, 31:299-490 (1982).

38. J.M. Williams, "Organic Superconductors" pp. 183-220, in "*Progress in Inorganic Chemistry*", Vol 33, S.J. Lippard, Ed., John Wiley & Sons, New York (1985).

39. J.M. Williams, H.H. Wang, T.J. Emge, U. Geiser, M.A. Beno, P.C.W. Leung, K.D. Carlson, R.J. Thorn, A.J. Schultz, and M-H. Whangbo, "Rational Design of Synthetic Metal Superconductors", pp. 51-218, in "*Progress in Inorganic Chemistry*", Vol 35, S.J. Lippard, Ed., John Wiley & Sons, New York (1987).

40. J.M. Williams and K. Carneiro, "Organic Superconductors: Synthesis, Structure, Conductivity, and Magnetic Properties", pp. 249-296, in "*Advances in Inorganic Chemistry & Radiochemistry*", Vol 29, Academic Press, Inc. (1985).

41. J.S. Miller, A.J. Epstein, "One-Dimensional Inorganic Complexes", pp. 1-151, in "*Progress in Inorganic Chemistry*", Vol 20, S.J. Lippard, Ed., John Wiley & Sons, New York (1976).

42. F. Wudl, "From Organic Metals to Superconductors: Managing Conduction Electrons in Organic Solids", *Accounts Chem. Res.*, 17:227-232 (1984).

43. N.J. Doyle, J.K. Hulm, C.K. Jones, R.C. Miller, and A. Taylor, *Phys. Lett.*, 26A:604-5 (1968).

44. T.B. Reed, M.D. Banus, M. Sjöstrand, and P.H. Keesom, *J. Appl. Phys.*, 43:2478-9 (1972).

45. M.D. Banus, *Mater. Res. Bull.*, 3:723-734 (1968).

46. G.F. Hardy and J.K. Hulm, *Phys. Rev.*, 93:1004-1016 (1954).

47. J.K. Hulm, M.S. Walker, and N. Pessall, *Physica*, 55:60-68 (1971).

48. A.M. Okaz and P.H. Keesom, *Phys. Rev.*, B 12:4917-4928 (1975).

49. J.K. Hulm, C.K. Jones, R.A. Hein, and J.W. Gibson, *J. Low Temp. Phys.*, 7:291-307 (1972).

50. H.R. Khan, Ch. J. Raub, W.E. Gardner, W.A. Fertig, D.C. Johnston, and M.B. Maple, *Mater. Res. Bull.*, 9:1129-36 (1974).

51. M.L. Cohen, *Phys. Rev.*, 134 A:511-521 (1964).

52. M.L. Cohen, *Rev. Modern Phys.*, 36:240-243 (1964).

53. M.L. Cohen, in Vol. 1, "*Superconductivity*", R.D. Parks, Ed., M. Dekker, New York, pp. 615-664 (1969).

54. J.F. Schooley, W.R. Hosler, E. Ambler, M.L. Cohen, and C.S. Koonce, *Phys. Rev. Lett.*, 14:305-7 (1965).

55. A.D. Wadsley, "*Non-Stoichiometric Compounds*", L. Mandelcorn, Ed., Academic Press, New York, p. 134 (1964).

56. Ch.J. Raub, *J. Less-Common Metals*, 137:287-295 (1988).

57. T.E. Gier, D.C. Pease, A.W. Sleight, and T.A. Bither, *Inorg. Chem.*, 7:102-103 (1968).

58. J.E. Ostenson, H.R. Shanks, and D.K. Finnemore, *J. Less-Common Metals*, 62:149-153 (1978).

59. A.R. Sweedler, Ch.J. Raub, and B.T. Matthias, *Phys. Lett.*, 15:108-9 (1964).

60. J.P. Remeika, T.H. Geballe, B.T. Matthias, A.S. Cooper, G.W. Hull, and E.M. Kelly, *Phys. Lett.*, 24A:565-566 (1967).

61. H.R. Shanks, *Solid State Commun.*, 15:753-756 (1974).

62. A.R. Sweedler, J.K. Hulm, B.T. Matthias, and T.H. Geballe, *Phys. Lett.*, 19:82 (1965).

63. P.E. Bierstedt, T.A. Bither, and F.J. Darnell, *Solid State Commun.*, 4:25-26 (1966).

64. R.J. Bouchard and J.L. Gillson, *Inorg. Chem.*, 7:969-972 (1968).

65. A.R. Aristimuno, H.R. Shanks, and G.C. Danielson, *J. Solid State Chem.*, 32:245-247 (1980).

66. F.F. Hubble, J.M. Gulick, and W.G. Moulton, *J. Phys. Chem. Solids*, 32:2345-2350 (1971).

67. L.H. Cadwell, R.C. Morris, and W.G. Moulton, *Phys. Rev. B*, 23:2219-2223 (1981).

68. R.K. Stanley, R.C. Morris, and W.G. Moulton, *Phys. Rev. B*, 20:1903-1914 (1979).

69. M.R. Skokan, W.G. Moulton, and R.C. Morris, *Phys. Rev. B*, 20:3670-3677 (1979) and *Phys. Rev.* Abstracts, 7, No. 1, p. 5 (1976).

70. M.J. Sienko, *Adv. Chem*, 39:224-236 (1963).

71. A.R. Sweedler, Ph.D. Thesis, Stanford Univ., (1969)., (as quoted by D.R. Wanlass and M.J. Sienko in *J. Solid State Chem.*, 12:362-369 (1975).

72. R.A. Beyerlein, A.J. Jacobson, and L.N. Yacullo, *Mater. Res. Bull.*, 20:877-886 (1985).

73. Y. Saito, T. Maruyama, and A. Yamanaka, *Thermochim. Acta*, 1150:199-205 (1987).

74. D.E. Cox and A.W. Sleight, *Solid State Commun.*, 19:969-973 (1976).

75. D.E. Cox and A.W. Sleight, *Acta Crystallogr.*, B35:1-10 (1979).

76. G. Thornton and A.J. Jacobson, *Acta Crystallogr.*, B34:51-354 (1978).

77. E.T. Shuvaeva and E.G. Fesenko, *Kristallografiya*, 14:1066-1068 (1970). *Soviet Phys.-Crystallogr.*, 14:926-927 (1970).

78. C. Chaillout, A. Santoro, J.P. Remeika, A.S. Cooper, G.P. Espinosa, and M. Marezio, *Solid State Commun.*, 65:1163-1169 (1988).

79. C. Chaillout, J.P. Remeika, A. Santoro, and M. Marezio, *Solid State Commun.*, 56:829-831 (1985).

80. C. Chaillout and J.P. Remeika, *Solid State Commun.*, 56:833-835 (1985).

81. Y. Khan, K. Nahm, M. Rosenberg, and H. Willner, *Phys. Stat. Sol.*, 39A:79-88 (1977).

82. A.M. Gabovich and D.P. Moiseev, *Sov. Phys. Usp. (Engl. Trans.)*, 29:1135-1150 (1986).

83. C.W. Chu, T.H. Lin, M.K. Wu, P.H. Hor, and X.C. Jin, in *"Solid State Physics under Pressure"*, pp. 223-228, S. Minomura, Ed., Terra Scientific Publ. Co. (1985).

84. J.B. Clark, F. Dachille, and R. Roy, *Solid State Commun.*, 19:989-991 (1976).

85. C.W. Chu, S. Huang, and A.W. Sleight, *Solid State Commun.*, 18:977-979 (1976).

86. R.J. Cava, B. Batlogg, G.P. Espinosa, A.P. Ramirez, J.J. Krajewski, W.F. Peck Jr., L.W. Rupp, Jr., and A.S. Cooper, *Nature*, 339:291-293 (1989).

87. L.F. Mattheiss, E.M. Gyorgy, and D.W. Johnson, Jr., *Phys. Rev. B*, 37:3745-3746 (1988).

88.a) R.J. Cava, B. Batlogg, J.J. Krajewski, R. Farrow, L.W. Rupp, Jr., A.E. White, K. Short, W.F. Peck, and T. Kometani, *Nature*, 332:814-816 (1988).

88.b) L.F. Schneemeyer, J. K. Thomas, T. Siegrist, B. Batlogg, L.W. Rupp, R.L. Opila, R.J. Cava, and D.W. Murphy, *Nature*, 335: 421-423 (1988).

89. S. Kondoh, M. Sera, K. Fukada, Y. Ando, and M. Sato, *Solid State Commun.*, 67:879-881 (1988).

90. N.L. Jones, J.B. Parise, R.B. Flippen, and A.W. Sleight, *J. Solid State Chem.*, 78:319-321 (1989).

91. D.G. Hinks, B. Dabrowski, J.D. Jorgensen, A. W. Mitchell, D.R. Richards, Shiyou Pei, and Donglu Shi, *Nature*, 333: 836-838 (1988).

92. J.E. Schirber, B. Morosin, and D.S. Ginley, *Physica C*, 157:237-239 (1989).

93. D.C. Johnston, *J. Low-Temp. Phys.*, 25:145-175 (1976).

94. J.H. Lin, T.H. Lin, and C.W. Chu, *J. Low-Temp. Phys.*, 58:363-369 (1985).

95. Y. Ueda, T. Tanaka, K. Kosuge, M. Ishikawa, and H. Yasuoka, *J. Solid State Chem.*, 77:401-40 (1988).

96. S. Foner and E.J. McNiff, Jr., *Solid State Commun.*, 20:995-998 (1976).

97. R.W. McCallum, D.C. Johnston, C.A. Luengo, and M.B. Maple, *J. Low-Temp. Phys.*, 25:177-193 (1976).

98. R.N. Shelton, D.C. Johnston, and H. Adrian, *Solid State Commun.*, 20:1077-1080 (1976).

99. U. Roy, A. Das Gupta, and C.C. Koch, *IEEE Trans. Magnetics*, (Mag-13):836-837 (1977).

100. A.H. Mousa and N.W. Grimes, *J. Mater. Sci.*, 15:793-795 (1980).

101. Hk. Müller-Buschbaum, "Oxometallates with Planar Coordination." *Angew. Chem.*, Int. Ed. Engl., 16:674-687 (1977).

102. J.B. Goodenough, G. Demazeau, M. Pouchard, and P. Hagenmuller, *J. Solid State Chem.*, 8:325-330 (1973).

103. G. Demazeau, C. Parent, M. Pouchard, and P. Hagenmuller, *Mater. Res. Bull.*, 7:913-920 (1972).

104. A.W. Webb, E.F. Skelton, S.B. Qadri, E.R. Carpenter, Jr., M.S. Osofsky, R.J. Soulen, and V. LeTourneau, *Phys. Letters A*, 137:205-7 (1989).

105. A.G. Nord and P. Kierkegaard, "Statistics of Divalent-Metal Coordination Environments in Inorganic Oxide and Oxosalt Crystal Structures.", *Chemica Scripta*, 24:151-8 (1984).

106. S. Åsbrink and L.-J. Norrby, *Acta Crystallogr.*, B26:8-15 (1970).

107. C. Michel and B. Raveau, *J. Solid State Chem.*, 43:73-80 (1987).

108. B. Grande, Hk. Müller-Buschbaum, and M. Schweizer, *Z. Anorg. Allg. Chem.*, 428:120-124 (1977).

109. A. Riou, *Bull. Soc. Fr. Min. Crist.*, 97:405-410 (1974).

110. L.W. Finger, *Amer. Mineral.*, 70:197-199 (1985).

111. R.D. Shannon and C. Calvo, *Acta. Cryst.*, B29:1338-1345 (1973).

112. L. Er-Rakho, C. Michel, and B. Raveau, *J. Solid State Chem.* 73:514-519 (1988).

113. J. Chenavas, J.C. Joubert, M. Marezio, and B. Bochu, *J. Solid State Chem.*, 14:25-32 (1975).

114. M. Marezio, P.D. Dernier, J. Chenavas, and J.C. Joubert, *J. Solid State Chem.*, 6:16-20 (1973).

115. M. Foëx, A. Mancheron, and M. Line, *Compt. rend. Acad. Sci.* 250:3027-8 (1960).

116. J.M. Longo and P.M. Raccah, *J. Solid State Chem.*, 6:526-531 (1973).

117. J.B. Goodenough, J.M. Longo, and P.M. Raccah, Lincoln Labs., research conducted between 1952-1976.

118. P. Lehuede and M. Daire, *Compt. rend. Acad. Sci.* Paris, 276C:1011-3 (1973).

119. M. Arjomand and D.J. Machin, *J. Chem. Soc.*, Dalton, 1055-1061 (1975).

120. M. Arjomand and D.J. Machin, *J. Chem. Soc.*, Dalton, 1061-1066, (1975).

121. I.S. Shaplygin, B.G. Kakhan, and V.B. Lazarev., *Z. Neorg. Khim.*, 24:1478-148 (1979), or *Russ. J. Inorg. Chem. Engl. Transl.* 24, #6:820-824 (1979).

122. A.A. Zakharov, V.B. Lazarev., and I.S. Shaplygin, *Z. Neorg. Khim.*, 29:788-79 (1984), or *Russ. J. Inorg. Chem. Engl. Transl.* 29, #3, 454-456, (1984).

123. N. Nguyen, J. Choisnet, M. Hervieu, and B. Raveau, *J. Solid State Chem.*, 39:120-127 (1981).

124. S. Saez Puche, M. Norton, and W.S. Glaunsinger, *Mater. Res. Bull.*, 17:1429-1435 (1982).

125. K.K. Singh, P. Ganguly, and C.N.R. Rao, *Mater. Res. Bull.*, 17:493-500 (1982).

126. T. Kenjo and S. Yajima, *Bull. Chem. Soc. Japan*, 46:1329 1333, (1973).

127. P. Ganguly and C.N.R. Rao, *J. Solid State Chem.*, 53:193 216, (1984).

128. P. Ganguly and C.N.R. Rao, *Mater. Res. Bull.*, 8:405-412 (1973).

129. J.B. Goodenough, *Mater. Res. Bull.*, 8:423-432 (1973).

130. A.M. George. I.K. Gopalakrishnan, and M.D. Karkhanavala, *Mater. Res. Bull.*, 9:721-726 (1974).

131. T. Kenjo and S. Yajima, *Bull. Chem. Soc. Japan*, 50:2847-2850 (1977).

132. H.N. Migeon, F. Jeannot, M. Zanne, and J. Aubry, *Rev. Chim. Miner.*, 13:440-445 (1976).

133. R. Kipka, and Hk. Müller-Buschbaum, *Z. Naturforsch.*, 32b:121-123 (1977).

134. M.T. Weller, and D.R. Lines, *J. Chem. Soc., Chem. Commun.*, 484-485, (1989).

135. Hk. Müller-Buschbaum, and W. Wollschläger, *Z. anorg. allg. Chem.*, 414:76-80 (1975).

136. O. Schmitz-DuMont and H. Kasper, *Monat. Chem.*, 96: 506-515 (1965).

137. N. Nguyen, L. Er-Rakho, C. Michel, J. Choisnet and B. Raveau, *Mater. Res. Bull.*, 15:891-897 (1980).

138. N. Nguyen, J. Choisnet, M. Hervieu, and B. Raveau, *J. Solid State Chem.*, 39:120-127 (1981).

139. N. Nguyen, F. Studer, and B. Raveau, *J. Phys. Chem. Solids*, 44, #5:389-400 (1983).

140. C. Michel and B. Raveau, *Rev. Chim. Min.*, 21:407-425 (1984).

141. L. Er-Rakho, C. Michel, J. Provost, and B. Raveau, *J. Solid State Chem.* 37:151-156 (1981).

142. C. Michel, L. Er-Rakho, M. Hervieu, J. Pannetier, and B. Raveau., *J. Solid State Chem.*, 68:143-152 (1987).

143. W.I.F. David, W.T.A. Harrison, R.M. Ibberson, M.T. Weller, J.R. Grasmeder, and P. Lanchester, *Nature*, 328:328-329 (1987).

144. B.T. Matthias, *Physics Today*, 23-28, August (1971).

145. J.G. Bednorz and K.A. Müller, *Phys. Rev. Lett.*, 52:2289-2292 (1984).

146. C. Michel, L. Er-Rakho, and B. Raveau, *Mater. Res. Bull.*, 20:667-671 (1985).

147. K. Alex Müller and J. Georg Bednorz, "*The Discovery Of A Class Of High-temperature Superconductors*", *Science*, 237: 1133-1139 (1987).

148. J.G. Bednorz and K.A. Müller, *Z. Phys. B, Cond. Matter*, 64:189-193 (1986).

149. J.G. Bednorz, M. Takashige, and K.A. Müller, *Europhys. Lett.*, 3:379-385 (1987).

150. S. Uchida, H. Takagi, K. Kitazawa, and S. Tanaka, *Jpn. J. Appl. Phys.*, 26:L1-L2 (1987).

151. H. Takagi, S. Uchida, K. Kitazawa, and S. Tanaka, *Jpn. J. Appl. Phys.*, 26:L123-L124 (1987).

152. K. Kishio, K. Kitazama, N. Sugii, S. Kanbe, K. Fueki, H. Takagi, and S. Tanaka, *Chem. Lett.*, 429-432 (1987).

153. S. Uchida, H. Takagi, K. Kitazawa, and S. Tanaka, *Jpn. J. Appl. Phys.*, 26:L151-L152 (1987).

154. S. Uchida, H. Takagi, K. Kitazawa, and S. Tanaka, *Jpn. J. Appl. Phys.*, 26:L196-L197 (1987).

155. Shoji Tanaka, "Research on High-T_c Superconductivity in Japan", *Physics Today*, 53-57 December (1987).

156. C.W. Chu, P.H. Hor, R.L. Meng, L. Gao, Z.J. Huang, and Y.Q. Wang, *Phys. Rev. Lett.*, 58:405-407 (1987).

157. R.J. Cava, R.B. van Dover, B. Batlogg, and E.A. Rietman, *Phys. Rev. Lett.*, 58:408-410 (1987).

158. M.K. Wu, J.R. Asburn, C.J. Torng, P.H. Hor, R.L. Meng, L. Gao, Z.J. Huang, Y.Q. Wang, and C.W. Chu, *Phys. Rev. Lett.*, 58:908-910 (1987).

159. R.M. Hazen, L.W. Finger, R.J. Angel, C.T. Prewitt, N.L. Ross, H.K. Mao, C.G. Hadidiacos, P.H. Hor, R.L. Meng, and C.W. Chu, *Phys. Rev.*, B 35:7238-7241 (1987).

160. J.J. Capponi, C. Chaillout, A.W. Hewat, P. Lejay, M. Marezio, N. Nguyen, J.L. Soubeyroux, J.L. Tholence, and R. Tournier, Europhys. *Lett.*, 3:1301-1307 (1987).

161. G. Calestani and C. Rizzoli, *Nature*, 328:606-607 Aug. 13, (1987).

162. H. Steinfink, J.S. Swinnea, Z.T. Sui, H.M. Hsu, and J.B. Goodenough, *J. Amer. Chem. Soc.*, 109:3348-3353 (1987).

163. T. Siegrist, S. Sunshine, D.W. Murphy, R.J. Cava, and S.M. Zahurak, *Phys. Rev.*, B35:7137-7139 (1987).

164. Y. Le Page, W.R. McKinnon, J.M. Tarascon, L.H. Green, G.W. Hull, and D.M. Hwang, *Phys. Rev.*, B35:7245-7248 (1987).

165. W.I.F. David, W.T.A. Harrison, J.M.F. Gunn, O. Moze, A.K. Soper, P. Day, J.D. Jorgensen, M.A. Beno, D.W. Capone II, D.G. Hinks, I.K. Schuller, L. Soderholm, C.U. Segre, K. Zhang, and J.D. Grace, *Nature*, 327:310-312 May 28, (1987).

166. M.A. Beno, L. Soderholm, D.W. Capone II, er, D.G. Hinks, J.D. Jorgensen, I.K. Schuller, C.U. Segre, K. Zhang, and J.D. Grace, *Appl. Phys. Lett.*, 51:57-60 (1987).

167. H. Fjellvag, P. Karen, and A. Kjekshus, *Acta Chem. Scand.*, A41:283-293 (1987).

168. A. Ourmazd, J.A. Rentschler, J.C.H. Spence, M. O'Keeffe, R.J. Graham, D.W. Johnson, Jr., and W.W. Rhodes, *Nature*, 327:308-310 May 28 (1987).

169. J.B. Goodenough and J.M. Longo, "Crystallograpic and magnetic properties of perovskite and perovskite-related compounds." in Landolt-Börnstein's Numerical Data and Functional Relationships in *Science and Technology*, New Series, K.-H. Hellwege, Ed., Group III/Vol. 4a, Springer Verlag, Berlin, (pp. 126-314) (1970).

170. F.S. Galasso, "*Structure and Properties of Inorganic Solids*", Chapter 7, pp. 162-210, Pergamon Press, N.Y. (1970).

171. D.W. Murphy, S. Sunshine, R.B. van Dover, R.J. Cava, B. Batlogg, S.M. Zahurak, and L.F. Schneemeyer, *Phys. Rev. Lett.*, 58:1888-1890 (1987).

172. P.H. Hor, R.L. Meng, Y.Q. Wang, L. Gao, Z.J. Huang, J. Bechtold, K. Forster, and C.W. Chu, *Phys. Rev. Lett.*, 58: 1891-1894 (1987).

173. E.M. Engler, V.Y. Lee, A.I. Nazzal, R.B. Beyers, G. Lim, P.M. Grant, S.S.P. Parkin, M.L. Ramirez, J.E. Vazquez, and R.J. Savoy, *J. Am. Chem. Soc.*, 109:2848-2849 (1987).

174. G. Xiao, F.H. Streitz, A. Gavrin, Y.W. Du, and C.L. Chien, *Phys. Rev.*, 35B:8782-8784 (1987).

Part II
Structural and Preparative Chemistry

2
The Complex Chemistry of Superconductive Layered Cuprates

*Bernard Raveau, Claude Michel,
Maryvonne Hervieu, Daniel Groult*

The high T_c superconductive copper oxides form a very large family which is very attractive for many physicists owing to the apparent ease of synthesis of those materials. For this reason a tremendous number of compositions have been investigated since the beginning of the race to high critical temperature in 1987. Most of the results which were published correspond to mixtures of compounds due to the fact that the chemistry of these phases is more complex than expected by people who are not familiar with nonstoichiometry in oxides. The purity of those mixed valence copper oxides is closely related to their method of synthesis, which influences strongly the oxygen stoichiometry especially for $YBa_2Cu_3O_{7-\delta}$, but also the existence of extended defects for instance in thallium cuprates. Moreover, the structure of a great number of these oxides is complicated by incommensurability effects. This chapter tries to give an overview of all these problems in connection with the superconducting properties of these materials.

1.0 THE STRUCTURAL PRINCIPLES

More than twenty different cuprates have been found to exhibit superconducting properties. All of them are characterized by a bidimensional character of the copper oxygen framework, i.e. their structure is formed of copper oxygen superconductive layers separated

Complex Chemistry of Superconductive Layered Cuprates   107

by insulating layers.

The synthesis of these compounds is based on two properties:

- The great ability of the perovskite structure AMO_3 and of the rock salt type structure AO to adapt to each other, forming intergrowths $(AMO_3)_m(AO)_n$, as shown, for instance in titanates $(SrTiO_3)_mSrO$ (n=1) (1).

- The property of copper to take coordinations smaller than six, and especially pyramidal and square planar coordinations which favour again the bidimensional character within the perovskite layers.

Moreover, the mixed valence of copper, Cu(II) - Cu(III), is absolutely necessary for the delocalization of holes in the copper oxygen framework, leading to semi-metallic or metallic properties.

The general formulation of these oxides, $(ACuO_{3-x})_m(AO)_n$, reflects for each of them the number m of copper layers which form each perovskite slab, and the number n of AO layers which form each rock salt type slab (the AO layers which lie at the boundary of the perovskite slabs and rock salt type slabs can only be counted as for 1/2). Thus all these oxides (2-34) can be represented by the symbol [m,n] in which m,n will be integral numbers. In most of these oxides one observes for one compound only one m and n value, corresponding to single intergrowths.

The [1,n] oxides (Figure 1) which correspond to the formulation $A_{n+1}CuO_{3+n}$ are all characterized by single $[CuO_3]_\infty$ perovskite layers involving only CuO_6 octahedra. These oxides which exhibit the lowest T_c's — 7K to 50K — are obtained for n=1, i.e., La_2CuO_4 type oxides (Figure 1a); n=2, i.e., $Tl_{1-x}Pr_xSr_{2-y}Pr_yCuO_5$ (Figure 1b); and n=3 i.e., for $Tl_2Ba_2CuO_6$ and $Bi_2Sr_2CuO_6$ (Figure 1c).

The [2,n] oxides (Figure 2), which can be formulated $A_{n+1}Ca LnCu_2O_{5+n}$ exhibit generally higher T_c's — 60K to 85K — and are all characterized by double $[Cu_2O_5]_\infty$ layers of corner-sharing CuO_5

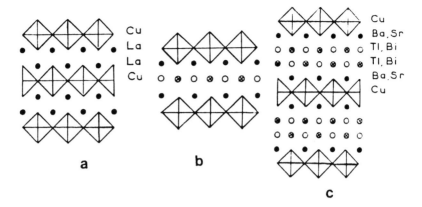

Figure 1: Schematic drawings of [1, n] oxides: (a) La_2CuO_4 type oxides (n=1); (b) $Tl_{1-x}Pr_xSr_{2-y}Pr_yCuO_5$-type oxides (n=2); (c) $Tl_2Ba_2CuO_6$-type oxides (n=3).

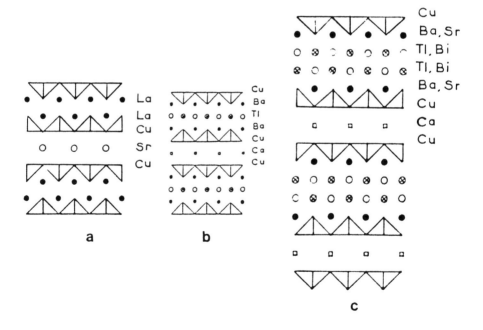

Figure 2: Schematic drawings of [2, n] oxides: (a) $La_{2-x}A_{1+x}Cu_2O_6$-type oxides (n=1) (A=Ca, Sr); (b) $TlBa_2CaCu_2O_7$-type oxides (n=2); (c) $Tl_2Ba_2CaCu_2O_8$-type oxides (n=3).

pyramids. Curiously the member n=1 which exhibits single rock salt type layers, $La_{2-x}A_{1+x}Cu_2O_6$ (A=Ca,Sr) (Figure 2a) does not superconduct. The members n=2 (Figure 2b) can be either superconductive as for $TlBa_2CaCu_2O_7$ (T_c=60K), $TlSr_2CaCu_2O_7$ (T_c=50K) and $Tl_{0.5}Pb_{0.5}Sr_2CaCu_2O_7$ (T_c=85K) or not superconductive as for $TlBa_2LnCu_2O_7$ (Ln=Pr, Y, Nd). The n=3 members (Figure 2c) which exhibit thallium or bismuth bilayers, are superconductive for $Tl_2Ba_2CaCu_2O_8$ (T_c=105K) and $Bi_2Sr_2CaCu_2O_8$ (T_c=85K), but tend to lose their superconducting properties by introduction of lead or of univalent thallium as shown for the oxides $Bi_{2-x}Pb_xSr_2Ca_{1-x}Y_xCu_2O_8$, and $Tl^{III}_{2-x}Ba_{1+x}Tl^I_xLnCu_2O_8$ (Ln=Pr, Nd, Sm).

The highest values of T_c are observed for the [3,n] oxides whose structure consists of triple copper layers built up from two $[CuO_{2.5}]_\infty$ pyramidal layers and one $[CuO_2]_\infty$ layer of corner-sharing CuO_4 square planar groups (Figure 3). These oxides, $A_{n+1}(Ca, Ln)_2 Cu_3O_{7+n}$, are represented for n=3 (Figure 3c) by $Tl_2Ba_2Ca_2Cu_3O_{10}$ (T_c=125K) and $Bi_{2-x}Pb_xSr_2Ca_2Cu_3O_{10}$ (T_c=110K) and for n=2 (Figure 3b) by $TlBa_2Ca_2Cu_3O_9$ (T_c=120K) and $Tl_{0.5}Pb_{0.5}Sr_2Ca_2Cu_3O_9$ (T_c=120K). The member n=1 (Figure 3a) is represented by $PbBaYSr Cu_3O_8$ which does not superconduct; nevertheless about 1% of the volume of this latter phase can become superconductive by replacing partially yttrium by calcium.

The [4,n] oxides, which are represented by the formulation $A_{n+1}(Ca,Ln)_3Cu_4O_{9+n}$ are characterized by quadruple copper layers involving two pyramidal $[CuO_{2.5}]_\infty$ copper layers and two $[CuO_2]_\infty$ layers of corner-sharing CuO_4 square planar groups (Figure 4). Only two oxides have been isolated in this series, $TlBa_2Ca_3Cu_4O_{11}$ ($T_c\approx$-108K) and $Tl_2Ba_2Ca_3Cu_4O_{12}$ ($T_c\approx$115K) which correspond to n=2 (Figure 4a) and to n=3 (Figure 4b) respectively. The [4,1] oxide has not yet been isolated.

The 92K superconductors $LnBa_2Cu_3O_7$ (35) belong also to this family (Figure 5). They are indeed characterized by the absence of rock salt type layers and an infinite number of copper layers involving $[CuO_2]_\infty$ rows of corner-sharing CuO_4 square planar groups and $[CuO_{2.5}]_\infty$ pyramidal layers. They can be represented by the symbol [0,∞].

The superconductive oxides $Pb_2Sr_2Ca_{1-x}Y_xCu_3O_8$ (36) and $Pb_{2-x}Bi_xSr_2Ca_{1-y}Y_yCu_3O_8$ (37) which exhibit a zero resistance at temperatures ranging from 46K to 79K respectively do not seem to

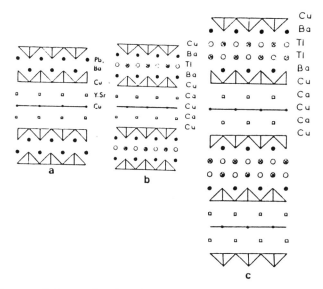

Figure 3: Schematic drawings of [3, n] oxides: (a) PbBaYSr Cu_3O_8-type oxides (n=1); (b) $TlBa_2Ca_2Cu_3O_9$-type oxides (n=2); (c) $Tl_2Ba_2Ca_2Cu_3O_{10}$-type oxides (n=3).

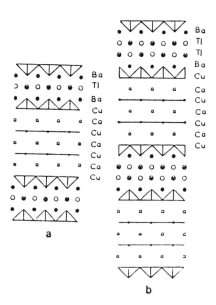

Figure 4: Schematic drawings of [4, n] oxides: (a) $TlBa_2Ca_3Cu_4O_{11}$-type oxides (n=2); (b) $Tl_2Ba_2Ca_3Cu_4O_{12}$-type oxides (n=3).

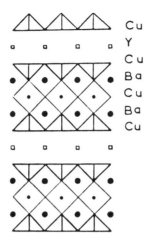

Figure 5: Idealized structure of the $LnBa_2Cu_3O_7$ superconductors.

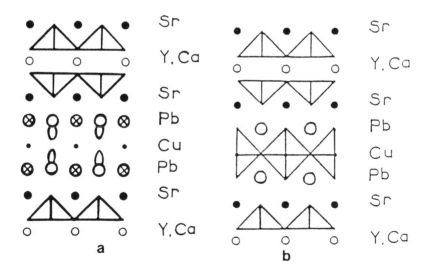

Figure 6: Idealized structures of $Pb_2Sr_2Y_xCa_{1-x}Cu_3O_8$-type oxides (a) and of the hypothetical $Pb_2Sr_2YCu_3O_{10}$ compound (b).

have a place at first sight in the above classification. In fact they correspond to a non integral m value (m=1.5) and to n=1 as will be shown from the analysis of their structure. This framework (Figure 6a) can be described as a stacking of double pyramidal copper layers like [2,n] oxides, and of CuO_2 sticks like in $YBa_2Cu_3O_6$, and of single rock salt type layers $[(Pb,Sr)O]_\infty$. Another description consists in the consideration of the hypothetical compound "$Pb_2Sr_2YCu_3O_{10}$" whose structure (Figure 6b) would be formed of identical rock salt layers and double copper layers and would differ by the existence of simple octahedral copper layers. Such a structure would correspond to an intergrowth of the [2,1] oxide and of the [1,1] oxide, and thus would be a double intergrowth. In fact the $Pb_2Sr_2CaCu_3O_8$-type oxides belong to this family since they are derived from this latter hypothetical structure by elimination of the oxygen atoms of the basal planes of the octahedra of the single perovskite layers. Thus, they can be described as a double intergrowth [1,1 | 2,1] in the above nomenclature leading to a non integral m value, i.e. [1.5,1]. These latter results also confirm that multiple intergrowths corresponding to non integral m or n values in the nomenclature [m,n] can be predicted.

$YBa_2Cu_4O_8$ (Figure 7a), the 80K superconductor which has been observed for the first time as a defect by high resolution electron microscopy in $YBa_2Cu_3O_{7-\delta}$ samples (38) and which has been synthesized as a pure phase by Bordet et al. (39), cannot be represented by the symbol [m,n]. Nevertheless it is closely related to those structures. It corresponds indeed to a sharing phenomenon similar to that observed in tungsten suboxides (40), leading to the formation of double rows of edge-sharing CuO_4 square planar groups. Both structures $YBa_2Cu_3O_7$ and $YBa_2Cu_4O_8$ can coexist in the same crystal so that a series of superconductive phases can be predicted, with the formula $(YBa_2Cu_3O_7)_n(YBa_2Cu_4O_8)_n$, for which the first member $Y_2Ba_4Cu_7O_{15}$ (n=n'=1) (Figure 7b) has been recently isolated (41).

The superconducting properties of $Nd_{2-x}Ce_xCuO_4$ (T_c=24 K) are very different from the other layered cuprates, since they involve electrons instead of holes (42). The structure of this phase which belongs to the Nd_2CuO_4-type (43) is closely related to La_2CuO_4. Both structures (Figure 8) exhibit identical position of the metallic atoms and of two oxygen atoms out of four forming $[CuO_2]_\infty$ layers of corner-sharing CuO_4 square planar groups. The different positions of the two oxygen atoms in Nd_2CuO_4 leads to a structure (Figure 8a),

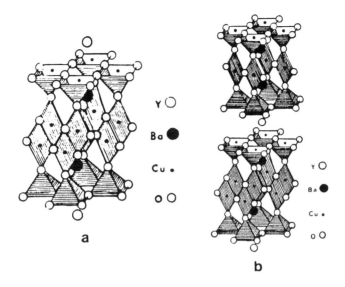

Figure 7: Schematic drawings of the structures of $YBa_2Cu_4O_8$ (a) and of $Y_2Ba_4Cu_7O_{15}$ (b).

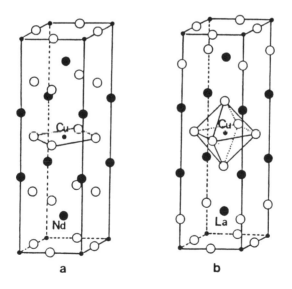

Figure 8: Idealized structure of the Nd_2CuO_4-type oxides (a) compared to that of the La_2CuO_4-type oxides (b).

that can be described as built up from double fluorite type layers $[Nd_2O_4]_\infty$ sharing the faces of their NdO_8 cubes. The analysis of this structure shows that the fluorite type structure can be adapted to the perovskite and rock salt type structures. The layered cuprates $Tl_{1+x}A_{2-y}Ln_2Cu_2O_9$ (A=Sr, Ba) open a new route to the investigation of superconductors although they do not superconduct (44). The structure of these oxides (Figure 9) can be described as an intergrowth of three types of layers: pyramidal layers derived from the perovskite structure, double rocksalt-type layers and double fluorite type layers. The double fluorite-type layers which are occupied by lanthanides are similar to those observed in Nd_2CuO_4. The rocksalt-type layers are similar to those observed in $TlBa_2CaCu_2O_7$; they are built up from two sorts of AO layers noted A1 and A2. The A1-layers correspond to the thallium monolayers observed in $TlBa_2CaCu_2O_7$, and are mainly occupied by Tl(III). The A2-layers are common to the perovskite pyramidal and to the rocksalt-type layers; i.e., they correspond to the $[BaO]_\infty$ layers in $TlBa_2CaCu_2O_7$. These latter layers are mainly occupied by barium and strontium but they exhibit a significant partial occupancy by Tl(I).

2.0 OXYGEN NON-STOICHIOMETRY AND METHODS OF SYNTHESIS

The ability of copper to take various coordinations as well as the great flexibility of the perovskite structure combine to allow significant deviations in oxygen stoichiometry that do not really change the structure but dramatically affect the superconducting properties.

It is indeed now well established that the La_2CuO_4-type oxides and $YBa_2Cu_3O_7$-type compounds must be prepared under an oxygen flow and especially annealed at 400°C in order to avoid the formation of oxygen vacancies. On the other hand, heating the bismuth cuprates in an oxygen flow destroys their superconducting properties likely owing to the partial oxidation of Bi(III) into Bi(V), thus these oxides must be prepared in air. In the same way, superconductive lead cuprates can only be synthesized in a relatively reducing atmosphere, i.e., in a nitrogen flow containing less than 2% O_2. The thallium

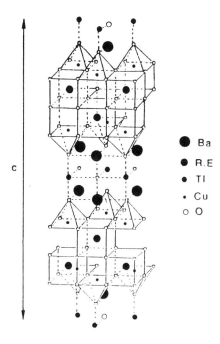

Figure 9: Idealized structure of the cuprates $Tl_{1+x}A_{2-y}Ln_2Cu_2O_9$ (A=Ba, Sr).

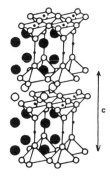

Figure 10: Schematic drawing of the insulating oxide $YBa_2Cu_3O_6$.

cuprates are also very sensitive to the oxygen pressure; the reduction of Tl(III) into Tl(I) tends to kill the superconductivity (45) so that these compounds are better prepared in evacuated ampoules using thallium trioxide and barium or strontium peroxides. This latter method also has the advantage of avoiding the volatilization of thallium oxide.

The $YBa_2Cu_3O_{7-\delta}$ oxide gives the most fascinating example of large deviation from oxygen stoichiometry - $0 \leq \delta \leq 1$ - which can affect the superconducting properties of this material (46)(47). The insulating limit compound $YBa_2Cu_3O_6$, which is obtained by heating the orthorhombic superconductive phase $YBa_2Cu_3O_7$ under argon, has a tetragonal symmetry. Its structure (Figure 10) is deduced from that of $YBa_2Cu_3O_7$ (Figure 5) by removing the oxygen atoms of the square planar groups located between the pyramidal layers. It results in pyramidal layers, in which the CuO_5 pyramids are only occupied by Cu(II), and layers of Cu^IO_2 sticks in which univalent copper exhibits the classical two-fold coordination.

The intermediate compositions - $0<\delta<1$ - raise the question of copper coordination. For instance the non-superconductive tetragonal phase $YBa_2Cu_3O_{6.3}$, obtained by quenching "$YBa_2Cu_3O_7$" in air, has its structure classically represented (Figure 11) as intermediate between $YBa_2Cu_3O_7$ and $YBa_2Cu_3O_6$, i.e. built up from identical pyramidal layers, but between these layers the oxygen vacancies are distributed at random in the basal planes of the CuO_6 octahedra. These results obtained either from single-crystal x-ray study or from x-ray or neutron diffraction data, correspond in fact to an average structure which does not reflect the actual copper coordination. Such an aleatory distribution of the oxygen vacancies would lead to three-fold coordination of copper which is not likely. It has not been possible to isolate single crystals of orthorhombic superconducting intermediate phases for a structure determination.

High resolution electron microscopy studies of several orthorhombic intermediate compositions reveal in fact an inhomogeneous distribution of oxygen in the crystals. For the oxides $YBa_2Cu_3O_{7-\delta}$, with $\delta=0.45$ and 0.37, which have a T_c of 60K, the electron diffraction investigations give evidence of streaks in all the recorded patterns. These streaks mainly lie along [100]* and [010]*, as shown in Figure 12. In the corresponding high resolution images, local superstructures are observed. Their principal mean directions that are

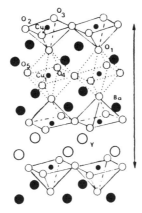

Figure 11: Structure of the tetragonal non-superconductive oxide $YBa_2Cu_3O_{7-\delta}$ ($\delta \approx 0.7$).

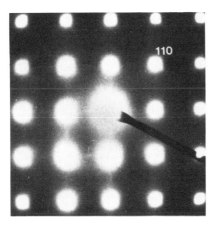

Figure 12: $YBa_2Cu_3O_{7-\delta}$ ($0.37 \leq \delta \leq 0.45$): typical [001] ED patterns showing streaks which lie mainly along [100]* and [010]*.

118 Chemistry of Superconductor Materials

Figure 13: $YBa_2Cu_3O_{7-\delta}$ (0.37≤δ≤0.45): 2 x a superstructure viewed along [010].

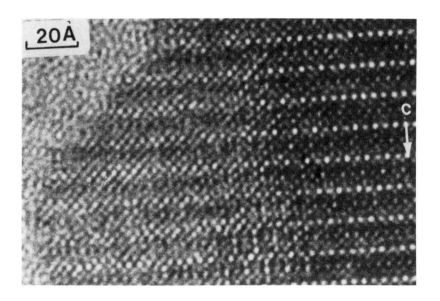

Figure 14: $YBa_2Cu_3O_{7-\delta}$ (0.37≤δ≤0.45): 2a x 2c superstructure ([010] image).

Complex Chemistry of Superconductive Layered Cuprates 119

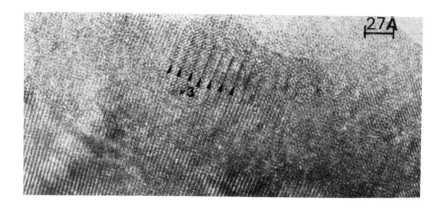

Figure 15: $YBa_2Cu_3O_{7-\delta}$ (0.37≤δ≤0.45): [001] image where 3 x a superstructures are observed.

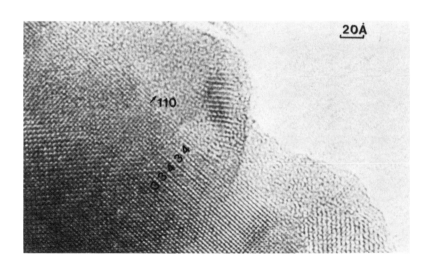

Figure 16: $YBa_2Cu_3O_{7-\delta}$ (0.37≤δ≤0.45): new periodicities along [110].

120 Chemistry of Superconductor Materials

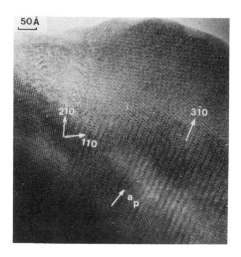

Figure 17: $YBa_2Cu_3O_{7-\delta}$ (0.37≤δ≤0.45): area of an orthorhombic crystal where superstructures are set up along [110], [2$\bar{1}$0] and [3$\bar{1}$0].

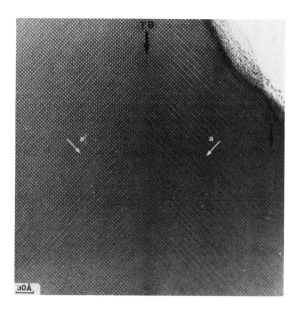

Figure 18: $YBa_2Cu_3O_{7-\delta}$ (0.37≤δ≤0.45): modulations of the contrast along a; they appear in orthorhombic crystals with directions which vary from one twinning domain to the other.

[100], [110], [210] and [310]. The favorable zone to characterize such phenomena is [001] but [100] and [010] were investigated too. Some typical examples are imaged in Figures 13 to 18.

A local 2 x a superstructure is observed in Figure 13, where the new periodicity is black arrowed; in this example, it extends over some 10 nm^2. A similar doubling of the a-parameter is observed in Figure 14, corresponding to a [010] image, but the phenomenon is complicated by a simultaneous doubling of the c-parameter. In such an area, the contrast consists in one bright dot out of two along the a-axis and this feature is translated (3.8 Å) in the adjacent perovskite triple layer, leading to a centered cell with 2a x 2c parameters. The a-parameter can be trebled too, as shown in Figure 15, where the new periodicity is limited by black arrows; eight supercells are then running on the edge of the crystal. The a-axis is undoubtedly the preferential direction for the establishment of such phenomena but others were observed too. A superstructure along [110] is observed in Figure 16, with the following sequence: - $4a\sqrt{2}$ - $3a\sqrt{2}$ - $4a\sqrt{2}$ - $3a\sqrt{2}$ - $3a\sqrt{2}$ -.

In other crystals, superstructures are set up throughout a larger area of the crystals (Figure 17) with shifting and modulated features. The new periodicities are established along [110], [2$\bar{1}$0] and [3$\bar{1}$0]; in the right part of the image, the [3$\bar{1}$0] direction is combined with local 2a superstructure leading to a 2a x $a\sqrt{10}$ supercell.

Besides all these localized phenomena, which take place in a more or less random way, a systematic contrast modulation appears in the crystals, the direction of which varies with the a-parameter, from one twinning domain to the other, as shown in Figure 18. In these images, a row of bright dots, parallel to the b-axis, alternates with a darker one, leading to a mean 2 x a periodicity.

This feature exhibits a similar periodicity to that observed on the [010] image of Figure 13 but differs as it extends throughout the whole crystal.

These phenomena suggest the existence of local variations of the oxygen content, giving rise to local ordering of the oxygens and oxygen vacancies in the central plane of copper atoms.

Many models have been proposed to explain oxygen and vacancy orderings in $YBa_2Cu_3O_{7-\delta}$, but most of them do not take into account the copper coordination, for which three-fold coordination is not likely. The general model which is proposed here is based on the two following points:

(i) Copper exhibits a usual coordination in those intermediate oxides, i.e. two-fold coordination for Cu(I) and 4-, 5- and 6-fold coordination for Cu(II)-Cu(III).

(ii) $YBa_2Cu_3O_{7-\delta}$ is one of the very rare oxides which is characterized by a disproportionation of Cu(II) into Cu(I) and Cu(III), explaining the fact that the compositions with $\delta \geq 0.50$ are still superconducting (48). It results in the coexistence in the same crystal of superconductive $YBa_2Cu_3O_7$ chains (or domains) involving the mixed valence Cu(II)-Cu(III) and insulating $YBa_2Cu_3O_6$ chains (or domains) involving Cu(I) and Cu(II). Such a disproportionation of copper is supported by x-ray absorption studies which clearly show that the Cu(I) content increases with δ and that Cu(III) should not be considered as a true state, but as a formal oxidation state, leading to holes in the oxygen band, often represented as Cu $3d^9\underline{L}$, where $\underline{L}$ corresponds to the ligand hole (49)(50).

With such assumptions, the multiplicity of the a-axis, n x a, implies that n x $[Cu^{III}O_2]_\infty$ rows of CuO_4 groups, parallel to b, alternate with a $[Cu^IO]_\infty$ row of Cu^IO_2 sticks, in a periodic way. A schematic model is drawn in Figure 19; in order to simplify the drawing, only copper and oxygen atoms of the central copper plane are represented. The 2a x b (Figure 19a) and 3a x b (Figure 19b) cells correspond to $O_{6.5}$ and $O_{6.66}$ formulations, respectively. These oxygen contents are in agreement with the nominal compositions of the samples. In the same way, the supercell 2a x 2c can be explained through a translation of the ordering from one triple layer to the adjacent one (Figure 20). It appears that the $[CuO_2]_\infty$ rows alternate with the $[CuO]_\infty$ rows both along a and c. The formulation is indeed $O_{6.5}$, as in model 19a, triple perovskite layers being similar.

The existence of new periodicities setting up along directions different from [001] implies a change of the copper coordination in one row, parallel to b. Such a change can be ensured, if we except three-fold coordination, either by a copper vacancy which can be occasionally encountered, but not in a systematic and periodic way, or by the interconnection of the rows. This interconnection can be ensured by an octahedron, a pyramid or a tetrahedron, all in agreement with the usual coordination of Cu(II) (or Cu(III)). A model is proposed for the supercell 2a x a$\sqrt{10}$ in Figure 21a; CuO_5 pyramids are lined up along the a-axis, with alternated positions of the vertex

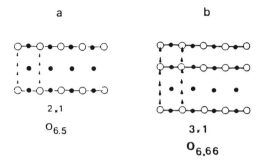

Figure 19: Schematic drawings of 2a x b (a) and 3a x b (b) superstructures involved for $YBa_2Cu_3O_{6.5}$ and $YBa_2Cu_3O_{6.66}$.

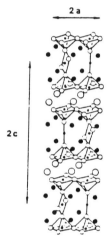

Figure 20: Schematic drawing of 2a x 2c superstructure corresponding to a double alternation of $[CuO_2]_\infty$ and $[CuO]_\infty$ chains.

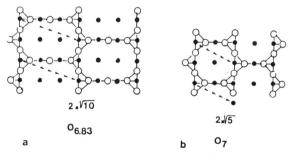

Figure 21: Schematic drawing of 2a x a$\sqrt{10}$ (a) and 2a x a$\sqrt{5}$ (b) supercells.

which ensure the connection of segments, parallel to b, built up from two CuO_4 square groups. This model can be generalized to various lengths of the segments of n x CuO_4 groups; the first one corresponds to a supercell of 2a x a$\sqrt{5}$ (along [210]), with n=1 (Figure 21b). The n value can range from n=1 (O_7) to n=∞ ($O_{6.5}$). The a- and b-parameters of the corresponding supercells are 2a x a$\sqrt{1+(n+1)^2}$ and the oxygen content, $O_{6.5+1/(n+1)}$.

Most of the crystals are covered with modulations (Figure 18) which can indeed be described through these mechanisms; indeed, they do not appear as regularly as proposed in the models (Figure 19). The variations along the a direction can be easily regarded as a way to accommodate a particular oxygen content and, in that case, $YBa_2Cu_3O_{7-\delta}$ could correspond to $[YBa_2Cu_3O_7]_{1-\delta}[YBa_2Cu_3O_6]_\delta$, i.e. to a ratio $(1-\delta)/\delta$ of $[CuO_2]_\infty$ chains and $[CuO]_\infty$ rows distributed all over the crystal. The contrast variations along one row are more difficult to interpret; they are observed in the thick areas of the crystals and can result from the superposition of differently ordered zones in the inner part, which would vary from one triple layer to the other along the c-axis. Moreover, the variations of the lattice parameters and of the Cu-O distances, due to the change in oxidation state of copper from Cu(II)-Cu(III) to Cu(I) are large enough to induce strong distortions in such mixed crystals and can explain that a uniform contrast is never observed throughout the matrix.

3.0 EXTENDED DEFECTS

The electron microscopy studies of the superconductive cuprates show that the different families differ from each other by the nature of their defect chemistry, in spite of their great structural similarities. For example, the La_2CuO_4-type oxides and the bismuth cuprates rarely exhibit extended defects, contrary to $YBa_2Cu_3O_7$ and to the thallium cuprates. The latter compounds are characterized by quite different phenomena.

3.1 Defects in $YBa_2Cu_3O_7$

The well known microtwinning phenomena that are inherent to the structure transition of orthorhombic $YBa_2Cu_3O_7$ to tetragonal

$YBa_2Cu_3O_6$ can be considered as extended defects. Such defects can induce a variation of the oxygen stoichiometry at the twin boundaries, as shown from the different models which can be proposed (Figure 22).

Besides this systematic microtwinning which is very close to a chemical twinning, three other types of extended defects are mainly observed which lead to a local variation of composition (51). For the first family of defects (Figure 23a), the simulations of the images have shown two different orientations of the structure at 90° to one another in the same crystal: area labelled 1 exhibits a contrast characteristic of a [100] image, whereas the area labelled 2 is characteristic of a [010] image. Thus an idealized model of the oriented domains can be proposed (Figure 23b) which corresponds to different orientations of the CuO_4 groups belonging to the $[CuO_2]_\infty$ layers. It is worth pointing out that, contrary to the twinned domains, these domains are characterized by a junction involving a juxtaposition of two different parameters "a" and "b", respectively, at the boundary. Consequently the domain interfaces are particularly disturbed as observed in the thickest part of the bulk. The other two families of defects result from two types of cationic disorders. For the first characteristic defect we observe clearly (Figure 24a) a variation of the spacing of the rows of spots: the rows of white spots move apart, implying a bending of the adjacent double-row and an extra row of small spots appears in the middle. Such a feature results from the formation of a double row of edge-sharing CuO_4 groups, forming a $YBa_2Cu_4O_8$-type defect (Figure 24b) described above. The second defect concerns the variations in the Y, Ba cation ordering (Figure 25a); the c/3 shifting of the fringes is observed in the top of the micrograph (arrowed). This defect is easily explained by a reversal of one barium and one yttrium layer from one part of the defect to the other while the second barium layer remains unchanged through the defect, as shown in the idealized model (Figure 25b). In that example, the defect extends over twelve triple layers $YBa_2Cu_3O_7$ with a clear contrast and the junction of the reverse layers is regularly translated in a direction perpendicular to the c-axis. A multiple shifting of the Y and Ba layers and the overlapping of the small domains lead to the formation of moiré patterns. It should be noted that this type of defect does not affect the O-Cu-O chains but implies a breaking in the bidimensionality of the structure.

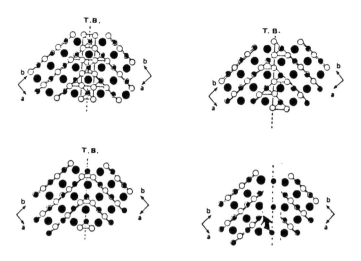

Figure 22: Idealized models of junction between twinning domains.

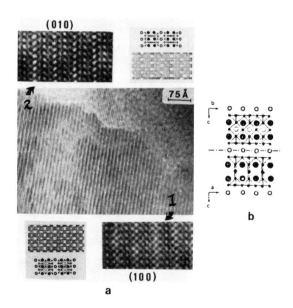

Figure 23: (a) HREM image of a crystal showing 90° oriented domains; enlargements are compared with the calculated images for [100] and [010] orientations. (b) Idealized drawing of the domains.

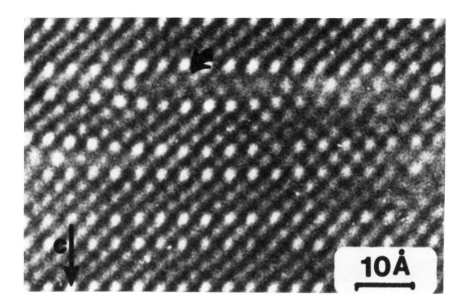

Figure 24a: HREM image of a characteristic defect, an extra row of atoms appears in the perovskite layer: such a defect is interpreted by the existence of a double row of edge sharing CuO_4 groups.

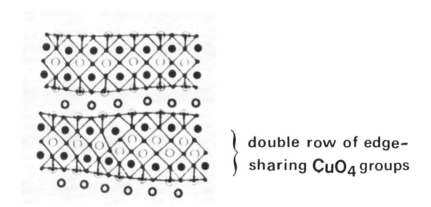

double row of edge-sharing CuO_4 groups

Figure 24b: Idealized model of the defect.

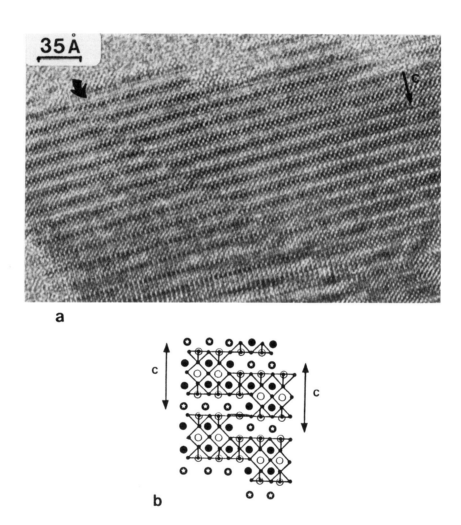

Figure 25: (a) Image of a defect dealing with a variation of the Y-Ba cation ordering: a c/3 shifting of the fringes is clearly observed. (b) Idealized model of the defect.

3.2 Intergrowth Defects in Thallium Cuprates.

It is now clear that the critical temperature of the layered cuprates depends on the number of copper layers forming the oxygen deficient perovskite slabs, and also on the thickness of the rock salt type layers. For instance, in the two series of thallium cuprates $TlBa_2Ca_{m-1}Cu_mO_{2m+3}$ and $Tl_2Ba_2Ca_{m-1}Cu_mO_{2m+4}$, T_c increases as m increases in each series up to m=3 and then decreases for m=4, and the thallium bilayer cuprates exhibit higher T_c's than the thallium monolayer cuprates. Thus it is clear that the formation of intergrowth defects, corresponding to a variation of either the rock salt type layers or of the perovskite layers will influence dramatically the superconductivity of these materials.

Curiously, the thallium cuprates often exhibit intergrowth defects contrary to the bismuth cuprates, whose structures are very similar.

Numerous extended defects affect the rock salt type layers. The thallium bilayer cuprates frequently exhibit defects involving thallium monolayers as shown in Figure 26a which shows a "$TlBa_2CaCu_2O_7$" defect in a matrix of the $Tl_2Ba_2CaCu_2O_8$ oxide. On the contrary, the formation of single rocksalt type layers in thallium monolayer cuprates, i.e. in a matrix involving double rocksalt type layers $[(TlO)(BaO)]_\infty$, as shown for $TlBa_2CaCu_2O_7$ (Figure 26b), is rarely observed. The formation of rock salt type layers larger than those of the matrix is sometimes observed. Figure 26c shows such a defect in a $Tl_2Ba_2CaCu_2O_8$ matrix in which the light contrast of the additional rows of white spots suggest that they correspond to CaO layers. The presence of thallium on the calcium sites as observed by x-ray diffraction, corresponds to a variation of the ordering of the cations. An example is shown in Figure 27, where a calcium layer is replaced by a thallium layer between two pyramidal $[CuO_{2.5}]_\infty$ layers of the perovskite slab. These various extended defects explain the thallium nonstoichiometry and also the possible calcium nonstoichiometry in these structures, and thus could influence their superconducting properties. These phases often present intergrowth defects corresponding to a variation of the perovskite layer thickness. For low m-values of the matrix, few intergrowth defects are generally observed and the m' value corresponding to the defective layer thickness is close to the nominal composition: m'=±1. Two images are

130 Chemistry of Superconductor Materials

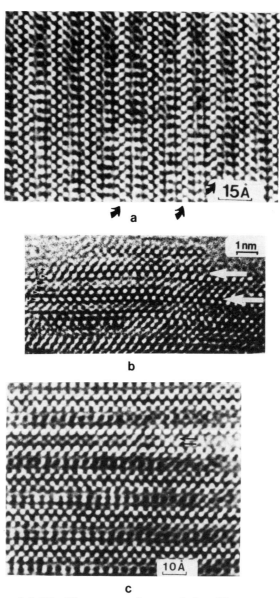

Figure 26: (a) Thallium monolayers ($n' = 2$) appear as defects in a matrix $Tl_2Ba_2CaCu_2O_8$ ($n=3$); (b) rare defect corresponding to the existence of a single rock salt type layer ($n' = 1$) in a $TlBa_2CaCu_2O_7$ matrix ($n=2$); (c) two additional CaO layers are intercalated between thallium layers ($n' = 5$) in a $Tl_2Ba_2CaCu_2O_8$ ($n=3$) matrix.

Complex Chemistry of Superconductive Layered Cuprates 131

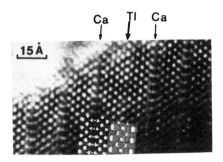

Figure 27: HREM image and calculated image of a defect corresponding to the replacement of a Ca layer by a Tl layer.

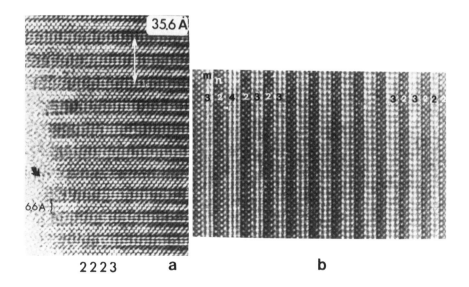

Figure 28: (a) Isolated defect corresponding to a double perovskite layer ($m'=2$) in a $m=3$ matrix (arrowed); (b) multiple intergrowth defects in a $m=3$ matrix: $m'=2$ and $m'=4$ members are observed.

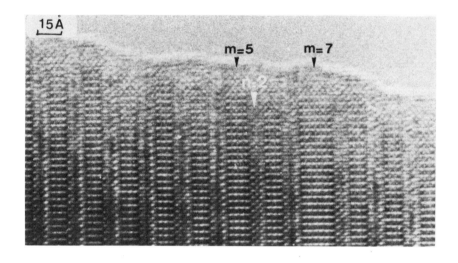

Figure 29: m'=5 and m'=7 layers are observed in a m=4 matrix ($Tl_2Ba_2Ca_3Cu_4O_{12}$).

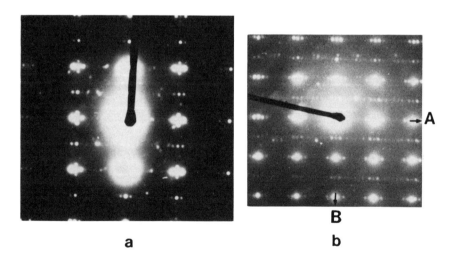

Figure 30: (a) $Bi_2Sr_2CaCu_2O_8$: E. D. patterns showing two systems of satellites due to the existence of two 90° oriented domains; (b) $Bi_2Sr_4Fe_3O_{12}$: E. D. patterns exhibiting satellites along the [100]* direction similar to those observed in the bismuth cuprates.

shown in Figure 28 : the first one (Figure 28a) corresponds to a "m'=2" isolated defect" and the second to more numerous "m'=2" and "m'=4" defects in a m=3 matrix ($Tl_2Ba_2Ca_2Cu_3O_{10}$) (Figure 28b). The defective layer can be significantly thicker than the perovskite layer of the nominal matrix, as m increases. Figure 29 shows an example of high m' defects (m=7) which appear in addition to m'=5 defects in a matrix m=4 ($Tl_2Ba_2Ca_3Cu_4O_{12}$). Thus it can be expected that the frequency of these defects and their deviation from the nominal m value will increase with m.

4.0 INCOMMENSURATE STRUCTURES AND LONE PAIR CATIONS

Lone pair cations exhibit external pairs of electrons which do not participate in the bonds but can influence dramatically the geometry of the structures (52). This is the case of cations like Bi(III), Pb(II) or Tl(I) whose $6s^2$ lone pairs have been shown to present an important stereochemical activity. Such cations which can be found in the rock salt type layers are capable of influencing the oxygen framework and may consequently affect the superconducting properties of the layered cuprates.

The first important point is that satellites in incommensurate positions are observed in all superconductive bismuth cuprates. They are directed along the [100]* or [010]* directions and can appear along two perpendicular directions due to the existence of domains at 90°, characterized by a perfect coherent interface (Figure 30a).

The substitution of iron for copper clearly shows that the presence of copper is not at the origin of such satellites: the isostructural oxides $Bi_2Sr_{3-x}Ca_xFe_2O_9$ (53) and $Bi_2Sr_4Fe_3O_{12}$ (54)(55) in which the FeO_6 octahedra replace the CuO_5 pyramids and CuO_4 square groups exhibit similar satellites on E.D. patterns (Figure 30b) and similar modulations on the [010] HREM images. The fact that these materials do not superconduct suggests that superconductivity is not really linked to the incommensurability of the structure. This particular behavior of the bismuth cuprates can be explained by the stereoactivity of the $6s^2$ lone pair of Bi(III). Such a lone pair, which tends to take the place of an anion, induces a distortion of the rock salt type layers. It results in a nonperiodic displacement of the Bi(III) ions in the layers and consequently of the oxygen atoms.

The latter hypothesis leads us to consider the lead cuprates in which Pb(II) is also characterized by a $6s^2$ lone pair. The replacement of bismuth by lead has been performed for $Bi_2Sr_2CaCu_2O_8$: the oxides $Bi_{2-x}Pb_xSr_2Ca_{1-x}Y_xCu_2O_8$ (56), isolated for $0 \leq x \leq 1$, superconduct for $x \leq 0.5$ ($T_c=80K$), whereas a drastic decrease of T_c is observed beyond this value, $BiPbSr_2YCu_2O_8$ being an insulator. The classical modulations of bismuth have disappeared. Nevertheless they are replaced by new types of incommensurate satellites, as shown for instance in Figure 31, where several types of E.D. patterns are observed in different areas of the same crystal. Moreover this incommensurability is observed independently of the presence of superconductivity. In the same way, E.D. patterns of the layered oxide $PbBaYSrCu_3O_8$ are all characterized by incommensurate satellites as seen for instance in Figure 32, which shows the superposition of two sorts of satellites oriented at 90°, accompanied by double diffraction phenomena. These observations reinforce the hypothesis of the role of the $6s^2$ lone pair, here of Pb(II), in the generation of incommensurability. On the contrary, the oxides $Pb_2Sr_2Ca_{0.5}Y_{0.5}Cu_3O_8$ and $Pb_{2-x}Bi_xSr_2Ca_{1-y}Y_yCu_3O_8$ (37) do not exhibit any incommensurability phenomena in spite of the presence of Bi(III) and Pb(II). The absence of satellites in these two oxides is easily explained by the lacunar character of the structure (Figure 6a): the $6s^2$ lone pairs extend towards the oxygen vacancies and consequently do not disturb the rock salt layers, so that no distortion, i.e., no modulation, is observed.

The case of the thallium cuprates is complex. The two series of superconductive oxides $TlA_2Ca_{m-1}Cu_mO_{2m+3}$ (A=Sr, Ba) and $Tl_2Ba_2Ca_{m-1}Cu_mO_{2m+4}$, exhibit only the trivalent state for thallium (57). Thus, the lone pair $6s^2$ of Tl(I) should not be involved in these two series of superconductive oxides. The electron microscopy observations confirm this point of view. Most of the authors (16)(29)(30)(33)(34)(58) do not observe satellites characteristic of an incommensurate structure. Nevertheless the existence of satellites was pointed out in the oxides $Tl_2Ba_2CaCu_2O_8$ and $Tl_2Ba_2Ca_2Cu_3O_{10}$ by Zandbergen et al. (59)(60). Further electron microscopy studies have resulted in a better understanding of the behavior of thallium cuprates.

The oxide $Tl_{0.5}Pb_{0.5}Sr_2CaCu_2O_7$, prepared for the first time by Subramanian et al. (34) was investigated by electron microscopy (35).

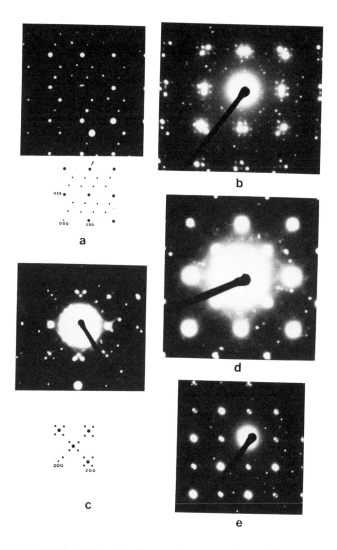

Figure 31: $BiPbSr_2YCu_2O_8$: Several types of satellites are observed in different areas of the same crystals ([001]): (a) satellites set up along a direction roughly parallel to [120]* (q=6.25); (b) bidimensional system of satellites resulting from the existence of misoriented areas and double diffraction phenomena, the angle between both systems is close to 100°; (c) satellites along [110]* with q=3.65; (d) previous satellites have disappeared but streaks remain along [110]*; (e) multisplitting of the spots and first satellites; they are indications of distortions and microtwins.

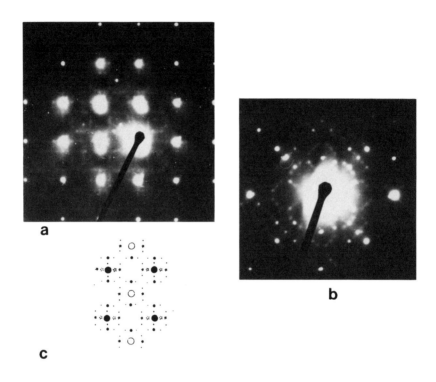

Figure 32: (a) Examples of satellites observed in [001] E. D. patterns of $PbBaYSrCu_3O_8$, [3,1] phase; (b) the extra spots are clearly visible when the zone is slightly deviated from the exact orientation; (c) schematic representation of the pattern, two sets of satellites are visible; they are characteristic of 90° oriented domains: basic spots at level 1/2 are represented as open circles, • ★ systems 1 and 2 (4 x d_{110}), · ☆ systems 3 and 4 (8 x d_{110}).

Its electron diffraction patterns exhibit numerous extra spots, contrary to the thallium cuprates without lead. Besides the simple superstructures, there exist complex arrangements of extra spots as shown for instance in Figure 33. Such satellites are very similar to those more recently observed for $Pb_2Sr_2Y_{0.5}Ca_{0.5}Cu_3O_8$ and do not correspond to an incommensurability of the structure in a single crystal. They can indeed be interpreted as the result of the misorientation of two single crystals, involving also double diffraction phenomena. Thus, in this oxide, the absence of incommensurability is quite in agreement with the absence of lone pair cations. The interatomic distances obtained from the single-crystal x-ray diffraction study confirm this point of view, since they imply short Tl-O and Pb-O distances only compatible with Tl(III) and Pb(IV).

The study of the thallium cuprates $Tl^{}_{2-x/3}Tl^I_{1-x}Ba_{1+x}LnCu_2O_8$ with Ln=Nd, Pr, Sm and x≈0.25 (45) sheds light on the role of the $6s^2$ lone pair of Tl(I) in the problem of incommensurability. This structure which belongs to the 2212-type (Figure 2b) can be formulated for the praseodymium phase $(Tl^{III}_{2-\epsilon})_{A1}$ $(Tl^I_{0.75}Ba_{1.25})_{A2}$ $PrCu_2O_8$ showing that the A1 sites of the classical thallium bilayers are only occupied by Tl(III) and slightly deficient, whereas the A2-sites corresponding to the barium layers are partly occupied by Tl(I). The important feature of those oxides, concerns the great similarity of their E.D. patterns with the bismuth cuprates. One indeed systematically observes satellites that are more clearly visible along zones which deviate from the ideal directions (Figure 34). The reconstruction of the reciprocal lattice (Figure 35) shows that the satellites run along a direction which is similar to that observed in bismuth cuprates, the wave vector being in the (110)* plane; however, their symmetry is different (tetragonal instead of orthorhombic) and the wave length of the modulation, close to six times d_{110} (2.78Å) is also different. Although they do not superconduct, these oxides are of capital importance since they show that the existence of satellites is closely related to the presence of the lone pair cation Tl(I), and that the presence of this cation in the 2212 structure may destroy the superconductivity.

The ability of univalent thallium to occupy the barium sites suggests a possible nonstoichiometry in the superconductive thallium cuprates involving a mixed valence of thallium, Tl(I)/Tl(III). Preliminary investigations (61), considering the possible formulations $Tl^{III}_2(Ba_{2-x}Tl^I_x)(Ca_{1-x}Tl^{III}_x)Cu_2O_8$ and $Tl^{III}_{2-x}(Ba_{2-x/2}Tl^I_{x/2})$

138 Chemistry of Superconductor Materials

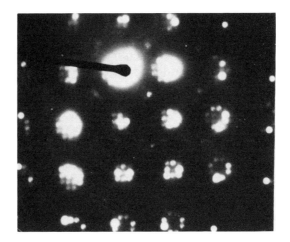

Figure 33: $Tl_{0.5}Pb_{0.5}Sr_2CaCu_2O_7$: [001] E. D. pattern, numerous extra spots are observed which can be interpreted as the result of the misorientation of two lamellae and double diffraction phenomena.

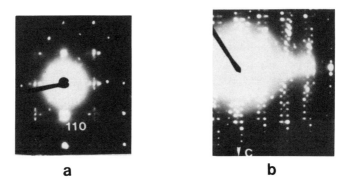

Figure 34: $Tl_{2.7}Ba_{1.25}PrCu_2O_8$: satellites are observed in (a) [001] and (b) [100] E. D. patterns

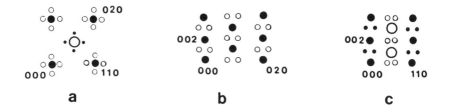

Figure 35: Reconstruction of the reciprocal lattice; (a) [001]*, (b) [100]* and (c) [110]*.

$(Ca_{1-x/2}Tl^{III}_{x/2})Cu_2O_8$ show that a pure phase is obtained for $0 \leq x \leq 0.20$, with T_c decreasing from 105K for x=0 to 95K for x=0.20. More spectacular is the electron diffraction study which shows that the sample x=0 which contains only Tl(III) ($Tl_2Ba_2CaCu_2O_8$) does not present any incommensurability phenomena whereas the sample x=0.1, which corresponds to the presence of Tl(I) exhibits incommensurate satellites (Figure 36) which do not exhibit the same orientation as those in the oxides $Tl^{III}_{2-x/3}Tl^{I}_{1-x}Ba_{1+x}LnCu_2O_8$ (45), but are absolutely identical to those described by Zandbergen et al. (59)(60) for the oxide to which they attributed the composition "$Tl_2Ba_2CaCu_2O_8$". These results reinforce the hypothesis of the important role of the $6s^2$ lone pair of Tl(I) in incommensurability.

In the same way the E.D. patterns of the layered cuprates $Tl_{1+x}A_{2-y}Ln_2Cu_2O_9$ (A=Sr, Ba) involving double fluorite layers exhibit incommensurate satellites lying along either <110>* (Figure 37a) or along <100>* (Figure 37b), with periodicities close to 4 x d_{110} (2.8Å) and 3 x a respectively. These latter observations confirm the mixed valence Tl(III) - Tl(I) and support again the role of Tl(I) lone pair in incommensurability.

5.0 CONCLUDING REMARKS

A route is opened to the synthesis of new high T_c superconductors. Up to now, it appears that the mixed valence of copper and the low dimensionality of the structure remain the two important factors for the research of new materials. However, if the mixed valence Cu(II) - Cu(III) is very promising, the discovery of the Nd_2CuO_4 superconductors doped with cerium (62) suggests a different mechanism. The nonstoichiometry phenomena, in those materials are very complex and the problems of oxygen nonstoichiometry and extended defects, especially the intergrowth phenomena, have of course a great influence on the superconducting properties and should be controlled carefully. The lone pair cations Bi(III), Pb(II) and Tl(I) are of interest since they induce the layered character of the structure. It is now clear that such cations are at the origin of incommensurability in those structures. Moreover, it is also evident that superconductivity is not really linked to this modulation of the structure. Particular attention should be paid to the thallium oxides, which

140 Chemistry of Superconductor Materials

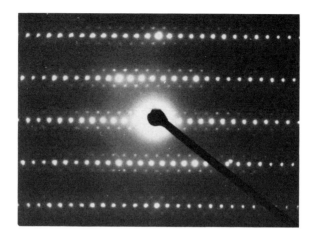

Figure 36: [001] E.D. pattern of $Tl_{2.2}Ba_{1.9}Ca_{0.9}Cu_2O_8$, satellites are clearly visible.

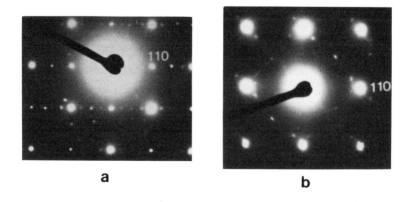

Figure 37: $Tl_{1+x}A_{2-y}Ln_2Cu_2O_9$: [001] E.D. patterns, (a) satellites along <110>* with a new periodicity close to $4 \times d_{110}$, (b) satellites along <100>* with periodicity $3 \times a$.

exhibit a very complex nonstoichiometry due to the mixed valence Tl(I)/Tl(III); the systematic absence of superconductivity in these latter oxides, in the presence of Tl(I), is not due to the $6s^2$ lone pair effect but might result from redox reactions which prevent the tranfer of holes from the thallium layers towards the copper oxygen layers.

6.0 REFERENCES

1. Wadsley, A.D., *Nonstoichiometric Compounds*, ed. L. Mandelcorn (1964) Academic Press.

2. Michel, C, Hervieu, M., Borel, M.M., Grandin, A., Deslandes, F., Provost, J. and Raveau, B., *Z. Phys.* B68:421 (1987).

3. Maeda, H., Tanaka, Y., Fukutomi, M. and Asano, T., *Jpn. J. Appl. Phys.* 27:L209 and L548 (1988).

4. Hervieu, M., Michel, C., Domengès, B., Laligant, Y., Lebail, A., Ferey, G. and Raveau, B., *Mod. Phys. Lett.* B2:491 (1988).

5. Torardi, C.C., Subramanian, M.A., Calabrese, J.C., Gopalakrishnan, J., Mc Carron, E.M.C., Morrissey, K.J., Askew, T.R., Flippen,R.B., Chowdhry, U. and Sleight, W., *Phys. Rev.* B38:225 (1988).

6. Tarascon, J.M., Le Page, Y., Barboux, P., Bagley, B.G., Greene, L.H., Mc Kinnon, W.R., Hull, G.W., Giroud, M. and Hwang, D.M., *Phys. Rev.* B37:9382 (1988).

7. Subramanian, M.A., Torardi, C.C., Calabrese, J.C., Gopalakrishnan, J., Morrissey, K.J., Askew, T.R., Flippen, R.B., Chowdhry, U. and Sleight, A.W., *Science* 239:1015 (1988).

8. Hervieu, M., Domengès, B., Michel, C. and Raveau, B., *Mod. Phys. Lett.* B2:835 (1988).

9. Bordet, P., Capponi, J.J., Chaillout, C., Chenavas, J., Hewat, A.W., Hewat, E.A., Hodeau, J.L., Marezio, M., Tholence, J.L. and Tranqui, D., *Physica* C 153-155:623 (1988).

10. Politis, C., *Appl. Phys.* A45:261 (1988).

11. Von Schnering, H.G., Walz, L., Schwartz, M., Beker, W., Hartweg, M., Popp, T., Hettich, B., Müller, P. and Kampt, G., *Angew. Chem.* 27:574 (1988).

12. Kajitani, T., Kusaba, K., Kikuchi, M., Kobayashi, N., Syono, Y., Williams, T.B. and Hirabayashi, M., *Jpn. J. Appl. Phys.* 27:L587 (1988).

13. Sunshine, S.A., Siegrist, T., Schneemeyer, L.F., Murphy, D.W., Cava, R.J., Batlogg, B., Van Dover, R.B., Fleming, R.M., Glarum, S.H., Nakahara, S., Farrow, R., Krajewski, J.J., Zahurac, S.M., Waszczak, J.V., Marshall, J.H., Marsh, P., Rupp, L.W. and Peck, W.F., *Phys. Rev.* B38:893 (1988).

14. Tarascon, J.M., Mc Kinnon, W.R., Barboux, P., Hwang, D.M., Bagley, B.G., Greene, L.H., Hull, G.W., Le Page, Y., Stoffel, N. and Giroud, M., *Phys. Rev.* B38:8885 (1988).

15. Kijima, N., Hendo, H., Tsuchiya, J., Sumiyama, A., Mizumo, M. and Oguri, Y., *Jpn. J. Appl. Phys.* 25:L821 (1988).

16. Torardi, C.C., Subramanian, M.A., Calabrese, J.C., Gopalakrishnan, J., Morrissey, K.J., Askew, T.R., Flippen, R.B., Chowdhry, U. and Sleight, A.W., *Science* 240:631 (1988).

17. Zandbergen, H.W., Huang, Y.K., Menken, M.J.V., Li, J.N., Kadouaki, K., Menovsky, A.A., Van Tendeloo, G. and Amelinckx, S., *Nature* 332:620 (1988).

18. Sheng, Z.Z., Hermann, A.M., El Ali, A., Almason, C., Estrada, J., Datta, T. and Matson, R.J., *Phys. Rev. Lett.* 60:937 (1988).

19. Sheng, Z.Z. and Hermann, A.M., *Nature* 332:55 (1988).

20. Hazen, R.M., Finger, L.W., Angel, R.J., Prewitt, C.T., Ross, N.L., Hadidiacos, C.G., Heaney, P.J., Veblen, D.R., Sheng, Z.Z., El Ali, A. and Hermann, A.M., *Phys. Rev. Lett.* 60:1657 (1988).

21. Parkin, S.S.P., Lee, V.Y., Engler, E.M., Nazzal, A.I., Huang, T.C., Gorman, G., Savoy, R. and Beyers, R., *Phys. Rev. Lett.* 60:2539 (1988).

22. Politis, C. and Luo, H., *Mod. Phys. Lett.* B2:793 (1988).

23. Maignan, A., Michel, C., Hervieu, M., Martin, C., Groult, D. and Raveau, B., *Mod. Phys. Lett.* B2:681 (1988).

24. Subramanian, M.A., Calabrese, J.C., Torardi, C.C., Gopalakrishnan, J., Askew, T.R., Flippen, R.B., Morrissey, K.J., Chowdhry, U. and Sleight, A.W., *Nature* 332:420 (1988).

25. Hervieu, M., Michel, C., Maignan, A., Martin, C. and Raveau, B., *J. Solid State Chem.* 74:428 (1988).

26. Martin, C., Michel, C., Maignan, A., Hervieu, M. and Raveau, B., *C. R. Acad. Sci. Fr.* 307:27 (1988).

27. Domengès, B., Hervieu, M. and Raveau, B., *Solid State Comm.* 68:303 (1988).

28. Parkin, S.S.P., Lee, V.Y., Nazzal, A.I., Savoy, R., Beyers, R. and La Placa, S., *Phys. Rev. Lett.* 61:750 (1988).

29. Hervieu, M., Maignan, A., Martin, C., Michel, C., Provost, J. and Raveau, B., *J. Solid State Chem.* 75:212 (1988).

30. Hervieu, M., Maignan, A., Martin, C., Michel, C., Provost, J. and Raveau, B., *Mod. Phys. Lett.* B2:1103 (1988).

31. Morosin, B., Ginley, D.S., Hlava, P.F., Carr, M.J., Baughman, R.J., Shirber, J.E., Venturini, E.L. and Kwak, J.F., *Physica* C 152:413 (1988).

32. Sheng, Z.Z. and Hermann, A.M., *Nature* 332:138 (1988).

33. Martin, C., Provost, J., Bourgault, D., Domengès, B., Michel, C., Hervieu, M. and Raveau, B., *Physica* C 157:460 (1989).

34. Subramanian, M.A., Torardi, C.C., Gopalakrishnan, J., Gai, P., Calabrese, J.C., Askew, T.R., Flippen, R.B. and Sleight, A.W., *Science* 242:249 (1988).

35. Capponi, J.J., Chaillout, C., Hewat, A.W., Lejay, P., Marezio, M., Nguyen, N., Raveau, B., Soubeyroux, J.L. and Tholence, J.L., *Europhysic Letters* 12:1301 (1988).

36. Cava, R.J., Batlogg, B., Krajewski, J.J., Rupp, L.W., Schneemeyer, L.F., Siegrist, T., Van Dover, R.B., Marsh,

P., Peck, W.F., Gallagher, P.K., Glarum, S.H., Marshall, J.H., Farrow, R.C., Waszczak, J.V., Hull, R. and Trevor, P., *Nature* 336:211 (1988).

37. Retoux, R., Michel, C., Hervieu, M. and Raveau, B., *Mod. Phys. Lett.* B 3:591 (1989).

38. Domengès, B., Hervieu, M., Michel, C. and Raveau, B., *Europhysic Letters* 4(2):211 (1987).

39. Bordet, P., Chaillout, C., Chenevas, J., Hodeau, J.L., Marezio, M., Karpinski, J. and Kaldis, E., *Nature* 334:596 (1988).

40. Raveau, B., *Proc. Indian Nat. Sci. Acad.* 52:67 (1986).

41. Chaillout, C., Bordet, P., Chenevas, J., Hodeau, J.L. and Marezio, M., *Solid State Comm.* 70:275 (1989).

42. Tokura, Y., Takagi, H. and Uchida, S., *Nature* 337:345 (1989).

43. Muller-Buschbaum, H.K. and Wollschläger, W., *Z. Anorg. Allg. Chem.* 414:76 (1975) and 4288:120 (1977).

44. Martin, C., Bourgault, D., Hervieu, M., Michel, C., Provost, J. and Raveau, B., *Mod. Phys. Lett.* B3(13):993 (1989).

45. Bourgault, D., Martin, C., Michel, C., Hervieu, M. and Raveau, B., *Physica* C 158:511 (1989).

46. Tarascon, J.M., Greene, L.H., Bagley, B.G., McKinnon, W.R., Barboux, P. and Hull, G.W., *Novel Superconductivity*, ed. A. Wolf, 705 (1987).

47. Cava, R.J., Batlogg, B., Shen, C.H., Rietman, E.A., Zahurak, S.M. and Wender, D., *Nature* 329:423 (1987).

48. Raveau, B., Michel, C., Hervieu, M. and Provost, J., *Physica* C 153-155:3 (1988).

49. Baudelet, F., Collin, G., Dartyge, E., Fontaine, A., Kappler, J.P., Krill, G., Itie, J.P., Jegoudez, J., Maurer, M., Monod, P., Revcolevschi, A., Tolentino, H., Tourillon, G. and Verdaguer, M., *Z. Phys.* B69:141 (1988).

50. Bianconi, A., De Santis, M., Di Cicco, A., Flank, A.M.,

Fontaine, A., Lagarde, P., Katayama-Yoshida, H., Kotani, A. and Marcelli, A., *Phys. Rev.* B38:7196 (1988).

51. Hervieu, M., Domengès, B., Michel, C. and Raveau, B., *Europhysic Lett.* 4:205 (1987) and 4:211 (1987).

52. Raveau, B., *Reviews on Silicon, Germanium, Tin and Lead Compounds* 6:287 (1982).

53. Hervieu, M., Michel, C., Nguyen, N., Retoux, R. and Raveau, B., *Europ. J. of Inorg. and Solid State Chem.* 25:375 (1988).

54. Retoux, R., Michel, C., Hervieu, M., Nguyen, N. and Raveau, B., *Solid State Comm.* 69:599 (1989).

55. Le Page, Y., McKinnon, W.R., Tarascon, J.M. and Barboux, P., *Phys. Rev.* B40:6810 (1989).

56. Retoux, R., Caignaert, V., Provost, J., Michel, C., Hervieu, M. and Raveau, B., *J. Solid State Chem.* 79:157 (1989).

57. Studer, F., Retoux, R., Martin, C., Michel, C., Raveau, B., Dartyge, E., Fontaine, A. and Tourillon, G., *Mod. Phys. Lett.* B3(7):1085 (1989).

58. Bourgault, D., Martin, C., Michel, C., Hervieu, M., Provost, J. and Raveau, B., *J. Solid State Chem.* 788:326 (1989).

59. Zandbergen, H.W., Van Tendeloo, G., Van Landuyt, J. and Amelinckx, S., *Appl. Phys.* A46:233 (1988).

60. Zandbergen, H.W., Groen, W.A., Mijlhoff, F.C., Van Tendeloo, G. and Amelinckx, S., *Physica* C 156:325 (1988).

61. Martin, C., Maigman, A., Provost, J., Michel, C., Hervieu, M., Tournier, R., and Raveau, B., *Physica* C, In Press (1990).

62. Akimitsu, J., Suzuki, S., Watanabe, M. and Sawa, H., *Jpn. J. Appl. Phys.* 27:L1859 (1988).

3

Defective Structures of $Ba_2YCu_3O_x$ and $Ba_2YCu_{3-y}M_yO_z$ (M = Fe, Co, Al, Ga, ...)

Anthony Santoro

1.0 INTRODUCTION

Superconductivity with $T_c \simeq 94K$ was discovered in a sample with nominal composition $Ba_{0.8}Y_{1.2}CuO_{4-y}$, prepared through solid state reaction of the appropriate amounts of Y_2O_3, $BaCO_3$, and CuO (1). Soon after this discovery, it was found in a number of laboratories that the reaction product was in fact a mixture of phases, one of which, identified as $Ba_2YCu_3O_x$ with $x \simeq 6.9$, was responsible for the superconducting properties of the material (2-6). From x-ray powder diffraction data, Cava et al. (2), determined an orthorhombic structure that could be described as an oxygen deficient perovskite with tripling of the c-axis caused by ordering of the Ba and Y atoms. Single-crystal x-ray data revealed without ambiguity the sites of the metal atoms in the unit cell, but, due to the presence of a large fraction of heavy atoms in the structure, did not permit one to determine with certainty the symmetry of the compound and the occupancies and locations of the oxygen atoms (7-9).

The structural determination of $Ba_2YCu_3O_x$, for $6.8 \leq x \leq 7.0$, was completed in several laboratories by Rietveld analysis of powder neutron diffraction data (10-15). The neutron diffraction experiments confirmed the space group Pmmm and the main structural features found by x-rays by Siegriest et al. (7), but revealed that some of the oxygen assignments made in the x-ray studies were not entirely correct. The refined structural parameters obtained in four of these neutron diffraction analyses are given in Table 1 (adapted from Reference 15) and the structure of $Ba_2YCu_3O_{7.0}$ is schematically illustrated in Figure 1, where it is compared with the parent structure of perovskite. Data collected at low temperatures (11-13)(16) revealed that no phase transitions take place in going from room temperature down to 5K.

2.0 DISCUSSION OF THE STRUCTURE OF $Ba_2YCu_3O_{7.0}$

The relationship between the structures of perovskite and $Ba_2YCu_3O_{7.0}$ illustrated in Figure 1, can be easily understood if we represent the sequence of layers in the two cases in the following way.

Perovskite ... $[(BX_2)_o \, (AX)_c] \, (BX_2)_o \quad (AX)_c \quad (BX_2)_o \quad (AX)_c \quad (BX_2)_o$ ...
$Ba_2YCu_3O_{7.0}$ ... $[(CuO)_o \, (BaO)_c \, (CuO_2)_o \quad (Y)_c \quad (CuO_2)_o \, (BaO)_c](CuO)_o$ ...
Approximate z/c   0   1/6   1/3   1/2   2/3   5/6   0

From this scheme it can be seen that the structure of $Ba_2YCu_3O_{7.0}$, when compared to that of perovskite, has two oxygen-deficient layers: the first, the CuO layer at z = 0, has the configuration of a BX_2 layer in which two oxygen atoms at the mid-points of opposite edges have been removed, and the other, the yttrium layer at z = 1/2, has the configuration of an AX layer in which all the oxygen atoms are missing.

The copper atoms are located on the two positions 1a (0,0,0) and 2q (0,0,z) of space group Pmmm. The first, Cu(1), at the origin of the unit cell, has four-fold planar coordination with two O(1) neighboring oxygen atoms along the c-axis at a distance of ~1.85Å and two O(4) atoms along the b-axis at a distance of ~1.94Å (precise inter atomic distances for various values of x are given in Table 3. The near-square $Cu-O_2$ units share one corner and form chains along the b-axis as shown in Figure 1. The second copper atom, Cu(2), has five-fold pyramidal coordination and is strongly bonded to two O(2) and two O(3) oxygen atoms with distances of ~1.93Å (along the b-axis) and ~1.96Å (along the a-axis), respectively. The fifth oxygen atom O(1) (the "apex" of the pyramid) is weakly bonded to Cu(2) with a distance of ~2.30Å along the c-axis. This weak Cu(2)-O(1) interaction confers a highly two-dimensional character to the CuO layers

TABLE 1. Results of Neutron Power Diffraction Rietveld Refinements for $Ba_2YCu_3O_x$.

Atom	Parameter	Beno et.al. (10)	Capponi et.al. (12)	Beech et.al. (13)	Cox et.al. (15)
Ba	z	0.1843(3)	0.1841(3)	0.1839(2)	0.1839(3)
	$B(A^2)$	0.54(5)	0.6(1)	0.65(5)	0.4(1)
Y	$B(A^2)$	0.46(4)	0.6(1)	0.56(4)	0.2(1)
Cu(1)	$B(A^2)$	0.50(5)	0.4(1)	0.55(4)	0.2(1)
Cu(2)	z	0.3556(1)	0.3549(3)	0.3547(1)	0.3546(2)
	$B(A^2)$	0.29(4)	0.5(1)	0.49(4)	0.3(1)
O(1)	z	0.1584(2)	0.1581(4)	0.1581(2)	0.1589(3)
	$B(A^2)$	0.67(5)	0.9(1)	0.78(5)	0.5(2)
	n	2.0	2.0	2.0	2.01(3)
O(2)	z	0.3773(2)	0.3779(4)	0.3779(2)	0.3783(3)
	$B(A^2)$	0.56(5)	0.1(1)	0.57(5)	0.5(2)
	n	1.89(2)	2.0	2.0	2.01(3)
O(3)	z	0.3789(3)	0.3777(5)	0.3776(2)	0.3780(3)
	$B(A^2)$	0.37(5)	0.3(1)	0.55(5)	0.4(2)
	n	2.0	2.0	2.0	2.02(3)
O(4)	$B(A^2)$	1.35(5)	2.4(3)	1.73(9)	1.6(3)
	n	0.92(2)	1.0	1.0	0.96(2)
O(5)	$B(A^2)$	-	-	-	1.6
	n	-	-	-	0.04(1)
	a(A)	3.8231(1)	3.8206(1)	3.8198(1)	3.8172(1)
	b(A)	3.8863(1)	3.8851(1)	3.8849(1)	3.8822(1)
	c(A)	11.6809(2)	11.6757(4)	11.6762(3)	11.6707(4)
composition: x		6.81(3)	7.00	7.00	6.96(2)
	R_N	4.86	7.15	4.84	3.41
	R_W	6.29	14.17	8.40	8.31
	R_E	3.51	10.84	5.42	4.15

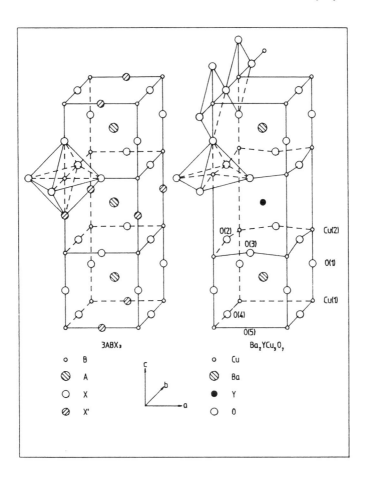

Figure 1: Structure of $Ba_2YCu_3O_{7.0}$ (right side) and its relationship to the structure of perovskite (left side). The atoms X' are the oxygen atoms of perovskite that do not exist in the structure of $Ba_2YCu_3O_{7.0}$. As a consequence of this elimination, the typical chains are formed on the basal plane of the superconductor, and the atoms Cu(2) assume five-fold, pyramidal coordination. The nomenclature of this figure is used in the discussions that follow.

perpendicular to the c-axis. The oxygen atoms O(2) and O(3) are almost exactly coplanar and are vertically displaced ~0.27Å from the plane of the copper atoms Cu(2), producing the characteristic buckling of the CuO layers, as indicated in Figure 1.

The ordering of the oxygen vacancies affects also the coordination of the other metal atoms. The yttrium atom is eight-fold coordinated and the coordination polyhedron is a square prism with Y-O distances of 2.41Å and 2.39Å. The barium atoms, on the other hand, are ten-fold coordinated and the coordination polyhedron can be described as a cuboctrahedron in which two oxygen atoms are missing (these atoms would be located on the sites indicated as O(5) in Figure 1. The Ba-O distances vary from 2.74Å to 2.98Å, with an average distance of 2.86Å.

2.1 Structural Changes as a Function of Oxygen Stoichiometry

All the structural determinations quoted in the previous sections were made on samples of $Ba_2YCu_3O_x$ with composition $6.8 \leq x \leq 7.0$. As shown in Table 1, for $x = 7.0$ all sites O(4) are filled. When $x < 7.0$ there must be oxygen vacancies in the structure and it was realized very soon that these are confined to the sites O(4) in the chains parallel to the b-axis (13). The presence of vacancies on the O(2) sites, reported by Beno et al. (10), and indicated in Table 1, has not been confirmed in any other diffraction experiment and will not be considered further in the present discussion.

It has been observed that the total oxygen stoichiometry decreases smoothly with increasing temperature (17)(18). The oxygen atoms that are removed from the structure are exclusively those located on the O(4) sites at (0,1/2,0). In situ neutron powder diffraction measurements (17) have shown that the positions O(5) at (1/2,0,0) are gradually filled as the temperature increases, and when the occupancies of the two sites become equal the symmetry of the structure changes from orthorhombic to tetragonal. The temperature of the transition depends on the partial pressure of the oxygen present in the atmosphere of the experiment, and is ~700°C in one atmosphere of oxygen and less than that at lower oxygen partial pressures. Continued heating at higher temperatures results in further loss of the oxygen from the sites O(4) and O(5) until the composition reaches the value $x = 6.0$. The oxygen stoichiometry at the transition is always x

≃ 6.5, so that the orthorhombic phase exists over the range $6.5 < x \leq 7.0$ and the tetragonal phase over the range $6.0 \leq x < 6.5$. However, if the oxygen is eliminated at sufficiently low temperatures, employing, for example, gettered annealing techniques as described by Cava et al. (19), the orthorhombic structure may be retained down to values of x as low as ~6.3.

Studies of the superconducting properties as a function of the oxygen content show that in the orthorhombic compound the value of T_c decreases as the oxygen is removed from the structure, and it becomes zero as the crystallographic transition is approached (19). Superconductivity has not been found in the tetragonal phase for any stoichiometry, and in fact, this phase is semiconducting (20)(21). The structure of the tetragonal phase has been refined by Rietveld analysis of neutron powder diffraction data (22) and by single-crystal x-ray diffraction (23). In Table 2 the refined parameters for the composition $Ba_2YCu_3O_{6.07}$ are given, and the structure is schematically illustrated in Figure 2.

It is interesting to determine the structural changes taking place in this system as a function of oxygen content. Some relevant interatomic distances for various values of the composition are given in Table 3. The coordination of the barium atoms decreases from ten-fold for x = 7.0 to eight-fold for x = 6.0. The data in Table 3 show that the distances Ba-O(1) and Ba-O(4) increase with decreasing oxygen content, and the distances Ba-O(2) and Ba-O(3) decrease. This means that the barium atoms move along the c-axis toward the bottom of the "cup" of oxygen atoms left from the cuboctahedron after atoms O(4) and O(5) are eliminated. The coordination of yttrium, on the contrary, does not change significantly, and the coordination polyhedron remains a rectangular prism over the entire range of composition. As we have mentioned, for x = 7.0 the Cu(1) atoms are four-coordinated and are located at the center of near-rectangular Cu-O$_2$ units connected by vertices and forming chains along the b-axis. For x = 6.0, these atoms are in two-fold coordination. As the oxygen is gradually removed from the structure, the distances Cu(1)-O(1) and Cu(1)-O(4) decrease continuously. The description of coordination of Cu(1) for intermediate compositions is more complex and we will discuss it later. The distance Cu(2)-O(1) increases when the oxygen content decreases and, correspondingly, the distance of Cu(2) from the average plane of the atoms O(2) and O(3) decreases. This means that

TABLE 2. Atom Positions, Occupancies and Agreement Indices for $Ba_2YCu_3O_6$.

Atom	Position	x	y	z	B	Occupancy
Ba	2h 4mm	1/2	1/2	0.1952(2)	0.50(4)	1.0
Y	1d 4/mmm	1/2	1/2	1/2	0.73(4)	1.0
Cu(1)	1a 4/mmm	0	0	0	1.00(4)	1.0
Cu(2)	2g 4mm	0	0	0.3607(1)	0.49(3)	1.0
O(1)	2g 4mm	0	0	0.1518(2)	1.25(6)	0.99(6)
O(2)	4i mm	0	1/2	0.3791(1)	0.73(4)	1.009(5)
O(4)	2f mmm	0	1/2	0	0.9	0.028(4)

$R_N = 3.89$ $R_P = 6.10$ $R_W = 7.95$ $R_E = 5.64$ $\chi = 1.41$

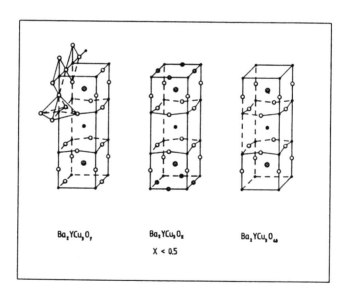

Figure 2: Comparison of the structures of $Ba_2YCu_3O_{7.0}$ (left side), $Ba_2YCu_3O_x$ (x < 0.5) (center) and $Ba_2YCu_3O_{6.0}$ (right side). The dotted atoms indicate partial occupation of the site.

TABLE 3. Relevant Bond Distances in $Ba_2YCu_3O_x$.

		x = 7	x = 6.8	x = 6.5	x = 6.06
Ba - O(1)	x4	2.7408(4)	2.7470(8)	2.7690(8)	2.7751(5)
Ba - O(2)	x2	2.984(2)	2.972(3)	2.930(4)	
					2.905(1)
Ba - O(3)	x2	2.960(2)	2.938(6)	2.902(4)	
Ba - O(4)	x2	2.896(2)	2.922(4)	2.956(2)	
Y - O(2)	x4	2.409(1)	2.403(3)	2.404(3)	
					2.4004(8)
Y - O(3)	x4	2.386(1)	2.389(2)	2.408(3)	
Cu(1) - O(1)	x2	1.846(2)	1.843(3)	1.795(3)	1.795(2)
Cu(1) - O(4)	x2	1.9429(1)	1.9428(1)	1.9374(1)	
Cu(2) - O(1)	x1	2.295(3)	2.323(4)	2.429(4)	2.469(2)
Cu(2) - O(2)	x2	1.9299(4)	1.9305(6)	1.9366(6)	
				1.9406(3)	
Cu(2) - O(3)	x2	1.9607(4)	1.9585(8)	1.9478(6)	

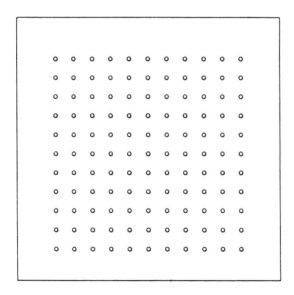

Figure 3: The basal plane of $Ba_2YCu_3O_{6.0}$. The circles indicate the Cu(1) atoms in two-fold coordination. The oxygen atoms above and below each copper atom are not indicated for clarity, and they will not be illustrated in any of the following figures showing the basal plane of the structure.

the coordination of Cu(2) tends to be more square-planar and less pyramidal with decreasing oxygen stoichiometry.

In practically all refinements of the structure of $Ba_2YCu_3O_x$, it has been found that the oxygen atoms O(4) have an unusually large temperature factor; e g., (14)(24). Low-temperature experiments (13)(16) have indicated the possible presence of static disorder. Anisotropic refinements of the structure showed that O(4) is highly anisotropic in the direction of the a- axis (24). Models with the atom O(4) split over the two positions (x,1/2,0) and (-x,1/2,0), rather than fixed at (0,1/2,0), resulted in a value of the coordinate x significantly different from zero and indicated that this atom may in fact be removed about 0.2Å from the b-axis. The same general conclusions were reached also by Francois et al. in their extensive study of the structure of $Ba_2YCu_3O_{6.91}$ and $Ba_2YCu_3O_{6.86}$ as a function of temperature (16).

2.2 Twinning, Twin Boundaries and Model of the Structure of $Ba_2YCu_3O_{7.0}$

As we have mentioned in the previous sections, the oxygen atoms involved in the oxidation and reduction of $Ba_2YCu_3O_x$ are those located on the basal plane of the structure at z=0 and labelled O(4) and O(5) in Figure 1. Therefore, in our discussion of structural models in the range of composition $6.0 \leq x \leq 7.0$, we need to consider only the atomic configuration on this plane and we may ignore all the other atoms in the unit cell. For x < 6.5, the compound is tetragonal with the symmetry of space group P4/mmm and the sites O(4) and O(5) are crystallographically equivalent. For x = 6.0, these sites are empty so that a layer of copper atoms is sandwiched between two BaO layers according to the scheme:

$Ba_2YCu_3O_6$... $[(Cu)_o\ (BaO)_c\ (CuO_2)_o\ (Y)_c\ (CuO_2)_o\ (BaO)_c](Cu)_o(BaO)_c$
z/c   0   1/6   1/3   1/2   2/3   5/6   0

The model for this composition is quite simple and the structure is schematically illustrated in Figure 3. For x > 6.5 the symmetry is orthorhombic Pmmm and sites O(4) and O(5) are not equivalent. When x = 7.0, sites O(4) are completely filled and sites O(5) are empty, and the model of the structure would be as simple as that illustrated for x = 6.0 if the presence of twinning would not introduce considerable complications in the configuration of the atoms on the basal plane. A realistic model cannot therefore be built without a knowledge of the twin laws operating in the system and of the arrangement of the oxygen atoms at the twin boundaries.

When the tetragonal oxide $Ba_2YCu_3O_x$ is cooled slowly, at the phase transition the symmetry changes from P4/mmm to Pmmm. The symmetry elements $(001)_{90°}$, $(110)_m$ and $(1\bar{1}0)_m$, present in the tetragonal structure, become pseudo-symmetry elements in the orthorhombic phase and twinning by pseudo-merohedry is possible (25). The existence of twin boundaries for x > 6.5 was detected early in transmission electron microscopy studies; e.g., (26)(27), but the complete analysis of twinning was carried out by Hodeau et al. by single-crystal x-ray diffraction techniques (28). These authors found that the general x-ray reflections are split into four spots, thus indicating the presence of four twinned individuals in the sample. These diffraction patterns could be indexed by assuming that two of the four individuals are related to each other by a reflection across the pseudo-mirror $(110)_m$ or $(1\bar{1}0)_m$, and the other two are generated from these by the operation $(001)_{90°}$ (rotation of 90° about the c-axis). The mismatch of the twin lattice due to the fact that a ≠ b, can be measured by means of the twin obliquity (29), $\epsilon = \arcsin((b^2 - a^2)/(b^2 + a^2))$, which, in this case, is a function of the oxygen stoichiometry and is of the order of ~1°. Transmission electron microscopy pictures showed that a movement of the twin walls takes place when the electron beam intensity is increased (28). This observation was interpreted by assuming that the heat treatment by the electron beam causes irreversible oxygen loss, and therefore, the motion of the twin walls indicates that there is a close relationship between twin boundaries and diffusion of oxygen in the sample.

Several models for the structure of the twin walls have been proposed (30, and ref. therein). The simplest of these is illustrated in Figure 4. The twin wall is formed by three consecutive diagonals. The copper atoms located on the central one have two-fold coordination with corresponding oxygen stoichiometry of x = 6.0. Those located on the two neighboring rows on each side of the central diagonal have three-fold coordination and oxygen stoichiometry x = 6.5. Within each domain, away from the boundary, the composition is x = 7.0 and copper has the usual four-fold planar coordination of

156   Chemistry of Superconductor Materials

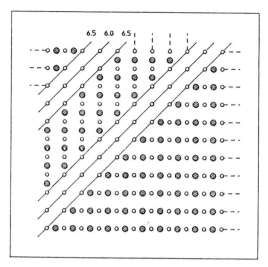

Figure 4: Possible twin boundary with copper atoms in two fold coordination (diagonal at the center of the boundary) and three-fold coordination (diagonals on the right and left side of each boundary). The composition corresponding to each diagonal is indicated. The shaded atoms are oxygen atoms and, as usual, the atoms above and below the basal plane are not indicated. The broken lines indicate the presence of oxygen atoms located outside the area represented by the figure.

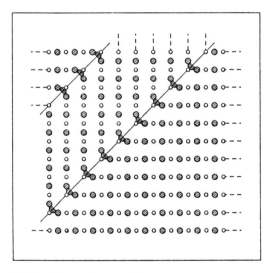

Figure 5: Twin boundary with copper atoms in distorted, tetrahedral coordination. In this model, the composition does not change at the boundary.

Defective Structures of $Ba_2YCu_3O_x$ and $Ba_2YCu_{3-y}M_yO_z$ 157

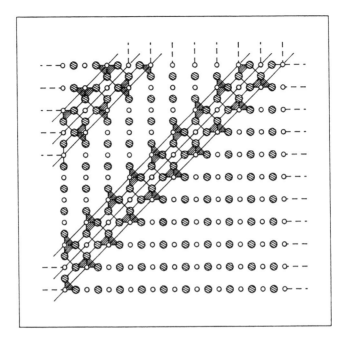

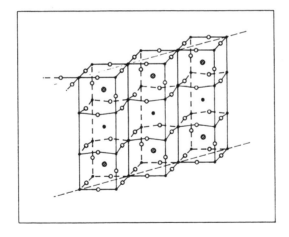

Figure 6: (a) Basal plane showing a possible twin boundary with copper atoms in five-fold coordination. Also in this model the composition does not change at the boundary. (b) Configuration of the unit cells at the boundary. Note that some oxygen atoms occupying the sites O(1) in the regular structure, have been relocated in positions O(5).

158   Chemistry of Superconductor Materials

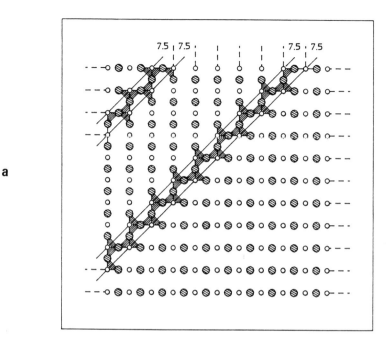

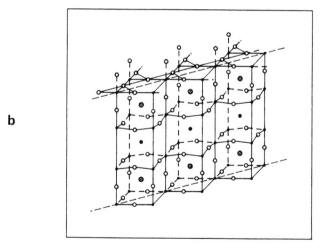

Figure 7: (a) Possible twin boundary with copper atoms in five-fold coordination. The composition at the boundary is 7.5. (b) Configuration of the unit cells at the boundary corresponding to the basal plane of (a).

Defective Structures of $Ba_2YCu_3O_x$ and $Ba_2YCu_{3-y}M_yO_z$ 159

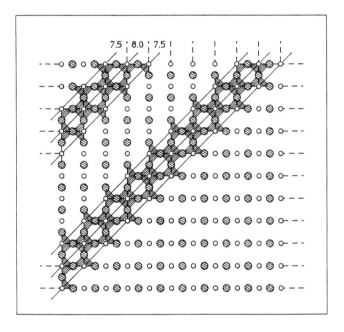

Figure 8: Twin boundary with copper atoms in five-fold and six-fold coordination. The composition of each of the diagonals forming the boundary is indicated in the figure.

the CuO layer. The domain model proposed by Ramakrishna et al. (31) is also oxygen deficient, with all the copper atoms at the boundary in three-fold coordination. These authors, however, ignore the presence of twinning, and in fact, the domain walls ("defect lamellae") are taken parallel to the a- and/or b- axes of the structure. In the model of Figure 5, the Cu(1) atoms are four-fold coordinated throughout the structure. However, while inside each domain the oxygen atoms have square-planar configuration, those at the boundaries form very distorted tetrahedra. The main feature of this model is that the oxygen stoichiometry does not change at the twin walls and that there are no copper atoms in three-fold coordination. Another configuration for which the composition does not change at the twin walls, is represented in Figure 6. In this case the copper-oxygen layer at z = 0 (basal plane) has composition CuO_2 instead of CuO, and the extra oxygen needed to increase the oxygen content is contributed by the two BaO layers located below and above the basal plane. In other words, we have here two structures of $Ba_2YCu_3O_{7.0}$, one inside the domains with the usual sequence

...$(BaO)_c[(CuO)_o(BaO)_c(CuO_2)_o(Y)_c(CuO_2)_o(BaO)_c](CuO)_o$...

and one at the boundary, with sequence

...$(BaO_{0.5})_c[(CuO_2)_o(BaO_{0.5})_c(CuO_2)_o(Y)_c(CuO_2)_o(BaO_{0.5})_c](CuO_2)_o$...

The oxygen atoms are removed from their usual positions so that the twinned domains are the pseudo-mirror images of each other, as required by the twin operations $(1\bar{1}0)_m$ or $(110)_m$, and the copper atoms at z = 0 are in four-fold planar and five-fold pyramidal coordination (Figure 6b). The models represented in Figures 7 and 8 have oxygen stoichiometry x > 7 at the twin boundaries. In the case of Figure 7, the copper atoms located on the two consecutive diagonal rows forming the twin wall have five-fold pyramidal coordination and the composition at the wall is x = 7.5. In the case of Figure 8, the twin wall is formed by three consecutive parallel diagonals. The one in the center has copper atoms with six-fold, octahedral coordination and composition x = 8.0, and the other two, copper atoms in five-fold, pyramidal coordination with stoichiometry x = 7.5.

Twin walls of the type illustrated in Figures 7 and 8 imply that oxygen stoichiometries with x > 7.0 could well exist, and that the upper limit of the composition range depends on the density of the twin boundaries present in the sample. Both of these conclusions are not supported by the experimental results obtained so far on this system. In addition, some of the oxygen needed to form the boundaries would be located on the sites O(5) and, as we have mentioned before, there is no evidence that these sites are partially occupied in orthorhombic $Ba_2YCu_3O_x$. It seems therefore reasonable to exclude these two models from any further consideration. Similarly, the model of Figure 6 can be excluded not only because it implies partial occupancy of sites O(5), but also because sites O(1) have never been found less that fully occupied. Furthermore, the formation of twin walls of this type would require large cooperative oxygen displacements that are rather unlikely to occur. Oxygen deficient twin boundaries such as those illustrated in Figure 4 predict that twinned samples having composition x = 7.0 cannot exist and that the twin boundary density of the material must decrease with increasing oxygen content. Again, there is no evidence supporting these conclusions. A further, important reason for rejecting an oxygen deficient model is that it contains a significant fraction of copper atoms in three-fold coordination. This configuration has never been found before for copper, and is less favorable energetically than either two-fold or four-fold coordination (32). In addition, nuclear quadrupolar resonance spectra have shown no evidence of three-fold coordinated copper in a sample of $Ba_2YCu_3O_{6.7}$ (33).

The model of Figure 5 contains oxygen-oxygen pairs across the interface with strong repulsive interactions (34)(35) rendering this type of boundary energetically unfavorable. This situation, however, can be compensated because the structure may assume unstable configurations that minimize the total boundary energy (36). In view of this possibility the model, which has been derived also in a number of electron microscopy studies (37)(38), will be adopted in the present discussion to represent the twinned structure of $Ba_2YCu_3O_x$ for the composition x = 7.0.

2.3 Oxygen Vacancy Ordering in $Ba_2YCu_3O_x$

One of the most important aspects of the structure of $Ba_2YCu_3O_x$ is the local arrangement of the oxygen atoms on the basal plane of the unit cell for compositions comprised between x = 6.0 and x = 7.0. X-ray and neutron powder diffraction methods usually reveal only the average structure, and, although we can measure rather precisely the occupancy factors of the oxygen sites with these techniques, we obtain no information about the configuration and the possible ordering of the atoms on a microscopic scale. For this reason electron diffraction and high resolution electron microscopy have been used extensively to study oxygen-deficient $Ba_2YCu_3O_x$.

Werder et al. (39) have observed diffuse scattering in electron diffraction patterns of samples of $Ba_2YCu_3O_{6.6-6.7}$ prepared with a gettering annealing technique. The diffuse streaks have maxima at Q = (h,k,l) + q with q = (1/2,0,0), (2/5,0,0) and (3/5,0,0). These results have been interpreted as due to short-range ordering of the oxygen vacancies. Clearly, grains with q = (1/2,0,0) have, in a short-range sense, every other row parallel to the b-axis totally vacant, so that the a-axis is doubled. The atomic configurations corresponding to q = (2/5,0,0) and (3/5,0,0) are more complex, and no interpretation for these cases has been given. The streaks have the shape of ribbons parallel to the reciprocal c* -axis, thus showing that there is no ordering along the c-axis. In these low-temperature annealed samples, diffuse scattering was observed only in the range of composition 6.6 $\leq$ x $\leq$ 6.7 and in samples with a high twin density of the order of 10^5 boundaries/cm^2.

These results have been confirmed by Fleming et al. (32) in a single-crystal x-ray scattering study of samples of composition x = 6.7. Although the single crystals used in the experiment were twinned, it was possible to determine the doubling of the a-axis. The coherence length associated with the diffuse peaks was estimated from the peak width to be of the order of four or five unit cells in the basal plane, and of one unit cell along the c-axis. This result shows that the chains are correlated in the basal plane, but uncorrelated along the c-axis. The ordering of the oxygen vacancies along the a-axis is plausible from the energetic point of view. In fact, let us consider the two ordering schemes represented in Figure 9 (32). If we assume that the oxygen-oxygen pair interaction energy along the a-axis (i.e., along the direction with no copper between the oxygen atoms) is negligible, then the configuration of Figure 9b has lower energy than that of Figure 9a because, as we have mentioned earlier, three-fold coordination of copper is less favorable than either two-fold or four-fold coordina-

Defective Structures of $Ba_2YCu_3O_x$ and $Ba_2YCu_{3-y}M_yO_z$ 163

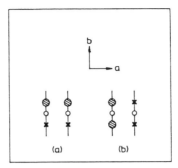

Figure 9: Ordering schemes of the oxygen atoms on the basal plane of the structure of $Ba_2YCu_3O_{6.5}$. Model (a) has copper in three-fold coordination. Model (b) shows vacancies concentrating on single chains and causing a short-range ordering along the a axis of the unit cell.

Figure 10: Basal plane of the structure of $Ba_2YCu_3O_{6.37}$. In this model the short-range doubling of the a axis is clearly visible. The number of Cu(1) in three-fold coordination, however, is rather high (~12% of the total).

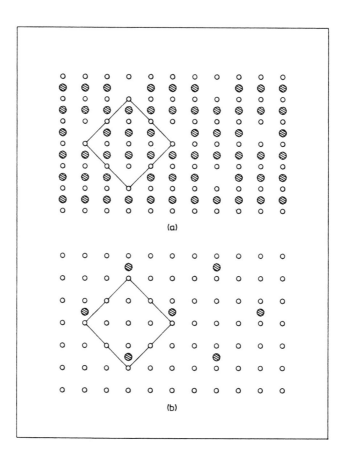

Figure 11: Models of the structures of $Ba_2YCu_3O_x$ for $x = 6.85$ (Figure a) and $x = 6.15$ (Figure b). Both structures are based on a unit cell of parameters $2\sqrt{2}a_c: 2\sqrt{2}a_c: 3a_c$ where a is the basic perovskite parameter. Also in this case some of the Cu(1) atoms are in three-fold coordination.

tion. Formation of random vacancies on the chains is not favored because they would produce large numbers of copper atoms in three-fold coordination and, as a consequence, the oxygen vacancies tend to concentrate on single chains (30)(32). This suggests that the oxygen deficient material will order preferentially along the a-axis with periodicities determined by oxygen stoichiometry.

Systematic electron diffraction studies of $Ba_2YCu_3O_x$ samples with different values of x have, in fact, shown that the oxygen vacancy ordering depends on the value of x (30). For $x \simeq 6.5$ the continuous diffuse streaks parallel to c^* have interceptions with the reciprocal (a^*, b^*) plane that can be indexed on a unit cell of parameters $2a_c \times a_c \times 3a_c$, where a_c is the basic parameter of cubic perovskite. The corresponding model is shown in Figure 10 showing a short range doubling of the a-axis, in agreement with previous results. For $x \simeq 6.15$ and $x \simeq 6.85$, the intersections on the (a^*, b^*) plane can be indexed on a unit cell of parameters $\sqrt{2}a_c \times \sqrt{2}a_c \times 3a_c$. The models corresponding to these two compositions are illustrated in Figure 11a and Figure 11b. The two atomic arrangements are clearly equivalent, in the sense that they have the oxygen atoms and the vacancies interchanged. More complex ordering schemes have been observed in samples slightly heated by the electron beam of the microscope. Such cases, however, are very complex and no models to describe them have been proposed.

The existence of vacancy ordering has also been predicted theoretically by ground-state calculations of ordered superstructures in the basal plane of $Ba_2YCu_3O_x$ (36)(40). This analysis, based on oxygen-oxygen pairs interactions, shows that, for $x = 7.0$, the most likely structures to be found in practice are: (i) a simple cell with plane-group symmetry P2mm (41), corresponding to the three-dimensional orthorhombic phase Pmmm; and (ii) a disordered structure with plane-group symmetry P4mm, corresponding to a three-dimensional tetragronal phase P4/mmm which would form by heating the orthorhombic $Ba_2YCu_3O_{7.0}$ at constant oxygen content. For $x = 6.5$ the theory predicts two double cells of planar symmetry P2mm with the a- or b- axis doubled. The first of these cases is the one discussed previously and determined by diffuse scattering in the electron diffraction patterns.

The existence of intermediate structures with various schemes of short-range oxygen ordering on the basal plane, indicates that the transition from the orthorhombic to the tetragonal phase is an order-disorder transition of second or higher order. Other results, however, support the view that the orthorhombic phase with x < 7.0 is unstable with respect to decomposition into a mixture of orthorhombic and tetragonal phases (42). High-resolution transmission electron microscopy studies (43)(44) have shown intergrowths of lamellae of the orthorhombic and tetragonal structures, having a thickness of several hundreds of Å and stacked alternately and almost regularly along the common (110) direction. Local variation of the lattice parameters of the a- and b-axes have also been detected (38)(44), and these results are consistent with the theoretical conclusions of Khachaturian et al. (45), postulating the existence of decomposed samples having a microscopic distribution of stoichiometric orthorhombic and tetragonal phases. Microstructural inhomogeneities have also been observed by high-resolution single-crystal x-ray diffraction (46) and have been related to the observed granular superconductivity showing a discrete distribution of critical temperatures for individual grains of the material (47).

If the coexistence of the two phases is the equilibrium condition for the system $Ba_2YCu_3O_x$, then the structural changes taking place at high temperature should be interpreted as a two-phase decomposition rather than an order-disorder transformation. The intermediate ordered phases described previously, then, have to be regarded as unstable or metastable transient states that occur during the low temperature decomposition of the material. The existence of such states during decomposition processes is not uncommon and Khachaturian and Morris (48) have in fact predicted the existence of a homologous series of transient ordered structures for $Ba_2YCu_3O_x$, having compositions $x = 7 - n/(2n + 1)$, where n is an integer. Thus, the type of ordering is a function of x, in agreement with previously discussed results (30), and, during decomposition, the structures of the decomposing phases pass sequentially through the series of configurations corresponding to the integral values of n and schematically illustrated in Figure 12.

In all the cases discussed so far, the short-range ordering of the oxygen atoms has a coherence length of four or five unit cells in the directions of the a- and b-axis, that is, the oxygen atoms are clustered in chains having a length of about 20Å. Although the three-fold coordinated copper atoms are much less numerous in this

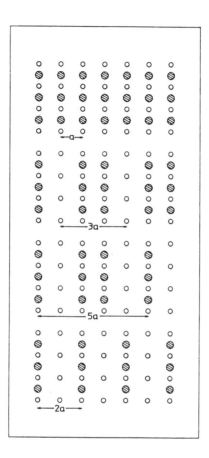

Figure 12: Short-range ordering of the oxygen atoms on the basal plane of the structure of $Ba_2YCu_3O_x$, as proposed by Khachaturian and Morris [48]. The composition corresponding to these ordering schemes is given by $x = 7-n/(2n + 1)$ with n = 0,1,2 and n → ∞.

168 Chemistry of Superconductor Materials

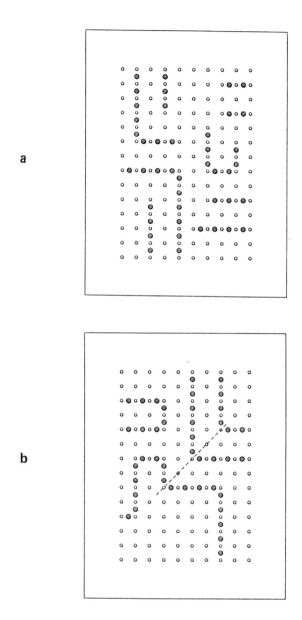

Figure 13: (a) In this figure, the model of Figure 10 has been modified to reduce the number of Cu(1) atoms in three-fold coordination with Cu(1) in tetrahedral coordination; (b) Total substitution; short-range ordering and stoichiometry are identical to those of Figure 10.

configuration than in a random distribution of vacancies, still their number is highly significant. For example, 12% of them have this unusual coordination in the model of Figure 10. This problem can be easily avoided if we accept the idea that first neighbor oxygen-oxygen pairs of the type illustrated in Figure 5 can exist, either as such, or in configurations that minimize the repulsive interactions. Taking again the model of Figure 10 as an example, it is possible to reduce the copper atoms in three-fold coordination from 12% to 8% by shifting just a few oxygen atoms from their original positions (Figure 13a.). Working on the same principle, it is not difficult to build a model in which no three-fold coordinated copper exists and in which all the characteristics pertinent to a short-range doubling of the a-axis are preserved (49). A model like the one illustrated in Figure 13b, for example, does not have any copper atoms in three-fold coordination and also explains how twin boundaries may form in the orthrhombic phase. In fact, for the composition of Figure 13b, the average symmetry of the material is tetragonal, and the interface indicated with a broken line may be considered the precursor of a twin boundary described by Iijima et al. as "incipient twinning" (50).

2.4 Mechanisms of Oxygen Elimination From the Structure of $Ba_2YCu_3O_x$

The short-range ordering of the oxygen atoms reviewed in the previous paragraph, is intimately related to the oxygen stoichiometry and to the mechanism by which oxygen can be added to, or eliminated from, the structure of $Ba_2YCu_3O_x$. A structural model for the reduction process has been recently proposed by Alario-Franco and Chaillout (51). These authors start with the model of Figure 11a based on a cell of sides $2\sqrt{2}a_c \times 2\sqrt{2}a_c \times 3a_c$ and corresponding to the composition $x = 6.875$, and gradually eliminate the oxygen atoms located on the same rows where vacancies were originally present, rather than at random, in order to minimize the number of copper atoms in three-fold coordination. At the composition $x = 6.5$, the structure with the short-range doubling of the a-axis is obtained, and the process is carried on until complete elimination of oxygen from the basal plane is achieved. The actual removal of oxygen is viewed first as a displacement of an atom from the site O(4) at (0,1/2,0) to the site (1/2,1/2,0) at the center of the mesh, and then as a diffusion

170 Chemistry of Superconductor Materials

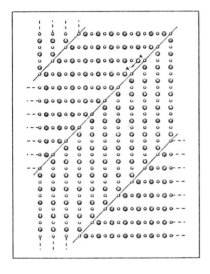 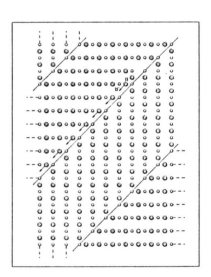

a b

Figure 14: Mechanism of oxygen elimination from the structure of $Ba_2YCu_3O_{7.0}$. (a) As an effect of temperature increase, atom A may jump into position A'; (b) As a consequence of this shift, atom B may jump into position B', C into C' etc., thus causing a correlated motion of the oxygen atoms terminating with the expulsion of half oxygen atom from the structure; (c) Atom C now may jump into positions C' or C" generating a second cascade
(continued)

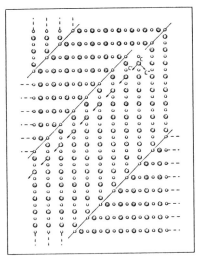

c

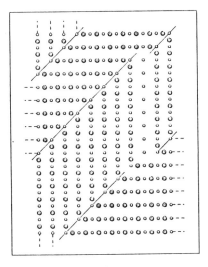

d

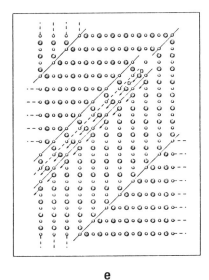

e

resulting in the elimination of one additional half atom of oxygen and the formation of a copper atom in two-fold coordination; (d) and (e) This process may be repeated until all the Cu(1) atoms of a row are in two-fold coordination. Note that, in this model, the elimination of oxygen is accompanied by a movement of the twin walls.

through the "tunnels", or channels, running along the b-axis through the positions (1/2,1/2,0) and (1/2,0,0).

A model of oxygen elimination in $Ba_2YCu_3O_x$, however, has to take into consideration, and be consistent with, a number of experimental results. First of all, it has to take into account the presence of twinning in the orthorhombic phase and explain the observed movement of the twin walls when the oxygen stoichiometry is varied, then it has to be consistent with a very fast diffusion process, and finally, it has to be based on an atomic configuration for which the three-fold coordinated copper atoms are present only in very small numbers or are totally absent. A possible mechanism that satisfies all these requirements is schematically illustrated in Figure 14 (49).[1] Let us assume that, initially, we have a twinned sample of composition x = 7.0, and let us suppose that the temperature of the sample is being gradually increased. In these conditions it may happen that one of the oxygen atoms located at a twin boundary (e.g., atom A in Figure 14a), being subjected to both increased thermal vibrations and to repulsive interactions with the corresponding atom across the wall, jumps into position A′. As a consequence of this motion, atom B (Figure 14b) has two nearest neighbors, rather than one, and, due to the resulting repulsion operated by these two atoms, may be ejected into position B′, thus initiating a chain of events, indicated with arrows in Figure 14b, ending with the elimination of a half oxygen atom from the grain. In the resulting configuration (Figure 14c) there is one copper atom in three-fold coordination, a situation that we assume to be unstable and that may be changed only if atom C moves into position C′ or C″. In either case, a second chain of jumps starts, ending again with the elimination of an additional half atom of oxygen and leaving one copper atom in two-fold coordination (Figure 14d). At this point atom D, which has, again, two nearest neighbors, will very likely move into position D′, and the process will continue as before until a whole row of two-fold coordinated copper atoms is formed (Figure 14e). In agreement with experimental results (28), the elimination of oxygen predicted by this model is accompanied by a continuous movement of the twin walls in the region of the grain

---

[1] In this figure, as well as in all the others representing the level at z = 0, the oxygen atoms below and above each copper atom are not shown for clarity.

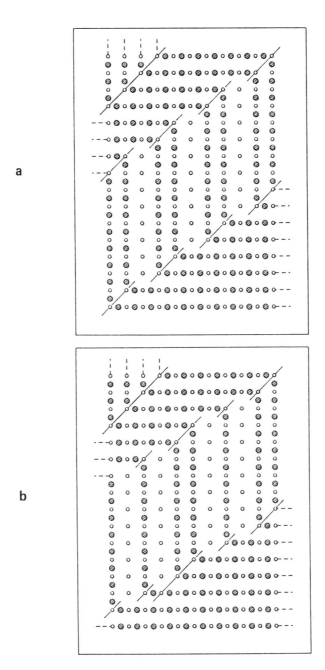

Figure 15: Short range ordering, according to Khachaturian and Morris [48], generated by the elimination mechanism illustrated in Figure 14.

involved in the reduction process. In addition, the cascade-like, correlated, sequence of atomic jumps postulated by the model to describe the loss of oxygen, can occur very rapidly, in agreement with the very fast diffusion process observed experimentally. If we eliminate with the same technique the oxygen from every third row in our model, we obtain the atomic configuration of Figure 15, which has one of the short-range ordering schemes predicted by Khachaturian and Morris (48) and, from this, it is possible to generate all the other structures of the homologous series postulated by these authors for $Ba_2YCu_3O_x$. It is worthwhile to note that in none of the structures generated by this method is there three-fold coordinated copper (Figures 14d, 14e, 15a, and 15b).

2.5 Metal Substitutions

Systematic substitution of trivalent rare-earth atoms for Y and of impurity transition metal atoms for Cu in $Ba_2YCu_3O_x$, and determination of the effects that the substituents have on the superconducting transition temperature of the resulting compounds, may provide a method for studying the possible mechanisms of superconductivity and for investigating which structural feature is relevant to high values of T_c. In addition to this, ceramic processing properties such as densification and microstructure, may depend on the presence and nature of dopants, and reactions with the substrates may modify the superconducting properties of the materials being used in the fabrication of these films. It is not surprising, therefore, that substituted and doped samples of $Ba_2YCu_3O_x$ have been investigated very intensely in the last three years.

The crystal chemistry of $Ba_2RCu_3O_x$ has been systematically studied by single-crystal and powder diffraction methods with R = La, Pr,... Yb, in addition to the conventional yttrium compound [(52)(53)(54) and references therein]. With the exception of La, Pr, and Tb, the substitution of Y with rare-earth metals has little or no effect on the superconductivity, with the values of T_c ranging from 87 to 95K. Also, a relatively small change is observed in the cell constants of these compounds. The La, Pr, and Tb-substituted materials are not superconductors. A detailed structural analysis of the Pr case (52) did not show any evidence of a superstructure or the presence of other differences with the atomic configuration of the yttrium prototype.

The possibility that Pr atoms may occupy some of the Ba sites has been taken into consideration, but no results for this type of substitution have been presented so far. As expected, the interatomic distances involving R atoms depend strongly on the values of the ionic radii of R. On the other hand, the distances that do not involve those atoms show only a weak or no dependence on the nature of R.

Far more complex from a crystallographic point of view are the compounds formed by substitution of Cu by impurity metal atoms. The complexity is due to the fact that the substituents may replace copper atoms located on two sets of non-equivalent sites, one, Cu(2), at $(0,0,z)$ and the other, Cu(1), on the basal plane of the unit cell at $(0,0,0)$, where the oxygen vacancies are also present. The general formula for this class of materials may be written as $Ba_2YCu_{3-y}M_yO_x$, and the most interesting cases studied so far are those where M = Fe, Co, Ni, Zn, Ga, and Al. Examples showing the stoichiometry, the type of substitution, and symmetry of these compounds are given in Table 4. The data in Table 4 show that some elements, such as Zn (and probably Ni) substitute preferentially copper atoms Cu(2), while others (e.g., Fe and Co) at first substitute Cu(1) and, when their concentration becomes large, begin to replace Cu(2). An additional interesting feature is that the oxygen stoichiometry may be larger than 7.0 and that the symmetry is not related to the oxygen content in the same way as in the undoped material. In all cases the value of T_c decreases with increasing amounts of impurities (58)(60), and such a decrease is much more pronounced when the substitution involves copper atoms of the Cu(2) sites than copper atoms of the Cu(1) sites. This observation has been taken as evidence that the integrity of the buckled CuO_2 planes at $(0,0,z)$ is much more important for sustaining high T_c superconductivity than that of the chains running along the b-axis of the structure (55).

The number of impurity atoms, the oxygen stoichiometry, and the symmetry of many of the compounds listed as examples in Table 4, provoke some thoughts about the configuration of the oxygen on the basal plane of the unit cell. In almost all cases, oxygen is incorporated in the structure above the limit $x = 7.0$, and neutron diffraction studies have shown that this excess is distributed over the O(4) and O(5) sites (55-58). Since copper should retain the square-planar coordination, we assume that the extra oxygen surrounds the M atoms so that their average coordination becomes larger than four. The possible

TABLE 4. Examples of Compounds Formed by Substitution of Copper by Impurity Metal Atoms.

	Cu(2) Sites	Cu(1) Sites	Oxygen Stoichiom	Symm.	Ref.
Ba_2Y	$[Cu(2)]_{1.82}Zn_{0.18}$	$[Cu(1)]$	$O_{6.80(6)}$	O*	(55)
Ba_2Y	$[Cu(2)]_2$	$[Cu(1)]_{0.76}Ga_{0.24}$	$O_{7.01(9)}$	T	(55)
Ba_2Y	$[Cu(2)]_2$	$[Cu(1)]_{0.84}Al_{0.16}$	$O_{7.0}$	T	(56)
Ba_2Y	$[Cu(2)]_2$	$[Cu(1)]_{0.77}Fe_{0.23(3)}$	$O_{7.13}$	T	(57)
Ba_2Y	$[Cu(2)]_{1.86}Fe_{0.14(5)}$	$[Cu(1)]_{0.69}Fe_{0.31(3)}$	$O_{7.15}$	T	(56)
Ba_2Y	$[Cu(2)]_2$	$[Cu(1)]_{0.78}Co_{0.22}$	$O_{7.04}$	T	(58)
Ba_2Y	$[Cu(2)]_{1.90}Co_{0.10}$	$[Cu(1)]_{0.27}Co_{0.73}$	$O_{7.30}$	T	(58)

(*) O: orthorhombic.   T: tetragonal

coordinations of the M atoms depend on the distribution of the oxygen atoms over the neighboring sites $O(4)$ and $O(5)$. If both these positions are occupied, then each atom M will be surrounded by two atoms $O(1)$, two $O(4)$ and two $O(5)$, forming a distorted octahedron. If the octahedra are randomly distributed and isolated, the required oxygen in excess over the composition $x = 7.0$ is in the ratio O:M = 2:1. This ratio decreases if the octahedra are grouped into clusters, and it would become 1:1 if all the Cu(1) sites were occupied by M atoms (in this case we would have $x = 8.0$). If only one site $O(5)$ is occupied by oxygen, the coordination of M is five-fold pyramidal. The excess oxygen needed for isolated pyramids is O:M = 1:1 and the oxygen stoichiometry would become $x = 7.5$ if all the sites $(0,0,0)$ were pyramidal. Finally, the M atoms can be four-coordinated by two $O(1)$, one $O(4)$, and one $O(5)$ forming a very distorted tetrahedron. Obviously, for this coordination there is no increase in the oxygen stoichiometry above $x = 7.0$. Thus the amount of the excess oxygen in the structure depends on the number of the substituting M atoms and on their coordination and distribution over the Cu(1) sites.

Models of the atomic configuration of $Ba_2YCu_{3-y}M_yO_x$ have been discussed in a number of cases. Miceli et al. (58) consider a value of $x = 6.93$ as the maximum oxygen concentration of the undoped $Ba_2YCu_3O_x$ (a value of $x < 7.0$ is invoked in order to allow the Ba atoms to bind not only to the $O(1)$ and $O(4)$ atoms, but also to the rest of the structure. Furthermore, values of x between 6.91 and 6.96 rather than 7.0 have been obtained in many neutron powder diffraction experiments using samples of $Ba_2YCu_3O_x$ annealed in oxygen). With this composition limit, the extra oxygen incorporated in the Co-substituted materials of Table 4 would be in a ratio of about O:Co = 1:2. This compositional relation has been explained with the possible existence of Co-O-Co groups, in which the extra oxygen atom could share one electron from each of the Co neighbors. The axial nature of these pairs may also cause some sort of short range orientational ordering. This model, however, does not explain the transition from orthorhombic to tetragonal symmetry that has been found to occur at very low concentrations of cobalt (61).

A more detailed analysis of the possible oxygen configurations in $Ba_2YCu_{3-y}M_yO_x$ has been made for M = Fe (57)(62), and the conclusions obtained for this case have been extended also to the Co-doped material. The values $x = 7.13$ and $x = 7.15$ of the samples in Table 4, correspond, approximately, to one extra oxygen for every two Fe atoms on the basal plane, and this O:Fe ratio would be even larger if we would adopt a composition limit x < 7.0, as proposed in reference (58). This means that the average coordination of the Fe atoms is larger than four. A possible model corresponding to the formula $Ba_2YCu(2)_{2.0}(Cu(1)_{0.77}Fe_{0.23})O_{7.13}$ is shown in Figure 16. In this configuration, the Fe atoms assume tetrahedral, pyramidal, and octahedral coordination, consistently with the results of Mossbauer spectroscopy (63), and are clustered along the diagonals of the unit cell, thus acting as a sort of twin walls. Electron microscopy results (57) indicate that this ordered clustering of the impurity atoms is not correlated along the c-axis and, therefore, crossings of domains may occur in this direction. Electron micrographs of samples with variable values of y (64), show for y < 0.06, twin lamellae similar to those obtained for the non-doped orthorhombic phase. For y > 0.06, these lamellae begin to develop along both of the (110) and (1$\bar{1}$0) directions, resulting in a cross-hatched pattern of bands having spacings much smaller than those for y < 0.06. The model of Figure 16 is not consistent with these results and does not explain why the symmetry becomes tetragonal at very low concentrations of the impurity atoms.

We may try to interpret the substitution of the copper atoms and the consequent rearrangement of the oxygen atoms on the basal plane of the unit cell of $Ba_2YCu_{3-y}M_yO_x$, by using the same mechanism adopted before to describe the elimination of oxygen from the structure of $Ba_2YCu_3O_x$ (49). Let us suppose that the impurity atom M has a strong preference for octahedral coordination (case of iron and cobalt, (65)). If the atom M substitutes one of the Cu atoms located on the twin boundaries, then the simplest way to arrange an octahedron around M is by moving the oxygen atoms A and A' of Figure 17a into positions B and B', respectively. This motion results in the configuration of Figure 17b in which one of the copper atoms has two-fold coordination. In this situation there is no change of the oxygen stoichiometry,, unless the correlated jumps indicated by the arrows take place, with consequent incorporation of one oxygen atom. In any case, this type of substitution does not cause a change of the twin boundaries or a reduction of the size of the orthorhombic domains in the grain. When the substitution occurs in the interior of a domain, however, the mechanism is very different. Let us consider

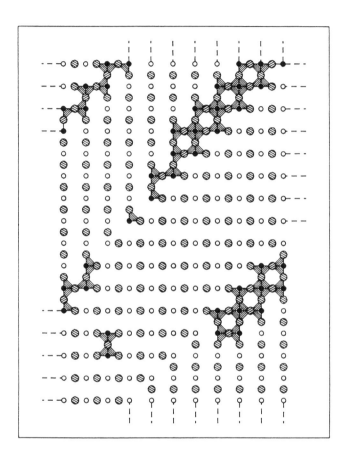

Figure 16: Model for the structure of $Ba_2YCu_{2.77}Fe_{0.23}O_{7.13}$ as proposed in ref. [57]. The impurity atoms (full circles) assume in this model tetrahedral, pyramidal, and octahedral coordination and are clustered along one of the diagonals of the unit cell.

180   Chemistry of Superconductor Materials

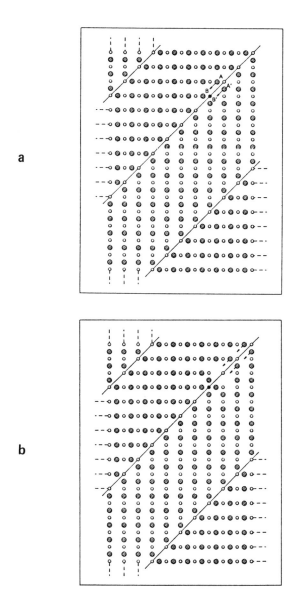

Figure 17: Possible mechanism for arranging octahedral coordination around an impurity atom (full circle) located on the twin boundary. (a) Atoms A and A' jump into B and B', respectively. (b) Resulting structure with unchanged stoichiometry and with one Cu(1) atom in two-fold coordination. The stoichiometry may be changed by shifting the atoms as indicated by the arrows.

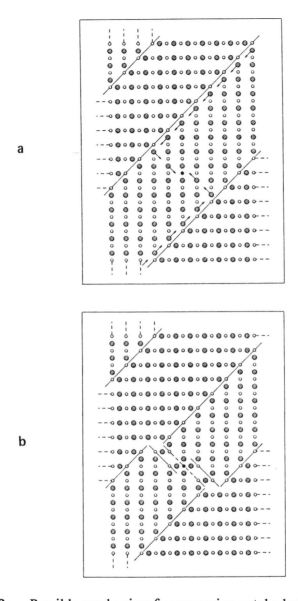

Figure 18: Possible mechanism for arranging octahedral coordination around an impurity atom located on the interior of a grain. (a) Shifts of the atoms are indicated by the arrows. (b) Resulting structure. With this mechanism the oxygen stoichiometry becomes larger than 7. The twin walls move as a consequence of the rearrangement of the oxygen atoms and cross-twinning becomes possible.

the case of Figure 18a. The six-fold coordination of the M atom may be obtained by shifting the oxygen atoms as indicated by the arrows, and now incorporation of oxygen is necessary to avoid the presence of three-fold coordinated copper. The configuration resulting from the rearrangement of the oxygen atoms, illustrated in Figure 18b, shows that the orthorhombic lamellae comprised between parallel twin boundaries are thinner than before and that a cross-hatched pattern, of the type observed in electron microscopy studies (64) is forming because the twin walls begin to develop in both the (110) and ($1\bar{1}0$) directions. More important, however, is that the model proposed here for the substitution of copper, shows that the size of an orthorhombic domain may be halved just by introducing one impurity atom, and it is possible to show that the domain structure on the basal plane can be disrupted by the presence of 3-5% of atoms M. This mechanism, therefore, may explain why the diffraction methods detect tetragonal symmetry at very low concentrations of M. The substitution in the interior of a domain may take place, however, also without incorporation of extra oxygen. The cooperative oxygen shifts necessary to accomplish this are indicated in Figures 19a and 19b. In this case one copper atom acts as an oxygen donor and remains in two-fold coordination, while the substituting atom becomes six-coordinated, or two of them assume five-fold, pyramidal coordination. This type of oxygen rearrangement may be used to explain symmetry and composition of substitution compounds such as $Ba_2YCu_{2.84}Al_{0.16}O_{7.0}$. The model illustrated in Figures 20a, b and c is an example in which all the impurity atoms are octahedrally coordinated and the composition is y = 0.11 and x = 7.0. Incorporation of oxygen, if needed, can be easily accomplished with the mechanism of correlated jumps illustrated in one of the previous sections. From Figure 20c it is obvious that the symmetry of the average structure is tetragonal even at low concentrations of the M ions, in agreement with experimental results; e g., (66-68). It is also clear that, according to our model, short-range order can easily occur in these structures.

Defective Structures of $Ba_2YCu_3O_x$ and $Ba_2YCu_{3-y}M_yO_z$ 183

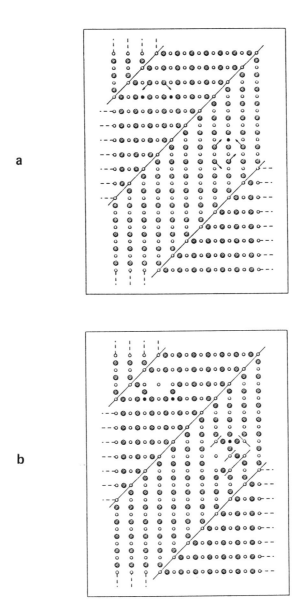

Figure 19: Possible mechanisms of formation of five-fold and six-fold coordination around impurity atoms without change of the oxygen stoichiometry. (a) Shifts of the atoms are indicated by the arrows. (b) Resulting structure.

184   Chemistry of Superconductor Materials

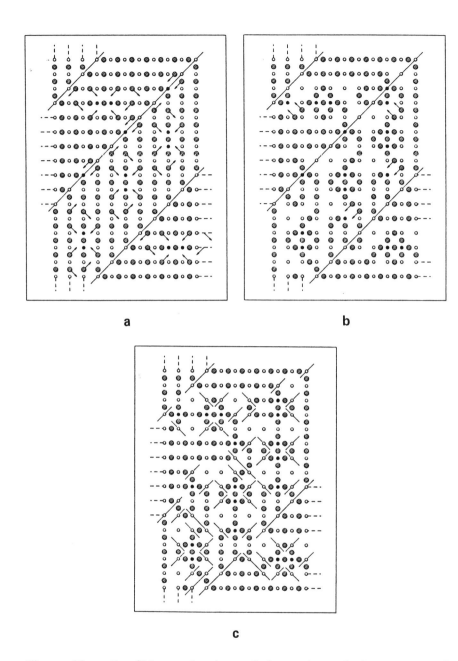

Figure 20: Possible mechanism of formation of the compound $Ba_2YCu_{2.89}M_{0.11}O_{7.0}$ with M in six-fold coordination. (a) Shifts of the atoms are indicated by the arrow. (b) Intermediate structure. (c) Resulting structure.

3.0 REFERENCES

1. M.K. Wu, J.R. Ashburn, C.J. Torng, P.H. Hor, R.L. Meng, L. Gao, Z.J. Huang, Y.Q. Wang, and C.W. Chu, *Phys. Rev. Lett.* 58 (9):908-910 (1987).

2. R.J. Cava, B. Batlogg, R.B. van Dover, D.W. Murphy, S. Sunshine, T. Siegrist, J.P. Remeika, E.A. Rietman, S. Zahurak, and G.P. Espinosa, *Phys. Rev. Lett.* 58 (16):1676-1679 (1987).

3. P.M. Grant, R.B. Beyers, E.M. Engler, G. Lim, S.S. Parkin, M.L. Ramirez, V.Y. Lee, A. Nazzal, J.E. Vazquez, and R.J. Savoy, *Phys. Rev.* B35 (13):7242-7244 (1987).

4. W.J. Gallagher, R.L. Sandstrom, T.R. Dinger, T.M. Shaw, and D.A. Chance, *Solid State Comm.* 63 (2):147-150 (1987).

5. D.G. Hinks, L. Soderholm, D.W. Capone II, J.D. Jorgensen, I.K. Schuller, C.U. Segre, K. Zhang, and J.D. Grace, *Appl. Phys. Lett.* 50 (23):1688-1690 (1987).

6. R. Beyers, G. Lim, E.M. Engler, R.J. Savoy, T.M. Shaw, T.R. Dinger, W.J. Gallagher, and R.L. Sandstrom, *Appl. Phys. Lett.* 50 (26):1918-1920 (1987).

7. T. Siegriest, S. Sunshine, D.W. Murphy, R.J. Cava, and S.M. Zahurak, *Phys. Rev.* B 35 (13):7137-7139 (1987).

8. R.M. Hazen, L.W. Finger, R.J. Angel, C.T. Prewitt, N.L. Ross, H.K. Mao, C.G. Hadidiacos, P.H. Hor, R.L. Meng, and C.W. Chu, *Phys. Rev.* B 35 (13):7238-7241 (1987).

9. Y. LePage, W.R. McKinnon, J.M. Tarascon, L.H. Greene, G.W. Hull, and D.M. Hwang, *Phys. Rev.* B 35 (13), 7245-7248 (1987).

10. M.A. Beno, L. Soderholm, D.W. Capone II, D.G. Hinks, J.D. Jorgensen, J.D. Grace, I.K. Schuller, C.U. Segre, and K. Zhang, *Appl. Phys. Lett.* 51 (1):57-59 (1987).

11. J.E. Greedan, A.H. O'Reilly, and C.V. Stager, *Phys. Rev.* B 35 (16):8770-8773 (1987).

12. J.J. Capponi, C. Chaillout, A.W. Hewat, P. Lejay, M. Marezio, N. Nguyen, B. Raveau, J.L. Soubeyroux, J.L. Tholence, and R. Tournier, *Europhys. Lett.* 3:1301- (1987).

13. F. Beech, S. Miraglia, A. Santoro, and R.S. Roth, *Phys. Rev.* B 35 (16):8778-8781 (1987).

14. W.I.F. David, W.T.A. Harrison, J.M.F. Gunn, O. Moze, A.K. Soper, P. Day, J.D. Jorgensen, D.G. Hinks, M.A. Beno, L. Soderholm, D.W. Capone II, I.K. Schuller, C.U. Segre, K. Zhang, and J.D. Grace, *Nature*, 327:310-312 (1987).

15. D.E. Cox, A.R. Moodenbaugh, J.J. Hurst, R.H. Jones, *J. Phys. Chem. Solids*, 49 (1):47-52 (1987).

16. M. Francois, A. Junod, K. Yvon, A.W. Hewat, J.J. Capponi, P. Strobel, M. Marezio, and P. Fischer, *Solid State Comm.* 66 (10):1117-1125 (1988).

17. J.D. Jorgensen, M.A. Beno, D. G. Hinks, L. Soderholm, K.J. Volin, R.L. Hitterman, J.D. Grace, I.K. Schuller, C.U. Segre, K. Zhang, and M.S. Kleefisch, *Phys. Rev.* B 36 (7):3608-3616 (1987).

18. G.S. Grader and P.K. Gallagher, *Adv. Cer. Mat.* 2:649- (1987).

19. R.J. Cava, B. Batlogg, C.H. Chen, E.A. Rietman, S.M. Zahurak, and D. Werder, *Phys. Rev.* B 36 (10):5719-5722 (1987).

20. P.K. Gallagher, H.M. O'Bryan, S.A. Sunshine, and D.W. Murphy, *Mat. Res. Bull.* 22:995 - (1987).

21. J.D. Jorgensen, B.W. Veal, W.K. Kwok, G.W. Crabtree, A. Umezawa, L.J. Nowicki, and A.P. Paulikas, *Phys. Rev.* B 36 (10):5731-5734 (1987).

22. A. Santoro, S. Miraglia, F. Beech, S.A. Sunshine, D.W. Murphy, L.F. Schneemeyer, and J.V. Waszczak, *Mat. Res. Bull.* 22:1007-1113 (1987).

23. P. Bordet, C. Chaillout, J.J. Capponi, J. Chenavas, and M. Marezio, *Nature* 327:687-689 (1987).

24. S. Miraglia, F. Beech, A. Santoro, D. Tran Qui, S.A. Sunshine, and D.W. Murphy, *Mat. Res. Bull.* 22:1733-1740 (1987).

25. G. Friedel, *Lecons de Cristallographie*, Paris: Berger-Levrault. Reprinted 1964, Paris: Blanchard (1926).

26. G. Van Tendeloo, H.W. Zandbergen, and S. Amelinckx, *Solid State Comm.* 63 (5):389-393 (1987).

27. M.M. Fang, V.G. Kogan, D.K. Finnemore, J.R. Clem, L.S. Chumbley, and D.E. Farrell, *Phys. Rev.* B 37 (4):2334-2337 (1988).

28. J.L. Hodeau, C. Chaillout, J.J. Capponi, and M. Marezio, *Solid State Comm.* 64 (11):1349-1352 (1987).

29. A. Santoro, *Acta Cryst.* A 30:224-231 (1987).

30. J.L. Hodeau, P. Bordet, J.J. Capponi, C. Chaillout, and M. Marezio, *Physica* C. 153-155:582-585 (1988).

31. K. Ramakrishna, B. Das, A.K. Singh, R.S. Tiwari, and O.N. Srivastava, *Solid State Comm.* 65 (8):831-834 (1988).

32. R.M. Fleming, L.F. Schneemeyer, P.K. Gallagher, B. Batlogg, L.W. Rupp, and J.V. Waszczak, *Phys. Rev.* B 37, (13):7920-7923 (1988).

33. W.W. Warren, Jr., R.E. Walstedt, G.F. Brennert, R.J. Cava, B. Batlogg, and L.W. Rupp, *Phys. Rev.* B 39 (1):831-834 (1989).

34. Z. Hiroi and M. Takano, *Solid State Comm.* 65 (12):1549-1554 (1988).

35. Z. Hiroi, M. Takano, and Y. Bando, *Solid State Comm.* 69 (3), 223-228 (1989).

36. D. de Fontaine, L.T. Wille, and S.C. Moss, *Phys. Rev.* B 36 (10):5709-5712 (1987).

37. C.S. Pande, A.K. Singh, L. Toth, D.U. Gubser, and S. Wolf, *Phys. Rev.* B 36, (10):5669-5671 (1987).

38. M. Sarikaya and E. Stern, *Phys. Rev.* B 37 (16):9373-9381 (1988).

39. D.J. Werder, C.H. Chen, R.J. Cava, and B. Batlogg, *Phys. Rev.* B 37 (4):2317-2319 (1988).

40. L.T. Wille and D. deFontaine, *Phys. Rev.* B 37 (4):2227-2230 (1988).

41. International Tables for Crystallography, D. Reidel Publishing Co; Dordrecht, Holland, p. 81 (1983).

42. A.G. Khachaturian and J.W. Morris, Jr., *Phys. Rev. Lett.* 59 (24):2776-2779 (1987).

43. Z. Hiroi, M. Takano, Y. Ikeda, Y. Takeda, and Y. Bando, *Jpn. J. Appl. Phys.* 27:L141- (1988).

44. Z. Hiroi, M. Takano, Y. Bando, Y. Takeda, and R. Kanno, *Physica* C 158:269-275 (1989).

45. A.G. Khachaturian, S.V. Semenovskaya, and J.W. Morris, Jr., *Phys. Rev.* B 37 (4):2243-2246 (1988).

46. H. You, J.D. Axe, X.B. Kan, S.C. Moss, J.Z. Liu, and D.J. Lam, *Phys. Rev.* B 37 (4):2301-2304 (1988).

47. X. Hai, R. Joynt, and D.C. Larbalestier, *Phys. Rev. Lett.* 58 (26):2798-2801 (1987).

48. A.G. Khachaturian and J.W. Morris, Jr., *Phys. Rev. Lett.* 61 (2):215-218 (1988).

49. A. Santoro. Unpublished results

50. S. Iijima, T. Ichihashi, Y. Kubo, and J. Tabuchi, *Jpn. J. Appl. Phys.* 26 (11):L1790-L1793 (1987).

51. M.A. Alario-Franco and C. Chaillout (1989)., *J. Solid State Chem.* in press.

52. Y. LePage, T. Siegrist, S.A. Sunshine, L.F. Schneemeyer, D.W. Murphy, S.M. Zahurak, J.V. Waszczak, W.R. McKinnon, J.M. Tarascon, G.W. Hull, and L.H. Greene, *Phys. Rev.* B 36 (7):3617-3621 (1987).

53. J.M. Tarascon, W.R. McKinnon, L.H. Greene, G.W. Hull, and E.M. Vogel, *Phys. Rev.* B 36 (1):226-234 (1987).

54. C.P. Poole, T. Datta, and H.A. Farach, *Copper Oxide Superconductors*, J. Wiley and Sons: New York (1988).

55. G. Xiao, M.Z. Cieplak, D. Musser, A. Gavrin, F.H. Streitz, C.L. Chien, J.J. Rhyne and J.A. Gotaas, *Nature* 332:238-240 (1988).

56. J. Stalick and J.J. Rhyne, private communication.

57. P. Bordet, J.L. Hodeau, P. Strobel, M. Marezio, and A. Santoro, *Solid State Comm.* 66 (4):435-439 (1988).

58. P.F. Miceli, J.M. Tarascon, L.H. Greene, P. Barboux, F.J. Rotella, and J.D. Jorgensen, *Phys. Rev.* B 37 (10):5932-5935 (1988).

59. Y. Oda, H. Fujita, H. Toyoda, T. Kaneko, T. Kohara, I. Nakada, and K. Asayama, *Jpn. J. Appl. Phys.* 26 (10):L1660-L1663 (1987).

60. A.M. Balagurov, G.M. Mironova, A. Pajaczkowska, and H. Szymczak, *Physica* C 158:265-268 (1989).

61. T. Kajitani, K. Kusaba, M. Kikuchi, Y. Syono, and M. Hirabayashi, *Jpn. J. Appl. Phys.* 26 (10):L1727-L1730 (1987).

62. J.L. Hodeau, P. Bordet, J.J. Capponi, C. Chaillout, J., Chenavas, M. Godinho, A.W. Hewat, E.A. Hewat, H. Renevier, A.M. Spieser, P. Strobel, J.L. Tholence, and M. Marezio. *Proceedings of the First Asia-Pacific Conference on High-T Superconductors*, Singapore, June 27-July 2 (1988).

63. M. Takano and Y. Takeda, *Jpn. J. Appl. Phys.* 26 (11):L1862-L1864 (1987).

64. Z. Hiroi, M. Takano, Y. Takeda, R. Kanno and Y. Bando, *Jpn. J. Appl. Phys.* 27 (4):L580-L583 (1988).

65. E. Takayama-Muromachi, Y. Uchida, and K. Kato, *Jpn. J. Appl. Phys.* 26 (12):L2087-L2090 (1987).

66. T. Siegriest, L.F. Schneemeyer, J.V. Waszczak, N.P. Singh, R.L. Opila, B. Batlogg, L.W. Rupp, and D.W. Murphy, *Phys. Rev.* 36, (16):8365-8368 (1987).

67. P.B. Kirby, M.R. Harrison, W.G. Freeman, I. Samuel, and M.J. Haines, *Phys. Rev.* 36 (16):8315-8319 (1987).

68. T.J. Kistenmacher, W.A. Bryden, J.S. Morgan, and K. Moorjani, *Phys. Rev.* B 36 (16):8877-8880 (1987).

4

Crystal Chemistry of Superconductors and Related Compounds

Anthony Santoro

1.0 INTRODUCTION

It is customary to describe the structures of superconductors in terms of known structural types. For example, the structure of $Ba_2YCu_3O_x$ has been related to that of perovskite, while the atomic arrangement of Ba- and Sr-doped La_2CuO_4 has been found to be isomorphous with that of K_2NiF_4. Similarly, the superconductors belonging to the systems $Bi_2Sr_2Ca_{n-1}Cu_nO_{2n+4}$ and $Tl_2Ba_2Ca_{n-1}O_{2n+4}$ have been compared with the Aurivillius phases, although these phases have some important features different from those found in the superconductors. These descriptions are useful for the purpose of classification, but they do not permit one to easily compare the various structures and to find common structural features, if they exist. As the number of superconductors found so far (and of materials structurally and chemically related to them) is rather large, it would be advantageous to discuss the crystal chemistry of these important compounds in terms of a coherent scheme providing not only a basis for the description of the crystal structures of the known materials, but also a starting point for the search of new compounds.

Fortunately, the known superconductors belong to a broad class of inorganic compounds in which the atoms have a layered configuration. Therefore, they can be classified and compared to one another by specifying the composition, the type and the sequence of the layers contained in the unit cell of the structure. This approach is not new. In a study of the homologous series $Sr_{n+1}Ti_nO_{3n+1}$, Ruddlesden and Popper (1)(2) describe the structures of the various compounds in the series in terms of perovskite-type slabs interleaved with layers containing Sr and O and having a rock salt configuration. This description was later applied to compounds of general formula A_2BX_4 (which include Ba- and Sr- doped La_2CuO_4) (3)(4), and extended to a larger class of superconductors of composition $mAO.nACuO_3$, where A is a large cation such as Ba, Bi, Ca, Sr, or Tl, and m and n are integers (5). In what follows we will generalize this treatment to the point of being able to give a unified description of the crystal chemistry of all the known superconductors.

2.0 DESCRIPTION OF LAYERED STRUCTURES

Oxide superconductors and related compounds can be easily described in terms of layers having composition AX and BX_2 (6). For example, in the structure of an idealized cubic perovskite ABX_3 (Figure 1a), the first layer perpendicular to the c-axis has composition BX_2, with the B atoms at the corners of a square mesh and the X atoms at the midpoints of the edges. This configuration may be represented with the symbol $(BX_2)_o$, indicating not only the composition of the layer, but also the choice of origin. A symbol like $(BX_2)_c$, therefore, represents the same layer with a different choice of origin, i.e., with the atom B at the center of the mesh and the X atoms at the midpoints of the edges. These two configurations can be derived from one another by shifting the origin by a vector $\vec{t} = 1/2(\vec{a} + \vec{b})$, i.e., we may write $(BX_2)_c = (BX_2)_{o(x+y)}$ and $(BX_2)_o = (BX_2)_{c(x+y)}$, where the subscript (x + y) indicates that the origin of a layer o or c is shifted by the vector $\vec{t} = 1/2(\vec{a} + \vec{b})$ to give a layer $(BX_2)_o$ or $(BX_2)_c$. The next layer along the c-axis has composition AX, with the atom A at the center of the mesh and the X atoms at the corners, a situation which may be expressed with the symbol $(AX)_c$. Also in this case we have $(AX)_c = (AX)_{o(x+y)}$ and $(AX)_o = (AX)_{c(x+y)}$ where $(AX)_o$ is a layer AX with A at the origin of the mesh and X at the center. The structure of perovskite, then, may be represented with the following sequence of layers:

$$... [(BX_2)_o(AX)_c](BX_2)_o(AX)_c ... \qquad (1)$$

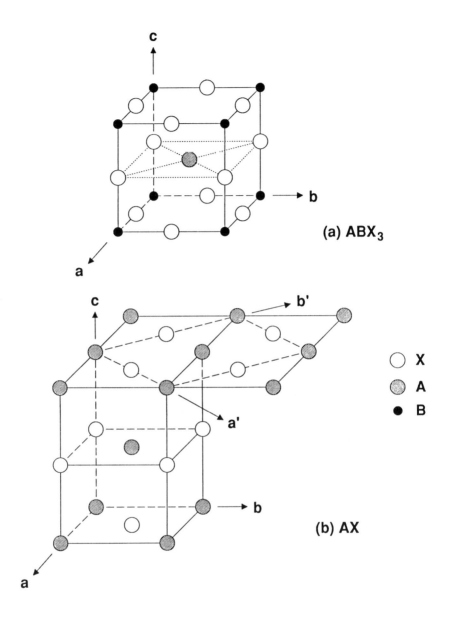

Figure 1: Crystal structures of (a) perovskite, and (b) rock salt. The axes a and b of rock salt are used in our description of the structure, and the axes a' and b' are those defining the conventional face-centered cell.

where the layers contained in one unit cell are enclosed in brackets. If the origin of the unit cell is shifted by (1/2, 1/2, 1/2), this representation becomes:

$$... [(AX)_o(BX_2)_c](AX)_o(BX_2)_c ... \qquad (1a)$$

In a similar way we may represent the structure of rock salt, illustrated in Figure 1b, with the sequence

$$... [(AX)_o(AX)_c](AX)_o(AX)_c ... \qquad (2)$$

The origin of an o or c layer may be shifted also by the vectors $t_x = (1/2)a$ or $t_y = (1/2)b$, and we may indicate this with the subscripts ox, oy, cx, and cy. It is easy to verify that

$$(BX_2)_{cx} = (BX_2)_{oy} \text{ and } (BX_2)_{ox} = (BX_2)_{cy} \qquad (3)$$

and

$$(AX)_{cx} = (AX)_{oy} \text{ and } (AX)_{ox} = (AX)_{cy} \qquad (3a)$$

In addition, the composition of the layers AX and BX_2 may vary, i.e., some or all of the atoms X may be missing. Layers of this sort may be called "defective" and they have compositions A and BX (the case of a layer of composition B is identical to A). All possible meshes of full or defective layers of type AX and BX_2 are illustrated in Figure 2.

A formulation of structures by layers, such as that represented by the sequence (1) or (2), allows one to derive in a simple way the coordination of the atoms. In the case of perovskite, for example, the atom A of a layer $(AX)_c$ is surrounded by twelve atoms X, four located at the corners of the same mesh, and eight at the midpoints of the edges of the $(BX_2)_o$ layers above and below $(AX)_c$. The coordination polyhedron is a cuboctahedron. In the case of the rock salt structure, the atom A of a layer $(AX)_c$ has coordination six, being surrounded by four atoms X at the corners of the same mesh, and by two atoms X at the center of the meshes $(AX)_o$ above and below $(AX)_c$. In this case the coordination polyhedron is an octahedron.

The relationship between perovskite, $Ba_2YCu_3O_{7.0}$, and

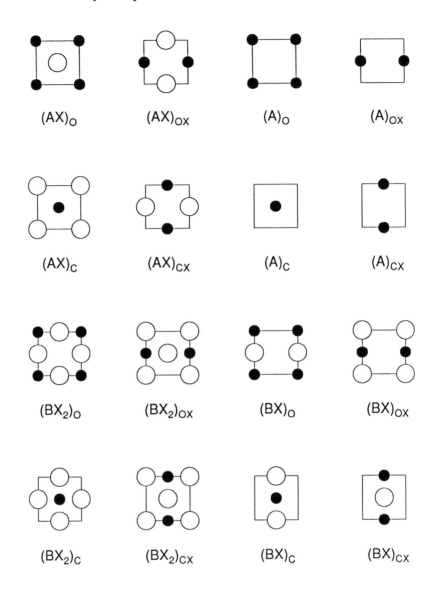

Figure 2: Possible meshes used to describe the layered structures of superconductors and related compounds.

$Ba_2YCu_3O_{6.0}$, is illustrated in Figure 3, where deviations from ideality due to distortions and minor displacements of the atoms are ignored. The three structures can also be easily understood and compared by representing them as sequences of layers, some of which are defective, according to the following scheme:

$[(BX_2)_o(AX)_c](BX_2)_o$ $(AX)_c$ $(BX_2)_o$ $(AX)_c$ $(BX_2)_o$ ... Perovskite

$[(CuO)_o(BaO)_c(CuO_2)_o$ $(Y)_c$ $(CuO_2)_o$ $(BaO)_c]$ $(CuO)_o$ ... $Ba_2YCu_3O_7$ (4)

$[(Cu)_o$ $(BaO)_c(CuO_2)_o$ $(Y)_c$ $(CuO_2)_o$ $(BaO)_c]$ $(Cu)_o$ ... $Ba_2YCu_3O_6$

In this case B = Cu and X = O, while A represents yttrium and barium atoms in an ordered sequence. This order, together with the particular distribution of the oxygen atoms, induce the tripling of the c-axis of the perovskite cell, clearly indicated in the scheme above. The structure of $Ba_2YCu_3O_x$ has two defective layers. The first is the yttrium layer in which all the oxygen atoms are missing. The second is the layer at z = 0 (basal plane) which has the configuration of a $(BX_2)_o$ layer in which two or all of the oxygen atoms at the mid-points of opposite edges have been removed to give $Ba_2YCu_3O_7$, or $Ba_2YCu_3O_6$, respectively.

The structure of La_2CuO_4 (K_2NiF_4-type) is illustrated in Figure 4 and is represented by the following scheme:

... $[(BX_2)_o$ $(AX)_c$ $(AX)_o$ $(BX_2)_c$ $(AX)_o$ $(AX)_c]$ $(BX_2)_o$ ... A_2BX_4

... $[(CuO_2)_o(LaO)_c(LaO)_o(CuO_2)_c(LaO)_o(LaO)_c](CuO_2)o$ ... La_2CuO_4 (5)

The description of this structure is more complicated than that of $Ba_2YCu_3O_x$. There are six layers in the unit cell of this structural type and they can be viewed in two quite different ways. In the first interpretation, we divide the six layers into two blocks of three layers each, the first being $(AX)_o(BX_2)_c(AX)_o$ and the second $(AX)_c(BX_2)_o(AX)_c$. These layers and these sequences are typical of perovskite and, therefore, in this description the structure is considered to be made of two perovskite blocks related to one another by a shift of origin of $\vec{t}$ = $(1/2)(\vec{a} + \vec{b})$. We may also regard the structure, however, as containing alternate blocks of perovskite (layers $(BX_2)_{o,c}$) and rock salt (layers $(AX)_{c,o}(AX)_{o,c}$). As before, the unit cell is made of two

196   Chemistry of Superconductor Materials

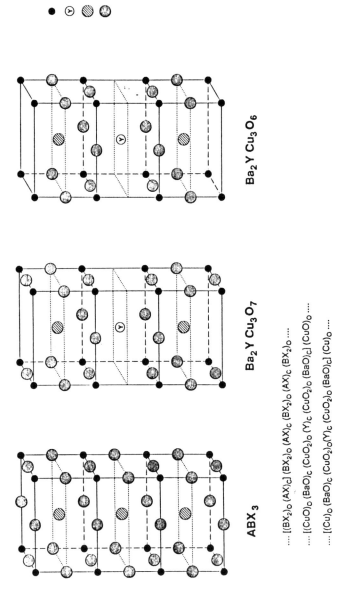

Figure 3: Comparison of the structures of perovskite (ABX_3), $Ba_2YCu_3O_7$, and $Ba_2YCu_3O_6$. The scheme at the bottom of figure shows the description of the three structures layer by layer.

.... |$(BX_2)_O$ $(AX)_C$| $(BX_2)_O$ $(AX)_C$ $(BX_2)_O$ ....

.... |$(CuO)_O$ $(BaO)_C$ $(CuO_2)_O$ $(Y)_C$ $(CuO_2)_O$ $(BaO)_C$| $(CuO)_O$ ....

.... |$(Cu)_O$ $(BaO)_C$ $(CuO_2)_O$ $(Y)_C$ $(CuO_2)_O$ $(BaO)_C$| $(Cu)_O$ ....

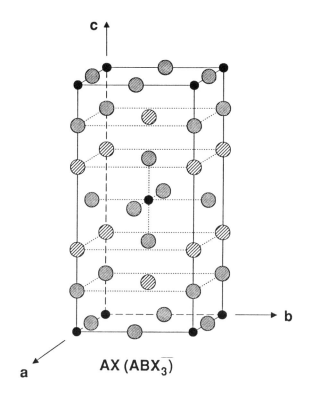

.... [(BX$_2$)$_O$ (AX)$_C$ (AX)$_O$ (BX$_2$)$_C$ (AX)$_O$ (AX)$_C$] (BX$_2$)$_O$ (AX)$_C$ ....

.... [(CuO$_2$)$_O$ (LaO)$_C$ (LaO)$_O$ (CuO$_2$)$_C$ (LaO)$_O$ (LaO)$_C$]   (CuO$_2$)$_O$ (LaO)$_C$ ....

(CuO$_2$)$_O$ (Ca)$_C$ (CuO$_2$)$_O$ .... (CuO$_2$)$_C$ (Ca)$_O$ (CuO$_2$)$_C$ .... (CuO$_2$)$_O$ (Ca)$_C$ (CuO$_2$)$_O$ ....

Figure 4: Structure of K_2NiF_4. The scheme at the bottom of figure represent the description by layers of $La_2Ca_{n-1}Cu_nO_{2n+2}$ for $n = 1$ and $n = 2$.

identical parts, related to one another by a shift t of the origin, one made by the layers ...$(BX_2)_o(AX)_c(AX)_o$ and the other by the layers ... $(BX_2)_c(AX)_o(AX)_c$ ...

This second interpretation is preferable to the first for at least two reasons. First, the A atoms are coordinated to nine X atoms (four on the same mesh $(AX)_{o,c}$, four on a $(BX_2)_{c,o}$ layer below or above, and one on another $(AX)_{c,o}$ layer above or below $(AX)_{o,c}$) giving a configuration that can be thought of as a sort of capped polyhedron, one half of which is cuboctahedral like in perovskite, and one half octahedral like in rock salt. The second reason is that the compound La_2CuO_4 is the end member of homologous series such as $La_{n+1}Cu_nO_{3n+1}$ and $La_2Ca_{n-1}Cu_nO_{2n+2}$ whose compounds have structures made of blocks of perovskite interleaved with blocks of rock salt. In fact, compounds belonging to the series can be generated from the structure of La_2CuO_4 by substituting each layer $(CuO_2)_{o,c}$ of La_2CuO_4 with blocks ... $(CuO_2)_{o,c}(LaO)_{c,o}(CuO_2)_{o,c}$ ... in $La_{n+1}Cu_nO_{3n+1}$ and with blocks ... $(CuO_2)_{o,c}(Ca)_{c,o}(CuO_2)_{o,c}$ ... in $La_2Ca_{n-1}Cu_nO_{2n+2}$. This configuration may be represented with the following scheme:

$[\{(CuO_2)_o \qquad \}\{(LaO)_c(LaO)_o\}\{(CuO_2)_c \qquad \}\{(LaO)_o(LaO)_c\}]$

$[\{(CuO_2)_o(LaO)_c(CuO_2)_o\}\{(LaO)_c(LaO)_o\}\{(CuO_2)_c(LaO)_o(CuO_2)_c\}\{(LaO)_o(LaO)_c\}]$

$[\{(CuO_2)_o(Ca)_c(CuO_2)_o\}\{(LaO)_c(LaO)_o\}\{(CuO_2)_c(Ca)_o(CuO_2)_c\}\{(LaO)_o(LaO)_c\}]$

where the blocks of perovskite and rock salt type structures are enclosed in braces. All the structures represented in the above scheme are made of two identical halves related to one another by a shift of the origin of $\vec{t} = (1/2)(\vec{a} + \vec{b})$, and, clearly, this property is common to all compounds in which the number of layers with the rock salt structure is even.

The compounds belonging to the series $(Bi,Tl)_2(Sr,Ba)_2Ca_{n-1}Cu_nO_{2n+4}$ are closely related to those belonging to $La_2Ca_{n-1}Cu_nO_{2n+2}$. The main difference between the two series is in the thickness and chemical nature of the blocks with the rock salt structure. More specifically, the blocks ... $(LaO)_{o,c}(LaO)_{c,o}$ ... of $La_2Ca_{n-1}Cu_nO_{2n+2}$ are replaced by blocks ... $(MO)_{o,c}(NO)_{c,o}(NO)_{o,c}(MO)_{c,o}$ ... in $M_2N_2Ca_{n-1}Cu_nO_{2n+4}$, where M = Ba,Sr and N = Tl,Bi. On the other hand, the configuration of the defective perovskite blocks is identical in the two cases, as shown by the scheme represented in Figure 5.

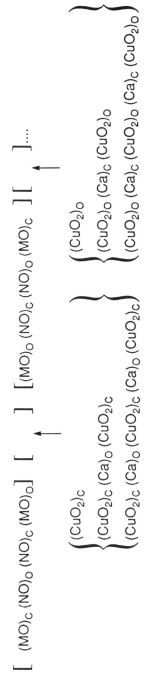

Figure 5: Layer by layer representation of the series $M_2N_2Ca_{n-1}Cu_nO_{2n+4}$ (M = Ba, Sr and N = Tl, Bi) for n = 1, 2 and 3.

2.1 Structural Types of Superconductors

Superconductors can be described and compared in a compact and logical way by deriving their structures from the basic atomic arrangement of perovskite. As we have mentioned earlier, the (BX_2) and (AX) layers of perovskite form a regular succession according to the scheme:

$$\ldots (BX_2)_{o,c}(AX)_{c,o}(BX_2)_{o,c}(AX)_{c,o}(BX_2)_{o,c} \ldots \quad (6)$$

Such regularity, however, can be altered by modifying the sequence in appropriate ways.

Let us consider first a layer (BX_2). Immediately before and after it, there must be a layer of type (AX), as two consecutive layers (BX_2) would generate an unreasonable atomic configuration with abnormally short oxygen-oxygen distances. This means that we may substitute a layer $(BX_2)_{o,c}$ only with a sequence

$$P = (BX_2)_{o,c} \cdot (n-1)[(AX)_{c,o}(BX_2)_{o,c}] \quad (7)$$

where this expression represents a sequence made by a layer $(BX_2)_{o,c}$ followed by n-1 bilayers $(AX)_{c,o}(BX_2)_{o,c}(n = 1,2,3,\ldots)$. This substitution does not produce anything different from expression (6), unless we introduce in the sequence defects or compositional changes. For example, if we substitute the layers $(BX_2)_o$ in expression (6) alternatively with

$$P = (CuO_2)_o(Y)_c(CuO_2)_o (n = 2) \text{ and } P' = (CuO)_o (n = 1) \quad (8)$$

we obtain the structure of $Ba_2YCu_3O_7$ ($(AX)_c = (BaO)_c$ in this case).

On the other hand, before and/or after a layer (AX) there may be a layer of type (BX_2) or one of type (AX). This last case is possible because (AX) is characteristic of both the perovskite and the rock salt structures and, therefore, is structurally coherent with both (BX_2) and (AX). A sequence $..(AX)_{c,o}(AX)_{o,c}...$ can consequently substitute the single layer $(AX)_{c,o}$, thus increasing the thickness of the rock salt monolayers (AX) present in the structure of perovskite. In this way, we may derive the structure of A_2BX_4 (K_2NiF_4 - type) from that of perovskite by substituting each $(AX)_{c,o}$ in expression (6) with bilayers

$(AX)_{c,o}(AX)_{o,c}$ (see expression (5)).

Following the above discussion, it is now possible to define a general structural type built with alternating blocks having the perovskite and the rock salt structures, according to the scheme:

$$...\{R\}\{P\}\{R\}\{P\}... \qquad (9)$$

where P is defined by expression (7), and

$$R = m \cdot (AX) \quad (m = 1,2,3,...) \qquad (10)$$

is a sequence of m layers $(AX)_{o,c}(AX)_{c,o}$ ... forming a block with the rock salt structure. We are now in a position to describe a fairly large number of superconductors and related compounds in terms of the general type represented by sequence (9).

2.2 Compounds with the Perovskite Structure

In these compounds, the blocks R of expressions (9) and (10) have m = 1, i.e., the rock-salt blocks are monolayers (AX). Important materials with the perovskite structure are represented in the scheme of Table 1, where they are compared to each other and to the general structural type discussed in the previous section.

In the system $Ba(Pb_{1-x}Bi_x)O_3$, the compound with x = 0.25 must be considered the first discovered ceramic material showing high-temperature superconductivity (7). Structure determinations have been carried out over the entire range of composition (8)-(11) and the refined parameters are presented in Table 2. Superconductivity in this system exists only for values of x between 0.05 and 0.35. The value of the critical temperature increases with x, reaches a maximum value Tc $\simeq$ 13K for x $\simeq$ 0.25, and then decreases. For x > 0.35, the material becomes a semiconductor.

The compound $NCuO_2$ (N = $Ca_{0.86}Sr_{0.14}$) is an insulator, but its structure, which is a simple defect perovskite made of layers (CuO_2) sandwiched between layers (N), can be considered as the parent structure of a large family of superconductors. The sequence $...(CuO_2)_{o,c}(N)_{c,o}(CuO_2)_{o,c}...$ is in fact one of the building blocks of many compounds considered in this review. The refined parameters for $NCuO_2$ are given in Table 3.

202   Chemistry of Superconductor Materials

Table 1:   Compounds with the perovskite structure.

	$\{P^{(i)}\}$	$\{R^{i)}\}$	$\{P^{(i+1)}\}$	$\{R^{(i+1)}\}$	
..[	$(LO_2)_0$	$(BaO)_c$	$(LO_2)_0$	$(BaO)_c$]	$BaLO_3$, $L = Pb_{1-x}Bi_x$ $x \approx 0.25$
..[	$(CuO_2)_0$	$(N)_c$]			$NCuO_2$, $N = Ca_{1-x}Sr_x$ $x \approx 0.15$
..[$(MO_2)_0(Y)_c$	$(MO_2)_0$	$(BaO)_c$]			$BaYM_2O_5$, $M = Fe_{0.5}Cu_{0.5}$ $BaY_{n-1}M_nO_{2n+1}$
..[$(CuO_2)_0(Y)_c(CuO_2)_0$	$(BaO)_c$	$(CuO_\delta)_0$	$(BaO)_c$		$Ba_2YCu_3O_{6+\delta}$ $Ba_2Y_{n-1}Cu_{n+1}O_{2n+2+\delta}$

- Square brackets enclose layers contained in one unit cell of the structure.
- The doubling of the c-axis in $Ba(Pb_{1-x}Bi_x)O_3$ is caused by shifts of the oxygen atoms in the layers (LO_2).

Table 2: Refined structural parameters of $Ba(Pb_{0.75}Bi_{0.25})O_3$ (Powder neutron diffraction data) (Ref. 12.)

Space group I4/mcm, z = 4[+]
a = 6.0496(1), c = 8.6210(2)Å[++]

Atom	Position[*]	x	y	z	B(Å^2)	Occupancy[**]
Ba	4b $\bar{4}$2m	1/2	0	1/4	0.81(3)	1.0
Pb	4c 4/m	0	0	0	0.30(2)	0.75
Bi	4c 4/m	0	0	0	0.30(2)	0.25
O(1)	4a 42	0	0	1/4	1.24(4)	1.0
O(2)	8h mm	0.2182(1)	0.7182(1)	0	1.17(2)	1.0

R_N = 4.94, R_P = 6.14, R_W 7.88, R_E = 5.04[o]

(+) Z is the number of formula units per unit cell.
(++) Figures in parentheses are standard deviations on the last decimal figure.
(*) The three symbols of an atomic position represent: (i) the multiplicity of the position; (ii) a letter that identifies the position; and (iii) the site symmetry. (See: International Tables for Crystallography, Vol. A, Reidel Publishing Co., 1983).
(**) The occupancy represents the fraction of sites that are occupied. The number of atoms in a position, then, is given by the occupancy, times the multiplicity of the position.
(o) For the definition of the agreement factors R see e.g., Ref. [13]. The oxygen atoms in this structure are shifted with respect to the positions they have in perovskite. As a consequence of this distortion there is a transformation of axes:

$$(\underline{a}\ \underline{b}\ \underline{c}) = (110/\bar{1}10/002)_p (\underline{a}\ \underline{b}\ \underline{c})_p$$

where $(\underline{a}\ \underline{b}\ \underline{c})_p$ is the column vector of the three vectors defining the unit cell of perovskite, and $(\underline{a}\ \underline{b}\ \underline{c})$ is the column vector of the unit cell vectors of $Ba(Pb_{0.75}Bi_{0.25})O_3$. In this compound the Ba atoms are twelve coordinated (distorted cuboctahedron) and the L atoms (L = $Pb_{0.75}Bi_{0.25}$) are octahedrally coordinated.

Table 3: Refined structural parameters of $(Ca_{0.86}Sr_{0.14})CuO_2$ (Single-crystal x-ray data) (Ref. 14)

Space group P4/mmm, Z = 1
a = 3.8611(2), c = 3.1995(2) Å

Atom	Position	x	y	z	B	Occupancy
Ca	1d 4/mmm	1/2	1/2	1/2	0.52(4)	0.86(?)
Sr	1d 4/mmm	1/2	1/2	1/2	0.52(4)	0.14(2)
Cu	1a 4/mmm	0	0	0	0.40(3)	1.0
O	2f mmm	0	1/2	0	0.7(2)	1.0

R_F = 3.5, R_W = 4.0

In this structure the (CuO_2) layers are flat (z(Cu) = z(O) = 0.0). The coordination of the Cu atoms is square-planar while the atoms N = $Ca_{0.86}Sr_{0.14}$ are eight-coordinated and the coordination polyhedron is a square prism. In this compound, the copper is formally 2+.

Table 4: Refined structural parameters of $BaYM_2O_5$ (M = $Fe_{0.5}Cu_{0.5}$). (Neutron power diffraction data (Ref. 15).

Space group P4mm, z = 1
a = 3.893(2), c = 7.751(3) Å

Atom	Position		x	y	z(*)	B	Occupancy
Ba	1a	4mm	0	0	0.0226(22)	0.82(6)	1.0
Y	1a	4mm	0	0	0.5112(3)	0.58(4)	1.0
$(CuFe)_1$	1b	4mm	1/2	1/2	0.2738(5)	0.47(11)	1.0
$(CuFe)_2$	1b	4mm	1/2	1/2	0.7387(6)	0.86(14)	1.0
O(1)	1b	4mm	1/2	1/2	0.0149(23)	1.68(8)	1.0
O(2)(**)	2c	mm	1/2	0	0.3331(9)	1.09(17)	1.0
O(3)	2c	mm	1/2	0	0.7028(9)	0.89(13)	1.0

R_N = 8.1, R_P = 17, R_E = 15

(*) Standard deviations for all atoms are given in the original paper. For this reason it is not clear how the origin in this polar space group was specified.
(**) Atoms O(2) and O(3) are labelled O(3) and O(4), respectively, in the original paper.

Every block of type P of the compound $BaYM_2O_5$ (M = $Fe_{0.5}Cu_{0.5}$) consists of three layers with sequence ..$(MO_2)_{o,c}(Y)_{c,o}$ $(MO_2)_{o,c}$..., and every block of type R is a monolayer (BaO). In $Ba_2YCu_3O_x$ (x = 6+δ, with $\delta \leq 1$), however, the situation is more complex because the blocks of type P are alternately made by the layers ...$(CuO_2)_{o,c}(Y)_{c,o}(CuO_2)_{o,c}$... and (CuO_δ), resulting in a longer c-axis (~11.7Å, against ~7.7Å for $BaYM_2O_5$). Both these compounds can be considered as members of homologous series of general formulas $BaY_{n-1}M_nO_{2n+1}$ and $Ba_2Y_{n-1}Cu_{n+1}O_{2n+2+\delta}$, where n, in both cases, is the same as that appearing in expression (7) and is related to the number of layers that constitute the block with the perovskite structure. For n = 3, for example, we would have for $Ba_2Y_{1-n}Cu_{n+1}O_{2n+2+\delta}$ the sequence:

... $[(CuO_2)_o(Y)_c(CuO_2)_o(Y)_c(CuO_2)_o(BaO)_c (CuO_\delta)_o(BaO)_c]$...

Compounds with n ≠ 2 have never been prepared, however. The refined parameters of $BaYM_2O_5$ are given in Table 4, and the structure of $Ba_2YCu_3O_x$ will be discussed in a separate section.

2.3 Compounds With Crystallographic Shear

In the structure of $Ba_2YCu_3O_7$, the layer (CuO) is of type (BX_2) and is defective, since one oxygen atom per mesh is missing (Figure 3). Immediately before and after (CuO), there are layers (BaO) of type (AX), and the sequence ... $(BaO)_c(CuO)_o(BaO)_c$ ... generates typical chains of cornersharing near-squares, with the copper atoms at the center (four-fold planar coordination), and the oxygen atoms at the corners. These chains run along the b-axis of the unit cell. A reasonable atomic configuration, however, can also be obtained if, after (CuO) there is a second layer $(CuO)_{oy,cy}$ (Figure 2), shifted with respect to the first by the vector $\vec{b}/2$. In this case the sequence becomes

...$(BaO)_c(CuO)_o(CuO)_{oy}(BaO)_{cy}$ ...

and the structure is obviously made of two parts related to one another by a crystallographic shear $\vec{b}/2$. A consequence of this configuration is the formation of double chains of edge-sharing near-squares with oxygen atoms at the corners and copper atoms at the center, in four-fold planar coordination as before. The geometry of the double chains, which are also parallel to the b-axis, is illustrated in Figure 6.

The first compound found to have a crystallographic shear of the type discussed above, has formula $Ba_2YCu_4O_8$. Its idealized structure is illustrated in Figure 7, where small shifts of the atoms from their ideal positions have been ignored. The structural parameters of this compound are given in Table 5. A comparison of Figures 3 and 7 shows that the structures of $Ba_2YCu_3O_7$ and $Ba_2YCu_4O_8$ are closely related, and the nature of this relationship is evident from the scheme of Figure 8. The unit cell of $Ba_2YCu_4O_8$ contains two blocks that have the structure of $Ba_2YCu_3O_7$. These two parts are connected together by one extra layer $(CuO)_o$ or $(CuO)_{oy}$ which causes the crystallographic shear with the shift of origin of $\vec{b}/2$. The presence of double chains with oxygen atoms bonded to three copper atoms, rather than two as in $Ba_2YCu_3O_7$, is probably the reason why the oxygen stoichiometry in $Ba_2YCu_4O_8$ is not variable, and this feature is indicated in the scheme of Figure 8 where the defective copper layers are written (CuO) rather than (CuO_δ).

The second compound with a crystallographic shear to be prepared and characterized has formula $Ba_4Y_2Cu_7O_{14+\delta}$. The refined parameters of its structure are given in Table 6 and the sequence of layers in the unit cell is shown in the scheme of Figure 8. This structure is composed of blocks made of two unit cells of $Ba_2YCu_3O_{6+\delta}$ and held together by extra layers $(CuO)_o$ or $(CuO)_{oy}$ which, as in the previous compound, cause the crystallographic shear and the consequent shift of origin. The structure can also be viewed as made of alternating blocks $Ba_2YCu_3O_{6+\delta}$ and $Ba_2YCu_4O_8$, in the ratio 1:1. Both these interpretations are indicated in Figure 8 and both are valid. The latter, however, is simpler and more amenable to generalization than the former. We may, in fact, write down the sequences of new hypothetical compounds by simply varying the ratio of the two building blocks, thus generating a homologous series of formula:

$$mBa_2YCu_3O_{6+\delta} \cdot nBa_2YCu_4O_8 \text{ or } Ba_{2(m+n)}Y_{m+n}Cu_{3m+4n}O_{6m+8n+m\delta}$$

Obviously, $Ba_2YCu_4O_8$ corresponds to the case m = 0, and $Ba_4Y_2Cu_7O_{14+\delta}$ to the case m = n = 1. In $Ba_4Y_2Cu_7O_{14+\delta}$ the double chains parallel to the b-axis alternate, along c, with single chains identical to those present in $Ba_2YCu_3O_{6+\delta}$. These single chains

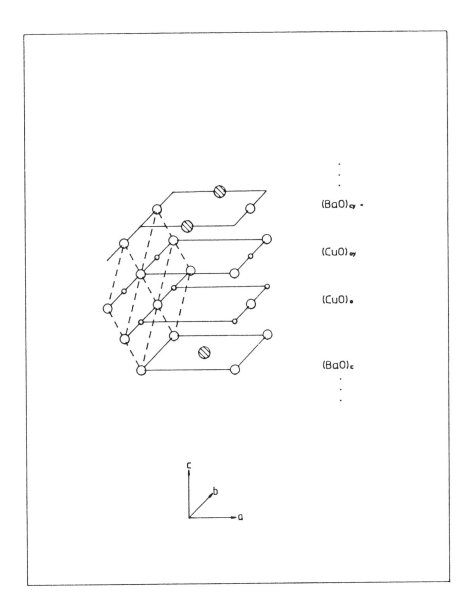

Figure 6: Double chains of edge sharing squares generated by a crystallographic shear $t = b/2$. The copper atoms are at the center of the squares and the oxygen atoms at the corners.

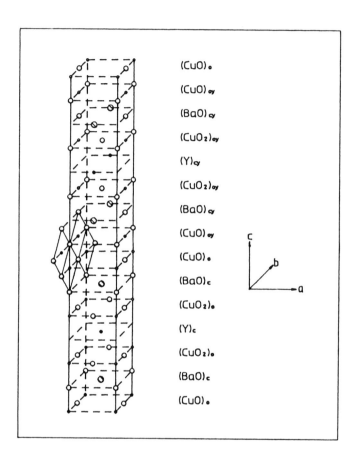

Figure 7: Schematic representation of the crystal structure of $Ba_2YCu_4O_8$.

Table 5: Atomic positional coordinates of $Ba_2YCu_4O_8$. (X-ray data from thin films) (Ref. 16).

Space group Ammm (n.65)[*], Z = 2
a = 3.86(1), b = 3.86(1), c = 27.24(6) Å

Atom	Position	x	y	z	$B(Å^2)$	Occupancy
Ba	4jmm	1/2	1/2	0.1347(5)		1.0
Y	2cmmm	1/2	1/2	0		1.0
Cu(1)	4imm	0	0	0.2135(8)		1.0
Cu(2)	4imm	0	0	0.0621(8)		1.0
O(1)	4imm	0	0	0.147(3)		1.0
O(2)	4jmm	1/2	0	0.049(4)		1.0
O(3)	4imm	0	1/2	0.057(3)		1.0
O(4)	4imm	0	1/2	0.216(3)		1.0

R_F = 4.0

[*] This is a different setting of the conventional space group Cmmm.
[**] An overall temperature factor for all atoms was used in this refinement. This compound is a superconductor with $T_c \approx$ 80K.

Figure 8: Representation of the sequence of layers in the compounds $Ba_2YCu_4O_8$ and $Ba_2Y_2Cu_7O_{14+\delta}$. The sequences of the hypothetical compounds $Ba_6Y_3Cu_{11}O_{22+\delta}$ and $Ba_6Y_3Cu_{10}O_{20+\delta}$ are also indicated. In this figure, $(S)_{c,o}$ represents the sequence $(S)_{c,o} \equiv (BaO)_{c,o}(Y)_{c,o}(CuO_2)_{o,c}$ so that $(CuO)_o(S)_c$ is $Ba_2YCu_3O_7$ (123) and $(CuO)_{oy}(CuO)_o(S)_c$ is $Ba_2YCu_4O_8$ (124).

Crystal Chemistry of Superconductors and Related Compounds   211

Table 6:   Refined structural parameters of $Ba_4Y_2Cu_7O_{14+\delta}$. (Single crystal x-ray data) (Ref. 17).

Space group Ammm, Z = 2
a = 3.851(1), b = 3.869(1), c = 50.29(2) Å

Atom	Position		x	y	z	U_{11}	U_{22}	U_{33}	Occup
Ba(1)	4j	mm	1/2	1/2	0.04310(3)	0.0141(5)	0.0147(5)	0.0004(4)	1.0
Ba(2)	4j	mm	1/2	1/2	0.18797(2)	0.0059(4)	0.0063(4)	0.0070(4)	1.0
Y	4j	mm	1/2	1/2	0.11545(4)	0.0003(6)	-0.0017(6)	0.0009(5)	1.0
Cu(1)	2a	mmm	0	0	0	0.042(3)	0.047(4)	0.020(2)	1.0
Cu(2)	4i	mm	0	0	0.08293(5)	0.0033(9)	0.0053(9)	0.0084(9)	1.0
Cu(3)	4i	mm	0	0	0.14831(5)	0.0027(9)	0.0045(9)	0.0071(9)	1.0
Cu(4)	4i	mm	0	0	0.23012(5)	0.011(1)	0.0048(9)	0.0015(8)	1.0
O(1)	4i	mm	0	0	0.0353(8)	0.010(9)	0.09(3)	0.10(3)	1.0
O(2)	4j	mm	1/2	0	0.0871(3)	0.5(2)(*)			1.0
O(3)	4i	mm	0	1/2	0.0865(3)	0.7(2)			1.0
O(4)	4j	mm	1/2	0	0.1430(3)	0.7(2)			1.0
O(5)	4i	mm	0	1/2	0.1432(3)	0.8(2)			1.0
O(6)	4i	mm	0	0	0.1937(2)	0.4(2)			1.0
O(7)	4i	mm	0	1/2	0.2328(3)	0.8(2)			1.0
O(8)	2b	mmm	0	1/2	0	0.6			0.10(7)
O(9)	2d	mmm	1/2	0	0	0.6			0.20(7)

R_F = 5.4

(*) Isotropic temperature factors were used for O(2) through O(9). This compound is a superconductor with $T_c \approx 40$ K.

(R)	(P)	(R˙)	(P˙)
[((LaO)ₐ (LaO)ₖ	) ((CuO₂)ₖ · (n-1)(LaO)ₖ (CuO₂)ₖ)	((LaO)ₑ(LaO)ₐ	) ((CuO₂)ₖ · (n-1)(Ca)ₐ (CuO₂)ₖ)]
[((LaO)ₐ (LaO)ₖ	) ((CuO₂)ₖ · (n-1)(Ca)ₖ (CuO₂)ₐ)	((LaO)ₑ(LaO)ₐ	) ((CuO₂)ₖ · (n-1)(Ca)ₐ (CuO₂)ₖ)]
[((PbO)ₐ (SrO)ₖ	) ((CuO₂)ₖ · (n-1)(Y)ₖ (CuO₂)ₐ)	((SrO)ₖ (PbO)ₐ	) ((CuO₄)))]
[((BaO)ₐ (TlO)ₖ (BaO)ₐ	) ((CuO₂)ₖ · (n-1)(Ca)ₐ (CuO₂)ₖ)]		
[((MO)ₐ (NO)ₖ (NO)ₐ (MO)ₑ)	((CuO₂)ₖ · (n-1)(Ca)ₖ (CuO₂)ₐ)	((MO)ₖ (NO)ₐ (NO)ₖ (MO)ₐ)	((CuO₂)ₖ · (n-1)(Ca)ₐ (CuO₂)ₖ)]

Figure 9:   Comparison of the structures of $La_{n+1}Cu_nO_{3n+1}$ (second line), $La_2Ca_{n-1}Cu_nO_{2n+2}$ (third line), $Pb_2Sr_2Y_{n-1}Cu_{n+1}O_{2n+4+\delta}$ (fourth line), $TlBa_2Ca_{n-1}Cu_nO_{2n+3}$ (fifth line), and $M_2N_2Ca_{n-1}Cu_nO_{2n+4}$ (sixth line). In the last homologous series M = Ba, Sr and N = Tl, Bi, respectively.

Table 7: Atomic coordinate of $(La_{1.85}Ba_{0.15})CuO_4$ at 295 K (Ref. 21) and 10 K (Ref. 22) (Neutron powder diffraction data).

Space group I4/mmm, Z = 2 at 295 K.
              Bmab , Z = 4 at 10 K.
a = 3.7873(1), c = 13.2883(3) A
a = 5.3430(3), b = 5.3479(3), c = 13.2504(2)

Atom	Position		x	y	z	B	Occup.
La + Ba	4e	4 mm	0	0	0.36063(9)		1.0
	8f	m	0	-0.0052(4)	0.3607(1)	0.15(2)	
Cu	2a	4/mmm	0	0	0		1.0
	4a	2/m	0	0	0	0.23(3)	
O(1)	4e	4mm	0	0	0.1828(2)		0.99(1)
	8f	m	0	0.0206(5)	0.1821(1)	0.90(3)	
O(2)	4c	mmm	0	1/2	0		1.02(1)
	8e	2	1/4	1/4	-0.0052(2)	0.42(3)	

R_N = NA R_P = NA R_W = 7.54 R_E = 4.38 X = 1.72
      3.68   5.87   7.95   4.71   1.69

Parameters in the first line of each entry refer to the refinement at 295 K and those in the second line refer to the refinement at 10 K.

Space group Bmab is a different setting of the conventional space group Cmca. Bmab is used here to permit direct comparison with the parameters of the room temperature phase.

The compound $(La_{1.85}Sr_{0.15})CuO_4$ has practically the same positional parameters as $(La_{1.85}Ba_{0.15})CuO_4$ and lattice parameters:

a = 3.7793(1),   c = 13.2260(3) A at 295K
a = 5.3240(1),   b = 5.3547(1),   c = 13.1832(1) at 10 K [ref. 24].

contain layers (CuO_δ), and for this reason the oxygen stoichiometry of this material is not fixed.

In Figure 8 are also shown the sequences of two hypothetical compounds, i.e. $Ba_6Y_3Cu_{11}O_{22+\delta}$, corresponding to the case m = 1, n = 2, and $Ba_6Y_3Cu_{10}O_{20+2\delta}$, corresponding to the case m = 2, n = 1. The structure of the first compound is made by blocks of two unit cells of $Ba_2YCu_4O_8$ alternating with blocks of one unit cell of $Ba_2YCu_3O_{6+\delta}$ and has approximate lattice parameters 3.9 x 3.9 x 39.0 $Å^3$, and that of the second by blocks of two unit cells of $Ba_2YCu_3O_{6+\delta}$ alternating with one unit cell blocks of $Ba_2YCu_4O_8$, with approximate lattice parameters 3.9 x 3.9 x 74 $Å^3$.

2.4 Compounds with the Rocksalt-Perovskite Structure

In this large class of materials the blocks R = m · (AX) with the rock-salt structure are made by two or more layers of type (AX) which may be identical to each other or have different chemical compositon. The blocks $P = (BX_2)_{o,c} \cdot (n-1) [(AX)_{c,o}(BX_2)_{o,c}]$ with the perovskite structure may have different values of n, and the layers (AX), sandwiched between layers (BX_2), may or may not be defective. The important homologous series with the rock salt-perovskite structure are listed in the scheme of Figure 9 where they are compared with each other and with the basic structure of perovskite.

System $(La_{2-x}M_x)Ca_{n-1}Cu_nO_{2n+2}$ (M = Ba, Sr): The discovery of superconductivity in this system was made on a sample consisting of a mixture of phases (18), and subsequently the superconducting compound was identified as $(La_{2-x}Ba_x)CuO_4$ (19)(20). The value of the critical temperature was found to be a function of x, with a maximum of ~35K for x ≃ 0.15. The material corresponding to this composition undergoes a phase transition from tetragonal to orthorhombic symmetry at about 180K. The detailed structure of $(La_{1.85}Ba_{0.15})CuO_4$ was determined by neutron powder diffraction (21)(22) and the results of these refinements are given in Table 7.

Substitution of Ba with Sr slightly increases the value of the critical temperature (T ≃ 40K for x ≃ 0.15) (23). Structural analyses of $(La_{1.85}Sr_{0.15})CuO_4$ have shown that this compound has a phase transition at ~200K and that its structure is isomorphic with that of $(La_{1.85}Ba_{0.15})CuO_4$ at room temperature as well as at low temperature (24)-(26). The lattice parameters for this composition are given in

Table 7. The positional parameters are practically identical to those of the barium analog.

The coordination polyhedra around the copper atoms of $(La_{2-x}M_x)CuO_4$ are elongated square bipyramids. At the transition these undergo a small but significant rigid rotation which causes the buckling of the (CuO_2) planes. As mentioned before, the atoms $(La_{2-x}M_x)$ are nine-coordinated and the shape of the coordination polyhedron is that of a capped square antiprism, more specifically, half of it is cuboctahedral like in perovskite, and half octahedral like in the rock-salt structure. No oxygen deficiency has been detected in these compounds. This means that charge compensation when Sr and Ba substitute La must be accomplished entirely by oxidation of copper from Cu^{2+} to Cu^{3+}.

Compounds of this homologous series with $n > 1$ exist and have been analyzed by neutron diffraction methods. More specifically, precise structural determinations of $La_{1.9}Ca_{1.1}Cu_2O_6$ (27)(28) and $La_{1.9}Sr_{1.1}Cu_2O_6$ (27) have been carried out and have shown that the sequence of layers, their composition, and the general distribution of the atoms in the unit cell are entirely consistent with the scheme of Figure 9. The structural parameters of the calcium compound are given in Table 8. We may write the sequence of layers for these two compounds as

$$...[(RO)_o(RO)_c(CuO_2)_o(R')_c(CuO_2)_o(RO)_c(RO)_o(CuO_2)_c(R')_o(CuO_2)_c]...$$

In $La_{1.9}Ca_{1.1}Cu_2O_6$, the calcium tends to occupy preferentially the site R', which is eight-coordinated, and the lanthanum the site R, which is nine-coordinated. In $La_{1.9}Sr_{1.1}Cu_2O_6$, on the other hand, the strontium seems to concentrate on site R. This distribution of the atoms in the structure may be understood considering the ionic radii of the cations. The ionic radii of Sr^{2+} for eight-fold and nine-fold coordination are larger than those of La^{3+} (1.26 and 1.31 Å versus 1.16 and 1.22 Å, respectively), and, for this reason, we should expect that the Sr^{2+} cations will prefer the nine-coordinated sites and displace La^{3+} into the eight-fold coordinated sites R'. On the other hand, the ionic radii of Ca^{2+} for eight- and nine-fold coordination, are smaller than those of La^{3+} (1.12 and 1.18 Å versus 1.16 and 1.22 Å, respectively) and, therefore, it is not surprising to find calcium concentration on sites R' in $La_{1.19}Ca_{1.1}Cu_2O_6$.

Crystal Chemistry of Superconductors and Related Compounds 215

Table 8: Structural parameters of $La_{1.9}Ca_{1.1}Cu_2O_6$. (Neutron powder diffraction data) (Ref. 27).

Space group I4/mmm, Z = 2
a = 3.8248(1), c = 19.4286(5)Å

Atom	Position		x	y	z	B(Å^2)	Occ.
La(1)	2a	4/mmm	0	0	0	1.0(1)	0.06(2)
Ca(1)	2a	4/mmm	0	0	0	1.0(1)	0.94(2)
La(2)	4e	4 mm	0	0	0.17578(9)	0.47(5)	0.94(2)
Ca(2)	4e	4 mm	0	0	0.17578(9)	0.47(5)	0.06(2)
Cu	4e	4 mm	0	0	0.58503(9)	0.32(4)	1.0
O(1)	8g	mm	0	1/2	0.08230(9)	0.74(5)	1.005(8)
O(2)	4e	4 mm	0	0	0.7046(2)	1.52(8)	1.01(1)
O(3)	2b	4/mmm	0	0	1/2	0.7	0.03(1)

R_N = 4.58, R_P = 5.96, R_W = 8.54, R_E = 4.34, χ = 1.97

Table 9: Structural parameters of $Pb_2Sr_2YCu_3O_8$. (Neutron powder diffraction data) (Ref. 30).

Space group Cmmm, Z = 2
a = 5.3933(2), b = 5.4311(2), c = 15.7334(6)

Atom	Position		x	y	z	B(Å^2)	Occ.
Pb	4l	mm	1/2	0	0.3883(2)	0.77(5)	1.0
Sr	4k	mm	0	0	0.2207(2)	0.53(7)	1.0
Y	2a	mmm	0	0	0	0.73(9)	1.0
Cu(1)	2d	mmm	0	0	1/2	0.96(9)	1.0
Cu(2)	4l	mm	1/2	0	0.1662(2)	0.59(6)	1.0
O(1)	4l	mm	1/2	0	0.2514(3)	0.97(7)	1.0
O(2)	16r	1	0.051(1)	0.074(1)	0.3849(4)	1.0(2)	1.0
O(3)	8m	2	1/4	1/4	0.0918(2)	1.20(7)	1.0

R_N = 6.24, R_P = 7.66, R_W = 9.74, R_E = 4.60, χ = 2.12

In La_2CuO_4 the coordination of copper is square-bipyramidal. The replacement of the (CuO_2) layers of La_2CuO_4 with blocks $(CuO_2)(R')(CuO_2)$ to produce $R_2R'Cu_2O_6$ has the effect of changing the polyhedra surrounding the copper atoms into square pyramids in which the Cu-O distances within the (CuO_2) layer are much shorter than the Cu-O distance with the oxygen atom at the apex of the pyramid. The existence of eight-coordinated cations in this homologous series is possible only for $n \geq 2$ and the coordination polyhedra are square prisms.

Neither the calcium nor the strontium compound is a superconductor. A possible reason may be that the oxidation state of copper is equal, or very close, to 2+.

System $Pb_2Sr_2Y_{n-1}Cu_{n+1}O_{2n+4+\delta}$: From a crystallographic point of view this system is perhaps the most interesting, not only because the layers of the rock-salt block are chemically different, but also because there are two perovskite-type blocks alternating along the c-axis, one $(CuO_2)_o \cdot (n-1)(Y)_c(CuO_2)_o$, and the other made by a monolayer (CuO_δ) $(0 \leq \delta \leq 1)$. Superconductivity near 70K was discovered in copper oxides based on $Pb_2Sr_2Y(Ln_{n-x}M_x)Cu_3O_{8+\delta}$, where Ln is a lanthanide element, and M = Ca, Sr (29). A precise determination has been carried out on the non-superconducting prototype compound $Pb_2Sr_2YCu_3O_8$ and the structural parameters are given in Table 9 (30).

If we represent this structure as

$...[(PbO)_o (SrO)_c (CuO_2)_o (Y)_c (CuO_2)_o (SrO)_c (PbO)_o (Cu)_c] (PbO)_o...$

we see that yttrium is sandwiched between two (CuO_2) layers and has eight-fold, square-prismatic coordination. The copper atoms of the layers (CuO_2) have five-fold, square pyramidal coordination, like that present in many of the compounds previously reviewed. The layers (PbO) are located between (SrO) and (Cu) layers and, as a consequence, the lead atoms are surrounded by five oxygen atoms at the vertices of a pyramid. The coordination polyhedron of the Sr atoms is a nine-fold capped square antiprism, similar to that of La in $La_2Ca_{n-1}Cu_nO_{2n+2}$ with $n \geq 2$. The layers (Cu) are located between two (PbO) layers, and the copper atoms have linear two-fold coordination similar to that present in $Ba_2YCu_3O_6$. Since the oxygen atoms on the (PbO) planes are distributed over the general (x,y,z) sites, instead of being located on the symmetrical sites (0,0,z) (see Table 9), the bonds O-Cu-O are not linear unless the two occupied oxygen sites on the (PbO) planes above and below (Cu) are diagonally opposite.

From a structural point of view, $Pb_2Sr_2YCu_3O_8$ can be oxidized to $Pb_2Sr_2YCu_3O_9$ with the layers (Cu) becoming (CuO). The formation of the oxidized compound, with a layer (CuO) in its structure, would make possible the existence of compounds with crystallographic shears of the type discussed before in the case of $Ba_2YCu_3O_7$. No oxidation experiments have been carried out so far, however. No other possible compounds of this homologous series have been reported so far.

System $TlBa_2Ca_{n-1}Cu_nO_{2n+3}$: Three compounds of this homologous series have been prepared and characterized, i.e. those with n = 1,2, and 3 (31). In each of these the rock-salt block is made of three layers having the sequence $(BaO)_o(TlO)_c(BaO)_o$, with the Tl atoms in octahedral coordination, and the Ba atoms in the nine-fold coordination provided by a capped square antiprism and typical of cations at the boundary of a rock-salt and a perovskite block. The difference between the three compounds lies in the thickness of the perovskite blocks, these being a monolayer (CuO_2) for n=1 and a sequence $(CuO_2)_o(Ca)_c(CuO_2)_o(Ca)_c(CuO_2)_o$ for n = 3. Electrical measurements show that the compound with n = 1 is a semiconductor while the other two are superconductors with T_c values of 65-85K and 100 - 110K for n = 2 and n = 3, respectively. Approximate positional parameters for the compound $TlBa_2Ca_2Cu_3O_9$ are given in Table 10. From these data, and from the scheme of Figure 9 we see that, when the number of layers in the rock-salt block is odd, the structure does not have two equal halves related to one another by a shift of the origin of $\vec{t} = 1/2\,\vec{a} + 1/2\,\vec{b}$ and, as a consequence, the parameter c is shorter than in the other case and the unit cell tends to be primitive rather than centered.

System $N_2M_2Ca_{n-1}Cu_nO_{2n+4}$ (N = Tl, Bi; M = Ba, Sr): Numerous compounds belonging to this system have been synthesized and characterized crystallographically. In the case of $Tl_2Ba_2Ca_{n-1}Cu_nO_{2n+4}$, the structures have been solved and refined from x-ray single crystal data for n = 1 (32), n = 2 (33), and n = 3 (34), and from neutron powder diffraction data for n = 2 and n = 3 (35). Similar studies have been carried out on the system $Bi_2Sr_2Ca_{n-1}Cu_nO_{2n+4}$ for n = 1 (32) and for n = 2 (36)-(40).

Table 10: Approximate positional parameters of $TlBa_2Ca_2Cu_3O_9$. (X-ray powder diffraction data) (Ref. 31).

Space group P4/mmm, Z = 1
a = 3.8429(16), c = 15.871(3) Å

Atom	Position		x	y	z
Tl	1a	4/mmm	0	0	0
Ba	2h	4mm	1/2	1/2	0.176
Ca	2h	4mm	1/2	1/2	0.397
Cu(1)	1b	4/mmm	0	0	1/2
Cu(2)	2g	4mm	0	0	0.302
O(1)[*]	2e	mmm	0	1/2	1/2
O(2)	4i	mm	0	1/2	0.304
O(3)	2g	4mm	0	0	0.132
O(4)	1c	4/mmm	1/2	1/2	0

[*] Oxygen positions were calculated from known coordination polyhedra in related crystal structures.
No temperature and occupancy factors are given in the original paper.

Table 11: Refined structural parameters of $Tl_2Ba_2CaCu_2O_8$. (Powder neutron diffraction data) (Ref. 35).

Space group I4/mmm, Z = 2
a = 3.8559(1), c = 29.4199(10) Å

Atom	Position		x	y	z	B	Occ.
Tl	4e	4mm	1/2	1/2	0.2129(1)	0.8(1)	0.96(1)
Ba	4e	4mm	0	0	0.1210(1)	0.6(1)	
Ca	2a	4/mmm	0	0	0	0.0(2)	1.09(2)
Cu	4e	4mm	1/2	1/2	0.0536(1)	0.1(1)	
O(1)	8g	mm	0	1/2	0.0526(1)	0.4(1)	
O(2)	4e	4mm	1/2	1/2	0.1455(1)	0.8(1)	
O(3)	16n	m	0.603(2)	1/2	0.2803(2)	0.5(3)	0.25(1)

$R_N = 4.1$, $R_W = 9.0$, $R_E = 6.0$

Table 12: Refined structural parameters of $Tl_2Ba_2Ca_2Cu_3O_{10}$ at 150K. (Neutron powder diffraction data) (Ref. 35).

Space group I4/mmm, Z = 2
a = 3.8487(1), c = 35.6620(15)

Atom	Position		x	y	z	B	Occ.
Tl	4e	4mm	1/2	1/2	0.2195(1)	0.8(2)	0.95(2)
Ba	4e	4mm	0	0	0.1444(2)	0.5(2)	
Ca	4e	4mm	0	0	0.0454(2)	0.1(2)	
Cu(1)	2b	4/mmm	1/2	1/2	0	0.3(2)	
Cu(2)	4e	4mm	1/2	1/2	0.0886(1)	0.1(1)	
O(1)	4c	mmm	1/2	0	0	0.6(2)	
O(2)	8g	mm	1/2	0	0.0878(1)	0.8(1)	
O(3)	4e	4mm	1/2	1/2	0.1650(2)	0.9(2)	
O(4)	16n	m	0.601(4)	1/2	0.2750(2)	-0.5(5)	0.23(5)

$R_N = 4.8$, $R_W = 10.2$, $R_E = 6.1$

All the structures belonging to this system have the sequence represented in the scheme of Figure 9. The rock-salt slab is made of four layers of type (AX), followed by slabs of variable thickness having the perovskite structure. Since the number of layers (AX) is even, every one of these structures is made of two identical halves shifted by $\vec{t} = 1/2\vec{a} + 1/2\vec{b}$. The Tl-Ba system is isomorphous with the Bi-Sr system. However, compounds $Bi_2Sr_2Ca_{n-1}Cu_nO_{2n+4}$ have superstructures whose atomic configurations have not been completely clarified.

The basic difference between the structures with different values of n is in the number of consecutive (CuO_2) planes in the perovskite blocks. The value of the critical temperature is a function of n. For example, in the Tl-Ba system T_c is ~85K, ~110K, and ~125K for n = 1, 2, 3, respectively. The compound with n = 5 has not been prepared as a bulk phase, but the value of T_c determined for this material on a microscopic level appears to be of the order of 140K (35).

Positional parameters for $Tl_2Ba_2Ca_{n-1}Cu_nO_{2n+4}$ are reported in Tables 11 and 12 for n = 2 and n = 3.

3.0 REFERENCES

1. S.N. Ruddlesden and P. Popper, *Acta Cryst.* 10:538-539 (1957).

2. S.N. Ruddlesden and P. Popper, *Acta Cryst.* 11:54-55 (1958).

3. J.M. Longo and P.M. Raccah, *J. Sol. State Chem.* 6:526-531 (1973).

4. A.H. Davies and R.J.D. Tilley, *Nature* 326:859-861 (1987).

5. D.M. Smyth, Ceramic Superconductors II, *Research Update.* Publ.: American Ceramic Society, Westerville, Ohio (1988).

6. A. Santoro, F. Beech, M. Marezio, and R.J. Cava, *Physica C* 156:693-700 (1988).

7. A.W Sleight, J.L. Gillson, and P.W. Bierstedt, *Solid State Comm.* 17:27 (1975).

8. D.E. Cox and A.W. Sleight, *Proceedings Conf. on Neutron Scattering*, Ed. R.M. Moon, Gatlinburg, Tennessee, June 6-10, (1976).

9. D.E. Cox and A.W. Sleight, *Solid State Comm.* 19:969-973 (1976).

10. G. Thornton and A.J. Jacobson, *Acta Cryst.* B 34:351-354 (1978).

11. D.E. Cox and A.W. Sleight, *Acta Cryst.* B 35:1-10 (1979).

12. C. Chaillout, personal communication.

13. H.M. Rietveld, *J. Appl. Cryst.* 2:65-71 (1969).

14. T. Siegrist, S.M. Zahurak, D.W. Murphy and R.S. Roth, *Nature* 334:231-232 (1988).

15. L. Er-Rakho, C. Michel, P. Lacorre, and B. Raveau, *J. Sol. State Chem.* 73:531-535 (1988).

16. P. Marsh, R.M. Fleming, M.L. Mandich, A.M. DeSantolo, J. Kwo, M. Hong, and L.J. Martinez-Miranda, *Nature* 334:141-143 (1988).

17. P. Bordet, C. Chaillout, J. Chenavas, J.L. Hodeau, M. Marezio, J. Karpinski, and E. Kaldis, *Nature* 334:596-598 (1988).

18. J.G. Bednorz and K.A Mueller, *Zeit. Phys.* B 64, 189 (1986).

19. S. Uchida, H. Takagi, K. Kitazawa, and S. Tanaka, *Jpn. J. Appl. Phys.* 26:L1 (1987).

20. H. Takagi, S. Uchida, K. Kitazawa, and S. Tanaka, *Jpn. J. Appl. Phys.* 26:L123 (1987).

21. J.D. Jorgensen, H.B. Schuettler, D.G. Hinks, D.W. Capone,II, K. Zhang, and M.B. Brodsky, *Phys. Rev. Lett.* 58 (10):1024-1027 (1987).

22. A. Santoro, S. Miraglia, and F. Beech, unpublished results (1987).

23. R.J. Cava, R.B. vanDover, B. Batlogg, and E.A. Rietman, *Phys. Rev. Lett.* 58 (4):408-410 (1987).

24. R.J. Cava, A. Santoro, D.W. Johnson, and W.W. Rhodes, *Phys. Rev.* B 35 (13):6716-6720 (1987).

25. M. Francois, K. Yvon, P. Fischer, and M. Decroux, *Sol. State Comm.* 63 (1):35-40 (1987).

26. P. Day, M. Rosseinsky, K. Prassides, W.I.F. David, O. Moze, and A. Soper, *J. Phys. C.* 20:L429-L434 (1987).

27. A. Santoro, R.J. Cava, and F. Beech, *Mat. Res. Soc. Symp. Proc.* 166:187-192 (1990).

28. F. Izumi, E. Takayama-Muromachi, Y. Nakai, and H. Asano, *Physica C* 157:89-92 (1989).

29. R.J. Cava, B. Batlogg, J.J. Krajewski, L.W. Rupp, L.F. Schneemeyer, T. Siegrist, R.B. van Dover, P. Marsh. W.F Peck, P.K. Gallagher, S.H. Glarum, J.H. Marshall, R.C. Farrow, J.B. Waszczak, R. Hull, and P. Trevor, *Nature* 336:211 (1988).

30. R.J. Cava, M. Marezio, J.J. Krajewski, W.F. Peck, A. Santoro, and F. Beech, *Physica C* 157:272-278 (1989).

31. S.S.P. Parkin, V.Y. Lee, A.I. Nazzal, R. Savoy, R. Beyers, and S.J. LaPlaca, *Phys. Rev. Lett.* 61 (6):750-753 (1988).

32. C.C Torardi, M.A. Subramanian, J.C. Calabrese, J. Gopalakrishnan, E.M. McCaron, K.J. Morrissey, T.R. Askew, R.B. Flippen, U. Chowdhry, and A.W. Sleight, *Phys. Rev. B* 38 (1):225-231 (1988).

33. M.A. Subramanian, J.C. Calabrese, C.C. Torardi, J. Gopalakrishnan, T.R. Askew, R.B. Flippen K.J. Morrissey, U. Chowdhry, and A.W. Sleight, *Nature* 332:420-422 (1988).

34. C.C. Torardi, M.A. Subramanian, J.C. Calabrese, J. Gopalakrishnan, K.J. Morrissey, T.R. Askew, R.B. Flippen, U. Chowdhry, and A.W. Sleight, *Science* 240:631-634 (1988).

35. D.E. Cox, C.C. Torardi, M.A. Subramanian, J. Gopalakrishnan, and A.W. Sleight, *Phys. Rev. B* 38 (10):6624-6630 (1988).

36. M.A. Subramanian, C.C. Torardi, J.C. Calabrese, J. Gopalakrishnan, K.J. Morrissey, T.R. Askew, R.B. Flippen, U. Chowdhry, and A.W. Sleight, *Science* 239:1015-1017 (1988).

37. P. Bordet, J.J. Capponi, C. Chaillout, J. Chenavas, A.W. Hewat, E.A. Hewat, J.L. Hodeau, M. Marezio, J.L. Tholence, and D. TranQui, *Physica C* 156:189-192 (1988).

38. S.A. Sunshine, T. Siegrist, L.F. Schneemeyer, D.W. Murphy, R.J. Cava, B. Batlogg, R.B. van Dover, R.M. Fleming, S.H. Glarum, S. Nakahara, R. Farrow, J.J. Krajewski, S.M. Zahurak, J.V. Waszczak, J.H. Marshall, P.Marsh, L.W. Rupp, Jr., and

W.F. Peck, *Phys. Rev. B* 38 (1):893-896 (1988).

39. J.M. Tarascon, Y. LePage, P. Barboux, B.G. Bagley, L.H. Greene, W.R. McKinnon, G.W. Hull, M. Giroud, and D.M. Hwang, *Phys. Rev. B* 37 (16):9382-9389 (1988).

40. E.A Hewat, M. Dupuy, P. Bordet, J.J. Capponi, C. Chaillout, J.L. Hodeau, and M. Marezio, *Nature* 333:53-54 (1988).

5

Crystal Growth and Solid State Synthesis of Oxide Superconductors

Lynn F. Schneemeyer

1.0 OVERVIEW

The aim of the study of materials is to establish correlations between the structures and properties of compounds. Such studies are motivated both by the intrinsic interest in new materials and by the technological relevance of many of the materials being studied. In the case of high temperature superconductors, unique behavior such as zero resistance conductivity offers hope for important future uses for these materials. High quality samples are essential for establishing intrinsic behavior. The aim of solid state synthesis is to produce homogeneous single phase samples of known stoichiometry and composition. This chapter will describe standard techniques for solid state synthesis of compounds in polycrystalline form and for the growth of single crystals. Special techniques for synthesis of samples containing volatile constituents will also be presented.

2.0 SOLID STATE SYNTHESIS

Compounds can be prepared either in polycrystalline form, often referred to as ceramics, or as single crystals. The synthesis of oxides in polycrystalline form is typically straightforward relative to single crystal growth. For the growth of single crystals of complex oxides, a unique growth system appropriate to the chemistry and properties of each desired compound must be developed. In contrast, standard techniques for the synthesis of ceramics are known, and such samples are suitable for a wide variety of physical measurements. In this section we describe standard approaches to the synthesis of oxide ceramics, focussing in particular on the synthesis of known oxide superconductors.

2.1 Oxide Synthesis

The standard approach to the synthesis of oxides in polycrystalline form is the direct reaction of a mixture of metal oxide starting materials at high temperature. The ratios of the starting materials will control the stoichiometry of the product provided the volatilities of the starting materials are relatively low. Syntheses of oxides containing volatile components are discussed in greater detail in the following sections.

While a binary oxide can be used as a starting material, some of these are reactive compounds that are difficult to handle. For example, BaO reacts rapidly with CO_2 in the air to produce barium carbonate. Other oxides, including BaO, are hygroscopic. Thus, compounds which are unreactive towards CO_2 and H_2O at room temperature, but decompose to oxides plus volatile gases at elevated temperatures, are often used as oxide precursors. In the case of barium, $BaCO_3$ is a stable compound under ambient conditions but decomposes upon heating to $\approx 950°C$ according to the reaction

$$BaCO_3 \rightarrow BaO + CO_2$$

Other compounds which are used in the synthesis of barium oxides include $Ba(NO_3)_2$ and BaO_2.

There are several other important considerations in choosing starting materials. First, the purity must be sufficient. A material that is 99% pure is 1% other compounds. Impurities may have profound effects on the properties of materials such as lowering the T_c of a superconductor. Also, the composition indicated on the label may not be what is found in the bottle. For instance, La_2O_3 readily converts to $La(OH)_3$ upon exposure to moisture, and a number of hydrates gain or lose water depending on the ambient humidity. Such changes in composition can produce weighing errors that affect results. Appropriate characterization of starting materials such as powder X-ray diffraction studies should be considered. For hydrates, thermal gravimetric analysis (TGA) can often readily provide information on the hydration state as well as the temperature at which water is lost.

High temperatures are generally needed in solid state synthesis to improve reaction rates and to facilitate solid state diffusion. Solid state diffusion is typically very slow. Thus, mechanical grinding steps are important to homogenize the sample and encourage complete reaction. It is important to realize, however, that some phases decompose at elevated temperatures. For example, $Ba_2YCu_3O_7$ is unstable above about 1050°C (1) and the related phase, $Ba_2YCu_4O_8$ is only stable to 860°C in one atmosphere of oxygen (2). Thus, efforts to prepare these phases require a balance between the heat put in to speed the reaction kinetics and the stability limits of the desired phase.

During reactions at high temperatures, the ceramic sample must be in contact with a container. Commonly used containers include alumina or zirconia boats or crucibles or noble metal foil-lined ceramic boats. It is important to be aware of possible reactions between the material being synthesized and the container which may be a source of foreign ions. For example, Al^{+3} ions may be incorporated if alumina crucibles are employed. Forming the material into a pellet minimizes the surface area and helps limit reactions with containers. Both alumina and zirconia crucibles and boats can be cleaned with mineral acid washes and reused.

Many of the oxide superconductors have relatively limited thermal stability. Strategies for the successful synthesis of such phases include the use of oxide precursors such as nitrates which react at relatively low temperatures, and the use of small particle size precursors and extensive grinding steps to promote reaction rates. These strategies have been important in the successful solid state syntheses of the superconducting phases, $Ba_2YCu_4O_8$ (2)(3) and $BaPb_{1-x}Sb_xO_3$ (4).

A number of low temperature routes to the preparation of oxides have begun to receive attention. In many cases these involve the decomposition of organometallic precursors such as oxalates. These alternative ceramic approaches will be discussed elsewhere in this volume.

2.2 Preparation of Samples Containing Volatile Constituents

Syntheses of oxides in ceramic form usually require reaction times of hours or days at relatively high temperatures. If any of the reactants are volatile, this combination of temperature and time often results in loss of these reactants from the system, wreaking havoc with stoichiometry control. A number of oxide precursors of interest in high T_c superconductors have appreciable volatilities including KO_2, PbO_2, Bi_2O_3, Tl_2O_3, and Tl_2O. The gaseous species of thallium oxide is Tl_2O and at 600°C, $p(Tl_2O) = 6.8 \times 10^{-4}$ atm for $p(O_2) = 0.01$ atm (5).

The toxicities of these heavy metal oxides, particularly thallium oxide, are sufficient that appropriate precautions in their handling should be observed (6)(7). Indeed, in general, it is prudent to avoid incorporation of foreign substances into the body. Substances can enter the body in three principal ways. Dust can be inhaled. Particles can be ingested. For instance, particles can be transferred to the mouth on fingertips or in a cup of coffee being drunk in the lab. Finally, some substances, notably thallium oxides, can enter the body through the skin. Thus, proper handling of materials during synthesis includes the use of well-ventilated fume hoods for grinding and high temperature reactions steps.

The usual strategies for preparing homogeneous samples of known stoichiometry when one or more constituent is volatile is to maximize the atomic scale mixing of constituents and minimize reaction times and temperatures.

Precursor Techniques: A homogeneous sample containing all of the non-volatile constituents is prepared first. Stoichiometry is easily maintained in the precursor and reasonable reaction times and temperatures can be employed. Generally, the precursor will be a mixture of phases, however. Then, the precursor and volatile constituents are formed into an intimate mixture and are subjected to a brief thermal treatment to form the desired phase.

The use of a prereacted precursor is the typical approach to the preparation of ceramics of the Tl-containing superconductors (8). For example, to prepare the 120K Tl phase, $Tl_2Ba_2Ca_2Cu_3O_{10}$, or 2223-Tl, an alkaline earth cuprate precursor which is a mixed phase sample of overall stoichiometry, $Ba_2Ca_2Cu_3O_x$ is prepared by standard ceramic techniques. An intimate mixture of the precursor and a stoichiometric amount of thallium oxide are formed into a pellet. The pellet is wrapped in Au foil to prevent reaction between the Tl-containing pellet and the container (usually an SiO_2 ampoule) during high temperature heat treatment. The pellet is heated at 900°C for 5-20 minutes until complete reaction is verified by powder X-ray diffraction.

Use of Sealed Tubes: Stoichiometry control is also enhanced when a closed system is used for the reaction. However, the material used to house the reaction must be inert to the reactants. Most commonly used for this purpose are gold or silver tubes. While noble metals are expensive, the metal can be recycled, minimizing the long-term cost. Seals can be welded, but a suitably tight seal can generally be formed by folding the end twice and crimping. Sealed gold tubes have been employed in the synthesis of Tl-based superconductors (9) and sealed silver tubes in the synthesis of $Ba_{1-x}K_xBiO_3$ superconductors (10).

Control of Oxygen Partial Pressure to Alter Constituent Volatility: The volatilities of some oxides are a strong function of temperature and oxygen partial pressure. Thus, reasonable control of stoichiometry can be obtained by choosing an oxygen partial pressure where the volatility of a reactant is limited. For example, the volatility of potassium oxide, KO_2, is high in air at 700°C but low in N_2 atmosphere at that temperature. The synthesis of $Ba_{1-x}K_xBiO_3$ (11) can then be carried out using BaO, KO_2 and Bi_2O_3 as starting materials by carrying out an initial reaction at 700°C under N_2 atmosphere. The oxygen stoichiometry is then adjusted to give a superconducting sample by annealing at 400°C in 100% O_2, conditions where the potassium vapor pressure remains low.

2.3 Ceramic Characterization

The quality of a ceramic sample is a function of the degree to which it consists of the desired product. An essential tool for the characterization of a polycrystalline sample is powder X-ray diffraction. The powder pattern is a fingerprint of the sample. For a sample to be declared single phase, all low angle peaks (below $\approx 60°2\theta$ for $CuK\alpha$ radiation) which are above the noise must be accounted for. Powder X-ray diffraction is often unable to see impurity phases present below the ~5% level. Visual inspection (using a microscope) can also reveal useful information about sample homogeneity since impurity phases may, for example, have different colors than the majority phase.

3.0 BULK CRYSTAL GROWTH

3.1 Introduction to the Growth of Single Crystals

Single crystals are of critical importance to the study of complex solids such as the new high temperature superconductors. Single crystals are free of grain boundary effects and allow measurements of physical properties as a function of crystal orientation.

Standard techniques for the growth of bulk single crystals are:

1.	Melt growth	Useful only for congruently melting phases.
2.	Sintering, grain growth	Generally produces very small crystals.
3.	Vapor phase transport	Requires volatile constituents and similar transport rates for the various constituents.
4.	Float zone	Requires specialized equipment. Detailed knowledge of the phase diagram is not required.
5.	Flux growth	Selection of flux and growth conditions must be tailored to the chemistry of a particular compound.

There are a number of excellent books on single crystal growth (12)(13). This section of this chapter will briefly outline the first four of the above techniques, then focus mainly on flux growth, the approach most commonly employed for the growth of superconducting single crystals. Finally, the techniques applied to the growth of the known high T_c phases will be reviewed.

Melt Growth: Melt growth techniques are useful for congruently melting phases, those in which the composition of the melt is the same as the single phase solid produced upon solidification. This is best diagnosed in a crystal growth experiment in which a relatively large fraction (50-70% or more) of the melt is crystallized. Then, by analyzing the top and bottom of the crystal, very small differences in composition are magnified and the true congruent composition can be determined. Note that a congruent composition is not necessarily a stoichiometric composition.

Crystals of congruently melting phases can be grown by directional solidification techniques such as the Bridgeman-Stockbarger technique (Figure 1) in which the sample in a crucible with a narrow end is passed through a temperature gradient. Solidification begins at the narrow end of the crucible, and in favorable situations, only one or a few nuclei persist to the end on the narrow region so that a sample with large single crystal domains is grown. Other melt growth techniques include Czochralski growth and gradient freeze growth.

Complex materials such as the high T_c superconducting oxides typically do not melt congruently and thus the above approach cannot be applied to the growth of single crystals. A possible exception among the superconducting cuprates is the bismuth strontium cuprate having a single two-dimensional Cu-O layer, $Bi_{2+x}Sr_{2-y}CuO_{6\pm\delta}$ (14) which may melt congruently at some composition. The congruently melting composition may not, however, correspond to the superconducting composition.

Sintering, Grain Growth: Sintering is the heating of a pressed polycrystalline sample. In sintering, a sample is held at a temperature just below its melting temperature. As a consequence of solid state diffusion, residual strain and orientation effects, larger crystallites grow at the expense of smaller ones. Small crystals suitable for single crystal X-ray diffraction studies and some physical properties studies can sometimes be obtained this way. The first studies of $Ba_2YCu_3O_7$ were carried out on large grains from sintered samples (15-17).

Vapor Phase Transport: Chemical vapor transport is a technique in which polycrystalline starting materials react with a gaseous reactant present in the system to form exclusively gaseous products. These gaseous species then travel to another place in the

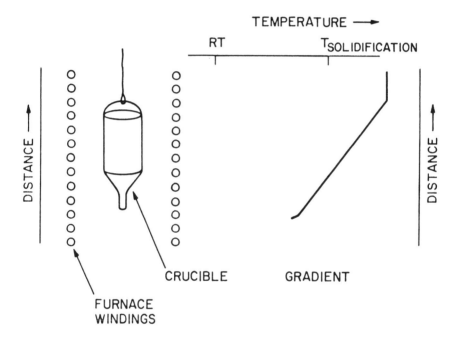

Figure 1: Schematic representation of furnace arrangement for Bridgeman-Stockbarger growth.

system where they decompose to yield the original compounds. It is important that the volatile species have similar transport rates to maintain the appropriate stoichiometry. The principles of vapor phase transport along with many examples of crystal grown are given in a series of articles by Nitsche (18-21) and by Shaefer (22)(23). Unfortunately, volatile alkaline earth species are rare, thus this technique has not been successfully applied to high temperature superconductors. However, single crystals of tenorite, CuO, have been grown using halides as transport agents (24).

Float Zone: The float zone technique is a melt growth technique which does not require the use of a crucible. Thus sample contamination by the crucible is not a problem. A vertical sample is heated with a laser, focused light source, or other heat source producing a molten zone of liquid held in place by surface tension. This molten zone can be seeded and oriented or single crystal sample produced. This technique has been effectively applied to the growth of $Bi_2Sr_2CaCu_2O_8$ by Feigelson and coworkers (25)(26). Since grain boundaries have detrimental effects on critical current densities in high T_c oxides, the preparation of long samples, effectively wires, that are free of grain boundaries is of interest.

3.2 Flux Growth

Overview of Flux Growth: Flux growth is high temperature solution growth. The term flux, taken from soldering, refers to a substance used to reduce the temperature for the growth of the desired phase. The flux may have different composition than the desired crystal, such as PbO used for the growth of La_2CuO_4 (27), or may contain one or more of the constituents of the desired phase, a situation referred to as a eutectic or non-stoichiometric melt, such as $BaCuO_2$-CuO mixtures used for the growth of $Ba_2YCu_3O_7$ (28)(29). In solution growth including flux growth, crystals are formed as the solubility of component A, the desired phase, in the melt, B, is reduced, usually by cooling the melt as illustrated in Figure 2. Solution methods permit crystal growth at a lower temperature than that required for growth from a pure melt. As a relatively low temperature technique, flux growth is useful for the growth of:

1. Incongruently melting phases.

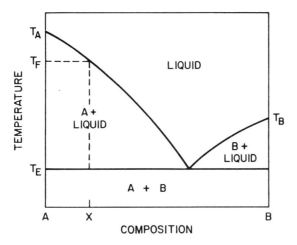

Figure 2: Binary eutectic phase diagram.

Table I. Growth techniques for superconducting oxides.

Compound	T_c (K)	Crystal Growth Technique	Ref.
$NbO_{1.0}$	1.61	?	
Rb_xWO_3	5-7	Molten salt Electrochemistry	30
$Li_{0.9}Mo_6O_{17}$	2	Gradient Flux	31
$SrTiO_3$	1	Verneuil technique	32
$LiTi_2O_4$	13	?	
$Ba(Pb,Bi)O_3$	13	Hydrothermal Flux - PbO/BaO	33,34 35
$Ba(Pb,Sb)O_3$	3.5	Flux - PbO	4
$Ba_{1-x}K_xBiO_3$	30	KOH Flux	36
$(La,Sr)CuO_4$	40	Eutectic melt Flux - $Li_4B_2O_5$ Flux - PbO	37 38 39,40
$Ba_2YCu_3O_7$	92	Eutectic melt	28,29
$Bi_2Sr_2Ca_nCu_{1+n}O_{6+2n}$	10, 85, 115	Eutectic melt Flux - alkali chloride Float Zone	41 42 25,26
$Tl_2Ba_2Ca_nCu_{1+n}O_{6+2n}$	80, 107, 125	Eutectic melt	43
$Pb_2Sr_2YCu_3O_8$	≈80	PbO Flux	44,45

2. Materials with volatile constituents.

3. Highly refractory materials.

Flux growth approaches have been invaluable for the growth of crystals of the new oxide superconductors as indicated in Table I, and should continue to yield single crystal samples as new superconductors are discovered. A variation on the flux growth approach is top-seeded solution growth. Top-seeding has been used to obtain large crystals of La_2CuO_4 (38-40).

Choosing a Suitable Flux: A wide variety of fluxes have been employed for the growth of oxide crystals. These are summarized in Table II. A suitable flux must have:

1. Reasonable solubility for the desired phase.

2. Relatively low melting point. (Crystals must grow below the decomposition temperature of the desired phase.)

3. Minimal substitutional doping of the desired phase.

Other features such as low volatility, low reactivity with container materials, high purity, and low toxicity are also desirable.

Fluxes vary in characteristics such as degree of covalency, acid-base characteristics and redox behavior. While chemical reasoning may provide some insights into appropriate choices of suitable fluxes, an empirical approach to the problem is usually necessary to produce crystals of a new compound.

Beyond the initial choice of a flux material, successful growth of crystals also requires appropriate choices of growth parameters including:

Melt composition	Often different from the ideal stoichiometry
Concentration	80 weight % Flux : 20 weight % Phase, a typical place to start

Table II. Summary of standard fluxes.

Flux System	Application to High T_c Superconductors	Considerations	Ref.
Pb-based	La_2CuO_4	Pb substitutes for Cu No superconductivity	27
Bi-based	—		
Borates	La_2CuO_4-$Li_2B_4O_7$	Li incorporation No superconductivity High Solubility for lanthanoid - But, Huge phase space	38
Vanadates	—		
Molybdates	—		
Tungstates	—		
Sulfates	—		
Phosphates	—		
Nitrates	—		
Hydroxides	La_2CuO_4 (Stacy) $Bi_{1-x}K_xBiO_3$	Low T H_2O content crucial Atmosphere matters	46 36
Halides			
- Fluorides	—		
- Chlorides	—		
· NaCl	$Bi_2Sr_2CuO_6$ $Bi_2Sr_2CaCu_2O_8$	Low solubility for oxides	47 42

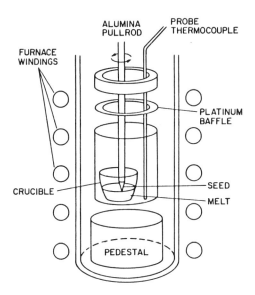

Figure 3: Schematic diagram of apparatus used for top-seeded solution growth.

236   Chemistry of Superconductor Materials

T_{MAX} Below T_{Decomp} of the desired phase

Soak time   Equilibration time, start short, ≈1 hr

Cooling rate   Relatively fast, 10-20°C/hr, for screening.

Top-seeded Solution Growth: In top-seeding, crystal growth takes place on a seed crystal that is slightly cool relative to the melt. The seed is immersed in the solution and is rotated slowly. The advantage of top seeding is control of nucleation, and, in the case of the superconducting cuprates, crystals that are relatively free of adhering flux. A schematic diagram of the experimental arrangement for top seeding is shown in Figure 3.

Containers: The crucible material of choice for oxide crystal growth has traditionally been platinum. However, copper oxide-based melts, particularly those containing alkaline earths, are reactive towards platinum producing various complex palatinates. Both platinum and gold crucibles are degraded as a result of alloying with copper extracted from the melts. Silver is relatively stable to attack by copper oxide-based melts but is too low melting to be very useful. Crucibles fabricated from metal alloys such as 5% gold/95% platinum are also used for crystal growth and offer advantages such as increased melting temperatures or decreased softness.

Ceramic crucibles are also used for oxide crystal growth. There are two considerations concerning the use of ceramic crucibles. First, they must have limited porosity in order to contain the melt during growth. Also, the purity of the ceramic, particularly with respect to binders used during ceramic fabrication, must be sufficient to minimize in advertent contamination of the product crystals. Because ceramic crucibles such as alumina are inexpensive and readily available, they can be very valuable during the early screening phase of a crystal growth problem. Table III lists common metal and ceramic crucible materials and their melting temperatures.

3.3 Growth of Superconducting Oxides

$La_{2-x}AE_xCuO_4$ AE = Alkaline Earth: Large crystals of the 30K superconductor, $Ba_{0.15}La_{1.85}CuO_4$, having the K_2NiF_4 structure

Table III. Common crucible materials.

Material	Melting Point (°C)[a]	Comments
Pt	1772	Standard choice for oxide growth
Au	1064	
Ag	962	
Ir	2410	
Ni	1455	
Al_2O_3	2072	Readily available in high purity and a variety of shapes
ZrO_2	2715	Comes stabilized with Ca or Y
MgO	2852	
SnO_2	1630	
ThO_2	3220	Radioactive

[a] Handbook of Chemistry and Physics, 65th ed., R. C. Weast, ed., CRC Press, Inc., Boca Raton, Florida, 1984.

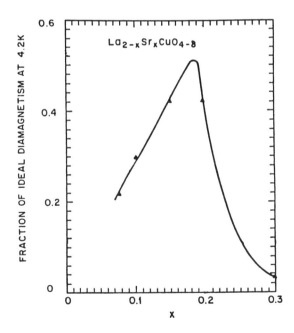

Figure 4: Superconducting volume fraction vs. x in $La_{2-x}Sr_xCuO_4$ (52).

(48)(49) and its 40K analog, $La_{1.85}Sr_{0.15}CuO_4$ (50)(51), with volumes approaching 1 cm^3 were first grown from CuO-based melts (37). Top seeding techniques have also been applied successfully in this system (38-40) using CuO- and $Li_4B_2O_5$-based melts. Crystals have also been grown from PbO-based melts (27). Observation of high superconducting transition temperatures in any of these crystals has, however, proven frustrating. Both lithium and lead dope the crystals and these impurities have detrimental effects on superconductivity. In the case of the CuO-based melts, the alkaline earth constituent has a low distribution coefficient, thus making incorporation of the alkaline earth element at appropriate levels difficult. There is evidence that superconductivity is favored at a fairly precise alkaline earth substitutional-doping level. The superconducting volume fraction in the solid solution is a maximum near $La_{1.85}Sr_{0.15}CuO_4$ as shown in Figure 4 (52). Additionally, superconductivity in the La_2CuO_4-based materials is sensitive to the precise oxygen stoichiometry while the oxygen diffusion in this phase is slow, making post growth anneals difficult. Some superconducting crystals have been prepared, but T_c's are generally low, <20K, and often, superconductivity is confined to the surface of the sample. Several puzzles remain to be answered concerning these materials, including the precise dependence of T_c on oxygen stoichiometry and on alkaline earth concentration.

$Ba_2YCu_3O_7$: Grains of the 91.5K superconductor, $Ba_2YCu_3O_7$, large enough for studies such as X-ray diffraction and magnetic characterization can be prepared by sintering near 1000°C (15-17). In the case of $Ba_2YCu_3O_7$ and related phases, the commonly used fluxes for high temperature solution growth examined so far either fail to yield perovskite-phase product crystals or show limited solubility for these compounds. Flux growth from a CuO-rich eutectic has proven useful for the preparation of crystals of $Ba_2YCu_3O_7$ of suitable size and quality for research purposes (28)(29). Figure 5 shows the main features of pseudo-ternary phase diagram for the $BaO-CuO-YO_{1.5}$ system at 975-1000°C in air, with a region of partial melting delineated by the dotted line in the CuO-rich region of the phase diagram (1). Crystals with surface areas approaching 1 cm^2 and thicknesses on the order of 100μm have been grown by slow cooling of certain of these partially melting compositions. Choice of an appropriate crucible material remains problematical, with Au and ZrO_2 giving the most satisfactory results so far.

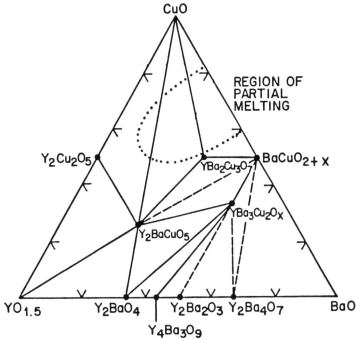

Figure 5: BaO-CuO-YO$_{1.5}$ ternary phase diagram (1) ($\approx$950-1000°C). Various compounds and compositions require various oxygen activities so that a complete rigorous representation of the diagram should include an additional dimension related to P$_{O2}$.

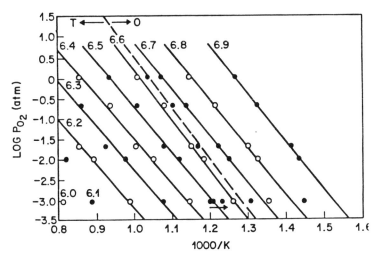

Figure 6: Oxygen pressure as a function of temperature for various values of x in Ba$_2$YCu$_3$O$_x$. The tetragonal (T) -orthorhombic (O) transition at x $\sim$ 6.63 is shown (54).

The oxygen stoichiometry, which strongly influences a variety of properties including T_c, can be controlled by annealing under set conditions of temperature and oxygen partial pressure as shown in Figure 6 (53)(54). Oxygen annealed crystals are, however, usually microtwinned in the *a-b* plane as a result of a tetragonal to orthorhombic phase transition. Several groups have recently reported thermo-mechanical techniques for detwinning $Ba_2YCu_3O_7$ crystals (55-57).

There is reasonable evidence that there are two bulk T_c's in $Ba_2YCu_3O_{7-\delta}$ rather than a continuous variation in T_c with δ (58). Figure 7 shows data of Cava et al. for T_c vs x in $Ba_2YCu_3O_x$. The plateaus near x = 7 and x = 6.7 correspond to bulk superconducting phases with T_c's of 92 and $\approx$60K respectively. We have recently verified this result for single crystals as well (59). Evidence of long range order with periodicities related to oxygen stoichiometry has been obtained from diffraction experiments (60)(61). For example, $Ba_2YCu_3O_{6.7}$ crystals show diffuse diffraction peaks indicating a cell doubling along the c axis and are superconducting at $\approx$60K (62). These orderings arise because oxygen atoms are not removed from the Cu-O chains at random, but rather, empty chains and full chains form ordered structures.

Superconductors in the Bi-Sr-Ca-Cu-O System: The Bi-Sr-Ca-Cu-O system is reported to contain several structurally-related superconducting phases. In the absence of calcium, a $\approx$10K superconducting phase having one infinite two-dimensional Cu-O layer can be prepared (63)(64). The Bi-Sr-Cu-O phase diagram (Figure 8) is surprisingly complicated (65)(66). Structurally similar non-superconducting phases can also be prepared (67-69). The superconducting phase has approximate composition, $Bi_{2+x}Sr_{2-y}CuO_{6+\delta}$, and is found in the Sr-rich portion of a region of solid solution (70). Single crystals of the superconducting phase have been grown in sealed gold tubes (71)(72) and from alkali chloride melts (47).

The 85K superconductor, $Bi_{2.2}Sr_2Ca_{0.8}Cu_2O_{8+\delta}$ (73-76), is also an incongruently-melting phase which is grown in single crystal form by flux techniques. Copper oxide-based fluxes yield research quality crystals (41). Alkali chloride melts are also useful for the growth of single crystals of this and related phases (42) and provide all of the advantages of high temperature solution growth including control of solute composition and concentration and of growth parameters such

Crystal Growth and Solid State Synthesis of Oxide Superconductors 241

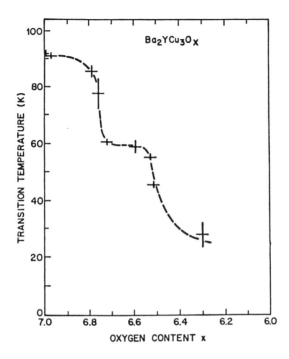

Figure 7: Superconducting critical temperature as a function of x in $Ba_2YCu_3O_x$ (58).

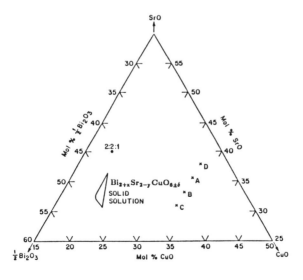

Figure 8: A portion of the $BiO_{1.5}$-SrO-CuO phase diagram (47).

as soak time and temperature. Both eutectic melt-grown and alkali halide melt-grown crystals of $Bi_{2.2}Sr_2Ca_{0.8}Cu_2O_{8+\delta}$ are very thin (a few μm) platelets. Some of these grown from alkali chloride melts are pictured in Figure 9. Recently, Feigelson and coworkers have used float zone techniques to produce single crystal fibers of $Bi_{2.2}Sr_2Ca_{0.8}Cu_2O_{8+\delta}$ 8mm in length and 300μm thick with fiber dimensions determined by the growth rate (22)(23).

A higher temperature, $T_c \approx 110K$, phase in the Bi-Sr-Ca-Cu-O system has also been reported (78-80). This phase contains three infinite two-dimensional Cu-O layers. However, lead appears to play an important role in stabilizing this compound in phase pure form (75)(79)(80). Single crystals grown by eutectic melt growth techniques have been reported (77) but the phase purity of these crystals with respect to intergrowths containing various numbers of Cu-O planes is a concern.

Superconductors in the Tl-Ba-Ca-Cu-O System: Superconducting transition temperatures have attained their maximum values to date in phases in the Tl-Ba-Ca-Cu-O system (8). Forming a homologous series of compounds of nominal composition, $Tl_2(Ba,Ca)_{1+n}Cu_nO_{6+2n}$, these phases show superconductivity onsets of 80K for n=1, 110K for n=2, and 125K for n=3. The Tl-Ba-Ca-Cu-O superconductors are structurally-similar to the Bi-Sr-Ca-Cu-O phases as is discussed in detail elsewhere in this book. Small single crystals of the various members of this series suitable for X-ray diffraction studies and some studies of physical properties have been obtained by slow cooling off-stoichiometric (normally Tl-O rich) mixtures contained in gold tubes (9)(43). Crystals are small black platelets similar in appearance to $Ba_2YCu_3O_7$ crystals and distinctly different from the micaceous $Bi_{2.2}Sr_2Ca_{0.8}Cu_2O_{8+\delta}$ crystals.

Crystal growth of these compounds is complicated by the high volatility of thallium oxides and thallium-containing compounds at elevated temperatures and the toxicity of thallium. Also, the similarity in structures leads to problems controlling phase purity and samples which appear to be single crystals based on their morphology can be shown to be complicated intergrowths by X-ray diffraction studies.

$Pb_2Sr_2M_{1-x}Ca_xCu_3O_8$: Compounds of formula $Pb_2Sr_2M_{1-x}Ca_xCu_3O_8$, where M = lanthanoid, form a new class of superconductors with T_c's as high as 80K (44)(81). The structure contains double layers of infinite planes of corner-shared CuO_5 pyramids interleaved

Crystal Growth and Solid State Synthesis of Oxide Superconductors 243

Figure 9: Photograph of $Bi_{2.2}Sr_2Ca_{0.8}Cu_2O_{8+\delta}$ crystals grown from alkali chloride fluxes (42).

by a lanthanoid/alkaline earth solid solution separated by Pb-O/Cu/-Pb-O planes containing linearly-coordinated Cu(1) atoms and is pictured in Figure 10. Single crystals are grown from PbO melts (45). A precursor containing the alkaline earth, lanthanoid and copper constituents is prepared using standard ceramic techniques. This precursor is then combined with lead oxide in a 1:1 weight ratio, placed in a high density Al_2O_3 crucible, and heat treated in a 1% oxygen atmosphere. The behavior with respect to oxygen is complicated as indicated in Figure 11 (82). Under higher oxygen partial pressures, the phase undergoes an irreversible decomposition to a tripled-perovskite related phase similar to $Ba_2YCu_3O_7$. The 1% O_2 atmosphere is necessary to maintain the desired phase during synthesis. Crystals with dimensions $1 \times 1 \times 0.1$ mm^3 are readily obtained, but display a range of T_c's depending on the thermal treatment following the growth portion of a given run. The relationship between T_c and δ for $Pb_2Sr_2M_{1-x}Ca_xCu_3O_{8+\delta}$ samples has been examined and suggests a maximum in T_c for $\delta = 0$ (83). However, control of δ in single crystal samples remains difficult. This phase has an interesting substitutional chemistry. A portion of the linearly-coordinated copper can be substituted by silver with little change in T_c (84).

$Nd_{2-x}Ce_xCuO_4$: The first electron-doped high temperature superconductor is $Nd_{2-x}Ce_xCuO_4$ (85). This phase which is of interest to provide additional clues to the mechanism of superconductivity is grown from CuO-rich melts similar to those used for the growth of La_2CuO_4-based phases (86)(87). The end member is readily grown, but addition of cerium to the melt raises the melting temperature which increases the volatility of CuO. Large rectangular plates of Nd_2CuO_4 and its Ce-doped variant have been grown. Top-seeded solution growth has also recently been reported (88).

Anionic substitution of F for O in Nd_2CuO_4 also results in bulk superconductivity with T_c's as high as 27K (89). To obtain superconductivity in these crystals, a complicated post-annealing procedure is required. So far, the fraction of the crystal which becomes superconducting appears to be relatively small (90).

$Ba_{1-x}K_xBiO_3$: A non-copper containing compound with a T_c of 30K, $Ba_{1-x}K_xBiO_3$ is of much interest in the study of oxide superconductors (10)(91). Unlike the cuprate phases which are layered, $Ba_{1-x}K_xBiO_3$ is an ideal cubic perovskite. Most of the work on $Ba_{1-x}K_xBiO_3$ has been carried out on ceramics. Recently, small

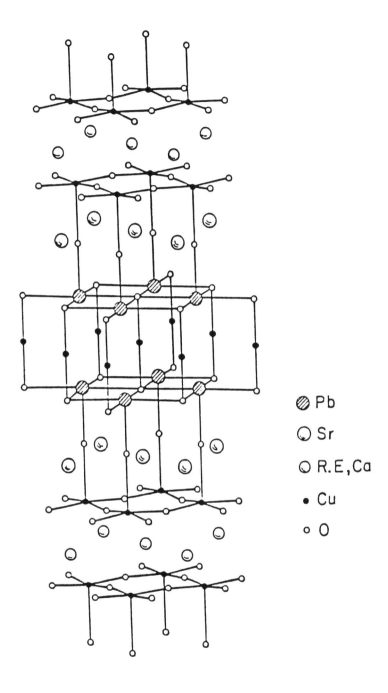

Figure 10: Structural representation of $Pb_2Sr_2MCu_3O_8$ (44).

246   Chemistry of Superconductor Materials

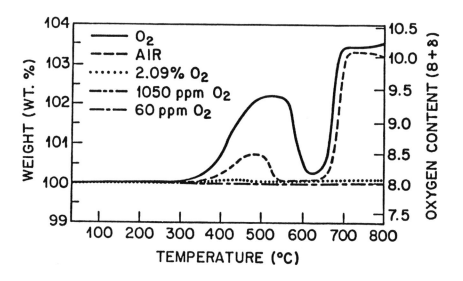

Figure 11: Thermogravimetric curves for $Pb_2Sr_2YCu_3O_{8.00}$ at $2°C min^{-1}$ in various atmospheres (82).

crystals were grown from dry KOH melts (36) following reports of the growth crystals which contained potassium but not at sufficient levels to produce superconductivity (92). These superconducting crystals are pictured in Figure 12. Apparently, nucleation occurs just above the freezing point of KOH. Hence, most crystals are of similar size and are small, < 0.1 mm in diameter.

3.4 Characterizing Single Crystals

Analysis: Unlike solid state synthesis, the stoichiometry of a newly grown crystal is unknown. Also, first attempts to grow crystals often produce relatively small samples. A useful tool for gaining information on the composition and stoichiometry of new crystalline samples is a Scanning Electron Microscope (SEM) equipped with an energy dispersive X-ray (EDX) spectrometer. Particularly when combined with appropriate standards, such a set-up can quickly provide valuable information on the number and kinds of phases obtained from a crystal growth experiment. Typical spectrometers are only sensitive to elements heavier than neon although "windowless detectors" can also give information on oxygen content. The accuracy of EDX is typically 3-5%.

Rutherford Backscattering Spectroscopy (RBS) can also give compositional information. It is a highly surface sensitive technique requiring specialized equipment and is best for samples containing elements with Z's that differ by several atomic numbers. The accuracy of RBS is controlled by counting statistics but is usually about 5%.

X-ray Diffraction: Structural characterization of single crystal samples can reveal both the atomic arrangement and the composition of a sample. The unit cell size and symmetry alone can quickly establish whether a crystal is a known material or a new phase. A detailed discussion of X-ray diffraction studies of the known superconductors is the topic of another chapter in this book.

Superconductivity: Superconductivity plays two roles in the study of superconducting oxides. Single crystals of high T_c phases are characterized to determine parameters including T_c, transition width, and superconducting volume fraction. Also, superconductivity can be employed as a handle in the search for new superconducting phases or for new approaches to the growth of existing phases.

248   Chemistry of Superconductor Materials

a.

b.

Figure 12: Photograph of 30.5K superconducting $Ba_{0.6}K_{0.4}BiO_3$ crystals grown from KOH flux (36).

Magnetic measurements and transport property studies are described in detail in other parts of this book. A few comments concerning such measurements on single crystals are noted here. First, the known superconducting cuprates are anisotropic materials. As the growth habit of compounds usually reflects their underlying structure, the superconducting cuprates tend to grow as thin platelets. This shape anisotropy has consequences for measurements, especially those to determine superconducting volume fraction. Depending on sample orientation relative to the field, a demagnetizing factor must be applied. The magnitude of the demagnetizing factor is, however, unknown for most sample geometries although it can be estimated. Failure to consider the demagnetizing factor can lead to serious errors in the size of the Meissner fraction.

The shape anisotropy also complicates efforts to determine the resistivity anisotropy. An extension of the Montgomery method must be employed for anisotropic materials (93). Such characterization has been carried out for $Bi_{2.2}Sr_2Ca_{0.8}Cu_2O_{8\pm\delta}$ (94).

In favorable cases, superconductivity is a sensitive and useful screen for desired crystals. Near zero field microwave absorption (95) can be used to examine very small samples of only a few micrograms. This technique has value in the search for new superconducting phases. In early attempts to identify the superconducting phase in the Pb-Sr-Y-Cu-O system, both superconducting and non-superconducting crystals were obtained. Individual crystals were examined for superconductivity using near-zero field microwave absorption. Then X-ray diffraction was used to establish the structure and stoichiometry of the superconducting phase.

However, there are limitations to the use of superconductivity as a screen for new superconductors. Experience with various of the known high T_c phases has taught us the sensitivity of superconductivity to doping. For, example T_c depends critically on oxygen stoichiometry in $Ba_2YCu_3O_7$. Thus, interesting phases might easily be overlooked.

ACKNOWLEDGEMENT

I am grateful to D. W. Murphy and R. J. Cava for helpful suggestions, and S. A. Sunshine for a critical reading of the manuscript.

4.0 REFERENCES

1. R.S. Roth, J.R. Dennis, and K.C. Davis, *Advan. Ceram. Mater.*, 2:303 (1987).
2. R.J. Cava, J.J. Krajewski, W.F. Peck, Jr., B. Batlogg, L. W. Rupp, Jr., R.M. Fleming, A.C.W.P. James, and P. Marsh, *Nature*, 338:328 (1989).
3. R.J. Cava, J.J. Krajewski, W.F. Peck, Jr., B. Batlogg, and L.W. Rupp, Jr., *Physica C* 159, 461 (1989).
4. R.J. Cava, B. Batlogg, G.P. Espinosa, A.P. Ramirez, J.J. Krajewski, W. F. Peck, Jr., L. W. Rupp, Jr., and A.S. Cooper, *Nature*, 339:291 (1989).
5. D. Cubicciotti and F. J. Keneshea, *J. Phys. Chem.* 71:808 (1967).
6. *"The Merck Index"*, 9th ed., M. Windholz, ed., Merck and Co, Rahway, 1976.
7. *"Casarett and Doull's Toxicology"*, 2nd ed., J. Doull, C.D. Klaassen, and M.O. Amdur, eds., MacMillian Publ. Co., NY, 1980.
8. Z.Z. Sheng and A.M. Hermann, *Nature* 332:138 (1988).
9. M. A. Subramanian, J. C. Calabrese, C. C. Torardi, J. Gopalakrishnan, T. R. Askew, R. B. Flippen, K. J. Morrissey, U. Chowdhry, and A. W. Sleight, *Nature*, 332:420 (1988).
10. R. J. Cava, B. Batlogg, J. J. Krajewski, R. Farrow, L. W. Rupp, Jr., A. E. White, K. Short, W. F. Peck, Jr., and T. Kometani, *Nature*, 339:291 (1989).
11. D. G. Hinks, B. Dabrowski, J. D. Jorgensen, A. W. Mitchell, D. R. Richards, S. Pei, and D. Shi, *Nature*, 333:836 (1988).
12. D. Elwell and H. J. Scheel, *"Crystal Growth from High-Temperature Solutions"*, Academic Press, New York, 1975.
13. R. A. Laudise, *"The Growth of Single Crystals"*, Prentice-Hall, NJ, 1970.
14. S. A. Sunshine, private communication.
15. R. J. Cava, B. Batlogg, R. B. van Dover, D. W. Murphy, S. Sunshine, T. Siegrist, J. P. Remeika, E. A. Rietman, S. Zahurak, and G. P. Espinosa, *Phys. Rev. Lett.*, 58:1676 (1987).

16. T. R. Dinger, T. K. Worthington, W. J. Gallagher, and R. L. Sandstrom, *Phys. Rev. Lett.* 58:2687 (1987).
17. G. W. Crabtree, J. Z. Liu, A. Umezawa, W. K. Kwok, C.H. Sowers, S.K. Malik, B.V. Veal, D.J. Lam, B. Brodsky, and J.W. Downey, *Phys. Rev. B* 36:4021 (1987).
18. R. Nitsche, H. U. Bolsterli, and M. Lichtensteiger, *J. Phys. Chem. Solids*, 21:199 (1961).
19. R. Nitsche and D. Richman, *Zeit. fur Elektrochemie* 66,709 (1962).
20. R. Nitsche, *Z. Krist.* 120:1 (1964).
21. R. Nitsche, *J. Phys. Chem. Solids* (Supp.) 1:28 (1967).
22. H. Schaefer, *"Chemical Transport Reactions"*, Academic Press, New York, 1964.
23. H. Schaefer, T. Grofe, and M. Trenkel, *J. Solid State Chem.*, 8:14, (1973).
24. A. Wold, *Proceedings of the Materials Research Society Symposium* on High Temperature Superconductors, San Diego, CA, April 24-28 (1989).
25. R. S. Feigelson, D. Gazit, D. K. Fork, and T.H. Geballe, *Science* 240:1642 (1989).
26. D. Gazit and R. S. Feigelson, *J. Crystal Growth* 91:318 (1988).
27. G. P. Espinosa, A. S. Cooper, and J. P. Remeika, private communication.
28. L. F. Schneemeyer, J. V. Waszczak, T. Siegrist, R. B. van Dover, L. W. Rupp, B. Batlogg, R. J. Cava, and D. W. Murphy, *Nature*, 332:601 (1987).
29. D. L. Kaiser, F. Holtzberg, B. A. Scott, and T. R. McGuire, *Appl. Phys. Lett.* 51:1040 (1987).
30. Ch. J. Raub, A. R. Sweedler, M. A. Jensen, S. Broadston, and B. T. Matthias, *Phys. Rev. Lett.* 13:746 (1964).
31. W. H. McCarroll and M. Greenblatt, *J. Solid State Chem.* 54:3087 (1984).
32. K. Nassau and A. E. Miller, *J. Crystal Growth* 91:373 (1988).
33. S. Hirano and S. Takahashi, *J. Crystal Growth* 78:408 (1986).

34. S. Hirano and S. Takahashi, *J. Crystal Growth* 79:219 (1986).

35. B. Batlogg, J. P. Remeika, R. C. Dynes, H. Bartz, A. S. Cooper, and J. P. Garno, in *"Superconductivity in d- and f-band Metals"*, W. Buckel and W. Weber, eds., (Kernforschungszeutrum Karlsruhe, 1982), p. 401.

36. L. F. Schneemeyer, J. K. Thomas, T. Siegrist, B. Batlogg, L.W. Rupp, R. L. Opila, R. J. Cava, and D. W. Murphy, *Nature* 335:421 (1988).

37. Y. Hidaka, Y. Enomoto, M. Suzuki, M. Oda, and T. Murakami, *J. Crystal Growth* 85:581 (1987).

38. P. J. Picone, H. P. Jenssen, and D. R. Gabbe, *J. Crystal Growth* 85:576 (1987).

39. P. J. Picone, H. P. Jenssen, and D. R. Gabbe, *J. Crystal Growth* 91:463 (1988).

40. C. Chen, B. E. Watts, B. M. Wanklyn, P. A. Thomas, and P. W. Haycock, *J. Crystal Growth* 91:659 (1988).

41. M. Hikita, T. Iwata, Y. Tajima, and A. Katsui, *J. Crystal Growth* 91:282 (1988).

42. L. F. Schneemeyer, R. B. van Dover, S. H. Glarum, S. A. Sunshine, R. M. Fleming, B. Batlogg, T. Siegrist, J. H. Marshall, J. V. Waszczak, and L. W. Rupp, *Nature*, 332:422 (1988).

43. D. S. Ginley, B. Morosin, R. J. Baughman, E. L. Venturini, J. E. Schirber, and J. F. Kwak, *J. Crystal Growth* 91:456 (1988).

44. R. J. Cava B. Batlogg, J. J. Krajewski, L. W. Rupp, L. F. Schneemeyer, T. Siegrist, R. B. van Dover, P. Marsh, W. F. Peck, P. K. Gallagher, S. H. Glarum, J. H. Marshall, R. C. Farrow, J. V. Waszczak, R. Hull, and P. Trevor, *Nature* 336:211 (1988).

45. L. F. Schneemeyer, R. J. Cava, A. C. W. P. James, P. Marsh, T. Siegrist, J. V. Waszczak, J. J. Krajewski, W. P. Peck, Jr., R. L. Opila, S. H. Glarum, J. H. Marshall, R. Hull, and J. M. Bonar, *Chem. of Mater.*, to be published.

46. W. K. Ham, G. F. Holland, and A. M. Stacy, *J. Am. Chem. Soc.* 110:5214 (1988).

47. L. F. Schneemeyer, J. V. Waszczak, R. M. Fleming, S. Martin, A. T. Fiory, and S. A. Sunshine, *Proceedings of*

the *Mater. Res. Soc. Conf.* on High Temperature Superconductors, San Diego, CA, April 23-28, 1989.

48. J. G. Bednorz and K. A. Muller, *Z. Physik* B64:189 (1986).

49. G. Takagi, S. Uchida, K. Kitazawa, and S. Tanaka, *Japan J. Appl. Phys.* 26:L231 (1987).

50. R. J. Cava, R. B. van Dover, B. Batlogg, and E. A. Rietman, *Phys. Rev. Lett* 58:408 (1987).

51. J. M. Tarascon, L. H. Greene, W. R. McKinnon, G. W. Hull, and T. H. Geballe, *Science* 235:1373 (1987).

52. R. M. Fleming, B. Batlogg, R. J. Cava, and E. A. Reitman, *Phys. Rev.* 35 7191 (1987).

53. P. K. Gallagher, H. M. O'Bryan, S. A. Sunshine, and D. W. Murphy, *Mater. Res. Bull.* 22:995 (1987).

54. P. K. Gallagher, *Adv. Ceram. Mater.* 2:632 (1987).

55. H. Schmid E. Burkhardt, N. B. Sun and J.-P. Rivera, *Physica C* 157:555 (1989).

56. D. L. Kaiser, F. W. Gayle, R. S. Roth, and L. J. Swartzendruber, *J. Mater. Res.* 4:745 (1989).

57. T. Hatanaka and A. Sawada, *Japn. J. Appl. Phys.* 28:L794 (1989).

58. R. J. Cava, B. Batlogg, A. P. Ramirez, D. Werder, C. H. Chen, E. A. Rietman, and S. M. Zahurak, *Proc. Mater. Res. Soc., Symp.* 99:16 (1987).

59. S. H. Glarum, L. F. Schneemeyer, and J. V. Waszczak, *Phys. Rev. B*, to be published.

60. C. H. Chen, D. J. Werder, L. F. Schneemeyer, P. K. Gallagher, and J. V. Waszczak, *Phys. Rev. B* 38:2888 (1988).

61. R. Beyers, B.T. Ahn, G. Gorman, V.Y. Lee, S.S.P. Parkin, M.L. Ramirez, K. P. Roche, J.E. Vazauez, T.M. Gur, and R.A. Huggins, *Nature* 340:619 (1989).

62. R. M. Fleming, L. F. Schneemeyer, P. K. Gallagher, B. Batlogg, L. W. Rupp, and J. V. Waszczak, *Phys. Rev. B* 37:7920 (1988).

63. C. Michel, M. Hervieu, M. M. Borel, A. Grandin, F. Deslandes, J. Provost, and B. Raveau, *Z. Phys. B* 68:421 (1987).

64. J. Akimitsu, A. Yamazaki, H. Sawa, and H. Fujiki, Jpn, J. Appl. Phys. 26:L2080 (1987).
65. R. S. Roth, C. J. Rawn, J. D. Whitler, C. K. Chiang, and W. K. Wong-Ng, J. Am. Ceram. Soc. 72:395 (1989).
66. B. C. Chakoumakos, P. S. Ebey, B. C. Sales and E. Sonder, J. Mater. Res. 4:767 (1989).
67. M. Onoda and M. Sato, Solid State Commun. 67:799 (1988).
68. G. Xiao, M. Z. Cieplak, and C. L. Chien, Phys. Rev. B 38:11824 (1988).
69. M. T. Casais, C. Cascales, A. Castro, M. de Pedro, I. Rasines, G. Domarco, J. Maza, F. Miguelez, J. Ponte, C. Torron, J. A. Veira, F. Vidal, and J. A. Campa, Proc. European MRS, Strasbourg, France, Nov. 8-11, 1988.
70. A. MacKenzie, E. Marseglia, I. Marsden, C. Chen and B. Wanklyn, Physica C, to be published.
71. C. C. Torardi, M. A. Subramanian, J. C. Calabrese, J. Gopalakrishnan, E. M. McCarron, K. J. Morrissey, T. R. Askew, R. B. Flippen, U. Chowdhry, and A. W. Sleight, Phys. Rev. B 38:225 (1988).
72. P. Strobel, K. Kelleher, F. Holtzberg and T. Worthington, Physica C 156:434 (1988).
73. H. Maeda, Y. Tanaka, M. Fukutomi, and T. Asano, Jap. J. Appl Phys. 27:L209 (1988).
74. R. M. Hazen, C. T. Prewitt, R. J. Angel, N. L. Ross, L. W. Finger, C. G. Hadidiacos, D. R. Veblen, P. J. Heaney, P. H. Hor, R. L. Meng, Y. Y. Sun, Y. Q. Wang, Y. Y. Xue, Z. J. Huang, L. Gao, J. Bechtold, and C. W. Chu, Phys. Rev. Lett. 60:1174 (1988).
75. S. A. Sunshine, T. Siegrist, L. F. Schneemeyer, D. W. Murphy, R. J. Cava, B. Batlogg, R. B. van Dover, R. M. Fleming, S. H. Glarum, S. Nakahara, R. Farrow, J. J. Krajewski, S. M. Zahurak, J. V. Waszczak, J. H. Marshall, P. Marsh, L. W. Rupp, and W. F. Peck, Phys. Rev. B 38:893 (1988).
76. J. M. Tarascon, Y. LePage, P. Barboux, B. G. Bagley, L. H. Greene, W. R. McKinnon, G. W. Hull, M. Giroud, and D. M. Hwang, Phys. Rev. B 38:896 (1988).

77. M. A. Subramanian, C. C. Torardi, J. C. Calabrese, J. Gopalakrishnan, K. J. Morrissey, T. R. Askew, R. B. Flippen, U. Chowdhry, and A. W. Sleight, *Science*, 239:1015 (1988).
78. J. L. Tallon, R. G. Buckley, P. W. Gilberd, M. R. Presland, I. W. M. Brown, M. E. Bowden, L. A. Christian, and R. Goguel, *Nature* 333:153 (1988).
79. U. Endo, S. Koyama, and T. Kawai, *Japn. J. Appl Phys.* 27:L1476 (1988).
80. M. Takano, J. Takada, K. Oda, H. Kitaguchi, Y. Miura, Y. Ikeda, Y. Tommii and H. Mazaki, *Japn. J. Appl. Phys.* 27:L1041 (1988).
81. M. A. Subramanian, J. Gopalakrishnan, C. C. Torardi, P. L. Gai, E. D. Boyes, T. R. Askew, R. B. Flippen, W. E. Farneth, and A. W. Sleight, *Physica C*, 157:124 (1989).
82. P. K. Gallagher, H. M. O'Bryan, R. J. Cava, A. C. W. P. James, D. W. Murphy, W. W. Rhodes, J. J. Krajewski, W. F. Peck, J. V. Waszczak, *Chem. of Mater.* 1:277, (1989).
83. E. M. Gyorgy, H. M. O'Bryan and P. K. Gallagher, Chem. of Mater., submitted.
84. A. C. W. P. James, D. W. Murphy, *Chem. of Materials*, 1:169 (1989).
85. T. Tokura, H. Takagi, and S. Uchida, *Nature* 337:345 (1989).
86. Y. Hidaka and M. Suzuki, *Nature* 338:635 (1989).
87. J.-M. Tarascon, E. Wang, L. H. Greene, B. G. Bagley, G. W. Hull, S. M. D'Egidio, and P. F. Miceli, *Phys. Rev. B*, to be published.
88. A. Cassanho, D. R. Gabbe and H. P. Jenssen, *J. Crystal Growth* 96:999 (1989).
89. A. C. W. P. James, S. M. Zahurak, and D. W. Murphy, *Nature* 338:6212 (1989).
90. C. H. Chen, D. J. Werder, A. C. W. P. James, D. W. Murphy, S. Zahurak, R. M. Fleming, B. Batlogg and L. F. Schneemeyer, *Phys. Rev. Lett.*, to be published.
91. L. H. Mattheiss, E. M. Gyorgy and D. M. Johnson, *Phys. Rev. B* 37:3745 (1988).
92. J. P. Wignacourt, J. S. Swinnea, H. Steinfink and J. B. Goodenough, *Appl. Phys. Lett.* 53:1753 (1988).

93. H. C. Montgomery, *J. Appl. Phys.* 42:2971 (1971).
94. S. Martin, A. T. Fiory, R. M. Fleming, L. F. Schneemeyer and J. V. Waszczak, *Phys. Rev. Lett.* 60:2194 (1988).
95. S. H. Glarum, J. H. Marshall and L. F. Schneemeyer, *Phys. Rev. B* 37:7491 (1987).

6

Preparation of Bismuth- and Thallium-Based Cuprate Superconductors

Stephen A. Sunshine and Terrell A. Vanderah

1.0 INTRODUCTION

Superconductivity in the Bi-Sr-Cu-O system at relatively low temperatures was first reported by Michel, et al. (1). Addition of Ca to this system led to the report of superconducting transitions between 80 and 110 K (2). These discoveries led to frantic attempts in a number of laboratories to prepare single-phase samples of the phases responsible for superconductivity. Study of single crystals revealed that these transitions are due to three different superconducting phases that are often represented by the formula $Bi_2Sr_2Ca_{n-1}Cu_nO_{2n+4}$; n = 1, 2, and 3. This represents an idealized formula for the phases involved and schematic drawings of the ideal structures are given in Figure 1.

Within weeks after superconductivity was observed above 77 K in the Bi-Sr-Ca-Cu-O system, Sheng and Hermann (3) reported a record-high T_c value of 120 K for a Tl-Ba-Ca-Cu-O phase. The results were quickly confirmed (4)-(9) and the existence of a large family of thallium-containing superconducting oxides was revealed (10).

A number of chemical reviews of cuprate superconductors have included the bismuth and thallium families (11)-(14). Reviews focussing on the structural chemistry of these two series are also available (15),(16). On the thallium cuprates, an overview of structural studies has appeared (17), and a detailed review of

258 Chemistry of Superconductor Materials

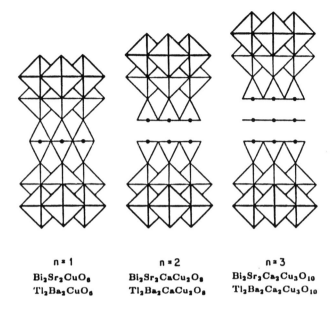

Figure 1: Schematic representation of the ideal structures of the homologous series of Bi- and Tl-based superconductors showing the coordination polyhedra about Bi or Tl (octahedra) and Cu (octahedra, square pyramids, or square planes).

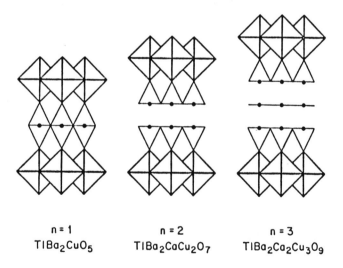

Figure 2: Schematic representation of the ideal structures of the thallium monolayer cuprate superconductors.

synthesis, structural chemistry, and electronic behavior will soon be available (18).

Two series of thallium-containing cuprate superconductors have been synthesized with the following **ideal** general formulas (10):

A. $Tl_2Ba_2Ca_{n-1}O_{2n+4}$

B. $TlA_2Ca_{n-1}Cu_nO_{2n+3}$; A = Ba, Sr

These structurally related series are intergrowths of oxygen-deficient perovskite layers (with "n" denoting the number of Cu-O layers per perovskite slab) and rock salt-type layers. These structures are cousins of the Aurivillius family of oxides (11) and are also related to the Ruddlesden-Popper-type phases (15). The compounds in Series A adopt the same ideal structures as those of the Bi family of superconductors which are depicted in Figure 1; the structures of the series B phases are shown in Figure 2. The two series differ in the thickness of the rock salt portion of the structure. In series A, the rock salt "slab" is three layers thick (2TlO + BaO); in series B, two layers thick (TlO + BaO). Series A is referred to as having "thallium bilayers"; series B, "thallium monolayers". The layer structures of these compounds lead to characteristic c-axis lengths of the unit cells. Although the unit cells all tend to be large and the X-ray powder patterns rather complex, the low-angle (002) or (001) reflections can often be used to sort out product phases; routine X-ray powder diffraction scans should therefore be started at 5 (or less) degrees two-theta in order to record this characteristic region (see Chapter 13).

It is common practice to refer to these compounds using numerical acronyms corresponding to the metal stoichiometries. The decoding scheme and T_c values are collected in the following table (13)(21)-(23); the lead-substituted phases are included but not generally referred to with acronyms.

Some of the compounds in Table 1 have been prepared as single crystals and/or well-crystallized single-phase samples, while others have been obtained in multi-phase products with structural information provided by electron microscopy (13). Compounds that have been well-characterized in single-phase polycrystalline form

Table 1
Bismuth- and Thallium-Based Superconducting Cuprates[1]

Acronym	Ideal Formula	$T_c(K)$
2201	$Bi_2Sr_2CuO_6$	10
2212	$Bi_2Sr_2CaCu_2O_8$	90
2223	$Bi_2Sr_2Ca_2Cu_3O_{10}$	110
2234	$Bi_2Sr_2Ca_3Cu_4O_{12}$	90
2201	$Tl_2Ba_2CuO_6$	90
2212	$Tl_2Ba_2CaCu_2O_8$	110
2223	$Tl_2Ba_2Ca_2Cu_3O_{10}$	122
2234	$Tl_2Ba_2Ca_3Cu_4O_{12}$	119
1201	$TlBa_2CuO_5$	non-SC
1212	$TlBa_2CaCu_2O_7$	80
1223	$TlBa_2Ca_2Cu_3O_9$	110
1234	$TlBa_2Ca_3Cu_4O_{11}$	122
1245	$TlBa_2Ca_4Cu_5O_{13}$	110
	$Tl_{0.5}Pb_{0.5}Sr_2CuO_5$	non-SC
	$(Tl,Pb)Sr_2CaCu_2O_7$	90
	$(Tl,Pb)Sr_2Ca_2Cu_3O_9$	122

[1] Recently, two analogous series of bismuth- and thallium-based cuprate series have been reported with the formula $Bi_2Sr_2(Ln_{1-x}Ce_x)_2Cu_2O_{10-y}$; Ln = Sm, Eu, or Gd (19). In the thallium series Bi,Sr is replaced by Tl,Ba. The bismuth series is superconducting with T_c values below 30 K, while the thallium compounds were not superconductors. The structure is similar to the bilayer Aurivillius-like thallium and bismuth families, except that a fluorite-like layer $(Ln_{1-x}Ce_x)O_2$ separates the CuO_2 sheets instead of calcium ions. A very similar series with thallium monolayers instead of bilayers has also been found, but was not superconducting (20).

and/or as single crystals include: the bismuth family, n=1,2,3 [2201, 2212, 2223] (13); the thallium bilayer phases, n=1,2,3 [2201, 2212, 2223] (13); the thallium monolayer phases, n=2,3 [1212, 1223] (13)(24); and the Pb-substituted thallium phases (25). The highest superconducting transition temperatures found to date, above 120K, are exhibited by the n=3 thallium bilayer phase (13), the n=4 thallium monolayer phase (20), and $(Pb_{0.5}Tl_{0.5})Sr_2Ca_2Cu_3O_9$ (25). It is now generally accepted that lower T_c values are found for phases with more than three or four Cu-O layers (21)(26).

The bismuth- and thallium-based cuprate superconductors differ in a number of aspects, most of which seem attributable to the chemical difference between trivalent bismuth and thallium: Bi^{3+} possesses a lone pair ($6s^2$) in its valence shell, in contrast to the noble gas configuration of Tl^{3+}. As can be seen in Figure 1 (see also Chapter 13), the ideal structures of the n=1-3 (Tl,Bi) bilayer phases are identical. However, the chemical nature of Tl^{3+} and Bi^{3+} result in different deviations from the ideal structures: Tl^{3+} behaves as if it were somewhat too small for its site, whereas Bi^{3+} possesses the lone pair that must be structurally accommodated (7). Figure 3 (7) depicts some of the differences between the average structures of the thallium and bismuth 2201 phases. Displacements within the Bi-O layers result in the formation of complex incommensurate superstructures which are discussed elsewhere. The formation of these superstructures, which involve the bonding between two adjacent Bi-O sheets, may be related to the failure to synthesize the bismuth monolayer analogs of the thallium compounds. On average, as can be seen in Figure 3, an unusually long Bi-O bond develops between the two Bi-O layers (3.2 Å), whereas the corresponding bond connecting the Tl-O sheets is much shorter (2.0 Å) (7). The weak Bi-O intersheet bonds likely cause the different morphologies of the bismuth and thallium compounds. The bismuth phases form extremely thin, micaceous platelets that tend to delaminate: cleavage between the Bi-O sheets yields charge neutral sections (7). The morphology of the thallium compounds is also platy, but not micaceous. The weak bonding along the c-axis of the bismuth compounds may also contribute to their stronger propensity to form intergrowth products, with sections of varying n-values in a single grain. The c-axis of the 2201 bismuth phase in Figure 3 is considerably

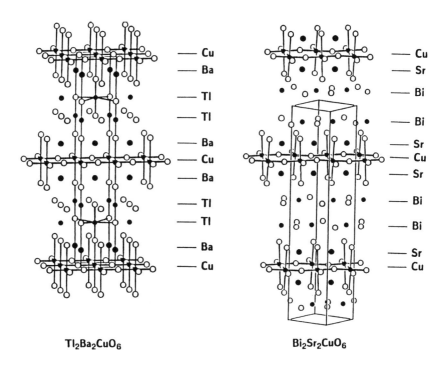

Figure 3: Average structures of $Tl_2Ba_2CuO_6$ and $Bi_2Sr_2CuO_6$ as determined by single-crystal X-ray diffraction (7).

longer than that of the thallium analog, despite very similar ionic radii of (Tl,Ba) and (Bi,Sr) (7). This could indicate that the thallium phases are also more tightly coupled electronically along the c-axis, thus leading to the higher T_c values observed for the thallium family.

The layer-type structures and chemical nature of the constituents of the bismuth and thallium-based cuprate superconductors - notably the lone-pair stereochemistry of Bi^{3+}, variable valence of copper, and considerable exchange among some of the cation sites - combine to make structural non-ideality, nonstoichiometry, and phase intergrowth the rule rather that the exception in these families of materials. These features, as well as the probable metastability of the phases (and possibly all high-temperature oxide superconductors), also contribute to the difficulties typically encountered in preparing single-phase samples with reproducible properties and compositions.

We are now acutely aware of the importance of the details of sample preparation; e.g., mixing, integrity of reactants, container, furnace design and sample placement therein, sample heating and cooling rates, actual vs. programmed temperatures, thermal gradients, and sample storage. Unfortunately, these details of sample preparation are not always known, and when known, are not always included in published reports - a situation that will hopefully self-correct in the future. Those attempting to repeat literature reports will most likely need to vary such details in their own particular experimental situation before success is attained. Hence, a considerable investment of time and effort should be expected.

This chapter presents an overview of our understanding of phase relationships and a summary of synthetic techniques for the synthesis of phase-pure superconducting samples in the bismuth- and thallium-based families of high T_c cuprate superconductors.

2.0 SYNTHETIC METHODS

Many of the high T_c ceramic superconductors can be prepared by similar solid state techniques. In the Bi-Sr-Ca-Cu-O system Bi_2O_3, $SrCO_3$, $CaCO_3$ or $Ca(OH)_2$, and CuO are often used as the starting materials. These powders are combined in appropriate ratios, ground in an agate mortar, and heated in a high density alumina crucible at 810-820°C for 3-5 hours. It is important that this initial heat treatment is carried out below the melting point of Bi_2O_3 ($\approx$825°C) to allow for reaction of the Bi_2O_3 before melting. The samples are cooled to room temperature, ground, pressed into pellets, and heated at 830°C for 3-5 hours. This process is repeated, raising the temperature by 10-15°C, until no change is observed in the X-ray powder diffraction pattern. Typically, 4-6 heatings up to 900°C are required to reach equilibrium. While these latter reactions can be performed in high density alumina, the possible reaction of the samples with Al_2O_3 makes silver or gold a better container material.

Other starting materials such as SrO, CaO, $Sr(NO_3)_2$, and $Ca(NO_3)_2 \cdot 4H_2O$ can be used in place of the carbonates. However, the oxides readily react with CO_2 and H_2O from the air to form carbonates and hydroxides while $Ca(NO_3)_2 \cdot 4H_2O$ often contains an unreliable amount of water of hydration. These starting materials should be analyzed prior to use to determine the actual SrO or CaO content. This can be done by thermogravimetric analysis or by heating a known amount of material in air to 950°C and quickly weighing the cooled sample to determine the weight loss due to CO_2, H_2O, and NO_2 evolution. Slow heating rates are necessary when the nitrates are used as starting materials since these compounds melt at low temperatures (550-600°C) and can splatter during NO_2 evolution.

The volatility of the reactants is a concern in some solid state syntheses. This may be a slight problem in bismuth cuprates synthesized at high temperatures because Bi_2O_3 has an appreciable vapor pressure at these temperatures (i.e. 900-950°C); however, chemical analyses of samples of bismuth-based superconductors before and after reactions at temperatures up to 900°C indicate no detectable loss of bismuth. This problem is much more severe in the case of thallium chemistry.

The primary difficulty in preparing the thallium-based cuprate superconductors lies in the toxicity and volatility of the reactant Tl_2O_3 and its decomposition products. Above 600°C, the following redox-vaporization process is well under way and would lead to substantial loss of reactant in an open system, although the magnitude of the partial pressure that would form in a closed system is not high enough to pose technical difficulties in containment (27):

$$Tl_2O_{3(s)} \rightarrow Tl_2O_{(g)} + O_{2(g)} \;;$$

small amounts of gaseous elemental Tl are also formed (27). Tl_2O_3 melts at 717°C and its conversion to Tl_2O is complete at 875°C in air (28). Reactions to prepare the complex oxides containing thallium should therefore be carried out in sealed containers that do not react with Tl_2O_3. In addition to problems with volatility, soluble thallium compounds are particularly toxic because they are erage work period (29); early symptoms of thallium-poisoning include hair loss. Safety considerations in handling thallium compounds should include the following: 1) gloves should be worn at all times to avoid skin contact; 2) powders should be handled as much as possible in a well-ventilated hood to avoid dust inhalation; 3) use of Tl_2O, which is soluble, should be avoided in favor of the sesquioxide; 4) reactions should be carried out in sealed containers in furnaces placed in chemical fume hoods; and 5) chemical waste must be isolated and disposed of properly. In flow-through systems, volatile thallium constituents can be safely isolated by exiting the flow gases through traps containing basic peroxide solutions; insoluble Tl_2O_3 precipitates.

The following sections will outline specific methods for the synthesis of Bi- and Tl-based cuprate superconductors. Because the synthetic methods and historical evolution of the compounds are different, the bismuth and thallium families are described separately.

3.0 SYNTHESIS OF Bi-BASED CUPRATE SUPERCONDUCTORS

3.1 Single Cu-O Layer Phase [2201], $T_c=10$ K

The most detailed studies to date have involved the n=1 system, i.e., the calcium-free system in which superconductivity was first detected near 20 K (1). This system can be described by a pseudo-ternary phase diagram. Equilibrium phase fields have been proposed by several groups (30)-(34). The equilibrium phase diagram for the Bi-Sr-Cu-O system as determined by Roth et al. is given in Figure 4 (30). The samples for this study were prepared from Bi_2O_3, CuO, and $SrCO_3$ or $Sr(NO_3)_2$ using the heating sequence described in section 2.0 above. The surprising result of this study was the isolation of two distinct phases close to the ideal composition $Bi_2Sr_2CuO_6$ (Figure 5). Phase A has a composition very close to $Bi_2Sr_2CuO_6$ although up to 10% Cu deficiency may occur (35). Phase B actually occupies a solid solution region described by the formula $Bi_{2-x}Sr_{2-y}CuO_{6\pm\delta}$. Importantly, Phase A is a semiconductor, although its composition is closest to the idealized $Bi_2Sr_2CuO_6$. Phase B shows signs of superconductivity between 6 and 10 K for samples at the Sr-rich end of this solid solution. The solid solution, as determined by Roth et al. (30), is bound by the compositions (a) $Bi_{2.17}Sr_{1.83}CuO_{6\pm\delta}$ (b) $Bi_{2.45}Sr_{1.76}CuO_{6\pm\delta}$, and (c) $Bi_{2.19}Sr_{1.39}CuO_{6\pm\delta}$ (Figure 5). The solid solution tolerates both Sr and Cu deficiencies although different limits of the solid solution have been reported by various groups (30)-(34). For example, Ikeda et al. (34) gives the limits as $Bi_{2+x}Sr_{2-x}Cu_{1+y}O_z$ with $0.1 < x < 0.6$ and $0 < y < x/2$, suggesting stability for excess Cu not Cu deficiency. The origin of these discrepancies is unclear. All researchers agree, however, that superconductivity is only observed in samples near the Sr-rich end of the solid solution. Furthermore, most groups report compositions for superconducting samples that are outside of some, if not all, of the solid solution ranges that have been reported. The superconducting fraction (as determined by Meissner measurements) is typically quite small which raises questions about the true nature and thermodynamic stability of the superconducting phase in the Bi-Sr-Cu-O system.

3.2 Cu-O Double Layers [2212], T_c=80 K

The idealized formula $Bi_2Sr_2CaCu_2O_{8\pm\delta}$ represents the next step in the "homologous series" of Bi-Sr-Ca-Cu-O superconductors. This phase has been reported to have a superconducting transition between 80 and 95 K. Because of the added complexity of the fourth element, Ca, no complete equilibrium phase diagram for the Bi-Sr-Ca-Cu-O system has been published. A myriad of partial

Bismuth- and Thallium-Based Cuprate Superconductors   267

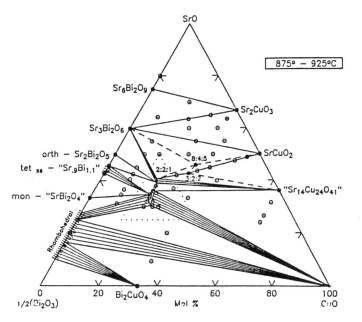

Figure 4: Equilibrium phase diagram for the Bi-Sr-Cu-O system in air at 875-925°C (30).

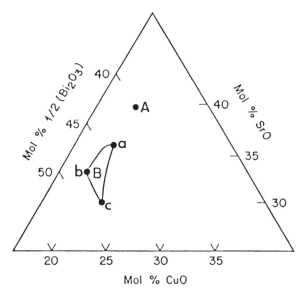

Figure 5: Portion of the Bi-Sr-Cu-O phase diagram near $Bi_2Sr_2CuO_6$ (30).

phase diagrams and synthetic techniques for the preparation of this superconducting phase have been reported.

The synthesis of the n=2 phase is similar to that described above for the n=1 system. In general, stoichiometric portions of oxides and carbonates (or nitrates) are ground and reacted in a high density alumina crucible at temperatures between 800 and 820°C. The product is crushed and pressed into a pellet for further heat treatments at 840-870°C. The firing temperature must be raised incrementally to prevent melting of the sample.

Initial studies reported the existence of a solid solution $Bi_2Sr_{2-x}Ca_{1+x}Cu_2O_{8\pm\delta}$ with $0.25 \leq x \leq 1.0$ (36)(37). These limits do not include the ideal formula $Bi_2Sr_2CaCu_2O_{8\pm\delta}$ as was noted early on (38). Recent work by Ono (39) indicates that the 80 K phase can be synthesized over a wider range of compositions described by the formula $Bi_2Bi_x(Sr_{2-y}Ca_{1+y})_{1-x}Cu_2O_y$ for $0.0 \leq x \leq 0.2$ and $0.0 \leq y \leq 0.5$ (Figure 6). Thus, this phase can be synthesized at a Ca/Sr ratio of 0.5 in the compound $Bi_{2.1}Sr_{1.93}Ca_{0.97}Cu_2O_y$. These results are consistent with the work in the n=1 system which suggests that Bi on the Sr site stabilizes the superconducting phase.

Alternate approaches to the standard synthetic methods described above have also been reported. Sastry et al. (40) report the synthesis of $Bi_2Sr_2CaCu_2O_{8\pm\delta}$ from a prereacted ceramic of "$Sr_2CaCu_2O_y$" and Bi_2O_3. The carbonates or oxides of Sr, Ca, and Cu are first reacted in appropriate ratios at 950°C to form a mixed-phase material. This is then combined with a stoichiometric amount of Bi_2O_3, thoroughly mixed, heated to 927°C for 3-5 minutes, and cooled to room temperature. The resulting black powder is ground well, pressed into a pellet, and heated again for 3 minutes at 927°C. A final heat treatment at 852°C for 2 hours with furnace-cooling resulted in a single-phase sample.

A similar route to $Bi_2Sr_2CaCu_2O_{8\pm\delta}$ developed by Beltran et al. (41) involves the reaction of Bi_2CuO_4, $CaCO_3$, $SrCO_3$, and CuO in appropriate ratios. These materials are well mixed and heated to 860°C in an alumina boat under flowing O_2 for 16 hours. The reacted powder is ground, compressed into pellets, and sintered at 880°C for 20 hours and furnace-cooled to room temperature. This technique reportedly also led to single-phase samples.

An alternate technique which eliminates the physical mixing

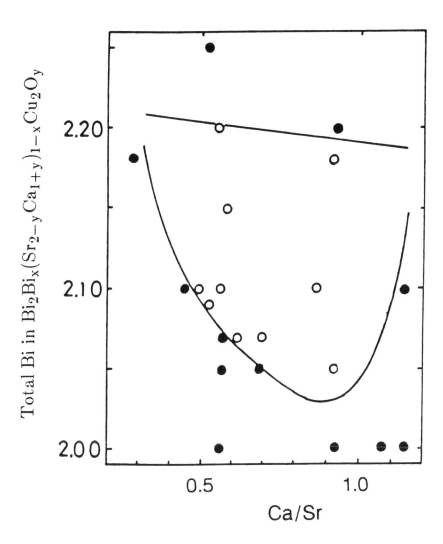

Figure 6: Single-phase region (open circles) for the 80 K superconductor in the Bi-Sr-Ca-Cu-O system (39).

necessary in the above reactions is the chemical mixing of salt solutions. One method involves the spray drying of nitrate solutions (36). Solutions of known molarity are prepared by dissolving the nitrates in dilute nitric acid (Bi, Ca, and Cu) or water (Sr). The desired compositions are mixed by volume and then sprayed through a quartz tube lining a furnace. The nitrates decompose on the walls of the quartz tube. This technique allows for the synthesis of a wide range of compositions in a relatively short time.

The nitrates can also be precipitated from solution in the form of the citrate salts (42). This also leads to intimate mixing and eliminates the need for a spray drying apparatus. For example, 8 mmol $Bi(NO_3)_3 \cdot 5H_2O$, 4 mmol $Ca(NO_3)_2 \cdot 4H_2O$, 8 mmol $Sr(NO_3)_2$, 8 mmol $Cu(NO_3)_2 \cdot 3H_2O$, and 4.62 g of citric acid are dissolved in 40 ml distilled water. The pH of the solution is adjusted to 6.65 by the addition of concentrated NH_4OH solution and 3.5 ml ethylene glycol is added. This solution is stirred and heated to 100-120°C for 1 hour to form a viscous mixture which is then pyrolyzed at 350°C for 1/2 hour to decompose the organic constituents and inorganic salts. At these temperatures, a vigorous exothermic reaction takes place and a spontaneous combustion is initiated by the nitrate salts. [**CAUTION: This exothermic reaction can be explosive if not carefully controlled. Appropriate precautions should be in place before this reaction is attempted.**]

The residue consists of homogeneous flakes. This residue is precalcined for 10 minutes at 500°C, ground, and then calcined for 8 hours at 800°C. This powder is then pressed into a pellet and sintered at 840°C for 4 hours. This method produces highly crystalline samples in spite of relatively low reaction temperatures and short reaction times.

3.3 Triple Cu-O Layers [2223], T_c=110 K

A phase in the Bi-Sr-Ca-Cu-O system with a superconducting transition temperature near 110 K was apparent in many early mixed-phase samples (2)(38). Superconducting onsets were often near 110 K although zero resistance was rarely achieved above 85 K. While chemical analysis, TEM, EDX, and single-crystal X-ray diffraction suggested a composition $Bi_2Sr_2Ca_2Cu_3$

$O_{10\pm\delta}$ for the high-T_c component, attempts to synthesize a single-phase ceramic at this composition were unsuccessful. This led to further studies on the dependence of the fraction of the 110 K phase on starting composition, reaction temperatures, and annealing conditions. Several studies have reported the enhancement of this phase by heating samples very near the melting point (875-885°C) (2)(38). This method has proven successful in samples with nominal compositions $Bi_{2.5}Sr_2Ca_2Cu_4O_x$ (43) and $Bi_4Sr_3Ca_3Cu_4O_x$ (38). Still other work has suggested that low temperature anneals improve the fraction of the 110 K phase (44). To date, samples containing a 110 K superconductor in the Bi-Sr-Ca-Cu-O system are multiphase.

It was recognized early on that small amounts of Pb enhanced the formation of the 110 K phase (45). This has led to significant efforts to synthesize a Pb-stabilized 110 K superconductor. Partial phase diagram studies in air (46) and in reduced partial pressures of O_2 (47) have been reported. The work of Sasakura et al. (46) reports the partial phase diagram in air for samples prepared from Bi_2O_3, PbO, $CaCO_3$, and CuO. Powders in the appropriate ratios were mixed and heated at 795°C for 15 hours, reground, and pressed into pellets. These pellets were sintered in air at 858°C for 90 hours and reground. New pellets were pressed, sintered a second time at 858°C for 65 hours, then cooled at 1-2.5°C/min to room temperature. Samples of nominal stoichiometry $Bi_{1.68}Pb_vSr_xCa_yCu_zO_w$ with v = 0.28 or 0.32 were prepared. Single-phase samples are formed over a fairly wide range of compositions as is depicted in Figure 7. For $Bi_{1.68}Pb_{0.32}Sr_{1.73}Ca_yCu_zO_w$ single-phase samples are formed for $1.75 \leq y \leq 1.85$ and $2.65 \leq z \leq 2.85$. These samples show a transition to zero resistance above 105 K and exhibit a substantial Meissner fraction.

Synthesis from the nitrates at reduced partial pressure (1/13 atm O_2) leads to a slightly different phase diagram (47). Samples were prepared by dissolving Bi_2O_3, PbO, $Sr(NO_3)_2$, $Ca(NO_3)_2 \cdot 4H_2O$, and CuO in nitric acid to insure intimate mixing. This solution is stirred and heated until dry, resulting in a light blue solid. This material was decomposed at 800°C for 30 minutes. The resulting powder was pressed into a pellet and heated at 828-843°C for 36-130 hours in 1/13 atm O_2 pressure, cooled slowly to 750°C, and then cooled in pure O_2 to room temperature.

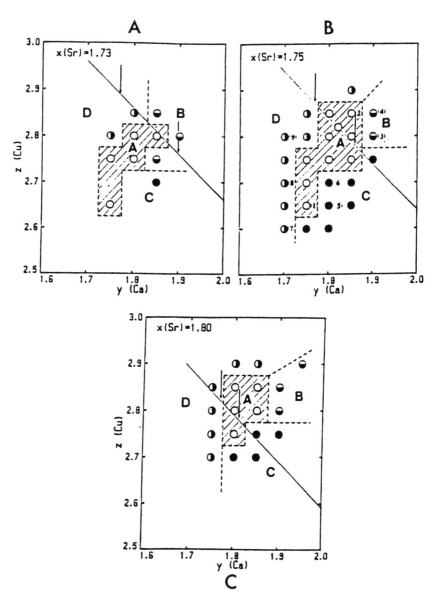

Figure 7: Composition diagrams for samples of nominal composition $Bi_{1.68}Pb_vSr_xCa_yCu_zO_w$. In (A), v=0.28 and x=1.73, for (B) v=0.32 and x=1.75, and for (C) v=0.32 and x=1.80. The dashed area (marked A) indicates where single-phase 110 K superconductors are prepared (46). Regions marked B, C, and D contain 110 K phase + Ca_2CuO_3, 110 K + 80 K + Ca_2CuO_3, and 110 K + 80 K phase, respectively.

Samples prepared with the general stoichiometry $Bi_{1.84}Pb_{0.34}Sr_xCa_y Cu_zO_w$ were found to be single-phase in the region $1.87 \leq x \leq 2.05$, $1.95 \leq y \leq 2.1$, and $3.05 \leq z \leq 3.2$ (Figure 8). Again, zero resistance at T > 105 K was exhibited within this range of compositions.

It should be clear from the above sections that the idealized formula $Bi_2Sr_2Ca_{n-1}Cu_nO_{2n+4}$ with n = 1, 2, and 3 does not accurately describe the actual stoichiometry of single-phase ceramic samples. Substantial cation intersite substitution and/or vacancies probably occur in these materials. The role of these defects in the stabilization of the superconducting phases and their contribution to the superconducting properties is still being investigated.

4.0 SYNTHESIS OF THALLIUM-BASED CUPRATE SUPERCONDUCTORS

4.1 Preparations in Air or Lidded Containers

The earliest syntheses of the 2212 and 2223 thallium-based cuprate superconductors were carried out by reacting stoichiometric amounts of Tl_2O_3, CaO, and a pre-reacted (920°C, 2 h) mixture of $BaCO_3$ and CuO (nominal composition $BaCu_3O_4$ or $Ba_2Cu_3O_5$) (3)-(5). The reaction mixtures were pressed into pellets, placed in hot furnaces at 880-950°C, heated for 2-5 min under flowing oxygen, and air-quenched. Since sealed containers were not used, product stoichiometries were not well controlled and the samples tended to be mixtures of the 2212 and 2223 phases with broad superconducting transitions. This group also reported a vapor-solid process that also produced mixtures of the 2212 and 2223 phases with broad transitions above 100 K (48): Pellets of different pre-reacted (925-950°C, 48 h) Ba-Ca-Cu-O mixtures were supported over platinum boats containing Tl_2O_3. This combination was then placed into a hot furnace at 900-925°C and heated for 3 min in flowing oxygen followed by furnace-cooling.

In another procedure (49), the reactants (Tl_2O_3, $Ba(NO_3)_2$, CaO, and CuO) were heated in two steps, the first in a lidded Pt crucible (1000°C, 15-25 min), and the second (as pellets wrapped in gold foil) in a sealed silica tube (865-905°C, 3-10 h) that had been flushed with oxygen. Products were multiphase, but a

274 Chemistry of Superconductor Materials

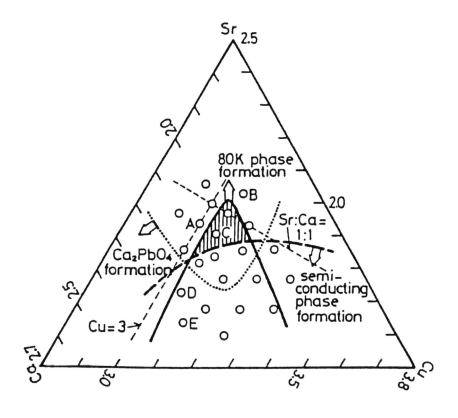

Figure 8: Partial phase diagram for phases of composition $Bi_{1.84}Pb_{0.34}Sr_xCa_yCu_zO_w$. The dashed area represents the single-phase region for samples of 110 K (n=3) material. The samples were prepared in 1/13 atmosphere O_2 (47).

starting composition of $Tl_{0.5}Ba_2Ca_2Cu_3O_y$ yielded (apparently) sufficient 2223 phase to give a superconducting onset temperature of 123 K with zero resistance at 121 K.

Millimeter-sized crystals of several of the thallium-based superconductors have been grown using a "pseudo-flux growth" method in Pt crucibles with tight-fitting lids (24)(50). Dry powder-mixtures of Tl_2O_3, BaO, CaO, and CuO of various starting metal ratios were placed in hot furnaces under flowing oxygen at 950°C, soaked 1 h, cooled to 700°C in 12.5 h and then to room temperature over 5-6 h; soak temperatures below 900°C did not produce crystals. In successful experiments, breakage of the reaction mass revealed "melt" pockets containing well-formed platelets 1-3 mm across and 0.01-0.20 mm thick. A starting Tl:Ba:Ca:Cu ratio of 4:3:1:5 yielded predominantly 2223 crystals; in other experiments, starting metal ratios of 1:1:1:2, 2:2:2:3, 2:2:1:2, 1:1:2:2, 2:5:1:2, 2:1:3:2, and 2:1:1:2 all yielded mixtures of crystals, some intergrown, of the 2212, 2223, and 1212 superconductors. In general, the T_c values for the crystals were somewhat lower than those observed in ceramic phases, an effect attributed to microscopic inhomogeneity and cation site disorder/vacancy formation.

All of these procedures involve heating Tl_2O_3 in non-sealed systems, and all are typified by superconducting product stoichiometries far different from the starting compositions. Again, in addition to safety concerns, little control of superconducting-phase composition and reproducibility of synthetic conditions is afforded by use of non-hermetically sealed reaction containers. The problem appears to be more complex than simply thallium reactant loss; factors related to reaction kinetics are most likely quite important for the preparation of these metastable phases.

4.2 Preparations in Hermetically Sealed Containers

Other researchers attempted improved control of product stoichiometry by wrapping reactant-mixture pellets (Tl_2O_3, BaO, CaO, and CuO) in gold foil and then sealing them in silica ampoules that had been flushed with oxygen (8)(51). Multiphase samples, however, were still obtained. Nominal starting metal compositions of 1:1:3:3 and 2:2:2:3 (Tl:Ba:Ca:Cu) resulted in the largest amounts of 2223 and 2212 phases, respectively; the samples were heated at 880°C for 3 h, followed by furnace-cooling over a 4 h period (51). Sufficient amounts of the 2212 and 2223 phases were formed to confirm bulk superconductivity at 108 and 120 K, respectively, and to elucidate the basic structural features and unit cells by combining electron and X-ray powder diffraction methods. Using the same synthetic procedure with different starting compositions, this group also obtained multiphase samples containing the thallium monolayer 1201, 1212, and 1223 compounds (8); the basic structural features of the series were again elucidated by combining electron microscopy and X-ray powder diffraction methods.

Syntheses of X-ray-diffraction pure bulk samples of the 1223 (9), 2223 (52), and 1212 (53) phases with T_c values of 120, 125, and 50-65 K, respectively, were reported by another group. In all cases, intimate reaction mixtures of stoichiometric quantities of Tl_2O_3, BaO_2, CaO, and CuO were pressed into pellets that were supported in alumina crucibles which were then sealed in evacuated silica ampoules. Heating rates were not given; samples were cooled slowly to room temperature. The products, heating temperatures, and soak times follow: 1223: 750-830°C, 5 h; 2223: 750-900°C, 3-12 h; 1212: 780-820°C, 5-6 h. For the 1212 sample, a final anneal under Ar raised T_c from 50 to 65 K. Sample microstructure was studied by electron microscopy, and the basic structural features and unit cells of the phases were elucidated by combining electron microscopy and X-ray powder diffraction methods. Using the same synthetic method, this group also prepared a number of phases with various degrees of cation substitution on the thallium and alkaline earth sites (10)(20)(23)(54).

Considerable single-crystal X-ray diffraction and powder neutron diffraction structural studies have been reported on samples prepared in sealed gold tubes (6)(7)(25)(55)-(58). Fifteen-inch long pieces of 5/8-inch-diameter pure gold tubing are first cleaned thoroughly with soap and rinsed with acetone. The open tubes are supported on a ceramic tile or brick and, using a torch with a moderate flame (somewhat cooler than that required for working pure silica), are preheated by passing the torch over the length in one direction only. The "annealed" tubing is then cut into 4-5 inch lengths with metal shears, and one end is crimped very flat with flat pliers. The crimp should be inspected using an optical microscope; it should be very flat with no folding or holes. The crimp is then melted shut by passing a hot flame over it in one direction only; the torch should be fitted with a fairly small nozzle and the flame temperature should be high enough to work silica. The tubes are then loaded with reactant mixture and the open end is crimped flat in a like manner. The fused end of the tube is placed in a beaker of water while the final crimp is melted shut with the torch flame. The tubing is not reused; the method reportedly works well with no leakage problems (59).

The gold tube containment procedure described above has been used in the following experiments to prepare a number of the thallium-based superconducting cuprates in single-crystal and pure or near-pure (according to X-ray powder diffraction) polycrystalline form from mixtures of Tl_2O_3, BaO_2, CaO_2, and CuO.*

1. In numerous initial experiments with various proportions of reactants heated at 850–915°C for 15 min to 3 h, most of the products were mixtures containing the 2212 and 2223 phases with superconductivity onset temperatures ranging from 100 to 127 K (6).

2. Essentially single-phase 2223 (T_c = 125 K with zero resistance at 122 K) was obtained by heating stoichiometric quantities of the reactants at 890°C for 1 h (6)(56); if this mixture was heated for more than 2 h or at temperatures above 900°C, a mixture resulted with 2212 as the major phase and 2223 as a minor component (56). The authors note that the temperature range in which the thallium-based superconductors can be formed without decomposition to barium cuprate phases is very narrow (6).

---

* Prior to use, BaO_2 and CaO_2 reagents should be analyzed thermogravimetrically to determine actual weight per cent metal.

3. Single crystals of the 2212 phase were grown from a 2:1:1:3 (Tl:Ba:Ca:Cu) copper-rich melt that was heated to 900-920°C, held 1 h, and cooled at 2°C/min (T_c = 110 K, narrow transition) (6).

4. Single crystals of the 2223 phase were grown from a 2:2:3:4 (Tl:Ba:Ca:Cu) starting composition that was heated at 920°C for 3 h and cooled to 300°C at 5°C/min (56); according to magnetic measurements the T_c values varied from 127 to 116 K in different preparations, with 125 K as a typically observed onset temperature.

5. Polycrystalline 2201 phase formed as the major and sole superconducting component when stoichiometric amounts of reactants were heated at 875°C for several hours (7); transitions were broad with T_c onsets of 84 K. Reheating at 900°C caused partial melting and an increase in the T_c onset value to 90 K (with zero resistance at 83 K); increased amounts of second-phase $BaCuO_2$ were observed.

6. Single crystals of the 2201 phase were grown from a starting composition 2:2:2 (Tl:Ba:Cu) (7); presumably the mixture was heated to 900°C, soaked several h, and slow-cooled. Flux exclusion indicated a T_c onset of 90 K with zero resistance at 83 K.

7. Single crystals of the 1223 thallium monolayer phase were grown from a copper-rich melt with molar composition 1:2:2:4 (Tl:Ba:Ca:Cu); the mixture was heated to 925°C, soaked 6 h, and cooled at 1°C/min (57). Magnetic flux exclusion experiments indicated a sharp T_c onset of 110 K.

8. A general summary of temperatures and heat-soak times that are favorable for the formation

of the different thallium bilayer superconductors follows (59): 2201: 875°C, 3 h; 2212: 900°C, 6 h; 2223: 890°C, 1 h; 2234: 860°C, 3 h; 2245: 880°C, 6 h - the 2234 and 2245 phases were observed by electron diffraction only.

The lead-substituted compounds, $Tl_{0.5}Pb_{0.5}Sr_2Ca_2Cu_2O_7$ and $Tl_{0.5}Pb_{0.5}Sr_2Ca_2Cu_3O_9$ were also synthesized in sealed gold tubing. Single-phase polycrystalline samples with somewhat broad superconducting transitions (T_c = 80-90 and 120 K, respectively) were obtained by reacting Tl_2O_3, PbO_2, CaO_2, SrO_2, and CuO at 850-915°C for 3-12 h (25). Single crystals of $Tl_{0.5}Pb_{0.5}Sr_2Ca_2Cu_3O_9$ were grown from a mixture of the above reactants in a molar ratio 1:1:2:3:4 (Tl:Pb:Sr:Ca:Cu) that was heated at 910°C for 6 h and then cooled at 2°C/min (25); flux exclusion measurements indicated a sharp superconducting transition at 118 K.

4.3 Preparations Under Flowing Oxygen

Bulk samples of the 2201, 2212, and 2223 thallium-based cuprates have been prepared by reacting Tl_2O_3, $BaCuO_2$, dry CaO, and CuO in stoichiometric amounts (60). The mixed reactants were pelletized, wrapped in gold foil, and heated under flowing oxygen at 890°C for 5 min for 2201, 905°C for 7 min for 2212, and 910°C for 7 min for 2223, followed by cooling at 10°C/min. The products were single-phase according to X-ray powder diffraction and the Rietveld profile refinement technique was used to determine the structural parameters for the three phases. Interestingly, the 2201 phase was not superconducting; zero-resistance was observed at 98 and 114 K for the 2212 and 2223 samples, respectively. This group also reported the preparation of bulk 2234 phase (26) using the same synthetic procedure described above; the starting composition was off-stoichiometry - 1:2:2:3 (Tl:Ba:Ca:Cu) - and was heated at 895°C for 50 min, followed by slow cooling. The product was a mixture of 2234, $BaCO_3$, $BaCuO_2$, and CaO. The T_c onset was 112 K with zero resistance at 108 K. The unit cell and basic structural features of the superconducting phase were determined by combining electron microscopy and X-ray powder diffraction methods.

Mixtures containing as major components the thallium monolayer phases 1212, 1223, 1234, and 1245 have been prepared by heating the thallium bilayer 2212 and 2223 phases at 890°C in oxygen for various times ranging from 4 to 10 h (22). Sufficient quantities of each of the four superconducting phases were obtained in separate products so that reasonably sharp superconducting transitions were observed. T_c values for the phases were determined by resistivity and susceptibility measurements, and the basic structural properties were elucidated using X-ray powder diffraction and high resolution electron microscopy.

Syntheses of near-single phases of the lead-substituted thallium monolayer phases with up to 6 Cu-O layers; i.e., Pb-doped 1212, 1223, 1234, 1245, and 1256, have been recently reported (21). Reactant mixtures of various proportions of Tl_2O_3, PbO, CaO, BaO_2, and CuO were pelletized, wrapped in gold foil, and sintered at 860-900°C under flowing oxygen for 10-30 h. The T_c value reached a maximum of 121 K for the 1234 compound and declined with further increase in the number of Cu-O layers. X-ray powder diffraction data for the different phases were refined using the Rietveld method and a consistent increase in the c-axis accompanied the increase in number of Cu-O layers.

The reports described above indicate that the thallium-containing cuprates can be prepared with reasonable degrees of purity under non-contained, flowing oxygen conditions. This somewhat surprising result is likely attributable to the relatively short heating times, the use of gold foil wrappings, and, possibly most importantly, a decrease in thallium volatility according to the principle of Le Chatelier (27). If the reactions are carried out in nonporous sleeves with exiting gases passed through multiple traps filled with basic peroxide solution, the method can be considered as adequately safe. (Furnaces should be placed in fume hoods).

5.0 CONCLUSION

A review of the synthetic methods used to prepare the bismuth and thallium families of cuprate superconductors has been presented. An overview of our current knowledge of phase relationships in the bismuth systems is also given; such studies of the thallium-containing compounds are yet to be reported. The importance of and the difficulties posed by the preparation of phase-pure samples with reproducible structures and chemical compositions cannot be understated; the elucidation of structure-property relationships that correlate with the appearance of superconductivity begins with sample integrity. The synthetic challenge is particularly acute in the case of the superconducting oxides because of their apparently inherent "unstable" nature, a situation reminiscent of the difficult syntheses of metal-insulator type metal oxides. Neophytes are urged to be patient, diligent, and well-equipped with knowledge of the literature.

While this chapter has focussed on synthesis of polycrystalline samples, other aspects of the bismuth and thallium cuprate superconductors are discussed elsewhere. An introduction to synthesis and crystal growth is given in Chapter 5, and a review of the crystal chemistry of the two families is presented in Chapter 2. The crystallographic data for these phases, including tables of calculated d-spacings and intensities for X-ray powder diffraction patterns, have been collected in Chapter 13. Characterization by electron microscopy, a particularly important technique because of the nature of the materials, is reviewed in Chapters 14 and 15.

6.0 REFERENCES

1. C. Michel, H. Hervieu, M.M. Borel, A. Grandin, F. Deslendes, J. Provost, and B. Raveau, *Z. Phys.* B68:421 (1987).

2. H. Maeda, Y. Tanaka, M. Fukutomi, and T. Assano, *Jpn. J. Appl. Phys.* 27:L209 (1988).

3. Z.Z. Sheng and A.M. Hermann, *Nature* 332:138 (1988).

4. R.M. Hazen, L.W. Finger, R.J. Angel, C.T. Prewitt, N.L. Ross, C.G. Hadidiacos, P.J. Heaney, D.R. Veblen, Z.Z. Sheng, A. El Ali, and A.M. Hermann, *Phys. Rev. Lett.* 60:1657 (1988).

5. L. Gao, Z.J. Huang, R.L. Meng, P.H. Hor, J. Bechtold, Y.Y. Sun, C.W. Chu, Z.Z. Sheng, and A.M. Hermann, *Nature* 332:623 (1988).

6. M.A. Subramanian, J.C. Calabrese, C.C. Torardi, J. Gopalakrishnan, T.R. Askew, R.B. Flippen, K.J. Morrissey, U. Chowdry, and A.W. Sleight, *Nature* 332:420 (1988).

7. C.C. Torardi, M.A. Subramanian, J.C. Calabrese, J. Gopalakrishnan, E.M. McCarron, K.J. Morrisey, T.R. Askew, R.B. Flippen, U. Chowdry, and A.W. Sleight, *Phys. Rev. B* 38:225 (1988).

8. S.S.P. Parkin, V.Y. Lee, A.I. Nazzal, R. Savoy, R. Beyers, S.J. La Placa, *Phys. Rev. Lett.* 61:750 (1988).

9. C. Martin, C. Michel, A. Maignan, M. Hervieu, and B. Raveau, *C. R. Acad. Sci. Paris* 307:27 (1988).

10. D. Bourgault, C. Martin, C. Michel, M. Hervieu, J. Provost, and B. Raveau, *J. Solid State Chem.* 78:326 (1989) and references therein.

11. C.N.R. Rao and B. Raveau, *Acc. Chem. Res.* 22:106 (1989) and references therein.

12. B. Raveau, C. Michel, M. Hervieu, and J. Provost, *Rev. Solid State Sci.* 2:115 (1988).

13. A.W. Sleight, *Science* 242:1519 (1988).

14. B. Raveau, C. Michel, M. Hervieu, J. Provost, and F. Studer in *Earlier and Recent Aspects of Superconductivity*; J.G. Bednorz and K.A. Mueller, Eds.; Springer-Verlag (1989).

15. D.P. Matheis and R.L. Snyder, *Powder Diffraction* in press.

16. B. Morosin, E.L. Venturini, D.S. Ginley, B.C. Bunker, E.B. Stechel, K.F. McCarty, J.A. Voight, J.F. Kwak, J.E. Schirber, D. Emin, R.J. Baughman, N.D. Shinn, W.F. Hammetter, D. Boehme, and D.R. Jennison in: *High Temperature Superconducting Compounds: Processing and Related Properties*; S.H. Whang and A. DasGupta, Eds.; The Minerals, Metals, and Materials Society, 1989.

17. R.B. Beyers, S.S.P. Parkin, V.Y. Lee, A.I. Nazzal, R.J. Savoy, G.L. Gorman, T.C. Huang, and S.J. La Placa, *IBM J. Res. Develop.* 33:228 (1989).

18. M. Greenblatt, S. Li, L.E.H. McMills, and K.V. Ramanujachary in *Studies of High-Temperature Superconductors*;

A.V. Narlikar, Ed.; Nova Science Publishers: N.Y.; in press.

19. Y. Tokura, T. Arima, H. Takagi, S. Uchida, T. Ishigaki, H. Asano, R. Beyers, A.I. Nazzal, P. Lacorre, and J.B. Torrance, *Nature* (1990).

20. C. Martin, D. Bourgault, M. Hervieu, C. Michel, J. Provost, and B. Raveau, *Modern Phys. Lett. B* 3:93 (1989).

21. H. Kusuhara, T. Kotani, H. Takei, and K. Tada, *Jpn. J. Appl. Phys.* 28:L1772 (1989).

22. S. Nakajima, M. Kikuchi, Y. Syono, T. Oku, D. Shindo, K. Hiraga, N. Kobayashi, H. Iwasaki, and Y. Muto, *Physcia C* 158:471 (1989).

23. C. Martin, D. Bourgault, C. Michel, J. Provost, M. Hervieu, and B. Raveau, *Eur. J. Solid State Inorg. Chem.* 26:1 (1989).

24. B. Morosin, D.S. Ginley, P.F. Hlava, M.J. Carr, R.J. Baughman, J.E. Schirber, E.L. Venturini, and J.F. Kwak, *Physica C* 152:413 (1988).

25. M.A. Subramanian, C.C. Torardi, J. Gopalakrishnan, P.L. Gai, J.C. Calabrese, T.R. Askew, R.B. Flippen, and A.W. Sleight, *Science* 242:249 (1988).

26. M. Kikuchi, S. Nakajima, Y. Syono, K. Hiraga, T. Oku, D. Shindo, N. Kobayashi, H. Iwasaki, and Y. Muto, *Physica C* 158:79 (1989).

27. D. Cubicciotti and F.J. Keneshea, *J. Phys. Chem.* 71:808 (1967).

28. *CRC Handbook of Chemistry and Physics*, 66th Edition; R. C. Weast, ed.; CRC Press, Inc.: Boca Raton, FL (1985) pp. B-38, B-151.

29. *The Merck Index*, 9th Edition; M. Windholz, ed.; Merck and Co., Inc.: Rahaway, NJ (1976) pp. 1194, MISC-31 and references therein.

30. R.S. Roth, C.J. Rawn, B.P. Burton, and F. Beech, *J. Res. NIST*, to be published.

31. B.C. Chakoumakos, P.S. Ebey, B.C. Sales, and E. Sonder, *J. Mater. Res.* 4:767 (1989).

32. M.T. Casais, C. Cascales, A. Castro, M. de Pedro, I. Rasines, G. Domarco, I. Maza, F. Miguelez, J. Ponte, C. Torron, J.A. Veira, F. Vidal, and J.A. Campa, *Proc. Eur. Mater. Res. Soc.* Strasbourg, France (1988).

33. J.A. Saggio, K. Sujata, J. Hahn, S.J. Hwu, K.R. Poeppelmeier, and T.O. Mason, *J. Am. Ceram. Soc.* 72:849 (1989).

34. Y. Ikeda, H. Ito, S. Shimomura, Y. Oue, K. Inabe, Z. Hiroi, and M. Takano, *Physica C* 159:93 (1989).

35. R.S. Roth, C.J. Rawn, and L.A. Bendersky, *J. Mater. Res.* 5:46 (1990).

36. G.S. Grader, E.M. Gyorgy, P.K. Gallagher, H.M. O'Bryan, D.W. Johnson, Jr., S. Sunshine, S.M. Zahurak, S. Jin, and R.C. Sherwood, *Phys. Rev.* B 38:757 (1988).

37. J.M. Tarascon, Y. LePage, P. Barboux, B.G. Bagley, L.H. Greene, W.R. McKinnon, G.W. Hull, M. Giroud, and D.M. Hwang, *Phys. Rev.* B 37:9382 (1988).

38. J.M. Tarascon, W.R. McKinnon, P. Barboux, D.M. Hwang, B.G. Bagley, L.H. Greene, G.W. Hull, Y. LePage, N. Stoffel, and M. Giroud, *Phys. Rev.* B 38:8885 (1988).

39. A. Ono, *Jpn. J. Appl. Phys.* 28:L1372 (1989).

40. P.V.P.S.S. Sastry, I.K. Gopalakrishnan, A. Sequeira, H. Rajagopal, K. Gangadharan, G.M. Phatak, and R.M. Iyer, *Physica C* 156:230 (1988).

41. D. Beltran, M.T. Caldes, R. Ibanez, E. Martinez, E. Escriva, A. Beltran, A. Segura, V. Munoz, and J. Martinez, *J. Less-Common Metals* 150:247 (1989).

42. N.H. Wang, C.M. Wang, H.C.I. Kao, D.C. Ling, H.C. Ku, and K.H. Lii, *Jpn. J. Appl. Phys.* 28:L1505 (1989).

43. T. Kajitani, M. Hirabayashi, M. Kikuchi, K. Kusaba, Y. Syono, N. Kobayashi, J. Iwasaki, and Y. Muto, *Jpn. J. Appl. Phys.* 27:L1453 (1988).

44. I. Iguchi and A. Sugishita, *Physica C* 152:228 (1988).

45. S.A. Sunshine, T. Siegrist, L.F. Schneemeyer, D.W. Murphy, R.J. Cava, B. Batlogg, R.B. van Dover, R.M. Fleming, S.H. Glarum, S. Nakahara, R. Farrow, J.J. Krajewski, S.M.

Zahurak, J.V. Waszczak, J.H. Marshall, P. Marsh, L.W. Rupp, Jr., and W.F. Peck, *Phys. Rev. B* 38:893 (1988).

46. H. Sasakura, S. Minamigawa, K. Nakahigashi, M. Kogachi, S. Nokanishi, N. Fukuoka, M. Yoshikawa, S. Noguchi, K. Okuda, and A. Yanase, *Jpn. J. Appl. Phys.* 28:L1163 (1989).

47. S. Koyama, U. Endo, and T. Kawai, *Jpn. J. Appl. Phys.* 27:L1861 (1988).

48. Z.Z. Sheng, L. Sheng, H.M. Su, and A.M. Hermann, *Appl. Phys. Lett.* 53:2686 (1988).

49. M. Eibschutz, L.G. Van Uitert, G.S. Grader, E.M. Gyorgy, S.H. Glarum, W.H. Grodkiewicz, T.R. Kyle, A.E. White, K.T. Short, and G.J. Zydzik, *Appl. Phys. Lett.* 53:911 (1988).

50. D.S. Ginley, B. Morosin, R.J. Baughman, E.L. Venturini, J.E. Schirber, and J.F. Kwak, *J. Crystal Growth* 91:456 (1988).

51. S.S.P. Parkin, V.Y. Lee, E.M. Engler, A.I. Nazzal, T.C. Huang, G. Gorman, R. Savoy, and R. Beyers, *Phys. Rev. Lett.* 60:2539 (1988).

52. M. Hervieu, C. Michel, A. Maignan, C. Martin, and B. Raveau, *J. Solid State Chem.* 74:428 (1988).

53. M. Hervieu, A. Maignan, C. Martin, C. Michel, J. Provost, and B. Raveau, *J. Solid State Chem.* 75:212 (1988).

54. D. Bourgault, C. Martin, C. Michel, M. Hervieu, and B. Raveau, *Physica C* 158:511 (1989).

55. D.E. Cox, C.C. Torardi, M.A. Subramanian, J. Gopalakrishnan, and A.W. Sleight, *Phys. Rev. B.* 38:6624 (1988).

56. C.C. Torardi, M.A. Subramanian, J.C. Calabrese, J. Gopalakrishnan, K.J. Morrissey, T.R. Askew, R.B. Flippen, U. Chowdry, and A.W. Sleight, *Science* 240:631 (1988).

57. M.A. Subramanian, J.B. Parise, J.C. Calabrese, C.C. Torardi, J. Gopalakrishnan, and A.W. Sleight, *J. Solid State Chem.* 77:192 (1988).

58. J.B. Parise, J. Gopalakrishnan, M.A. Subramanian, and

A.W. Sleight, *J. Solid State Chem.* 76:432 (1988).

59. M.A. Subramanian, private communication.

60. M. Kikuchi, T. Kajitani, T. Suzuki, S. Nakajima, K. Hiraga, N. Kobayashi, H. Iwasaki, Y. Syono, and Y. Muto, *Jpn. J. Appl. Phys.* 28:L382 (1989).

7

Synthesis of Superconductors Through Solution Techniques

Phillipe Barboux

1.0 INTRODUCTION

1.1 Ceramic Processing

Although the superconducting cuprates have high critical temperatures, their other superconducting properties such as critical currents and flux expulsion remain quite poor even after the large amount of research that has been made in this field (1).

It has been shown that such behavior can be related to the superconducting coherence length that is anisotropic and extraordinarily short, a few angstroms, much smaller than the characteristic thickness of any inhomogeneity such as a grain boundary in ceramics (2). As a result the critical currents are still much lower than required by any technological application using ceramics. Their large decrease in a magnetic field is characteristic of a weak link behavior that relates to the poor connections between grains (3). Thus, superconducting properties are very sensitive to the microstructure of ceramics. It has been already demonstrated that any texturing of the ceramics, such as grain alignment, that increases the grain boundary area may have a drastic effect on the superconducting properties (4)(5). Also, inhomogeneities in the ceramic resulting from too high a synthesis temperature or from an incomplete solid state reaction may be related to the occurrence of low critical currents

To date, conventional processing has failed to produce

ceramics of sufficient homogeneity, leading to best values of the critical current of the order of 10^2 A/cm^2 at 77K (6). More sophisticated synthesis techniques such as the so-called "Melt Textured Growth" process, where grains are aligned during cooling from the melt by the use of a thermal gradient, have achieved much better critical currents, one order of magnitude higher (4). But such processes have yet to be adapted to large scale production of materials.

Better control of the chemistry during synthesis should enhance resulting properties of the ceramics and the preparation of these superconducting oxides via solution techniques is therefore attractive.

1.2 Interest in Solution Techniques

Sol-gel process is the generic term for wet chemistry methods that start from mixture of molecular precursors and through condensation, deposition and heat treatment processes lead to a final multicomponent system (7-9). Some of the advantages are well known such as:

- Better homogeneity achieved through an intimate mixing at the molecular level in the solution

- Lowering of the synthesis temperature because of the increased reactivity of the mixed precursors.

- The synthesis of intermediary inorganic sols or polymers in solution whose rheological properties may be attractive for the film deposition techniques such as spin coating or the spinning of fibers.

However, the growing interest during the last ten years in the area of sol-gel processes has focussed primarily on silica glasses or ceramics made of materials such as Al, Zr, Ti whose many precursors form a versatile basis for sophisticated chemistry. Yet, there have been very few reports about the chemistry of solution precursors for elements such as Y (10)(11), Bi (12) or Cu (13) and to our knowledge, before 1987, none about their use in ceramic processing. Therefore, the majority of the works that have been described to date in the

Synthesis of Superconductors Through Solution Techniques 289

field of superconductivity relate to common salts such as carboxylates, hydroxides or nitrates.

The object of this chapter will be to review all the different precursors that have been used in the literature and then to describe all the processing difficulties encountered in the different synthesis steps.

1.3 The Different Solution Techniques

The simplest technique may be coprecipitation. In this method, a reagent is added to the stock solution that is destabilized and precipitated. Better mixing at a microscopic level is then achieved without mechanical grinding and mixing. Insoluble carboxylates such as citrates, oxalates and carbonates or hydroxides are the most suitable reagents.

Better control of the precipitation by an adjustment of the solution parameters such as pH, temperature, and/or amount of reagent may slow down or hinder the particle growth. Inorganic polymers of low molecular weight instead of large solid particles may be obtained this way. This results in gels or colloidal solutions (sols), which, because of their interesting rheological properties can be used for thin film deposition or fiber spinning. Weakly soluble carboxylates such as acetates or organometallics such as metal alkoxides appear the most appropriate precursors for such methods: polymerization of acetates may be controlled by the pH and the condensation of alkoxides by the addition of water to the solution.

Another procedure is to rapidly dry the solution at a rate that does not allow the elements to segregate. As an example, aerosols can be sprayed through a furnace heated at the reaction temperature of 900°C. This yields spherical submicronic particles that are most suitable to sinterable compacts. Another way is to spray the solution onto a substrate. This is a simple method of thick film deposition. Very ionic species such as nitrates dissolved in water can be sprayed onto a hot substrate. Long-chain carboxylates, soluble in volatile organic solvents, can be deposited as well, even at room temperature.

2.0 SYNTHESIS PROCEDURES

2.1 General Principles Of Synthesis

Conventional procedure: We will focus primarily on the fabrication of the most studied material, the 90K superconductor, $YBa_2Cu_3O_7$ (the "123" phase). When starting from a mixture of powders such as CuO, Y_2O_3 and $BaCO_3$, the reaction leading to the pure superconducting perovskite requires a high temperature, extended annealings and several grindings before homogeneity is obtained. This is compared to the 650°C that is necessary to crystallize a thin film obtained by the vacuum evaporation technique. Such a difference can be related to two reasons:

- The thermodynamic stability of the 123 phase as compared to that of the starting materials. For example $BaCO_3$ is unreactive and is only slowly decomposed by reaction below 900°C.

- The very slow diffusion rate in the solid state below 850°C. This kinetic limitation can be overcome in the case of thin films where intimate mixing of the cations at the atomic level has already been obtained during the evaporation procedure. Therefore, in this case crystallization of the correct phase can be obtained directly from the amorphous mixture, around 650°C for the optimized processes (14). This temperature may be expected to be the lowest that can be reached through a solution procedure where perfect mixing of all three cations together has been achieved. Indeed, this is the synthesis temperature used by Horowitz et al. in their precipitation method using hyponitrites (16).

Lowering the synthesis temperature: There are multiple reasons for lowering the synthesis temperature. First, it appears that no ceramic with good critical currents can be achieved without texturing or alignment of the grains parallel to each other. This can be done within thin films where the thickness is less or the same order as the largest grain size. Also, interaction with the substrate causes grain alignment.

For electronic applications, as is well known, ideal substrates would be alumina, silicon or silicon oxide. The different dilatation coefficients of the substrate and of the film may be a cause for cracks and dislocations if too high a reaction temperature is used. Moreover, at temperatures above 650°C only substrates such as MgO, $SrTiO_3$ and ZrO_2 can be used (15). In addition to their lesser interest, they still react with the superconducting materials since the large number of grain boundaries present in these ceramic-films promote rapid intergranular diffusion of the impurities from the substrate to the film. Substrate contamination is then larger than for crack-free evaporated films although the films derived from solution techniques are usually one order of magnitude thicker (17).

Also, in the case of ceramics a formidable grain growth happens as soon as the temperature is increased above 900°C-950°C. Instead of the desired sintering effect, this grain growth may cause a coarsening of the intergranular junctions. Low temperature synthesis may yield particles with small grain size that readily sinter.

Problems intrinsic to small grain sizes: One problem that is specific to the solution technique is the small grain size of the resulting powders. When low synthesis temperature is used, submicron powders are obtained whose particle size is the same order of magnitude as the characteristic London penetration length. In this case, flux exclusion can become very weak. Cooper et al. (19) have demonstrated the lowering of Meissner effect for particles of decreasing sizes made from an amorphous citrate method (18). These measurements (Figure 1) allowed the determination of the London penetration depth. Its temperature dependence was found to be stronger than expected from simple BCS theory. The same authors determined the London penetration length extrapolated at 0 K to be around 0.6 μm (19).

This is consistent with the result that Horowitz et al. have found on submicron particles obtained by a hyponitrite method: they are too small to give any diamagnetic behavior. However, the heat capacity measurement shows a break at 90 K that corresponds to about 90% of superconducting phase in the powder (16).

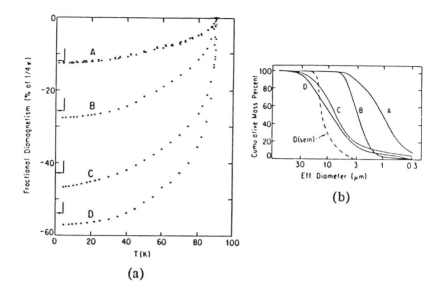

Figure 1: (a) Low field ac magnetization for samples of different particle size distribution. (b) Size distribution of the samples (After J.R. Cooper et al. (19)).

One of the advantages of the solution techniques is to obtain powders with very small grain sizes that do not, however, yield a good superconducting behavior precisely because of their particle size. Therefore, sufficient sintering has to be effected in order to eliminate the possibility of flux penetration between grains; the whole ceramic will then be diamagnetic instead of having granular superconductivity. Since this may be very difficult to achieve, it may be necessary to heat the powders at higher temperatures in order to obtain grain growth.

2.2 The Different Solution Processes

Hydrolysis or complexation: The following very schematic reactions describe the most common representation of a solution process, starting from a stable stock solution of some metal complexes to which reagents are added. The first step consists in the substitution or addition of molecular groups in the coordination sphere of a cation M. Most common reactions are (9):

$M(OR)_m + HOH \longrightarrow MOH(OR)_{m-1} + ROH$
(hydrolysis of an alkoxide)

$M(H_2O)_m^{n+} + OH^- \longrightarrow MOH(H_2O)_{m-1}^{(n-1)+} + H_2O$
(complexation with a hydroxide ligand replacing water in an aqueous solution)

$M(H_2O)_m^{n+} + L^- \longrightarrow ML(H_2O)_{m-1}^{(n-1)+} + H_2O$
(complexation by a ligand such as a carboxylate).

Condensation or precipitation: The species that are so formed may be unstable and in a second step may condense upon elimination of molecules such as ROH or H_2O, thus leading to the formation of a hydroxide or oxide network. This step corresponds to an inorganic polymerization such as (9):

$=MOR + HOM= \longrightarrow =M-O-M= + ROH$
(formation of an oxide bridge)

$$=M\begin{array}{c}OH\\OH_2\end{array} + \begin{array}{c}_2HO\\HO\end{array}M= \longrightarrow =M\begin{array}{c}OH\\HO\end{array}M= + 2\ H_2O$$

(formation of a hydroxo bridge)

This process leads to the precipitation of the oxides, hydroxides or carboxylates if we replace OH^- by L^- in the last reaction. If this polycondensation can be stopped or if it is slow enough, a colloidal solution is obtained that has rheological properties useful for application purposes.

As a matter of fact, the major difficulty in applying the solution techniques to the superconducting cuprates is the choice of precursors, since the low charges of Cu, Y, Ba and their relatively large ionic radii make it difficult to find versatile soluble precursors.

2.3 The Different Precursors

Alkoxides: Yttrium and barium alkoxides can be obtained by direct reaction of the metal with alcohols (10)(11). They readily hydrolyze in the presence of traces of water. Homogeneous precipitation with copper alkoxides such as copper methoxide or ethoxide, that are rather insoluble (13), is therefore difficult.

However, copper alkoxides with longer chains appear to be more soluble in their parent alcohol. S. Shibata et al. (20) have used the n-butoxides of Y, Ba and Cu dissolved in n-butanol and hydrolyzed with water. They obtain a precipitate of oxides that is composed of a very fine submicron powder that readily sinters starting above 250°C. However, the different reaction rates for the hydrolysis and the precipitation of the three different cations lead to cationic segregation.

Another way to dissolve the ethoxides was described by Uchikawa et al. (21) and used a mixture of water, diethylenetriamine and acetic acid, in which cations are both hydrolyzed and complexed. Upon solvent evaporation, very viscous thermoplastic gels are obtained from which fibers may be prepared. The unfired fibers contain 45% by weight inorganic materials. They yield good superconducting transitions once heat-treated.

The nonpolar copper ethoxide and methoxide may be substituted by more polar groups such as N,N-diethylethanolamine or butoxyethanol (16). The resulting alkoxides can be mixed with Y- and Ba-isopropoxides in tetrahydrofuran and hydrolyzed. Hydrolysis appears to be quite complete since the resulting precipitates are almost carbon-free. Thus, crystallization of the superconducting phase occurs at temperatures lower than 800°C.

Other attempts have used the cations dissolved in a chelating solvent such as methoxyethanol that forms alkoxide species (22)(23). When water is added to a solution of these elements in a ratio between 1 and 4 of H_2O/alkoxide, gels are obtained that are suitable for thin film deposition.

These studies are probably the first attempts in a field that may develop further in the future after discoveries of other new alkoxides.

Insoluble carboxylates: Although we should later conclude that this method leads only to ceramics having low critical currents, this has been probably the most common technique described up to now. Moreover, since it is one of the simplest approaches to the solution processes, we will review the different studies in order to outline all the difficulties that may be encountered during processing.

Coprecipitation of the metals is usually achieved from an aqueous solution of nitrates upon addition of anions such as carbonates, citrates, or oxalates (10)(24-27). First reports in this field have underlined the necessity to neutralize the pH of the solution in order to obtain complete precipitation of barium or strontium. Also, oxalate or citrate ligands may bind to two different cations. This should allow a better mixing at a microscopic level. However, care should be taken since some cations such as Y or La may precipitate as double salt complexes with alkaline ions that have been added to the solution as hydroxides in order to control the pH (24).

At low enough pH, precipitation does not occur. Citric acid diluted in water can then be used as a chelating agent for the cations while it may react with a polyalcohol such as ethyleneglycol to form a viscous polyester. The viscosity of these solutions allows thin film deposition by spin-coating or dip-coating. Interest in this method is high because it can be used as well for $YBa_2Cu_3O_7$ (28) or BiSrCaCuO (29) phases. Films having thickness on the order of a micron, sharp transition and very good orientation of the grains with the c axis perpendicular to the substrate have been obtained through this method.

Homogeneous precipitation: During the addition of the reagents, local pH or concentration gradients appear that induce inhomogeneity. To obtain highly uniform particles it may therefore be necessary to produce the reagent slowly in situ. Some homogeneous precipitation methods (30-32) have used the decomposition of urea between 90°C and 100°C in an aqueous solution of the mixed salts of Cu, Ba and Y. Carbonate and ammonia are slowly released:

$$CO(NH_2)_2 + H_2O \longrightarrow CO_2 + 2NH_3$$

The pH is thus slowly raised in the solution by the gradual release of ammonia; basic carbonates are precipitated. Although very uniform spherical particles are obtained (Figure 2), they do not have a homogenous cation distribution since upon the slow increase of pH, Cu is precipitated first as the nucleus of these particles, then Y (32). Ba can be precipitated only if large amounts of urea are used (31). Another procedure is to first precipitate Y:Cu particles then disperse them in a solution containing $Ba(NO_3)_2$ and urea in order to, by the same process, coat these particles with $BaCO_3$ (30).

Similar spherical particles have been obtained by slowly evaporating water from a solution of citric acid and Y, Ba, and Cu salts (33).

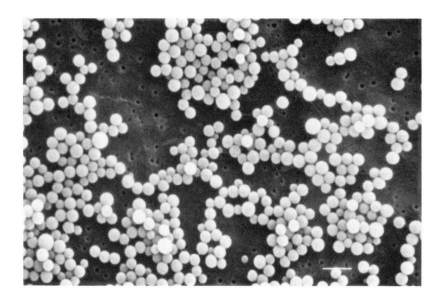

Figure 2: Controlled precipitation of spherical particles containing mixed Cu and Y. After F. Ribot et al. (32)

Weakly soluble carboxylates: Acetates are not strong enough ligands to yield precipitates alone. However, combined with hydroxides, they yield gels at pH around 7. The pH can be raised either by addition of a base such as ammonia (34)(35) or by use of barium hydroxide and colloidal yttrium hydroxide (35). The resulting gels have been used for film deposition or fiber processing(36).

Soluble carboxylates: The use of long chain carboxylates such as 2-ethylhexanoates (37)(38) or neodecanoates (38)(39) or of more ionic precursors like trifluoroacetates (40) enhances the solubility of the precursors in volatile solvents suitable for spin-coating. As an example, Y, Ba and Cu neodecanoates can be obtained from their acetates by reaction with ammonium decanoates or tetramethylammonium decanoate in the case of copper. The salts are all soluble in xylene or mixtures of xylene and pyridine (38). After spin-coating and drying the films at 150°C, microlithography can be performed on the resulting film by use of electron beam or ion beam in the same way as for a photoresist film (41). Unprocessed areas are then washed out by dissolution in the suitable solvent as shown in Figure 3. Typical resolution is 100 lines per mm for films of a characteristic thickness of 0.3 microns (41). Such inexpensive metaloorganic deposition techniques offer an interesting alternative to expensive vacuum evaporation.

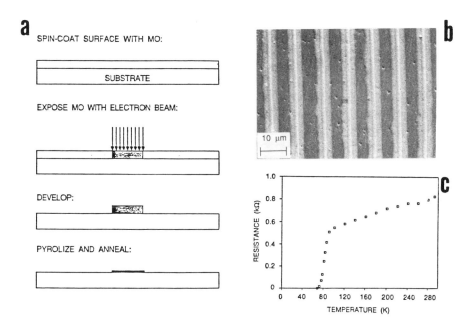

Figure 3: (a) Schematic of the lithography sequence on neodecanoate films; (b) Patterned lines after rapid thermal annealing; (c) Resulting film properties. After Mantese et al. (41) and Hamdi et al. (39)(40).

2.4 Thermal Processing

Ceramics: Heat treatment can be described as the following in the case of the 123 phase:

$$\text{precipitate} \xrightarrow{200°C-400°C} BaCO_3, CuO, Y_2O_3 \xrightarrow{800°C} 123T \xrightarrow{900°C} 123O$$

Most of the organics are usually removed between 200°C and 400°C by calcination, as shown by thermogravimetric and differential thermal analysis (TGA and DTA respectively, Figure 4a and 4b). The sharp peak observed in the DTA curve shows that, for this particular mixture of acetates and hydroxides, the reaction associated with the burning of organic materials may be violent and exothermic. In the case of thin films this may cause shrinkage or cracks. Moreover, a closer look at the plateau of the TGA curve above 600°C reveals that weight loss is never completed up to high temperatures. X-ray patterns reveal the presence of crystalline $BaCO_3$, CuO and Y_2O_3 (Figure 5).

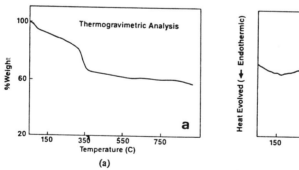
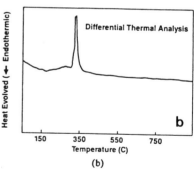

Figure 4: (a) TGA and (b) DTA traces for a solution method using acetates as the starting precursors (35).

Decomposition of the organics leads then to the separation of all elements into Cu and Y oxides and barium carbonate. This brings the synthesis parameters back to the problems encountered in the conventional solid state approach except that the heterogeneity may be over a submicron scale. Thus, reaction temperature may be slightly lowered (850°C instead of 900°C).

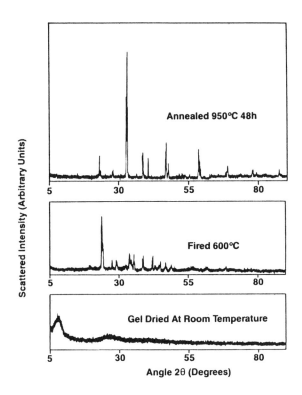

Figure 5: X-ray diffraction patterns using CuKα radiation of a xerogel dried at room temperature, fired at 600°C and annealed at 950°C. The peak observed at 600°C correspond to CuO, Y_2O_3 and $BaCO_3$. (After Reference 35.)

Some authors have noticed that if the ceramics are processed for a long time below 900°C, they yield a tetragonal semiconducting phase (denoted here as 123T) even after further annealing in oxygen at 500°C to recover full oxygen content. This tetragonal phase may then be explained by a cationic disorder in the (a,b) plane, related to poor crystallization (26)(27).

Below 900°C, for all the described processes, the powders are submicron. but the average particle size increases above this temperature. Then, grain coarsening competes with sintering (Figure 6). Such behavior does not promote good grain boundaries with thin intergranular interfaces that would accommodate the small superconducting coherence length. Thus, the resulting critical currents do not exceed 1000 A/cm^2. It seems that the best results are obtained at quite low temperatures (920°C) even though long annealing times are required for sintering (17).

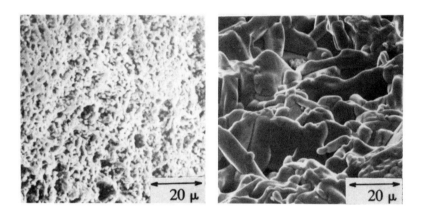

Figure 6: Scanning electron micrograph of a ceramic prepared from the acetate method and heated at (a) 920°C for 12h, (b) 950°C for 12h. (After Reference 17.)

Films: A way to overcome these difficulties, as already mentioned, is to synthesize textured films. Indeed, grain alignment has been generally observed in many techniques using spin-coating. As shown in Figure 7, X-ray patterns are characteristic of a crystalline orientation, with only one set of diffraction peaks observed. This may be explained by the fact that after heat treatment the films are usually composed of platelet-like grains whose diameter is roughly the same order as the film thickness. The platelets lay flat on the substrate with "c" axis perpendicular to the substrate.

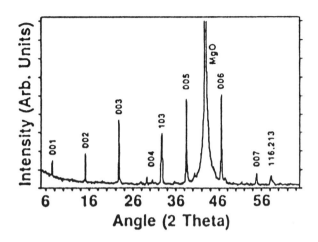

Figure 7: X-ray pattern of a $YBa_2Cu_3O_7$ film on MgO substrate. (After Reference 43.)

The best superconducting films are obtained on substrates such as $SrTiO_3$, MgO or ZrO_2. Some authors have succeeded in the preparation of films on Si, but using a MgO buffer that was first sputtered onto the substrate (38).

To our knowledge, no real superconducting film has been made without heating above 900°C. This may be related again to the poor reactivity of the intermediate phases. For example, Uchikawa et al. have found that, starting from precipitated carbonates, 40 min at 930°C are necessary to completely decompose $BaCO_3$ and form the pure superconducting superconductor (25). Therefore, there is a competition between the short time needed to minimize reaction with the substrate and the long time needed for phase crystallization and grain sintering. As a result, films that can be only heat-treated for 10-20 minutes at 920°C have a less metallic behavior and a larger superconducting transition width than ceramics made from the same gels but reacted for much longer times (43) (Figure 8).

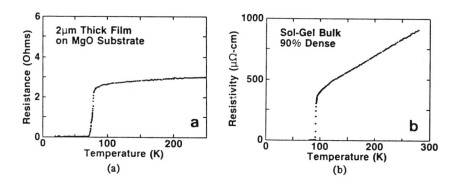

Figure 8: Comparison between the transport properties of (a) a thick film and (b) a ceramic obtained from the same gel.

Decreased reaction temperature: There are possibly a few ways to overcome these difficulties:

Manthiram and Goodenough (26) have noticed that they can lower the synthesis temperature of the $YBa_2Cu_3O_7$ phase to about 800°C under an argon atmosphere. Growth of the tetragonal $YBa_2Cu_3O_6$ phase is then favored and further oxidation at lower temperatures yields the $YBa_2Cu_3O_7$ superconducting phase.

Hamdi et al. (40) have used ultrarapid thermal annealing of the films at temperatures above 900°C for about 30s. The interesting feature of this work is that they obtain quite good metallic behaviors although starting from carbon-containing precursors. The rapid heating does not allow enough time in the temperature range 500°C-800°C for the cations to segregate. Therefore, a short annealing time (30s) is sufficient to obtain the orthorhombic phase.

Another possibility is also to use carbon-free precursors, as follows.

2.5 Carbon-Free Precursors

It appears that if homogeneous gels or coprecipitates can be obtained using carbon-based chemicals, they do not yield much better ceramics than conventional solid state synthesis. This is primarily attributed to the resulting carbonate-containing materials whose reactivity below 900°C is low. Therefore techniques based on carbon-free precursors or giving carbon-free precipitates like fully hydrolyzed alkoxides may help to overcome these difficulties.

Nitrates: All elements that belong to the composition of the superconducting cuprates form nitrates soluble in water or in acidic solutions. But, because of the low complexing ability of nitrate, the only application of these solutions is spraying. This method has been successively applied to $YBa_2Cu_3O_7$ (44-47), but also to $Bi_2(Sr,Ca)_3Cu_2O_8$ (48)(49) and $Tl_2Ba_2Ca_2Cu_3$ (49). Because of the increased reactivity compared to carbon-containing materials, films having good metallic properties can be obtained.

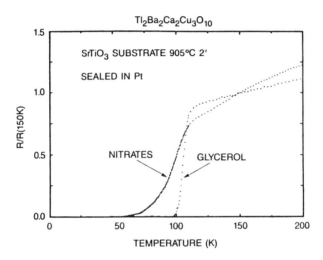

Figure 9: Resistive behavior for a film of $Tl_2Ba_2Ca_2Cu_3O_{10}$ made by spraying of nitrates. (After Reference 49.)

In the case of Bi and Tl phases only very short reaction times (of the order of minutes) are possible because of the evaporation of Bi and Tl. This may be compensated by using excess Bi and Tl in the solution or by modification of the solvent (mixture of water and glycerol) that enhances the solubility of Bi and allows the use of less nitric acid in the solution. A more homogeneous deposition results in a better product after heat treatment. Figure 9 shows a typical resistivity measurement for a thallium film that has, however, a very weak critical current (50 A/cm^2 at 77K).

Note also that for the Bi-based material, the same $Bi_2Sr_2Ca_1Cu_2O_8$ composition may result in the formation of other phases such as the $Bi_2Sr_2Cu_1O_6$ phase (that does not superconduct above 20K) if the films are processed at an intermediate temperature such as 500°C (48)(49) (Figure 10a). The 85K superconducting phase is obtained only if the films are quickly heated to temperatures of 850°C, just near the melting point, as shown in Figure 10b. Good crystalline orientation of the films with the c axis perpendicular to the substrate is always observed.

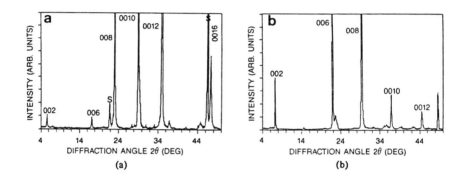

Figure 10: X-ray diffraction of films on $SrTiO_3$ with metal composition $Bi_4Sr_3Ca_3Cu_4$: (a) Directly heated to 850°C for 10 min; the peaks correspond to the $Bi_2Sr_2Ca_1Cu_2O_8$ phase. (b) First fired at 500°C then same as in (a); the peaks correspond to the phase $Bi_2Sr_2CuO_6$.

Nitrate solutions are also suitable for aerosol flow production of fine powders (50)(51). This technique yields small spherical particles that should be suitable for good sintering. However, because of the resulting small size of the superconducting grains, ceramics may have very poor flux expulsion properties if they are not thoroughly sintered (50).

Hydroxides: Precipitation of hydroxides may be done through addition of a base to an aqueous solution of salts. Yttrium and copper hydroxides have a similar precipitation range but for the less polarizing Ba cation, very high pH is necessary to start the precipitation. However, two processes have been described that increase the pH of a solution of nitrates in water by addition of either NaOH (52) or tetramethyl ammonium hydroxide (53). It appears that at 60°C and pH=13 all the barium has been precipitated. Another possibility is to start from soluble precursors such as trifluoroacetates in alcohol or to use a mixed acetone-water solution in which the nitrates are soluble but the hydroxides are not (54).

The hydroxides are decomposed below 600°C to finally yield the pure $YBa_2Cu_3O_{7-x}$ phase around 800°C, but reactivity is increased compared to organic precursors. This is demonstrated in Figure 11. The phase forms in a short time (20 min). It should therefore be very interesting to obtain homogeneous gels (53). They would allow the synthesis of thin films without excessive reaction with the substrate.

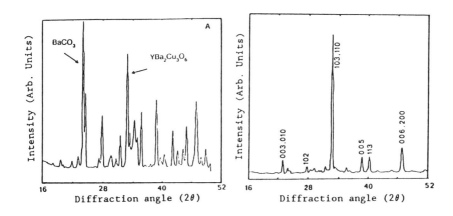

Figure 11: X-ray diffraction pattern of a precipitate that has been calcined at 850°C for 20 min under oxygen and quenched. (a) Starting from a precipitate of acetates (35). Peaks corresponding to the 123 phase can be seen mixed with peaks characteristic of $BaCO_3$. (b) Starting from hydroxides (54). The pure tetragonal 123 phase is observed.

3.0 CONCLUSION

There have been many solution techniques used to synthesize the superconducting perovskite-related phases, but no experiment has really achieved a substantial improvment of the synthesis, compared to conventional solid state reaction.

By using carbon-free precursors, the synthesis temperature could be significantly lowered but this did not result in any improvement of the properties since the particle size remained too small.

Interesting methods have demonstrated that thin films can be deposited and patterned using microlithography. However, there are still difficulties in thermal processing and the resulting films obtained by most of the solution techniques have wide transitions and low critical currents.

Solution chemistry remains attractive as an inexpensive technique useful for processing ceramics, fibers or coatings, even on a large scale not possible with evaporation techniques. However, the application of these solution techniques to multicomponent systems has rarely been attempted. A better basic understanding of all reaction steps from the solution to the ceramic is needed before a real improvement in the process will be possible.

4.0 REFERENCES

1. D.W. Murphy, D.W. Johnson Jr, S. Jin and R.E. Howard; *Science* 241:922 (1988).

2. B. Batlogg, A.P. Ramirez, R.J. Cava, R.B. Van Dover, E.A. Rietman, *Phys. Rev. B* 35:5343 (1987).

3. D.C. Larbalestier, S.E. Babcock, X. Cai, M. Daeumling, D.P. Hampshire, T.F. Kelly, L.A. Lavanier, P.J. Lee and J. Seuntjens. *Physica C* vol 153-155:1580 (1988).

4. S. Jin, T.H. Tiefel, R.C. Sherwood, M.E. Davis, R.B. Van Dover, G.W. Kammlot, R.A. Fastnacht and H.D. Keith, *Appl. Phys. Lett.* 52:2074 (1988).

5. D.E. Farrel, B.S. Chandrasekhar, M.R. Deguire, H.H. Fang, V.B. Kogan, J.R. Clem and D.K. Finnemore, *Phys. Rev. B* 36: 4025 (1987).

6. D.W. Johnson and G.S. Grader, *J. Am. Cer. Soc.* 71:C291 (1988).

7. H. Dislich, *Angew. Chemie Int. Ed.* 10:363 (1971).

8. B.E. Yoldas, *J. Mat. Sci.* 12:1203 (1977).

9. J. Livage, M. Henry and C. Sanchez, *Prog. Sol. State Chem.* 18:259 (1988).

10. K.S. Mazdiyasni, C.T. Lynch and J.S. Smith, *Inorg. Chem.* 5: 342 (1966).

11. L.M. Brown and K.S. Mazdiyasni, *Inorg. Chem.* 9:2783 (1970).

12. R.C. Mehrotra and A.K. Rai, *Indian J. Chem.* 4:537 (1966).

13. C.H. Brubaker and M. Wicholas, *J. Inorg. Nucl. Chem.* 27:59 (1965).

14. X.D. Wu, A. Inam, T. Venkatesan, C.C. Chang, E.W. Chase, P. Barboux, J.M. Tarascon and B. Wilkens, *Appl. Phys. Lett.* 52: 754 (1988).

15. T. Venkatesan, C.C. Chang, D. Dijjkamp, S.B. Ogale, E.W. Chase, L.A. Farrow, D.M. Hwang, P.F. Miceli, S.A. Scwhwarz, and J.M. Tarascon, *J. Appl. Phys.* 63:4591 (1988).

16. H.S. Horowitz, S.J. McLain, A.W. Sleight, J.D. Druliner, P.L. Gai and M.J. Vankavelaar, *Science* 243:66 (1989).

17. P. Barboux, J.M. Tarascon, L.H. Greene, G.W. Hull, B.W. Meagher and C.B. Eom, *Mater. Res. Soc. Symp. Proc.* 99:49 (1988).

18. B. Dunn, C.T. Chu, L.W. Zhou, J.R. Cooper and G. Gruner, *Adv. Ceram. Mat.* 2:343 (1987).

19. J.R. Cooper, C.T. Chu, L.W. Zhou, B. Dunn and G. Gruner, *Phys. Rev. B* 37:638 (1988).

20. S. Shibata, T. Kitagawa, H. Okazaki, T. Kimura and T. Murakami, *Jpn. J. of Appl. Phys.* 27:L53 (1988).

21. F. Uchikawa and J.D. Mackenzie, *J. Mat. Res.* 4:787 (1989).

22. S. Kramer, G. Kordas, J. McMillan, G.C. Hilton, and D.J. Van Harlingen, *Appl. Phys. Lett.* 156 (1988).

23. P. Ravindranathan, S. Komarneni, A. Bhalla, R. Roy and L.E. Cross, *J. Mat. Res.* 3:810 (1988).

24. H.H. Wang, K.D. Carlson, U. Geiser, R.J. Thorn, H.I. Kao, M.A. Beno, M.R. Monaghan, T.J. Allen, R.B. Proksch, D.L. Stupka, J.M. Williams, B.K. Flandermeyer and R.B. Poeppel, *Inorg. Chem.* 26:1474 (1987).

25. T. Itoh and H. Uchikawa, *J. Crystal Growth* 87:157 (1988).

26. A. Manthiram and J.B. Goodenough, *Nature* 329:701 (1987).

27. X.Z. Wang, M. Henry, J. Livage and I. Rosenman, *Solid State Com.* 64:2725 (1988).

28. Y.M. Chiang, D.A. Rudman, D.K. Leung, J.A.S. Ikeda, A. Roshko and B.D. Fabes, *Physica C* 152:77 (1988).

29. S. L. Furcone and Y.M. Chiang, *Appl. Phys. Lett.* 52:2180 (1988).

30. P.M. Kayima, S. Qutubuddin, *J. Mater. Sci. Letters.* 8:171 (1989).

31. R.S. Liu, C.T. Chang, and P.T. Wu, *Inorg. Chem.* 28:154 (1989).

32. F. Ribot, S. Krathovil and E. Matijevic, *J. Mater. Res.* 4:1123 (1989).

33. R. Sanjinés, K. Ravidranathan Thampi and J. Kiwi, *J. Am. Ceram. Soc.*, 71:C512 (1988).

34. H. Kozuka, T. Umeda, J. Jin, F. Miyaji and S. Sakka, *J. Ceram. Soc. Japan* 96:355 (1988).

35. P. Barboux, J.M. Tarascon, L.H. Greene, G.W. Hull and B.G. Bagley, *J. Appl. Phys.* 63:2725 (1988).

36. T. Umeda, H. Kozuka and S. Sakka, *Advanced Ceramic Materials* 3:520 (1988).

37. M.E. Gross, M. Hong, S.H. Liou, P.K. Gallagher and J. Kwo, *Appl. Phys. Lett.* 52:160 (1988).

38. H. Nasu, H. Myoren, Y. Ibara, S. Makida, Y. Nishiyama, T. Kato, T. Imura and Y. Osaka, *Jpn. J. of Appl. Phys.* 4:216 (1988).

39. A.H. Hamdi, J.V. Mantese, A.L. Micheli, R.C.O. Laugal, D.F. Dungan, Z.H. Zhang, and K.R. Padmanabhan, *Appl. Phys. Lett.* 51:2152 (1987).

40. A.H. Hamdi, J.V. Mantese, A.L. Micheli, R.A. Waldo, Y.L. Chen and C.A. Wong, *Appl. Phys. Lett.* 53:435 (1988).

41. J.V. Mantese, A.B. Catalan, A.H. Hamdi, A.L. Micheli and K. Studer-Rabeler, *Appl. Phys. Lett.* 53:526 (1988).

42. A. Gupta, R. Jagannathan, E.I. Cooper, E.A. Giess, J.I. Landman, and B.W. Hussey, *Appl. Phys. Lett.* 52:2077 (1988).

43. J.M. Tarascon, P. Barboux, L.H. Greene, B.G. Bagley, G.W. Hull, Y. LePage and W.R. McKinnon, *Physica C* 153-155:566 (1988).

44. A. Gupta, G. Koren, E.A. Giess, N.R. Moore, E.J.M. O'Sullivan and E.I. Cooper, *Appl. Phys. Lett.* 52:163 (1988).

45. A. Gupta and G. Koren, *Appl. Phys. Lett.* 52:665 (1988).

46. R.L. Henry, H. Lessof, E.M. Swiggard and S.B. Qadri, *J. Crystal Growth* 85:615 (1987).

47. M. Kawai, T. Kawai, H. Masuhira and M. Takahasi, *Jpn. J. of Appl. Phys.* 26:L1740 (1987).

48. E.I. Cooper, E.A. Giess and A. Gupta, *Material Letters* 7:5 (1988).

49. P. Barboux, J.M. Tarascon, F. Shokoohi, B.J. Wilkens and C.L. Schwartz, *J. Appl. Phys.* 64:6382 (1988).

50. T.T. Kodas, E.M. Engler, V.Y. Lee, R. Jacowitz, T.H. Baum, K. Roche and S.S.P. Parkin, *Appl. Phys. Lett.* 52:1622 (1988).

51. P. Odier, B. Dubois, M. Gervais and A. Douy, *Mat. Res. Bull.* 24:11 (1989).

52. B.S. Khurana, R.B. Tripathi, S.M. Khullar, R.K. Kotnala, S. Singh, K. Jain, B.V. Reddi, R.C. Goel, M.K. Das, *J. Materials Science* 8:234 (1989).

53. M. Fujiki, M. Hikita and K. Sukegawa. *Jpn. J. Of Appl. Phys.* 26:L1159 (1987).

54. I. Valente, P. Barboux, L. Mazerolles, D. Michel, R. Morimeau and J. Livage, in *Better Ceramics Through Chemistry* (C.J. Brinker, D.E. Clark and D.R. Ulrich eds.), Mat. Res. Soc., Pittsburgh (In Press).

8

Cationic Substitutions in the High T_c Superconductors

Jean Marie Tarascon and Brian G. Bagley

1.0 INTRODUCTION

The discovery by Bednorz and Müller (1) of superconductivity at 34K in the La-Ba-Cu-O system stimulated the search for new high transition temperature (T_c) cuprates. Superconductivity above liquid nitrogen temperature was subsequently established by Wu et al. (2) in the $YBa_2Cu_3O_7$ compound (so-called 123 or 90K phase) and more recently Bi- and Tl-based cuprates having T_c's greater than 100K, and as high as 125K, were uncovered (3-4).

A prominent feature common to these high T_c cuprates (deduced from band structure calculations (5)) is the existence of a strong hybridization between the Cu 3d and O 2p levels. Because of this strong hybridization, it is more appropriate to characterize the nature of the copper by its formal valence rather than by its oxidation state. The formal valence of Cu in these materials, which is centered around 2, can be monitored chemically. For instance, Cu can be oxidized (formal valence becoming greater than 2) or reduced (formal valence becoming less than 2) either by the uptake or removal of oxygen, respectively, or through a cationic substitution. For instance, superconductivity can be induced in the La_2CuO_4 system (so-called 40K or 214 phase) by replacing some of the trivalent La ions by divalent Sr ions thereby increasing the formal valence of the copper (6-8). Based on the observation that copper has a formal valence greater than 2 in all these high T_c Cu-based oxides, it was thought that such a value (greater than 2) was crucial for the occurrence of superconductivity. Recently, however, Tokura et al. (9) examined the other possibility and discovered superconductivity in the Nd_2CuO_4 system when some of the trivalent Nd ions are replaced by tetravalent Ce ions (i.e., lowering the Cu valence below 2) thus eliminating as a necessary criterion that the valence be greater than 2 for the occurrence of superconductivity in the cuprates. The crystal structures of $La_{2-x}Sr_xCuO_4$ (termed the T phase (10)), Nd_2CuO_4 (termed the T' phase (11)), the 123 phase (12) and the Bi-based cuprates of general formula $Bi_2Sr_2Ca_{1-n}Cu_nO_y$ (13-15) are shown in Figure 1. A structural feature common to all of these phases is the presence of CuO_2 planes but with the difference that the copper coordination within these planes can be four-fold (T' phase), five-fold (123 phase and bismuth n=2 phase) or sixfold (T phase or bismuth n=1 phase). These various coordinations for the copper play an important role in the choice of possible dopants. Throughout this review we will frequently refer to Figure 1, as structural considerations (e.g., crystallographic sites and their coordination) are useful in determining which substitutions have any chance for success.

Over the past three years, one common goal of chemists, physicists and theorists has been to determine the mechanism for superconductivity in these cuprates. Theories are based on experimental measurements done by physicists on materials prepared by chemists, and measurements made on poorly characterized samples can lead to unreliable data and thereby erroneous theories. Thus, as chemists, our primary role is twofold; 1) to optimize the processing conditions (ambient, cooling rate, annealing temperature) and determine other synthesis routes for the preparation of very pure and homogeneous samples such that the resulting measurements are reliable and 2) to synthesize new materials or perform chemical substitutions (cationic or anionic) on known systems so that general phenomenological trends can be determined that will help in establishing a viable theory.

The purpose of this chapter is not to cover the entire chemistry of the high T_c oxides but instead is to focus on one specific point: the study of these materials through cationic chemical substitutions in order to illustrate how this type of chemistry can aid in the understanding of some of the crystal-chemical aspects of these new high T_c

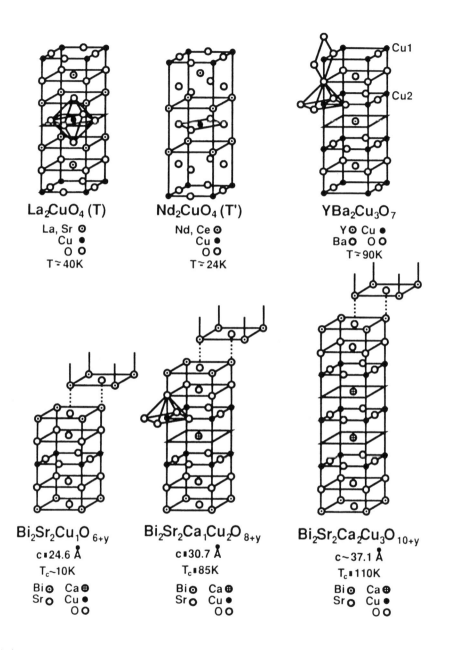

Figure 1: Structures of the high T_c cuprates are shown with the Cu coordination in the CuO_2 layers highlighted.

oxides and to experimentally answer some key questions. For example, one key question involves where superconductivity is confined within the cuprate structure or, more specifically, the relative importance of the Cu-O chains vs. CuO_2 planes in the 123 phase. Because of the short coherence length in these materials (very much shorter than that of the conventional metallic superconductors), sample homogeneity is a key issue. We address this point in the first part of the chapter with a comparison of several synthesis techniques. In section 3.0 we review the results for cationic substitutions at the Cu, at the rare earth, or at the divalent alkaline earth sites for most of the high T_c oxides and show how these substitutions affect their structural and physical properties. For clarity, the results for the 40K materials (La_2CuO_4 and Nd_2CuO_4), the 90K materials ($YBa_2Cu_3O_7$) and the Bi- and Tl-based cuprates are presented separately. Finally, in section 4.0, we discuss and analyze the data presented so that general trends are established and guidelines for the discovery of other cuprates can be proposed.

2.0 MATERIALS SYNTHESIS

The usual synthesis technique for the preparation of a cuprate superconducting ceramic consists of mixing and firing stoichiometric amounts of CuO, rare earth oxides and either a carbonate or nitrate as the precursor for the divalent alkaline earths (i.e., Ca, Sr or Ba). The firing temperatures, the time at temperature, the cooling rate and the ambient used depend strongly on the system being studied. Temperatures greater than 1100°C in air or oxygen are required for the synthesis of the 40K materials, whereas lower temperatures are used to synthesize the 123 (980°C) or Bi-based cuprate phases (840-880°C) using either air or oxygen. Barium nitrate decomposes at 600°C whereas the carbonate decomposes at 1450°C. Thus, using the nitrate (because of it's low melting point) a more complete reaction at lower temperatures is obtained than if a carbonate is used and this yields a more homogeneous sample (16). This is especially important for the preparation of 123 or Bi compounds because they are synthesized at (relatively) lower temperatures. To further improve homogeneity a sol-gel process, which allows mixing at the molecular level, was developed for the synthesis of doped or undoped 123 (17). Gels of the

123 are obtained by mixing appropriate amounts of copper acetate, barium hydroxide and barium acetate with colloidal $Y(OH)_3$ at a pH close to 7. The resulting gels can be dried to produce glassy-like materials which, after firing, yield a polycrystalline ceramic 123 with a T_c of 92K and a ΔT_c of 0.6K. The solution technique can give high quality ceramics, but for each system the processing parameters must be optimized in order to prepare a stabilized gel. For instance, we could not extend the above conditions to the preparation of Bi or Tl-based cuprate gels because of the poor solubility of Bi and Tl in solutions of neutral pH (18). The measurements reported in the following section were made on samples prepared with the solid state reaction technique using carbonates for the 40K materials, carbonates or nitrates for the Bi-based materials, and the 123 compound was made using either the nitrates as a precursor in a solid state reaction or by sol-gel.

3.0 RESULTS

3.1 $La_{2-x}Sr_xCuO_4$

In this system, the copper valence can be varied through a substitution for the trivalent La by divalent Ca, Sr, and Ba. Solid solutions $La_{2-x}M_xCuO_4$ have been determined to exist for these divalent substitutions from x=0 up to x=0.2 for M=Ca, 1.0 for M=Sr, and 0.2 for M=Ba. For the Sr substitution, up to x=0.3 the a lattice parameter decreases whereas the c-axis increases (Figure 2a). Beyond this concentration c decreases because of a simultaneous increase in the number of oxygen vacancies. The strontium substitution causes the disappearance of the antiferromagnetic ordering which exists in undoped La_2CuO_4 and also produces changes in the transport properties, as exemplified by the temperature dependence of the resistivity (Figure 3a) for several members of the Sr-doped series. As x increases the resistivity temperature dependence evolves from semiconducting-like to a superconducting behavior with a maximum T_c at 38K for x=0.15 (i.e., a formal valence of 2.15 per Cu atom). Above T_c there is a metallic-like linear variation of the resistivity with temperature. Finally, increasing x beyond 0.3 (data not shown) produces samples having metallic behavior but which are not super-

Cationic Substitutions in the High T_c Superconductors 315

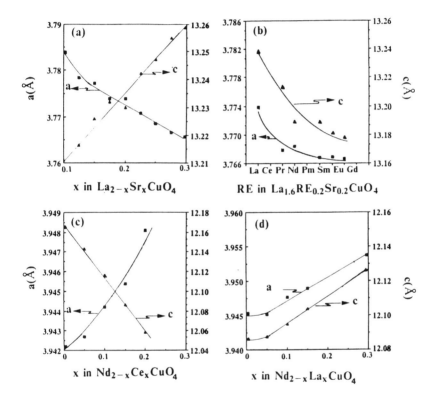

Figure 2: Variation of the tetragonal lattice parameters **a** and c as a function of x for $La_{2-x}Sr_xCuO_4$, $La_{1.6}RE_{0.2}Sr_{0.2}CuO_4$, $Nd_{2-x}Ce_xCuO_4$, and $Nd_{2-x}La_xCuO_4$ are shown in (a), (b), (c), and (d) respectively.

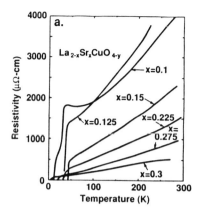
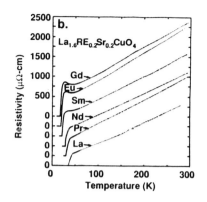

Figure 3: In (a) is shown the resistivity as a function of temperature and x for the $La_{2-x}Sr_xCuO_4$ series and in (b) the resistivity as a function of temperature and rare earth (RE) for the $La_{1.6}RE_{0.2}Sr_{0.2}CuO_4$ series.

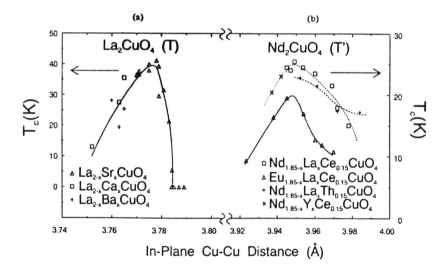

Figure 4: The variation of T_c as a function of the Cu-Cu distance (i.e., a-axis) is shown in (a) for several members of the $La_{2-x}A_xCuO_4$ series with A being Ca, Sr, and Ba (after Ref. 21) and in (b) for several members of the $Nd_{1.85-x}RE_xCe_{0.15}CuO_4$ series (RE= Y, La).

conductors. Chemical analyses (19), as well as Hall measurements (20), have shown that upon increasing the Sr content to x=0.3 the formal valence per Cu changes as (2+x) and deviates negatively from this value for larger x. This deviation results from the fact that, for a Sr content greater than 0.3, the system introduces oxygen vacancies so that the number of carriers is no longer a simple function of x. Similar measurements have been performed on Ca and Ba substituted systems and a maximum in T_c was also observed at x~0.15 indicating that there is a particular carrier concentration for maximum T_c in this system. An experimental determination of T_c as a function of the in-plane Cu-Cu distance (21-22) for the Ca, Sr and Ba doped systems indicates that there is also a particular Cu-Cu distance (3.786Å) for which T_c is a maximum (Figure 4).

To examine the importance of the magnetic nature of a particular rare earth (RE) on T_c in this system the La can be replaced by all the other rare earths. However, in order to maintain the T (La_2CuO_4) structure only a partial substitution can be achieved. A survey of the $La_{2-x}RE_xCuO_4$ solid solution has shown that the T phase (La_2CuO_4) is maintained only at low x (0<x<0.3) for the lightest rare earths up to Gd. A complete RE substitution leads to the T' phase (RE_2CuO_4) to be discussed later. To induce superconductivity in these phases some of the La must be substituted by Sr and it was observed that, independent of the rare earth (denoted in the following as RE) in each series, a maximum of T_c was obtained at a Sr content of 0.2. Thus we systematically studied the compositions $La_{1.6}RE_{0.2}Sr_{0.2}CuO_4$ with RE being Pr to Gd (23). Magnetic measurements performed on all the members of these series showed that the magnetic moments per rare earth agree with those expected for free trivalent ions. This indicates that the rare earth adopts the oxidation state +3 in these compounds. It was found (Figure 2b), as expected, that the a and c lattice parameters decrease with decreasing ionic radii of the substituted rare earth (i.e., in going from Pr to Gd). The effects on T_c for these various substitutions is shown in Figure 3b. Note that all the samples are superconductors and T_c decreases with increasing atomic number of the substituted rare earth such that T_c decreases from 39K for RE=La down to 18K for RE=Gd. Again, above T_c the resistivity varies linearly with temperature. Note that the depression in T_c is about the same for the compound doped with a Van-Vleck ion (J=0) rare earth (Eu) as for a compound doped with a strong magnetic rare earth (Gd). Thus the depression in T_c is not a magnetic effect, but is more likely a volume effect since it was found that the continuous decrease in T_c is correlated to a continuous decrease in the unit cell volume from 188.54Å^3 to 186.94Å^3 in going from La to Gd.

Further extension of this substitution work in the $La_{2-x}RE_x Sr_{0.2}CuO_4$ system led to the discovery of another phase at x=1, the so called T* phase, whose structure contains half of the T and half of the T' phases such that copper is five-fold coordinated (24-25). The T* phases, when synthesized under oxygen, exhibit superconductivity at 25K.

Tokura et. al., (9) recently studied other cationic substitutions in the T' phase and found, surprisingly, that superconductivity can be achieved in the $Nd_{2-x}M_xCuO_4$ system by first replacing the trivalent rare earth (Nd) not by a divalent ion but rather by a tetravalent ion (Ce or Th) and then by reducing the samples under nitrogen. Studies on polycrystalline samples of the $Nd_{2-x}Ce_xCuO_4$ system (26) have shown that a solid solution only exists up to x=0.2 and with increasing x the a-axis increases whereas the c-axis decreases (Figure 2c). Only samples within a narrow range of Ce composition (0.14<x<0.17) show superconductivity after the nitrogen anneal and the maximum in T_c(24K) occurs at x=0.15. Above T_c the resistivity exhibits semiconducting-like behavior. Single crystal studies (27-28) have confirmed this narrow range of x for which superconductivity can be achieved, but indicate that the semiconducting-like behavior above T_c is an extrinsic effect. Above T_c the resistivity of a crystal is metallic-like and linear from 150K to room temperature (Figure 5). The origin of the difference between polycrystalline ceramic and single crystal samples is beyond the scope of this chapter and is discussed elsewhere (28).

Markert et al. (29) have shown that there is nothing unique about Nd, as superconductivity was also found in $RE_2Ce_{0.15}CuO_4$ compounds with, for instance, RE being Pr, Sm and Eu. A depression in T_c by magnetic ions is usually observed in conventional superconductors. The substitution for Nd by these other rare earths (Pr, Sm, Eu) causes a slight depression in T_c, which is greater when the substituted rare earth is a J=0 Van Vleck ion (Eu) (ΔT_c=12K) rather than a magnetic ion (Pr or Sm) (ΔT_c=4K). This indicates that, as with the T phase materials, the magnetic nature of the rare earth ion does not affect T_c.

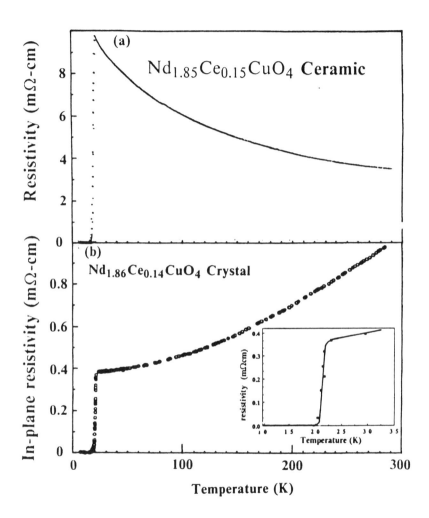

Figure 5: The resistivity as a function of temperature from 4.2 to 300K is shown in (a) for a superconducting $Nd_{1.85}Ce_{0.15}CuO_4$ ceramic and in (b) for a superconducting $Nd_{1.86}Ce_{0.14}CuO_4$ single crystal.

To study the T′ phase stability with respect to the T and T* phases four solid solution series were prepared and characterized (30). These four series consisted of; $Nd_{1.85-x}La_xCe_{0.15}CuO_4$, $Eu_{1.85-x}La_xCe_{0.15}CuO_4$, $Nd_{1.85-x}La_xTh_{0.15}CuO_4$ and $Nd_{1.85-x}Y_xCe_{0.15}CuO_4$. It was observed that upon increasing either the La or Y content the T′ phase transforms sharply to the T phase without any evidence of the T* phase, and the solubility limits for both the La and Y are increased by the presence of the Ce. Upon making these cationic substitutions (La or Y) the a and c lattice parameters behave similarly, either both increasing (Figure 2d) for La or decreasing for Y (30). In contrast, however, T_c decreases upon increasing the amount of the substituted element, with the exception of the Eu series for which a maximum in T_c is observed. By analogy to the observations made on the La_2CuO_4 system the variation of T_c for all the above phases were plotted as a function of the in plane Cu-Cu distance and, as with the Sr doped La system, there is a threshold Cu-Cu distance for which T_c is maximum (3.97Å). Thus, in this respect, there is a close similarity between the Sr-doped La and the Ce-doped Nd systems.

The weak dependence of T_c on magnetic ion substitutions suggests that superconductivity is not directly related to the La-O layers. Thus a $La_{1.85}Sr_{0.15}Cu_{1-x}M_xO_{4-y}$ series in which a magnetic (Ni) or a non-magnetic (Zn or Ga) ion was substituted for the Cu was studied (31-32). The resistivity temperature dependencies for several members of this series are shown in Figure 6. The important feature is that both the Ni and the Zn depress T_c rapidly such that compounds with values of x greater than 0.025 do not superconduct. This sharp depression in T_c suggests that superconductivity is confined to the CuO_2 plane. Surprisingly, however, T_c is depressed faster when M is non-magnetic (Zn) than magnetic (Ni), the opposite of what one would expect if these were conventional BCS superconductors. Magnetic measurements (33-34) indicate that a moment is induced by the Zn so that magnetic pair breaking must still be considered as a possibility. This would account for the faster decrease in T_c for the Zn-doped materials. Another possibility which could account for the faster decrease of T_c with Zn than with Ni is related to the local disorder within the CuO_2 planes, which is expected to be greater (at the same x) for the Zn than for the Ni samples since Ni forms the same structure over the entire solubility range whereas Zn does not.

Thus, from studies of cationic substitutions in the T and T′

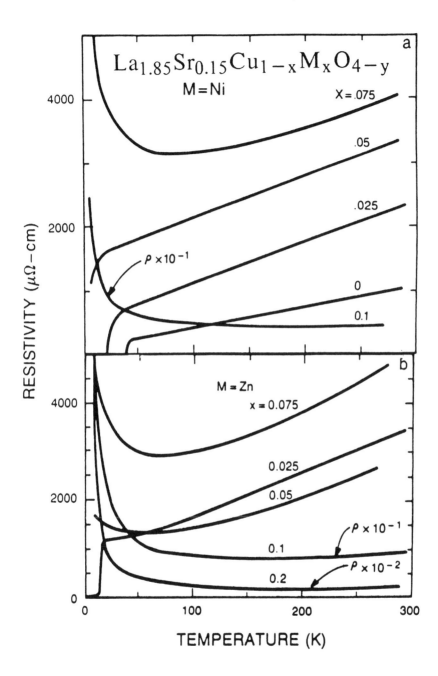

Figure 6: Resistivity as a function of temperature and x for the $La_{1.85}Sr_{0.15}Cu_{1-x}Ni_xO_{4-y}$ series (a) and for the $La_{1.85}Sr_{0.15}Cu_{1-x}Zn_xO_{4-y}$ series (b).

phases one finds: 1) that a substitution at the Cu site destroys T_c independent of whether the substituted element is magnetic or non-magnetic and 2) that a substitution at the rare earth site by another RE ion of the same oxidation state (+3) does not change the formal valence of copper, and thereby T_c, whereas a substitution for the RE by either a divalent (e.g., Sr) or a tetravalent (e.g., Ce) ion is marked by changes in the formal valence of the Cu and thereby T_c.

3.2 $YBa_2Cu_3O_7$

To understand the reasons why the 123 phase cannot be formed with Sr a study of the $YBa_{2-x}Sr_xCu_3O_7$ series was undertaken (35). X-ray analysis showed that single phase materials can be obtained from x=0.0 up to only x=1.4. Over this range of Sr substitution the tetragonal lattice parameters **a** and **c** decrease continuously with increasing x, and T_c also continuously decreases from 90K at x=0 to 78K at x=1.4. Beyond x=1.4 the materials are multiphase. However, extensive studies of the Sr based 123 system have shown that 3d metal impurities stabilize the fully Sr-doped 123 phase (36). The Sr substitution results in a reduction of the c-axis (Sr being smaller than Ba) and it is possible, therefore, that the repulsions between the CuO_2 planes and chains are less screened. Thus the chains and planes will tend to separate. The presence of a trivalent charge in one of these layers will compensate for this effect and thereby improve the stability of the Sr-doped 123 phase, as is observed experimentally.

Yttrium in the 123 phase can be replaced by most of the rare earths, exceptions being Ce and Tb (37). From magnetic measurements it was deduced that, in all these compounds, the rare earth is trivalent and therefore the RE substitution at the Y site does not affect the Cu valence. An exception is Pr for which the magnetic data suggests the coexistence of trivalent and tetravalent Pr ions. The fully RE-doped 123 phases, when prepared under the same conditions of temperature and ambient, are orthorhombic (an exception being again for Pr which is tetragonal) and their unit cell volumes decrease with decreasing ionic radii of the substituted RE (i.e., in going from La to Lu). Figure 7 shows that the substitution of a rare earth for Y has little effect on the superconducting and resistive properties of the 90K material. Above T_c the resistivity is metallic-like and linear with

Cationic Substitutions in the High T_c Superconductors 323

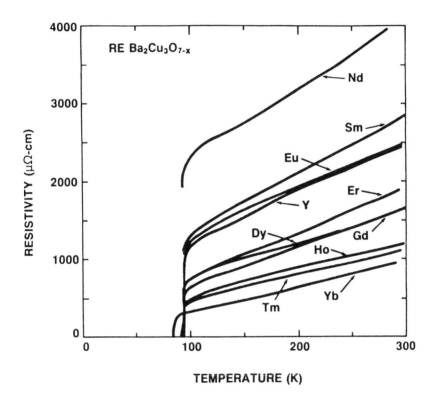

Figure 7: Resistivity as a function of temperature for $YBa_2Cu_3O_7$ and compounds for which a rare earth (RE) has been substituted for the Y.

temperature. With the exception of La and Pr (data not shown), the T_c's of the fully RE-doped samples lie between 87 and 95K independent of the magnetic nature of the substituted RE. This indicates there is little interaction between the RE and the charge carriers in these materials.

The coordination of Ba in the 123 compound is similar to that of La in the 214 phase. We showed that in the 214 material La can be replaced by Ba, thus we could expect the possibility of a RE substituting for Ba in the 123. This possibility was examined by studies of the $RE_{1+x}Ba_{2-x}Cu_3O_7$ series (38). In a $REBa_{2-x}La_xCu_3O_7$ series, for example, it was shown that single-phase materials can be obtained up to x=1. The La substitution results in an orthorhombic-tetragonal structural transition (denoted O-T) at x=0.4. The value of x at which this O-T transition occurs depends upon which rare earth is being substituted. The substitution of a trivalent ion for a divalent ion introduces a charge imbalance which, in this system, is compensated (for x < 0.4) by a reduction of the formal valence of Cu from 2.3 at x=0 to 2.2 at x=0.4. Over the same range of composition T_c decreases from 90K to less than 4.2K at x=0.4 which indicates a strong correlation between T_c and the formal copper valence. For x greater than 0.4, the La substitution induces an uptake of oxygen (an effect observed for several other cationic substitutions) which also compensates for a part of the charge imbalance.

In contrast to the 214 phase, the 123 phase contains both CuO_2 planes and CuO chains leading to two different Cu sites. The Cu in the planes (denoted Cu2) is five-fold coordinated whereas the Cu in the chains (denoted Cu1) is four-fold coordinated. To determine the relative role of chains vs. planes in these compounds, substitutions for Cu by 3d metals were performed and solid solutions of composition $YBa_2Cu_{3-x}M_xO_7$ (M being Al, Co, Fe, Ni, and Zn) were studied (39-41). Figure 8 shows the variation of the **a** and **b** lattice parameters for these solid solutions. Substitutions for Cu by Fe, Co, and Al result in an O-T transition (i.e., **a**=**b**) over the range of composition 0.05<x<0.1, whereas the cell remains orthorhombic for the Ni and Zn substituted compounds over the entire range of solubility. Al can only have an oxidation state of +3. Therefore the similar structural behavior for the Fe- and Co-doped materials to those that are Al-doped implies that both the Fe and Co are also trivalent in these compounds. This was confirmed, for Fe, by Mossbauer measurements (42). Zn is

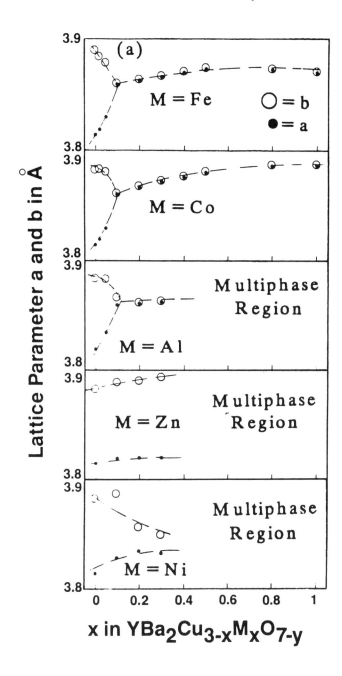

Figure 8: The unit-cell parameters **a** and **b** are shown as a function of x for the $YBa_2Cu_{3-x}M_xO_{7-y}$ series with M=Fe, Co, Al, Zn and Ni.

always a divalent ion and the similar composition dependence of the lattice parameters to those of the Ni-doped series suggest that Ni also enters these structures as a divalent ion. A debated question is: Which of the sites, Cu1 or Cu2, do these 3d metal substituted elements occupy? From simple chemical considerations, and using thermogravimetric analysis (TGA), we deduced that Co and Al occupy the Cu1 sites, that Fe occupies the Cu1 site at low concentrations and the Cu2 site at high concentrations, and finally that the divalent Zn and Ni ions occupy the Cu2 sites. The Cu1 site is square planar and it is well established that Fe^{+3} or Co^{+3} prefers an octahedral environment. This suggests that there may be an uptake of oxygen accompanying this cationic substitution. TGA measurements have indeed shown an oxygen content which increases with increasing doping content and which can be greater than 7. Neutron diffraction measurements (43-45) confirm both the uptake of oxygen and the assignments just discussed for the 3d metal sites but with some remaining ambiguity about the Ni and Zn. A differential anomalous x-ray scattering study of the 3d-ion occupancy in $YBa_2Cu_{3-x}M_xO_y$ with M=Fe, Co, Ni and Zn has been recently reported by Howland et. al., (46). They observed that the Fe and Co atoms predominantly occupy Cu1 sites whereas the Ni and Zn occupy both the Cu1 and Cu2 sites about equally.

The substitution for Cu by a 3d metal produces drastic changes in the physical properties of the 90K phase, as shown in Figure 9a where the T_c's (as determined by ac-susceptibility measurements) are plotted as a function of x for the various series. Independent of the magnetic nature of the substituted element, for the trivalent ions (Fe, Co, and Al) T_c remains constant and equal to 90K up to the O-T transition and then decreases continuously to less than 4.2K at x=0.5. Upon increasing x further, the compounds become semiconductors and simultaneously antiferromagnetism associated with the Cu ions develops. In contrast to this behavior, for the divalent substituted Ni and Zn ions T_c decreases markedly, even at low x.

How T_c varies with x and the maximum solubility range (x) for a specific 3d metal differs from group to group. These differences simply arise from the synthesis history of the samples (47). We illustrate this point in Figure 9b where we show the variation of T_c as a function of x for a series of samples prepared at 970°C and 920°C. Note that increasing the annealing temperature increases the T_c dependence upon x and also reduces the solubility range. This is also

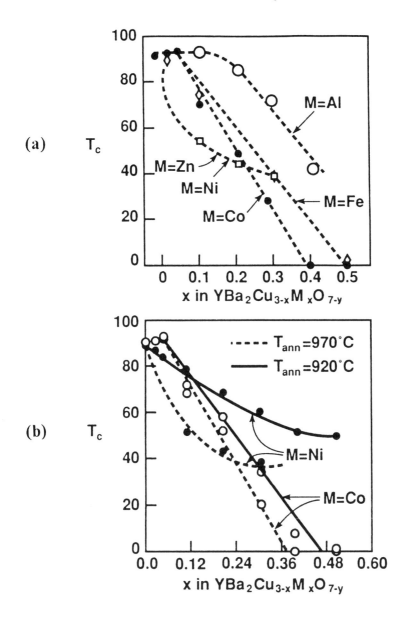

Figure 9: Superconducting critical temperatures, determined inductively, are reported as a function of x for the $YBa_2Cu_{3-x}M_xO_{7-y}$ series with M= Fe, Co, Al, Zn, and Ni in (a). In (b) a processing effect (annealing temperature) on T_c is shown for the Ni- and Co-doped 90K series.

consistent with our inability to substitute Ga for Cu in the 123 at temperatures of 970°C whereas others were able to achieve this substitution at 920°C. However, a general result is that (independent of the synthesis history) at small x we observe a sharper decrease in T_c for the divalent ions than for the trivalent ions. This implies that T_c is more affected when the doping occurs in the CuO_2 planes than in the CuO chains and therefore that superconductivity is confined to the CuO_2 planes.

To confirm this point, neutron diffraction measurements were performed on several members of the Co-doped 123 series in order to determine how bond lengths vary with Co content (48). An important result, shown in Figure 10, is that upon increasing the doping at the Cu1 site there is a decrease in T_c concomitant with a shortening of the Cu-O4 bond length (O4 being that oxygen bridging the Cu1 chains and Cu2 layers) and the Ba ion moving towards the CuO_2 planes. In undoped materials Ba shares its electrons between the Cu2 and Cu1 site containing layers. As the system is doped at the Cu1 site the barium moves towards the CuO_2 plane and shares more of its electrons with the Cu2 than the Cu1 layers thereby filling the electrical holes in the Cu2 layers that are necessary for superconductivity. This would result in a depression of T_c as is observed experimentally. Or viewed differently, as a result of doping at the Cu1 site the Ba moves towards the planes and O4 towards the chains such that the CuO_2 layers become isolated (i.e., there is no charge transfer between chains and planes) and the material becomes an insulator. Thus, from the 3d metal substitutions we unambiguously establish that superconductivity is confined to the CuO_2 layers and show that the CuO chains act as a carrier reservoir which provides the electronic coupling between the CuO_2 layers. This disestablished the early belief (simply based on the presence of Cu-O chains in the 90K phase and not in the 40K phase) that (1D) linear Cu-O chains were required to obtain higher T_c materials among the cuprates. The discovery of the Bi and Tl-based cuprates, as discussed next, completely contradicted the need for Cu-O chains for higher T_c's, since these materials do not contain Cu-O chains and exhibit even higher T_c's, greater than 100K.

3.3 $Bi_2Sr_2Ca_{n-1}Cu_nO_y$

The Bi system, which contains three phases of general formula

Cationic Substitutions in the High T_c Superconductors 329

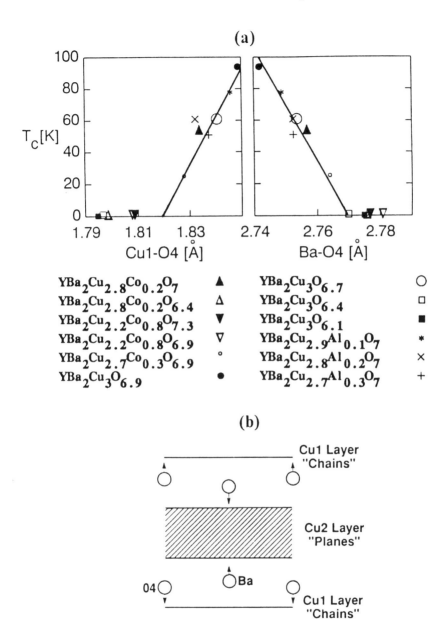

Figure 10: The variation of T_c with bond length is shown in (a) and in (b) is a schematic representation of the motion of the atoms (i.e., of the charge transfer) as the chemical doping at the Cu1 site proceeds. Note that this picture applies as well when the oxygen content in the material changes.

$Bi_2Sr_2Ca_{n-1}Cu_nO_y$ with n=1, 2 and 3 (49), provides another opportunity to test the trends established in the previous paragraphs dealing with the copper valence, low dimensionality structures, and magnetism. The structure of the Bi compounds can be viewed as a packing of CuO_2 layers interleaved with Ca layers and BiO layers and sandwiched between SrO layers and having, in addition, an incommensurate superstructure. The incommensurability of this structural modulation has prevented the complete determination of the structure and thereby the location of the extra oxygens. This, of course, impedes the understanding of the doping mechanism in these phases. To experimentally resolve this structural problem, compounds isostructural to the Bi cuprates but with a commensurate modulation are needed. This is another reason to study cationic substitutions in the Bi system. A complication with this system, compared to the 40K or 90K materials, is that there are three layered phases (n=1, 2 and 3) whose energy of formation is very close so it is difficult to obtain each phase free of stacking faults. This is particularly true for the n=3 phase which is always contaminated by n=2 faults (50). For this reason our chemical substitution studies have focused on the n=1 and n=2 phases. As an aid in selecting (based on crystal-chemical factors only) the kind of cationic substitutions that have any chance for success we go back to Figure 1 and emphasize the structural similarities between the Bi compounds and the previously discussed T, T' and 123 phases.

A striking similarity between the Bi and 123 phases is that the site occupied by Ca in the n=2 Bi phase is similar to that occupied by Y in 123 which suggests that Ca can be replaced by a RE in these materials. Single phase $Bi_2Sr_2Ca_{1-x}RE_xCu_2O_y$ materials with RE being most of the rare earths (exceptions are La, Pr, and Nd) were indeed synthesized for values of x ranging from 0 to 1 (51-53). Attempts to increase x (RE content) above 1 failed, which indicates that the rare earth substitution is limited to the Ca site. Magnetic measurements on the fully RE-doped Bi phases (x=1) have shown that the inverse susceptibility temperature dependence follows a Curie-Weiss law (Figure 11). From this data values of the magnetic moment per RE atom were determined and observed to be similar to those expected for free trivalent RE ions indicating, therefore, that the rare earths substituted for Ca in these materials exist as trivalent ions. Also deduced from the above fit were values of the paramagnetic Curie temperature (Θ_p) for each of the RE-doped Bi compounds. The value

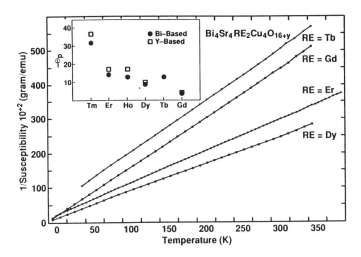

Figure 11: The inverse susceptibility temperature dependence is shown for several members of the $Bi_4Sr_4RE_2Cu_4O_{16+y}$ series. Data were collected in a field of 10kG. The inset shows a comparison of the Θ_p values for the RE-doped 90K and RE-doped Bismuth n=2 phases. Note the excellent agreement.

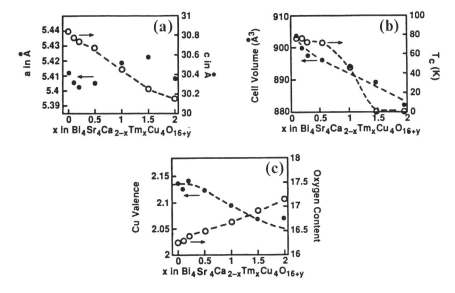

Figure 12: Lattice parameters (a), cell volumes and T_c's (b), copper valences and oxygen contents (c) are shown for compositions in the $Bi_4Sr_4Ca_{2-x}Tm_xCu_4O_{16+y}$ series.

of Θ_p reflects the magnetic interaction between neighboring atoms and depends strongly upon the crystallographic site occupied by the RE. The inset of Figure 11 shows the values of Θ_p obtained for both the RE-doped 123 and RE-doped Bi series. Note that the values are similar which confirms the crystallographic site similarity occupied by the RE in the 123 and in the Bi-based materials. Chemical analyses and TGA measurements have shown that, as a RE is substituted for Ca, additional oxygen is added to the structure with approximately one half oxygen added for each added rare earth (See Figure 12c for the Tm series). The formal valence of the copper, upon the RE substitution, remains constant and equal to 2.14 at low x (Figure 12c) and then, beyond x=0.5, drops continuously to reach 2.05 at x=2. Over the same composition range T_c remains constant at low x (Figure 12b) and then drops sharply around x=1 to become less than 4.2K at x=1.5. At x=2 the materials are semiconductors. The similarity of this curve to that depicting the Cu valence changes is striking and supports the general belief that T_c is related to the amount of Cu III. The fully rare earth substituted samples are semiconductors, even though the formal Cu valence is greater than 2. This indicates that for each system there is a threshold value for the copper formal valence for which T_c is maximum. We note also that the non-superconducting fully RE-doped Bi phases are isostructural (to the superconductors) and show a structural modulation that is non-commensurate. This clearly demonstrates, contrary to early conjecture, that in these materials the modulation has nothing to do with superconductivity.

Another structural similarity which exists between the crystallographic Sr site in the Bi phase, the Ba site in 123, and the La site in the T phase is that they are all nine-fold coordinated. We previously discussed how, in the 123 system, Ba can be replaced by a RE. This would again lead to the expectation that Sr can also be replaced by a RE in the Bi phase. Single-phase materials having compositions $Bi_2Sr_{2-x}La_xCuO_y$ with x=0 up to x=1 and RE=La, Pr and Nd were made and characterized (52,54). As previously noted these phases still show the structural modulation and for each added rare earth there is half an oxygen added to the structure. The compound evolves with increasing x from superconducting at 10K to non-superconducting, even though at x=1 the formal valence of the Cu is still greater than 2.

Finally we discuss a 3d metal substitution at the Cu site in the

Bi phases. We immediately recognize that in the n=1 Bi phase the Cu is six-fold coordinated as in the T phase, that in the n=2 phase the Cu is five-fold coordinated, and that in the n=3 phase the Cu is both five-fold and four-fold coordinated as in the 123 phase. From studies done on the 90K and 40K materials we know that the 3d metals do not (or barely so) substitute for five-fold coordinated Cu, but they do substitute for six-fold coordinated Cu as well as for square planar Cu. In the latter case the substitution was found to be possible only if accompanied by an uptake of oxygen such that the 3d metal ended up to be octahedrally coordinated. Thus, based on these observations, we first focus on the n=1 phase and indeed find that single phase $Bi_2Sr_2(M,Cu)O_y$ materials with M being Fe, Co and Mn can be prepared in bulk polycrystalline form or as single crystals using the corresponding 3d-metal oxide in excess (55-57). It is required that the atmosphere be more reducing as one moves down to the lightest 3d metal such that the $Bi_2(Sr,Ca)_2MnO_y$ phase forms only with a nitrogen ambient at temperatures of about 1000°C.

In the n=2 phase the Ca planes (free of oxygen) are sandwiched between the CuO_2 planes in which the Cu is five-fold coordinated. Fe or Co ions prefer to be six-fold coordinated. One way to achieve this is to replace the Ca by a Sr which will form a SrO layer thereby providing the extra oxygen required to produce a six-fold coordinated site. With this reasoning we simultaneously substituted 3d-metals (Fe, Co or Mn) for Cu and Sr for Ca. There was success in synthesizing the Bi phases having n=2 and 3 when M=Fe (58-60) and n=2 when M=Co (61), but with Mn neither the n=2 or n=3 phases could be synthesized (independent of the ambient or annealing temperature used) and only the n=1 phase could be formed. It is interesting to note that the n=1 phase becomes less stable as the ionic radius of the 3d metal decreases (in going from Mn to Fe), whereas the opposite is observed for the stability of the n=3 phase which is the most stable phase in the case of the Fe compound. The formulae of the newly synthesized phases obtained through cationic substitutions in the Bi compounds at the Cu sites are summarized in Table 1 together with the crystal data obtained from either x-ray powder diffraction experiments or single crystal analysis. A feature common to all these 3d substituted phases is that they are isostructural to the parent bismuth copper phases, but with the important difference that the structural modulation is commensurate. This commensurability

Table 1. Crystallographic data are presented for the new phases which resulted from a 3d-metal substitution at the copper site in the n=1, 2 and 3 Bi-based cuprates. The periodicity of the structural modulation (p) is given by bold numbers when obtained from single crystal x-ray studies, otherwise the values were obtained from TEM studies.

n =	Compound	p x a(Å)	b(Å)	c(Å)	V(Å)^3 subcell
1	$Bi_2Sr_2CuO_y$	5 x 5.379(5)	5.382(8)	24.51(4)	710.2
	$Bi_2Sr_2Cu_{0.5}Fe_{0.5}O_y$	4.2 x 5.412(4)	5.420(7)	23.92(3)	703.11
	$Bi_2Sr_2CoO_y$	4 x 5.4590(5)	5.4615(8)	23.450(4)	699.15
	$Bi_2Sr_2MnO_y$	4 x 5.451(2)	5.426(2)	23.613(8)	698.40
	$Bi_2Ca_2MnO_y$	4 x 5.343(2)	5.360(1)	23.151(2)	663.00
2	$Bi_2Sr_2CaCu_2O_y$	4.7 x 5.393(5)	5.393(5)	30.52(3)	887.66
	$Bi_2Sr_2SrFe_2O_y$	5 x 5.449(1)	5.4617(6)	31.690(4)	943.26
	$Bi_2Ba_2Ba_1Co_2O_y$	?	?	30.80(7)	?
	$Bi_2Sr_2SrCo_2O_y$	?	?	29.85(6)	?
	$Bi_2Ca_2CaCo_2O_y$	?	?	29.28(8)	?
3	$Bi_2Sr_2Ca_2Cu_3O_y$	? x 5.39(1)	5.39(1)	37.10(2)	1077.80
	$Bi_2Sr_2Sr_2Fe_3O_y$	4.9 x 5.418(1)	5.418(1)	37.10(1)	1089.33

Cationic Substitutions in the High T_c Superconductors 335

allowed the complete crystal structure to be solved for the n=2 Fe compound and for the n=1 Co and Mn phases. The solution, displayed in Figure 13, shows that the distortion associated with the modulation originates in the Bi-O planes and is associated with the presence of an extra row of oxygen atoms within the Bi layer (60). (This extra oxygen is undoubtedly present in the superconducting cuprates and certainly plays an important role in the anionic doping mechanism for these materials, but that is beyond the scope of this chapter.) These 3d-metal substituted phases, although they are non-superconducting, show interesting physical properties such as a novel magnetic behavior. The compounds with M=Co or Mn, for instance, are antiferromagnets but with their magnetic susceptibilities as a function of temperature (Figure 14a) peaking sharply and indicating a hidden ferromagnetism. In fact, $Bi_2Ca_2MnO_{6+y}$ (Figure 14b) behaves like a classic ferromagnet at low temperatures. These Mn-doped Bi compounds (n=1) can be grown as large crystals free of intergrowths (since higher n intergrowths do not exist for Mn, or are very difficult to make). Studies on such crystals may elucidate the magnetic behavior and thus, by extension, superconductivity in the cuprates since several theories have claimed that superconductivity and magnetism are related in these high T_c oxides.

4.0 DISCUSSION

From the study of cationic substitutions in the La, Y, and Bi-based cuprates a large amount of data has been obtained that we now try to systematize in order to develop the most prominent trends.

It is clear that the presence of CuO_2 planes is an essential feature required for superconductivity in these oxides. However, superconductivity can be achieved only if the chemical entity inserted between these planes allows them to electrically communicate. This is, for instance, the role played by the chains in the 123 or by the double Bi-O layers in the Bi compounds.

A substitution for Cu by Fe, Co or Mn in the $Bi_2Sr_2CuO_y$ phase produces a drastic change in the transport properties, in particular an electronic transition from superconducting to semiconducting for Fe or Co or to insulating for Mn. This indicates that the presence of 3d metal oxygen layers is not, by itself, sufficient for

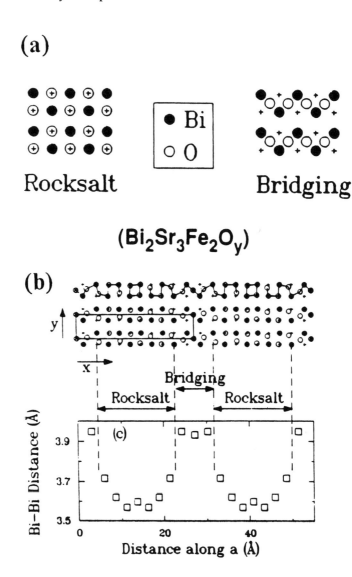

Figure 13: Two planar arrangements for the oxygen atoms, designated here rocksalt and bridging (perovskite), are shown in (a). The crosses refer to oxygens in rocksalt positions. In (b) are shown sections of the structure at z=0.06 through the Bi layer for the $Bi_2Sr_3Fe_2O_y$ phase (note that the Bi atoms form chains running along the **a** direction). The oxygen positions are shown in detail. Note that there is one oxygen inserted at the bridging position for every 10 rows of Bi atoms. In (c) the Bi-Bi distances are plotted with the same scale along x as in (b).

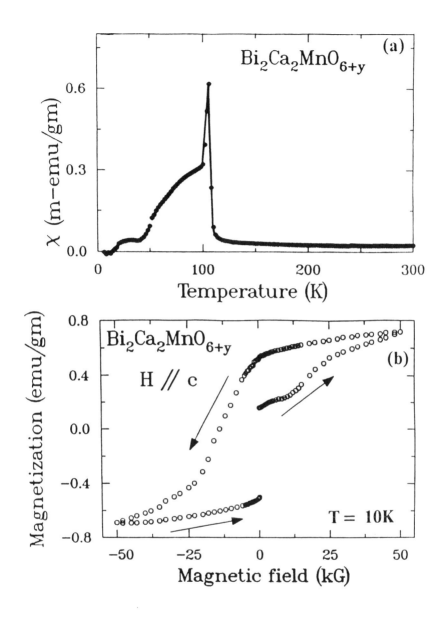

Figure 14: In (a) is shown the variation of the susceptibility with temperature for a $Bi_2Ca_2MnO_{6+y}$ crystal obtained while warming the sample in a field of 500G after having cooled the sample in zero field. The magnetization vs. field curve measured at 10K on the same sample is shown in (b).

superconductivity. In $Bi_2Sr_2CuO_y$ the Cu 3d levels mix with the O 2p levels such that the electrons collectively give rise to metallic behavior. In moving from Cu to those 3d-metals to the left in the periodic table, the energy difference between the 3d levels and O 2p levels progressively increases so that the resulting orbital overlap decreases producing a localization and thereby a semiconducting or insulating behavior. Thus, what is important for superconductivity is to have a 3d metal whose 3d states are strongly hybridized with the O 2p levels and that is the very special role played by Cu. We note that in the other classes of high T_c oxides (the bismuthates) there is also a strong hybridization between the Bi 6p and O 2p levels.

To better understand the dependence of T_c with respect to cationic substitutions we distinguish between those substitutions in the CuO_2 planes and those not in the CuO_2 planes.

We found that the T_c's of the out-of-CuO_2-plane substituted cuprates are nearly insensitive to any chemical substitutions (whether or not the substituted element is magnetic) which do not modify the average charge balance (e.g., RE for La in the 40K system, RE for Y or Sr for Ba in the 123 system). They are, however, extremely sensitive to those substitutions that create a charge imbalance and thereby modify the formal valence of the Cu through charge transfer (e.g., Sr for La in the 40K system, La for Ba in the 90K system or RE for Sr or Ca in the Bi system). Thus the formal valence of copper is very important for superconductivity. There is an optimum value for the Cu formal valence of copper which is different from system to system with a value of 2.15, 2.33, and $\cong 2.14$ for the La, Y, and Bi based cuprate phases. Any cationic substitution that leads to a deviation of the copper formal valence from the values just quoted results in a decrease in T_c. Furthermore, the non-superconducting behavior of the $Bi_2Sr_2RECu_2O_y$ compounds which contain 0.1 Cu (III), or of the non-superconducting $LaCuO_3$ phase which contains only Cu (III), indicates that the presence of Cu (III) in a cuprate does not a priori imply superconducting behavior.

In the $La_{1.85}Sr_{0.15}Cu_{1-x}M_xO_y$ (M=Ni, Zn) system T_c is destroyed for values of x greater than 0.025. The replacement of 0.025 Cu(2.15) by 0.025 Zn(II) or Ni will not change the formal valence of Cu by more than 0.004 if we assume that the Zn and Ni do not affect the oxygen content in these materials (as indicated by TGA measurements). However, even though no change in the formal copper valence is involved, the T_c of an in-CuO_2-plane substituted cuprate phase is extremely sensitive to any substitution and independent of the magnetic nature of the substituted element. This is a clear difference between these new high T_c oxides and the old conventional BCS superconductors which are extremely sensitive to magnetic impurities. Recall that in the 40K phase, for instance, the substitution of diamagnetic Zn for Cu induces a magnetic moment so that, at present, we cannot rule out the magnetic pair breaking argument as the origin of the depression of T_c. A large number of experiments, however, suggest that the presence of a local structural disorder is the more likely possibility for the depression in T_c that is observed as soon as the chemistry of the CuO_2 planes is affected.

In the superconducting La, Y, and Bi-based compounds the formal valence per Cu is greater than 2, and a linear relation between T_c and the number of Cu(III) (i.e., number of holes) for all of these oxides has been established by muon spin resonance (62). In these high T_c oxides the carriers are holes, and thus called p-type superconductors, and it was postulated that only cuprates with holes (oxidized Cu) as the carriers could be superconductors. Thus, observations in the Nd-Ce-Cu-O system in which superconductivity is achieved by reducing the Cu (i.e., by giving electrons) are very important. The Ce doped neodymium materials are called n-type superconductors. Recent Hall measurements on superconducting $Nd_{1.85}Ce_{0.15}CuO_y$ single crystals show a positive Hall coefficient and may be considered contradictory to the above distinction. However, we show next, by discussing Figure 2, that this distinction is unambiguous from a chemical point of view.

Figure 2 summarizes the variation of the tetragonal lattice parameters **a** and **c** for four series of compounds. Two series, $La_{1.6}RE_{0.2}Sr_{0.2}CuO_4$ and $La_{2-x}Sr_xCuO_4$, adopt the T crystal structure whereas the two other series, $Nd_{2-x}Ce_xCuO_4$ and $Nd_{2-x}La_xCuO_4$, adopt the T' structure. Note that, independent of whether it is the T or T' phase, when no charge transfer is involved both the **a** and **c** lattice parameters change in a similar way either increasing (Figure 2d) or decreasing (Figure 2b). In contrast, for the Sr-doped series or the Ce-doped series the **a** and **c** axes change in an opposite direction. For the Sr series **a** decreases and **c** increases whereas the opposite is observed for the Nd-doped series. The variation of the c-axis in this system can be understood based on the ionic radius of the dopant vs.

that of the substituent. In the RE-doped T-phase series, for instance, as the ionic radius of the substituent decreases one would expect a decrease in both the c and a axes as is observed. In the Sr-doped La phase the La is 8-fold coordinated. The ionic radii of an 8-coordinated La and an 8-coordinated Sr are 1.3Å and 1.45Å respectively and thus the replacement of the La by a larger ion, Sr, is expected to increase both the c and a axes. The data show that only c increases. In the T' phase the Nd is 8-fold coordinated and thus the replacement of Nd (1.25Å) by a smaller Ce ion (1.11Å) is expected to decrease both the c and a axes. Only a decrease of c is observed. Finally, when Nd is replaced by a larger La ion (1.25Å vs. 1.30Å) one would expect an increase of both the a and c axes as is observed experimentally. Thus, based on ionic radii considerations alone, one can explain satisfactorily the variation of the c axis for all the series and the variation of the a axis only for those series for which the hole or electron concentration is not affected. This indicates that charge transfer plays a key role in the variation of a for both the Sr-doped and Nd-based phases. It was shown by Whangbo et al. (63) that the in-plane Cu-O bonds have an antibonding character in the CuO_2 layer x^2-y^2 bands. The substitution for La^{+3} by Sr^{+2} corresponds to removing electrons from the x^2-y^2 bands. Thus, the removal of an antibonding electron would be expected to decrease the Cu-O bond length and thereby the a axis as is observed. In contrast, the replacement of Nd^{+3} by Ce^{+4} corresponds to adding electrons into the antibonding band orbital so that a lengthening of the Cu-O bond, and thereby of the a axis, is expected and is observed. Thus we have demonstrated that the variation of the a-axis (i.e., the in-plane Cu-Cu distance) is consistent with the T phase being doped by holes (p-type) and the T' being doped by electrons (n-type).

Finally we recall that in the Sr-doped system the maximum T_c is achieved for a Sr content of 0.15, which corresponds to a formal valence per Cu of 2.15 (i.e., 0.15 holes per Cu). In the Ce-doped system the maximum in T_c (24K) is achieved for a Ce content of 0.15, which corresponds to a formal valence per Cu of 1.85 (i.e., 0.15 electrons per Cu). Thus, a maximum T_c occurs in the n-type materials at the same electron concentration as the hole concentration in the p-type materials which produces a maximum T_c and indicates a doping symmetry. The symmetry between the n-type and p-type materials is further reflected in Figure 4 where we have shown that for each system the T_c dependence upon the in-plane Cu-Cu distance is similar, but with a difference that the critical distance at which T_c is maximum is different for the two systems. Thus, from a chemical point of view the n-type materials are identical to the p-types, but there is a significant electronic difference in that the carriers are electrons and not holes. The T' superconducting materials (n-type), to a certain extent, can be viewed as a mirror image of the superconducting T materials (p-type) which leads to the possibility of mirror images of the 123 and Bi phases. Thus far we have failed in this direction.

In summary, we demonstrated that cationic substitution studies done on known compounds can be a powerful tool to use to answer important crystal-chemical questions (e.g. the origin of the modulation in the Bi phases), to synthesize new phases (e.g., Fe, Co or Mn bismuth phases) and, to establish general phenomenological trends that will contribute to a better understanding of the factors that give rise to the existence of superconductivity in these new materials (the electron-type superconductors, for instance, stand as a serious test for new or existing theories). Although this chapter has only addressed cationic substitutions, the reader should be aware that important chemical issues are also addressed by looking at the anion sites in these materials (changes in oxygen content, fluorine substitution). Finally, we wish to leave the reader with the view that solid state chemistry is a powerful and fascinating science not only because of the novel materials with exciting properties that have been, or will be, produced but also because the results are essential for testing theoretical models.

ACKNOWLEDGEMENTS

We wish to thank P. Barboux, L. H. Greene, P. F. Miceli, W. R. McKinnon, Y. LePage, E. Wang, R. Ramesh, J. Barner, and J. H. Wernick for valuable discussions.

5.0 REFERENCES

1. G.Bednorz and K.A.Muller, *Z.Phys.* B64:189 (1986).
2. M.K.Wu, J.R.Ashburn, C.J.Torng, P.H.Hor, R.L.Meng, L.Gao, Z.J.Huang, Y.Q.Wang, and C.W.Chu, *Phys. Rev. Lett.* 58:908 (1987).
3. H.Maeda, Y.Tanaka, M.Fuksutumi, and T.Asano, *Japn. J. Appl. Phys.* 27:L209 (1988).
4. Z.Z.Sheng and A.M.Hermann, *Nature* 332:138(1988).
5. L.F.Mattheis and D.R.Hamman, *Solid State Commun.* 63:395 (1987).
6. K.Kishio, K.Kitazawa, S.Kanbe, N.Sugii, H.Takagi, S.Uchida, K.Fueki, and S.Tanaka, *Chem. Lett.* (Japan) 429 (1987).
7. R.J.Cava, R.B.van Dover, B.Batlogg and E.A.Rietman, *Phys.Rev.Lett.* 58:408 (1987).
8. J.M.Tarascon, L.H.Greene, W.R.McKinnon, G.W.Hull, and T.H.Geballe, *Science* 235:1373 (1987).
9. Y.Tokura, H.Takagi and S.Uchida, *Nature*, 337:345 (1989).
10. J.M.Longo, and P.M.Raccah, *J.Solid State Chem.* 6:526 (1973).
11. H.Muller-Buschbaum, *Angew. Chem. Int. Ed. Eng.* 16:674 (1977).
12. Y. LePage, W.R. McKinnon, J.M. Tarascon, L.H. Greene, G.W. Hull, and D.M. Hwang, *Phys. Rev.* B35:7249(1987).
13. J.M. Tarascon, Y. Le Page, P. Barboux, B.G. Bagley, L.H. Greene, W.R. McKinnon, G.W. Hull, M. Giroud and D.M. Hwang, *Phys. Rev.* B37:9382 (1988).
14. S.A. Sunshine et al., *Phys. Rev.* B38:893 (1988).
15. C.C. Torardi, M.A. Subramanian, J.C. Calabrese, J. Gopalakrishnan, K.J. Morissey, T.R. Askew, R.B. Flippen, U. Chowdry and A.W. Sleight, *Phys. Rev.* B38:2504 (1988).
16. J.M. Tarascon, P. Barboux, B.G. Bagley, L.H. Greene, W.R. McKinnon, and G. W. Hull in, "*Chemistry of High Temperature Superconductors*", edited by D.L. Nelson, M.S. Whittingham, and T.F. George, American Chemical Society, Washington DC, p 198 (1987).
17. P. Barboux, J.M. Tarascon, L.H. Greene, G.W. Hull and B.G. Bagley, *J. Appl. Phys.* 63:1768 (1987).

18. P. Barboux, J.M. Tarascon, F. Shokoohi, B.J. Wilkens and C.L. Schwartz, *J. Appl. Phys.* 64, 6382 (1988).
19. J.B. Torrance, Y. Tokura, A.I. Nazzal, A. Bezinge, T.C. Huang, and S.S. Parkin, *Phys. Rev. Lett.* 61:1127 (1988).
20. Z.Z. Wang, J. Clayhold, N.P. Ong, J.M. Tarascon, L.H. Greene, and G.W. Hull, *Phys. Rev.* B36:7222 (1987); P. Chaudari et al., *Phys. Rev.* B36:8903 (1987).
21. D.W. Murphy, L.F. Schneemeyer and J.V. Waszczak, in *"Chemistry of High Temperature Superconductors II"*, ed. T. F. George and D.L. Nelson, American Chemical Society, Washington DC, p. 315 (1988).
22. M.H. Whangbo, D.B. Kang and C.C. Torardi, *Physica C*, 158:371 (1989)
23. J.M. Tarascon, L.H. Greene, W.R. McKinnon, and G.W. Hull, *Solid State Commun.* 63:499 (1987).
24. E. Takayama-Muromavhi, Y. Matsui, Y. Uchida, F. Izumi, M. Onoda and K. Kato, *Jpn. J. Appl. Phys.* 27:L2283 (1988).
25. S.-W Cheong, Z. Fisk, J.D. Thompson and R.B. Schwarz, *Physica C* 159:407 (1989).
26. H. Takagi, S. Uchida, T. Tokura, *Phys. Rev. Lett.* 62:1197 (1989).
27. Y. Hidaka and M. Suzuki, *Nature* 388:635 (1989).
28. J.M. Tarascon, E.Wang, L.H. Greene, B.G. Bagley, G.W. Hull, S.M. D'Egidio, P.F. Miceli, Z.Z. Wang, T.W. Jing, J. Clayhold, D. Brawner and N.P. Ong, *Phys. Rev. B* 40:4494 (1989); J.M. Tarascon, E. Wang, L.H. Greene, R.Ramesh, B.G. Bagley, G.W. Hull, P.F. Miceli, Z.Z. Wang, D. Brawner and N.P. Ong, *Physica C* 162-164:285 (1989).
29. J.T. Markert, E.A. Early, T. Bjornholm, S. Ghamtay, B.W. Lee, J.J. Neumeier, R.D. Price, C.L. Seaman and M.B. Maple, *Physica C* 158:178(1989); J.T. Markert and B. Maple, *Solid State Commun.* 70:145 (1989).
30. E. Wang, J.M. Tarascon, L.H. Greene, G.W. Hull and W.R. McKinnon, *Phys. Rev. B* 41:6582 (1990).
31. J.M. Tarascon, L.H. Greene, P. Barboux, W.R. McKinnon, G.W. Hull, T.P. Orlando, K.A. Delin, S. Foner, and E.J. McNiff, *Phys. Rev. B* 36:8393 (1987).
32. G. Xiao, A. Bakhshai, M.Z. Cieplak, Z. Tesanovic and C.L. Chien, *Phys. Rev. B* 39:315 (1989).

33. L.H. Greene, J.M. Tarascon, B.G. Bagley, P. Barboux, W.R. McKinnon and G.W. Hull, Rev. *Solid State Science* 1:199 (1987).

34. J.M. Tarascon, P. Barboux, P.F. Miceli, L.H. Greene and G.W. Hull, *Phys. Rev. B* 37:7458 (1988).

35. J.M. Tarascon, L.H. Greene, B.G. Bagley, W.R. McKinnon, P. Barboux, and G.W. Hull, in "*Novel Superconductivity*", edited by S.A. Stuart and V. Z. Kresin, Plenum, New York, and London, 1987, p. 705.

36. S. Sunshine, L.F. Schneemeyer, T. Siegrest, D.C. Douglas, J. Waszczak, R.J. Cava, F.M. Gyorgy, and D.W. Murphy, *Chem. of Mat.* 1:331 (1989).

37. J. M. Tarascon, W. R. McKinnon, L. H. Greene, G. W. Hull and E. M. Vogel, *Phys. Rev. B* 36:226 (1987).

38. R.J. Cava, B. Batlogg, R.M. Fleming, S.A. Sunshine, A. Ramirez, E.A. Rietman, S.M. Zahurak, R.B. van Dover, *Phys. Rev. B* 37:5912 (1988).

39. J.M. Tarascon, P. Barboux, P.F. Miceli, L.H. Greene, and G.W. Hull, *Phys. Rev. B* 37:7458 (1988).

40. G. Xiao, F. H. Streitz, A. Gavrin, Y. W. Du and C. L. Chien, *Phys. Rev. B* 35:8782 (1987).

41. Y. Maeno, T. Tomita, M. Kyogoku, S. Awaji, Y.A. Oki, K. Hoshino, A.A. Minami and T. Fujita, *Nature* 328:512 (1987).

42. M. Eibschutz, M. Lines, J.M. Tarascon, and P. Barboux, *Phys. Rev. B* 38:2896 (1988).

43. P.F Miceli, J.M. Tarascon, L.H. Greene, P. Barboux, F.J. Rotella and J.D. Jorgensen, *Phys. Rev. B* 37:5932 (1988).

44. P. Bordet, J.L. Hodeau, P. Strobel, M. Marezio, and A. Santoro, *Solid State Commun.* 66:435 (1988).

45. G. Xiao, M.Z. Cieplak, D. Musser, A. Gavrin, F.H. Streitz, C.L. Chien, J.J. Rhyne and J.A. Gotaas, *Nature* 332:238 (1988).

46. R.S. Howland, T.H. Geballe, S.S. Laderman, A. Fischer-Colbrie, M. Scott, J.M. Tarascon and P. Barboux, *Phys. Rev. B* 39:9017 (1989).

47. J.M Tarascon, P. Barboux, L.H. Greene, B.G. Bagley, P.F. Miceli and G.W. Hull in, "*High Temperature Superconductivity: The First Two Years* " edited by R.M. Metzger, p. 199 (1988).

48. P.F. Miceli, J.M. Tarascon, L.H. Greene, P. Barboux, J.D. Jorgensen, J.J. Rhyne and D.A. Neumann, *Materials Research Soc. Symp. Proc.* Vol. 156:119 (1989).
49. J. M. Tarascon, W. R. McKinnon, P. Barboux, D. M. Hwang, B. G. Bagley, L. H. Greene, G. W. Hull, Y. LePage, N. Stoffel and M. Giroud, *Phys. Rev. B* 38:8885 (1988).
50. J.M. Tarascon, Y. LePage, L.H. Greene, B.G. Bagley, P. Barboux, D.M. Hwang, G.W. Hull, W.R. McKinnon and M. Giroud, *Phys. Rev. B* 38:2504 (1988).
51. J.M. Tarascon, P. Barboux, G.W. Hull, R. Ramesh, L.H. Greene, M. Giroud, M.S. Hegde and W.R. McKinnon, *Phys. Rev. B* 38:4316 (1989).
52. A. Manthiran, J.B. Goodenough, *Appl. Phys. Lett.* 53:420 (1988).
53. B. Chevalier, B. Lepine, A. Lelirzin, J. Darriet, J. Etourneau and J.M. Tarascon, *Mat. Sci. Eng. B* 2:277 (1989).
54. J. Darriet, C.J.P. Soethout, B. Chevalier and J. Etourneau, *Solid State Commun.* 69:1093 (1989).
55. J.M. Tarascon, P.F. Miceli, P.F. Barboux, D.M. Hwang, G.W. Hull, M. Giroud, L.H. Greene, Y. LePage, W.R. McKinnon, E. Tselepsis, G. Pleizier, M. Eibschutz, D.A. Neuman, and J.J. Rhynes, *Phys. Rev. B* 39:11587 (1989).
56. J.M. Tarascon, Y. LePage, W.R. McKinnon, E. Tselepsis, P. Barboux, B.G. Bagley and R. Ramesh, *Materials Research Soc. Symp. Proc.* Vol. 156:317 (189).
57. W.R. McKinnon, E. Tselepis, Y. LePage, S.P. McAlister, G. Pleizier, J.M. Tarascon, P.F. Miceli, R.Ramesh, G.W. Hull, J.V Waszczak, J.J. Rhyne and D.A. Neumann, *Phys. Rev. B* 41:4489 (1990).
58. B.G. Bagley, J.M. Tarascon, P. Barboux, Y. LePage, W.R. McKinnon, L.H. Greene, and G.W. Hull, The 1988 Fall Meeting of the MRS (Boston).
59. R. Retoux, C. Michel, M. Hervieu, N. Nguyen and B. Raveau, *Solid State Comm.* 69:599 (1988).
60. Y. Le Page, W. R. McKinnon, J. M. Tarascon and P. Barboux, *Phys. Rev. B* 40:6810 (1989).
61. J.M. Tarascon, R. Ramesh, P. Barboux, M.S. Hedge, G.W. Hull, L.H. Greene, and M. Giroud, *Solid State Commun.* (1989).

62. Y.J. Uemura et al., *Journal de Physique*, C8:2087 (1988).
63. M.-H. Whangbo, M. Evain, M. Beno, and J.M. Williams, *Inorg. Chem.* 26:1829 (1987).

9

The Chemistry of High T_c in the Bismuth Based Oxide Superconductors $BaPb_{1-x}Bi_xO_3$ and $Ba_{1-x}K_xBiO_3$

Michael L. Norton

1.0 INTRODUCTION

Rarely in the study of superconducting oxides does one encounter the beautifully colored materials observed among the bismuthates. They vary in color from red to yellow-gold to irridescent blue. These changes in color reflect the powerful changes in electronic properties brought about through chemical modification of these systems. They also belie the unexpected complexity of these "simple" perovskite materials.

It has been remarked that the bismuthate materials are so simple that, if we cannot understand the mechanism of superconductivity operative in them, then we have no hope of understanding the more complex cuprates. Throughout this chapter we will find that incontrovertible evidence for one mechanism or another has not been forwarded, despite considerable research on the part of many capable scientists. The available data concerning preparation, physical and electronic properties, and theories pertaining to these materials will be considered here, and their import for researchers pursuing the synthesis of these or even higher transition temperature (T_c) ceramic materials will be highlighted. We may begin our study of these materials by reviewing the idealized structures and stoichiometries, and their impact on the physical properties.

2.0 THEORETICAL UNDERPINNINGS OF BISMUTHATE RESEARCH

The simple or idealized perovskite structure is shown in Figure 1.

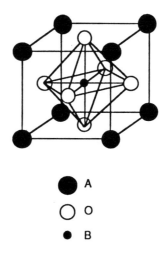

● A
○ O
● B

Figure 1: Representation of the ideal Perovskite structure.

This structure or a variant of it, is adopted by a large number of compounds with stoichiometries 1:1:3 or ABO_3. The compound is considered to be very simple in that the unit cell only contains 5 atoms, 1/8 of an A atom at each of the eight cell corners, one B atom at the center of the cell, and six halves of O atoms, one in the center of each face in the cell. The crystal generated from this unit cell is then considered in a polyhedral description as consisting of a regular array of BO_6 or MO_6 octahedra joined by sharing O atoms at the vertices. A two dimensional view of this metal-oxygen framework is presented in Figure 2.

In the two nonstoichiometric compounds which are the focus of this chapter, $BaPb_{1-x}Bi_xO_3$ and $Ba_{1-x}K_xBiO_3$, the A lattice sites are occupied by Ba or by a mixture of Ba and K, respectively. The B sites are then occupied by a mixture of Pb and Bi in $BaPb_{1-x}Bi_xO_3$, and by Bi only in $Ba_{1-x}K_xBiO_3$.

Let us consider the impact of doping (variation in x) in these

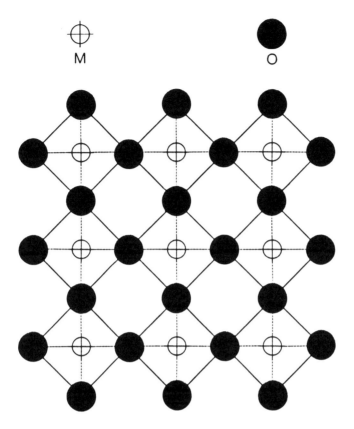

Figure 2: Metal-oxygen sublattice present in ideal perovskite.

materials. We may consider both families as originating from the doping of the bronze colored compound $BaBiO_3$. In this end member, normal chemical valence considerations would lead us to consider the Ba ion as +2, the O atoms as -2 leaving the Bi ion with a very high formal charge or valence of +4. Whereas this is a common oxidation state (although not exceedingly stable) for Pb, it is highly unstable for Bi. Bi's most stable cationic states are +3 and +5, where +5 is not notably stable. In order for the Bi ion to satisfy the requirements of +4 average valence, yet remain in stable oxidation states, two of these fictitious +4 ions must disproportionate into one +3 and one +5 ion. These two different bismuth ions must necessarily have different ionic radii, with the radius of the more reduced, and electronically polarizable +3 ion being the larger of the two. In our simple model of the perovskite unit cell we are now faced with a problem. When the occupants of the B site in two adjoining unit cells are not the same size, then the unit cells are not the same size. Thus, for $BaBiO_3$, we necessarily cannot have the simple perovskite structure (assuming the charge disproportionation discussed above). It was such reasoning that lead to the discovery that the unit cell structure of $BaBiO_3$ contains two sizes of Bi (1), a large and presumably electron rich ion, and a smaller, presumably highly electronegative ion. Of course the chemist's valence assignments do not and cannot represent actual electronic charges on the ions, especially in these highly covalent (or semicovalent) systems any more than the valence of +4 of carbon in a saturated hydrocarbon denotes its charge. The inequivalence of the two types of Bi ions are, however, well predicted by this chemical bookkeeping.

In a simplification of this structure we could view the structure of $BaBiO_3$ as consisting of a three dimensional array composed by alternately linking large and small octahedra at their corners. This material, even though it is a mixed valence compound, is not conducting at room temperature. An understanding of this property is essential to the understanding of the insulator to metal transition brought about by doping, and has great relevance to the superconducting properties of the materials as well. Of course conductivity involves the transfer of at least one electron from, in this case, one metal B site to another B site, through an intervening O site. The geometry diagramed in Figure 1 in which nearest neighbor metal ion coordination polyhedra are joined at vertices may be contrasted with other systems in which the octahedral polyhedra are joined at an edge or a face. Such edge or face sharing allows direct overlap of the metal orbitals, and conduction bands formed via this geometry have little if any nonmetal character. Some oxygen character of the conduction bands may be required for high T_c, and this may explain the central importance of the perovskite structure in this field. Electron transfer can require large excitation or activation energies if a large reorganization of the lattice is required to accommodate the new valence arrangement. In this particular example, a transfer of one electron from a large, electron rich metal site to a small, electrophilic ion would yield a transition from a configuration in which all 6 of the oxygens nearest the small ion would suddenly have to expand against the 6 neighboring octahedra. The associated compression of these neighboring octahedra would mean that their excess charge would have to move to nearby octahedra. Thus, a large perturbation of the lattice, viewed most simply as a local volume fluctuation, could only be stabilized by a complete reversal of the valencies of every Bi ion in the crystal. Although such fluctuations certainly occur locally, the observation of two sizes of ions by the neutron diffraction experiments (1) does indicate that the valency of the Bi ions is locked in. A static, periodic distortion of the lattice traps the charge carriers in their places. Because the charge alternates from one Bi to the next in a periodic manner, this end member is said to display a charge density wave (CDW) phase or structure.

The identification of the valence of the B site ion through the determination of the size of the octahedra, or the metal-oxygen bond length, leads directly to an understanding of the concept of electron-phonon coupling. Bismuth ions in small octahedral cages of oxygens must be low in electronic volume. Bismuth ions in the larger cages, i.e. displaying larger bond lengths, have a larger number of outer electrons. If we consider a vibrational mode in which all 6 oxygen atoms move in and out with respect to the central Bi ion, then we see the possibility for the octahedra surrounding a Bi ion to change from that appropriate for a +3 ion (large) to the small size appropriate for a +5 ion. This oscillation in size can be called a breathing mode, and, when these motions occur in a solid they are called phonons. If charge can flow into and out of the central ion position following this breathing motion, then the charge motion, which may be very fast (almost instantaneous) is correlated with the atomic motions, which are considered slow in comparison with electronic transition times. This correlation is the physical expression of electron phonon coupling. The above phonon type is only one of many available to the perovskite structure, but it well demonstrates the possibility of strong electron phonon interaction (EPI) in oxygen coordinated materials. In fact, we could consider the static CDW observed in $BaBiO_3$ to be the result of electron trapping by an overly strong electron phonon interaction. The static CDW could be called a frozen phonon, since it is not a freely oscillating oxygen breathing mode.

As was mentioned above, the motion of one electron in these systems affects the motion of other electrons in the system. Therefore, the electron - phonon interaction can be said to mediate electron

- electron interactions, and particularly pairing interactions in superconducting systems.

In the solid solution series $BaPb_{1-x}Bi_xO_3$, where Pb atoms are substituted for Bi atoms present at the B sites in the parent compound $BaBiO_3$, a transition from the CDW or semiconductor state to a metallic state is observed at approximately $x = .35$ or for approximately 65% Pb occupation of the B site in the ideal perovskite structure (2). Obviously Pb substitution has lead to a decrease in the self trapping (or carrier freeze out) in this material. One contribution that this lead substitution makes is that it disorganizes the charge disproportionated bismuth centers, thus weakening their phonon mediated physical interactions with each other. Aside from simply physically separating and insulating the Bi ions from each other, the Pb ions have another significant effect. Because on the average the majority of the neighboring B site ions about each Bi ion will be Pb, the average local lattice parameter will also be that of a +4 ion. This pseudo substrate or lattice strain effect could be expected to influence the two types of Bi in opposite manners: the +5 and +3 cells would both tend toward the average size of a +4 ion. This would decrease the atomic reorganizational energy required for electron transfer, and may be the source of delocalization or of the insulator to metal transition in this material. Equivalence of all Bi sites has been demonstrated in the superconducting composition with $x = .25$ via EXAFS studies (3). These Bi octahedra are not the same size as the Pb octahedra, however.

We may now consider the impact of doping or substituting K atoms into the Ba sites in the parent $BaBiO_3$ material. Since the total valence of the A + B sites must remain constant (+6), the valence of the B site ion (Bi) must compensate any loss of valence at the A site. For every K added into the lattice, one Bi must change to a higher valence state. Since +4 valence is not reasonable in the CDW phase, we must consider that for every two K ions inserted, one +3 valence Bi is converted into a +5 ion. K substitution then leads to a lattice which becomes more disordered as K concentration goes up, and again the self trapping energy must decrease. At a critical concentration, the material becomes metallic (the electrons become delocalized). Although the precise value of this composition has not been well determined, it is clear that the CDW or distortion amplitude changes rapidly as a function of K substitution (4). In view of the fact that the concentration of bismuth ions, which, as described above, have a very

Bismuth Based Oxide Superconductors $BaPb_{1-x}Bi_xO_3$ and $Ba_{1-x}K_xBiO_3$

strong electron-lattice interaction, is very high in $Ba_{1-x}K_xBiO_3$, it is understandable that it might display physical properties (i.e. a higher T_c) reflecting this stronger interaction.

We may now consider the most basic form of the relationship between lattice properties and one of the most important physical properties of the bismuthates, the superconducting T_c, as derived from the theories of Bardeen, Cooper and Schrieffer:

$$T_c \sim T_D \exp(-1/K_{ef})$$

where: T_c = superconducting transition temperature
T_D = Debye temperature of solid
K_{ef} = effective electron-electron attractive force (5).

From a consideration of this simple formula we can understand the purpose of much of the research to characterize the bismuthates, as well as the research aimed at discovering materials with higher transition temperatures. We can also understand why the bismuthates have been the center of considerable attention after their discovery since their properties (discussed below) seem to conform poorly to the theoretically expected parameters. From the formula we see that the transition temperature is proportional to the Debye temperature of a solid. This is meant to indicate that the higher the frequency of the lattice vibrational mode coupling to electron motion in the solid, the more times per second the electron-electron interaction is "renewed". This yields a stronger total electron-electron interaction, which means that it is more stable against disruption by thermally excited phonons or lattice vibrations. The chemist has two ways to control or adjust vibrational frequencies, since they are dependent upon both the spring constant (i.e. the bond strength), and the mass (i.e. the molecular weight) of a harmonic oscillator. Metal atoms with variable valence yet high charges can be used to produce strong bonds. Since lighter atoms have higher vibrational frequencies, materials containing O, N, C or B are favored candidates for displaying reasonably high T_c's, assuming that these are the atoms in the structure which provide for the electron-phonon coupling interaction through their vibrations. K_{ef} represents the exponential dependence of T_c on the coupling of electron motion with the mode driving superconductivity. The frequency, as well as the type of mode are the most important chemically manipulatable aspects of a superconducting material.

The discussion so far has been based upon the idealized perovskite structure. At this point we may consider the real crystal structures observed in these materials.

3.0 CRYSTALLOGRAPHY

The crystallography of these systems certainly reflects the fact that the electronic and bonding properties are much more complex than the simple models provided in the previous discussion.

An overview of the crystallographic findings in the $BaPb_{1-x}Bi_xO_3$ system has been published by Batlogg et al. (6). The structure changes as a function of x or mole fraction Bi. From the pure lead end member $BaPbO_3$ to $x = .05$, the compound displays an orthorhombic structure. From .05 to .35 the compound is a tetragonal superconductor with its highest T_c near $x = .25$. From the metal to insulator composition of $x = .35$ until $x = .95$ the material adopts an orthorhombic structure. All compositions from $x = .95$ including the pure bismuth end member $BaBiO_3$ are monoclinic. Such structural results have mostly been based upon powder diffraction data. Neutron diffraction has been a major force in this research both because codes for powder pattern fitting are at a high state of development, and because the oxygen scattering of neutrons is more significant than that of X-rays. It is, however, possible that greater disorder, perhaps due to fluctuations in the compositional homogeneity, is present in powder samples than in single crystal samples, especially near the very important phase boundaries present in such abundance in these phases. This provides strong motivation for the development of crystal growth techniques for these temperature sensitive phases. It may be noted that samples studied at different facilities may vary in oxygen or even Ba/Pb stoichiometry, and that, although the assignment of the tetragonal phase to all superconducting compositions is well supported by synchrotron studies (7), not all laboratories agree on the symmetries present in this composition range (8)(9).

The development of a low temperature crystal growth technique utilizing KOH as a flux by Wignacourt et al. (10) and his demonstration of its utility for studying the composition - structure relationships in the $Ba_{1-x}K_xBiO_3$ system has lead Schneemeyer et al (4) to the best structural determinations for single crystal samples. It should be noted that these crystals are reported to have large mosaic spreads, which is often indicative of compositional inhomogeneity. The structural results indicate that the $Ba_{1-x}K_xBiO_3$ material transforms from the monoclinic structure of the bismuth end member $BaBiO_3$ to an orthorhombic structure at very low mole fraction of K i.e. for x less than or equal to .05.

In this orthorhombic structure, superconductivity is not observed, and two inequivalent Bi sites are observed. A cubic, but non-superconducting specimen has been reported (10) for x = .13. In this structure, all Bi sites are equivalent, and a metallic or delocalized electronic structure would be expected to exist. This structure apparently persists into the superconducting compositions, where it has been observed for x = .374 (4) which displays a T_c of 30.5 K. A cubic phase has also been reported for x = .6, with T_c = 34K. The limit of this solid solution which may be at or beyond this x = .6 composition, may be dependent upon synthesis temperature and technique. Indications of phase disproportionation at these high x values have been published by Jones et al. (11).

The $Ba_{1-x}K_xBiO_3$ phase may not be stable under conditions used for high resolution transmission electron microscopy (12). This could impede studies of domain structures which may be present in the "single" crystals that structural research now revolves around.

In summary, the superconducting compositions appear to have a tetragonal structure for $BaPb_{1-x}Bi_xO_3$, and a cubic structure for $Ba_{1-x}K_xBiO_3$. Dopant atom ordering has not been reported in structural studies.

4.0 MATERIALS PREPARATION

4.1 Bulk Growth

Sleight's discovery of superconductivity at 13K in polycrystalline samples of $BaPb_{1-x}Bi_xO_3$ (13) ignited interest in this system.

The apparent thermal stabilities of the two families, $BaPb_{1-x}Bi_xO_3$ and $Ba_{1-x}K_xBiO_3$ are quite different. Under the normal conditions of ceramic synthesis, the oxidation states desired in $BaPb_{1-x}$-

Bi_xO_3 are quite readily achieved (14). The synthetic procedure reported by Khan et al (15) consists of mixing stoichiometric quantities of $BaCO_3$, PbO_2 and Bi_2O_3, firing for 24h at 800°C, remixing and refiring for 24h at 850°C, followed by pressing into pellets and air annealing for 24h at 930°C. Because of the fluxing action of both Bi and Pb oxides, a platinum liner for the alumina crucibles is used to avoid impurities and nonstoichiometry. This reaction has been studied using thermal analysis by Gilbert et al (16). The simplicity of powder production enabled immediate preliminary studies of Ba substitution by K, Cs, Rb, Na and Sr by Sleight (17), and later Sr (18)(19) and Sn (19) substitution were studied in more detail.

It was obvious from the beginning (20)(21) that powder synthesis is not so simple for $Ba_{1-x}K_xBiO_3$, which displays two types of thermal instability. First, the K content of $Ba_{1-x}K_xBiO_3$ grown in an open system is strongly affected by the time-temperature parameters of the sample preparation. Both high temperatures and long time firings can lead to products quite different from their nominal (original) stoichiometries. A second form of instability revolves around the oxidation state of Bi, which is mixed in the desired product. It has been shown that firing the material at temperatures above 450°C results in the reduction of the oxygen content of the material and thus of the Bismuth oxidation state (22). An optimized synthesis resulting in single phase materials maintaining nominal K content has been suggested (22). This consists of mixing stoichiometric quantities of BaO, KO_2, and dried Bi_2O_3 and firing in dry nitrogen gas at 700°C for 1 h. Ramping rates are not important since a sample of reduced oxygen content is the desired result. This material is then annealed in oxygen for 4h at 450°C, followed by a slow cooling to room temperature. A similar procedure for making dense ceramics has been published (23). The two step synthesis strongly resembles the preparative procedures used for $YBa_2Cu_3O_7$, where a low temperature anneal is required to reach stoichiometric oxygen contents. The thermal instability of these two high T_c materials is no coincidence, and represents the instability of the bonding states involved in the materials' cohesion and in their superconductivity.

4.2 Crystal Growth

There are two major types of crystal growth for these materials with variable oxidation states, synthetic crystallization and recrystallization. The techniques used for the preparation of $Ba_{1-x}K_xBiO_3$ fall into the former category, while those producing $BaPb_{1-x}Bi_xO_3$ fall into both categories. The methods used to produce $BaPb_{1-x}Bi_xO_3$ will be considered first, in the order of increasing difficulty of harvesting undamaged single crystals.

A hydrothermal technique operating at 450°C has been shown to yield samples with narrow transition temperature widths (24)(25). Growth times on the order of 3 days were found to yield crystals 1mm in diameter.

A higher temperature (~900°C) technique involves use of the molten metal halides as a flux (26). Growth from KCl under slow cooling conditions for times on the order of one week produced crystals on the order of 1mm in diameter (27). The small size of the product crystals in the above experiments is a result of the limited solubility of the product in the flux. The advantages include mild conditions of product separation, and ease of control of product stoichiometry.

The technique in which excess PbO_2 and Bi_2O_3 are used as a solvent has been shown to yield quite large crystals, almost 1.5 cm on an edge (28). Crystals grown on the surface are often [100] oriented, and may be used without freeing from the surrounding solidified melt. Smaller crystals can be obtained from the crucible by decantation, followed by highly destructive excavations. Although poor control of stoichiometry is possible utilizing this technique the high solubility and large product crystal size makes it the preparative method of choice for preparing $BaPb_{1-x}Bi_xO_3$ for many characterization methods.

Because of the instability of the high oxidation states in $Ba_{1-x}K_xBiO_3$, and the fact that oxidation processes (annealing methods) generally result in disintegration, or at least produce highly stressed crystals, low temperature methods should produce the best single crystalline samples for research. KOH, first used as a flux in this system to grow crystals with x = 0.13 by Wignacourt et al (10), is currently the flux of choice. The x range has been extended to ~0.374 displaying T_c of ~30K by Schneemeyer et al in a more detailed study of the growth and symmetry of this system (4). Even higher T_c's (34K) and higher K contents (x = 0.6) have been reported by Jones et al. (11). This technique utilizes dissolution of the component oxides in KOH. At this point in the synthesis of $Ba_{1-x}K_xBiO_3$, the average nominal oxidation state of Bi is raised from +3 (its state in Bi_2O_3) to "+4" via air or O_2 oxidation. The limited solubility of $Ba_{1-x}K_xBiO_3$ then drives crystal growth. Alternative interpretations of the precipitation phenomena observed have been forwarded (4). The synthesis of large crystals is usually slow because of the kinetics of air oxidation, and because of the large number of spontaneous nucleation sites available. These difficulties are ameliorated by use of low temperature molten salt isothermal electrochemical deposition, in which the oxidation process and average metal oxidation state are under potential control, and the nucleation area is readily limited (29).

4.3 Thin Film Preparation

Although bulk superconducting wire for superconducting magnets represents the largest commercial utilization of low T_c materials, it is expected that oxide materials will make their earliest impact as electronic materials in the form of thin films. Because $BaPb_{1-x}Bi_xO_3$ is the more mature material, many techniques for thin film preparation have been reported. The earliest studies (30-32) found that rf-diode sputtering from stoichiometric targets resulted in nonstoichiometric films unless relatively high background gas pressures were utilized. The rf-sputtered films reported in most studies require the use of nonstoichiometric targets, and post deposition annealing of the amorphous or microcrystalline films in oxidizing atmospheres enriched in PbO (33-38). Superconducting as deposited single crystalline thin films have been prepared by planar rf magnetron sputtering on $SrTiO_3$ substrates (39). In this study a thin molten flux layer of lead and bismuth oxides is suggested to aid epitaxial growth.

One technique which has produced thin, but not epitaxial films of $BaPb_{1-x}Bi_xO_3$, and which shows good promise is the use of laser evaporation methods (40). Since the compound is efficiently transported stoichiometrically from the target to the substrate at a high rate and does not require a vacuum, this method may be superior to sputtering techniques.

The thermal instability of $Ba_{1-x}K_xBiO_3$ compounds makes it desirable to utilize low temperature deposition techniques. Use of a three cell evaporation source coupled with oxygen, ionized by an electron cyclotron resonance source and accelerated toward the growing film and neutralized, has been reported by Enomoto et al. (41). This technique requires a substrate temperature of 500°C. A lower temperature growth of the related material (Rb,Ba)BiO$_3$, at 300°C, has been reported through the use of molecular beam epitaxy (42). Epitaxial thin films on MgO and SrTiO$_3$ have been obtained via both of these metal beam based techniques.

Techniques utilizing spray on of precursors or chemical vapor deposition offer considerable competition as cost effective methods of producing films and should be pursued.

5.0 PHYSICAL PROPERTIES

Because these materials are the first examples of highly oxidized nonstoichiometric ceramic oxide superconductors, the determination and optimization of the physical properties has been a major technical and scientific challenge. The observed properties, and the impact of chemistry on them are reviewed here.

5.1 Electrical Transport Properties

Although the superconducting properties of these materials are of major interest at this time, future applications of these materials may depend upon utilization of the dramatic changes in electrical properties accompanying stoichiometric changes in these solid solutions. Particularly interesting are the ferro-, ferri-, piezo- and paraelectric properties displayed by the insulating phases in $BaPb_{1-x}Bi_xO_3$, and presumably in $Ba_{1-x}K_xBiO_3$ (43-45). Epitaxial composites of these two materials could be expected to display complex properties uncharacteristic of any single composition.

In the region of superconducting compositions in $BaPb_{1-x}Bi_xO_3$, oxygen vacancies have been found to decrease T_c (46)(47). This has definite implications for preparation of high T_c material, but also indicates that the T_c may be tuned to any desired transition temperature below the maximum T_c by careful technological treatments.

One of the features first observed in $BaPb_{1-x}Bi_xO_3$, and now known as a hallmark of bulk ceramic superconductors is the appearance of grain boundary Josephson junctions. One of the most convincing demonstrations of this effect in bulk material, a reentrant superconducting resistive transition, has been reported by Lin et al (48). The bulk ceramic compound can be considered to consist of weakly coupled superconducting grains. The conductivity of the grain boundaries is a function of temperature, resulting in a transition of the resistive state of the material from superconducting to resistive as temperature is lowered while the majority of the material remains in the diamagnetic, superconducting state. Such transitions are not observed in all samples, but such grain boundaries are present in most samples of the material, and can strongly affect the critical property data obtained from macroscopic experiments. The destruction of these weak links, or a chemical explanation such as loss of oxygen may be responsible for the decrease in diamagnetic response observed after crushing $BaPb_{1-x}Bi_xO_3$ (49).

In view of the fact that these materials have relatively large coherence lengths, the critical current density obtained for a grain boundary free single crystalline film of $BaPb_{1-x}Bi_xO_3$ on $SrTiO_3$ [100] (39) of 2.5 X 10^5 A/cm^2 at 4.2K in zero magnetic field should be taken as a minimum. The large coherence length means that the electron-electron pairing interaction is active over relatively great lengths, allowing pairs to interact with lattice defects. Since the critical current in such materials is usually not an intrinsic property, but an extrinsic property, dependent upon the types and densities of material defects, much larger critical currents are expected of the higher T_c material $Ba_{1-x}K_xBiO_3$, especially after effective pinning mechanisms and defects are identified.

Activated, or hopping conductivity would, of course not be expected to be an intrinsic property of these metallic materials, as demonstrated for single crystalline films (39).

Compositional homogeneity may strongly impact the observed properties of bulk material. Often the fact that T_c is a sensitive function of composition (the maximum T_c for $BaPb_{1-x}Bi_xO_3$ is observed at 13K (13) for x = 0.25, and a T_c of 34K is found for x = .6 in $Ba_{1-x}K_xBiO_3$ (11)) means that the width of the resistive transition can be used to determine the uniformity of the product. Use of the resistive T_c onset to characterize these solid solutions, especially in bulk form can lead to errors due to stoichiometric nonuniformity. The results of Tomeno and Ando (50) who have reported that T_c remains constant at 29K for x values from 0.28 to 0.44 in $Ba_{1-x}Rb_xBiO_3$, are based upon magnetic data, and the disparate behavior of this compound series from that of $BaPb_{1-x}Bi_xO_3$ and $Ba_{1-x}K_xBiO_3$ is not readily explained.

5.2 Magnetic Properties

A comparison of the magnetic properties of $BaPb_{1-x}Bi_xO_3$ and $Ba_{1-x}K_xBiO_3$ has been published recently (51). The data were obtained through magnetic studies of single crystals rather than by monitoring the electrical resistivity of the materials as a function of magnetic field intensity. Therefore, these critical magnetic field determinations are free from the artifacts usually introduced by grain boundaries in measurements taken on bulk materials. Although the H_{c1} slopes for $BaPb_{1-x}Bi_xO_3$ vs $Ba_{1-x}K_xBiO_3$ are different: ~2.2 vs 4.5 Oe/K, the H_{c2} slopes appear to be quite similar and near 5.5KOe/K (51). Of course, the 20K difference in T_c between the materials yields much higher critical fields for $Ba_{1-x}K_xBiO_3$ at any given temperature.

5.3 Optical and Infrared Properties

The CDW stabilized energy gap, i.e. the disproportionation of Bi^{+4} into Bi^{+3} and Bi^{+5}, present in $BaBiO_3$ (52) gives rise to one of the most important optical features of these materials and its understanding is central to the development of a mechanism for high T_c in them. Early studies indicated a collapse of the gap for superconducting compositions in $BaPb_{1-x}Bi_xO_3$ (15). More recent studies indicate the coexistence of the gap and superconductivity indicating the possibility that the CDW, the oxygen breathing mode, and the electron phonon coupling may combine to yield the observed T_c's (53).

A second optical property allowing determination of electronic properties of these materials is the plasma frequency, or the onset of a collective excitation of the free or metallic electrons in a metallic material. The plasma frequency has been shown to increase with increasing doping and T_c in the metallic ranges of $BaPb_{1-x}Bi_xO_3$ (54) and in $Ba_{1-x}K_xBiO_3$ (41). Carrier mass (0.67 m_e) and carrier densities of $4.5 \times 10^2 cm^{-3}$ for $BaPb_{1-x}Bi_xO_3$ with x = .25 have been determined through analysis of the reflectivity and plasma edge (54). These results have been shown to be consistent with theoretical calculations of the optical properties by Mattheiss (55) and Sofo et al (56).

One of the most important features in the IR region of the spectrum should be the appearance of the superconducting energy gap. The higher transition temperature materials should display larger values for the gap if they are BCS superconductors. The gap observed for $BaPb_{1-x}Bi_xO_3$ has been shown to agree with the BCS prediction (57).

Raman spectra indicate that phonons are coupled to electronic states in $BaPb_{1-x}Bi_xO_3$ (53) and in $Ba_{1-x}K_xBiO_3$ (58). These studies do show a strong electron - phonon interaction is present in the superconducting phases, but do not prove that these modes are responsible for high T_c driven by a phonon - only mechanism.

Tunneling Spectroscopy has revealed electron - phonon coupling strengths for optical phonons in $Ba_{1-x}K_xBiO_3$ which suggest that phonon mediated coupling is responsible for superconductivity in this system (59).

5.4 Specific Heat

Measurement of the specific heat at T_c should yield quite informative data. Information can be gained about the binding energy of the electrons in the Cooper pairs as mediated by electron-phonon coupling (in the BCS theory).

The original specific heat experiments on $BaPb_{1-x}Bi_xO_3$ by Methfessel et al (60) immediately raised the prospect that an unusual mechanism was operative in this newly found system. Their finding of no heat capacity anomaly at T_c could actually have a number of possible interpretations, including: an impurity phase giving rise to superconductivity, a non-phonon mechanism, or some new form of conductivity.

More recent measurements have shown (61)(62) that the problem with the experiment was that the thermal effect at T_c is actually very small. In fact, after the true heat had been determined, research was (and still is) directed toward finding out why the T_c is so high (63). The Sommerfeld parameter or electronic specific heat parameter appears to be identical (1.5mJ/mol K^2) in $BaPb_{1-x}Bi_xO_3$ (63) and in $Ba_{1-x}K_xBiO_3$ (64). From specific heat measurements, then, $BaPb_{1-x}Bi_xO_3$ appears to be a strong coupling (63) and $Ba_{1-x}K_xBiO_3$ a weak coupling superconductor with a coupling parameter of ~0.35 (64). This result is surprising when one considers that the T_c of the latter is 20K higher than that of the former.

5.5 Pressure Effects

The dependence of T_c on pressure is studied for a variety of reasons. In a chemical sense, bond lengths are shortened, and orbital interactions are increased. The volume decrease leads in principle to a rise in carrier density. In reality, however, not only do vibrational frequencies change, but crystal structure and symmetry are often affected by high pressure. Numerous materials undergo semiconductor to metal phase transitions as a function of pressure. Increasing pressure can often be considered analogous to a decrease in temperature.

One may then expect that, in materials as complex as the bismuthates, the observed pressure effects are complex. The phase transformations apparently observed in $BaPb_{1-x}Bi_xO_3$ via Mossbauer (65) as a function of temperature can most likely also be driven by pressure. Some indication of these transitions can be observed in the very high pressure room temperature measurements of Clark et al. (66) on $BaPb_{1-x}Bi_xO_3$, and Sugiura and Yamadaya (67) on $BaBiO_3$.

In $BaPb_{1-x}Bi_xO_3$ with the optimum composition (x = 0.25), T_c is suppressed smoothly by the application of pressure up to 15 kbar (68). However, in $BaPb_{1-x}Bi_xO_3$ with x = .20, T_c rises with pressure at a rapid rate (10^{-4} K/bar) from 0 to 4 kbar (69). From 4 kbar to 20 kbar, T_c decreases uniformly at ~3 X 10^{-5} K/bar, as calculated from figure 2 in reference 69. The rate of decrease of T_c with pressure is the same for both compositions. Room temperature resistivity data taken as a function of pressure for the x = .2 material indicates an apparent electronic phase transition at 4 kbar, accompanied by a sharp rise in the resistivity of the sample. Such a rise indicates a stronger interaction between the electron current and a dissipative excitation. Although the T_c's observed in the two materials differ, the similarity of the slope of the dT_c/dP curves indicates that the superconductivity mechanism is the same for all compositions of $BaPb_{1-x}Bi_xO_3$. This study would also appear to support the contention that the x = .25 composition is the optimum material in the $BaPb_{1-x}Bi_xO_3$ system.

It is exciting that the results of studies on $Ba_{1-x}K_xBiO_3$ indicate a positive pressure dependence of T_c (70). Unfortunately, the range of the materials measured (T_c's = 16K to 25K) did not include the higher T_c (30K) materials, and the pressure coefficient of T_c drops quickly as the T_c of the material at zero pressure increases. The study of this coefficient as a function of x and T_c would indicate further directions toward higher T_c materials. A saddle point of this coefficient within the currently known x values would indicate that attempts to apply isotropic lattice pressure would not significantly raise T_c. The cubic $(Ba_{1-x}K_xBiO_3)$ materials may not be as susceptible to phase transformation as $BaPb_{1-x}Bi_xO_3$, suggesting that a larger range of pressure may be applied. In a phonon - only mechanism, this could increase vibrational frequencies, effecting significant increases in T_c before a maximum is observed.

5.6 Isotope Effects

In the first order of BCS theory, one would expect that the transition temperature one observes for a material is dependent upon the frequency of the phonon most strongly coupled to the ensemble of superconducting electrons. This would mean that by changing the mass of an important component atom in the system to a lower mass, while keeping the bond strength constant, one could raise T_c of a material because the phonon would now have a higher frequency of oscillation. While just such an effect has been found in numerous systems, it is not always observed, and usually is only rigorously calculated after the experimental result has been obtained. In the simple model, T_c should vary as M^{-a}, where M is the reduced mass of the phonon, and a is 0.5 for a purely phonon mechanism (71). If breathing modes of oxygen octahedra play a major role in the superconductivity as has been suggested (72), substitution of O^{18} for O^{16} should result in a decrease in T_c. It should be noted that the preparation of these materials is difficult, and homogeneity is seldom obtained in bulk samples. In view of these synthetic chemical difficulties, it is not surprising that a variety of results have been reported. For $BaPb_{1-x}Bi_xO_3$, a = .22 (73) and > .5 (71) have been reported. For $Ba_{1-x}K_xBiO_3$, isotope effects of 0.21 (73), 0.41 (74), and 0.35 (75) have been reported. An isotope effect is always observed, and any true isotope effect reflects the interaction of the phonon system with the superelectrons. Although the smaller isotope effects may be interpreted as supportive of non-phonon mechanisms, they may also be explained within the phonon mechanism (76). It does not appear that these measurements have settled the question of mechanism.

6.0 THEORETICAL BASIS FOR FUTURE BISMUTHATE RESEARCH

In contrast to the copper containing high temperature oxide superconductors, the bismuthates appear much simpler, possessing tetragonal ($BaPb_{1-x}Bi_xO_3$) or cubic ($Ba_{1-x}K_xBiO_3$) perovskite structures. Not only are the structures of the bismuthates less complex than those of the cuprates, but the electronic situation seems to be more comprehensible also. Because the ground state of the bismuthates is diamagnetic, rather than displaying the antiferromagnetism of the cuprates, the concern that a magnetic interaction is driving the superconductivity to such high temperatures may tentatively be ruled out (77). Even without the added complexity of invoking a magnetic mechanism to drive superconductivity in this system, one must consider whether the mechanism is all phonon, all nonphonon, or a mixture of both. At this time it cannot be said that a consensus has been reached on the mechanism operative in these systems. The evidence for and against these mechanisms has been discussed here. To some extent, the availability of higher quality samples, specifically chemically well characterized single crystals and thin films, would resolve some of the problems involved in sorting intrinsic from extrinsic properties when comparing the results of different researchers. At this point, two crucial questions, present from the first moment of research on these materials remain. "What is the nature of the charge carriers involved in High T_c?", and "What is the pairing mechanism?"

It would seem that the carrier identification would be readily resolved by experiment. In fact, numerous experiments have found the carriers to be electrons in $BaPb_{1-x}Bi_xO_3$ (78)(79), the carrier concentration and T_c readily scaling with Bi content until a CDW decreases the carrier density. This interpretation is well supported by a magnetic susceptibility study (80). Even though the parent compound $BaBiO_3$ is a hole type semiconductor (81), the carrier type has been shown to be n or electronic in character (75) in $Ba_{1-x}K_xBiO_3$.

In order to investigate possible similarity (82) to the more complex Cu containing p type systems, hole type carrier mechanisms are being investigated intensively, both experimentally (83) and theoretically (56), to discover hole subbands. The observed hole concentration does not, however, scale well with T_c (83). The electronic carrier concentrations observed have been quite low (84) and have brought about significant questions about why T_c is so high at such low densities. Actually, these questions are poorly founded, considering the experimental situation, where the maximum T_c has been found to scale inversely with carrier density (85) (allowing that each system has its optimum carrier density). The necessity of low carrier concentrations should be a recognized cornerstone of all high T_c research, in that electron trapping by a global phase transformation will be precipitated in systems with strong electron - phonon interactions when the carrier density is high. The associated loss of a phonon coupling mode necessarily decreases the T_c of the material. Of course, electronic screening effects are also reduced at low carrier concentrations.

The second question has been found to be unexpectedly hard to answer. These materials involve covalency effects to a much greater extent than the intermetallic superconductors, and band theory may not be ideally applicable to them. There are unresolved questions of low temperature phase transitions in $BaPb_{1-x}Bi_xO_3$ (65) which may have analogs in the $Ba_{1-x}K_xBiO_3$ system, although current X-ray diffraction studies have not observed them (86). Calculations concerning the dynamics of the electron - lattice interaction cannot be performed without precise structural information. Yet few experimental tools for obtaining structural dynamics data exist. Even the static structure of $BaPb_{1-x}Bi_xO_3$ and $Ba_{1-x}K_xBiO_3$ may have more short range order than has previously been suspected. A superlattice structure has been observed in $BaPb_{1-x}Bi_xO_3$ single crystals (25), long after it was inferred from Mossbauer studies (65). It has already been shown for the $BaPb_{1-x}Bi_xO_3$ system that electronic structure calculations lacking a realistically detailed structural model (87) do not model experimental reflectivity results (88) well. Even more recent calculations do not lead to a clear specification of the superconducting mechanism (89). Two of the ordered superlattice structures proposed for $Ba_{1-x}K_xBiO_3$ and used for electronic structure calculations (90) would introduce two dimensional character into the structures, thus bringing the bismuthates into closer analogy to the Cu superconductors.

The mechanisms proposed as responsible for high T_c in the bismuthates range from bipolarons (91) to purely phonon (92-97) to purely electronic. Favored electronic mechanisms utilize electronic excitations (73), excitons (89)(98) or plasma excitations. These materials may be of a transitional, or hybrid type, and a stronger signature of unconventional mechanisms may become evident in higher T_c materials. It was once thought possible to simply use T_c as an indicator of mechanisms, but, noting that a phonon-only mechanism has been predicted to yield a T_c of 230K for metallic hydrogen (99), we must look for other signatures.

Perhaps the best guideline to use in designing high T_c materials is to look for electron localization mechanisms, as suggested by Sleight (100). The random disorder suggested by Jurczek (101) for $BaPb_{1-x}Bi_xO_3$ may be responsible for its low Hall mobility of 3.0 $cm^2/V \cdot s$ (39) which is unusually small for a nontransition metal oxide conductor. The ferroelectric phase transition observed in the bismuth containing parent compound $BaBiO_3$ at 643K (102) certainly is a strong indicator that this system possesses a strong mechanism for charge localization and hopefully means that a higher T_c can be coaxed out of a related system.

From the above discussions, it is clear that the major contribution of theory in guiding future experimental work is the demand it places for high quality empirically determined data. Theory has rarely, if ever, contributed to a rise in T_c, and, without a clearly defined mechanism, the contributions of theory to the pursuit of high T_c remains low.

7.0 APPLICATIONS

The rather low critical current densities expected in bulk samples of these granular superconductor materials (103)(104), and their relatively low magnetic critical fields (50)(78)(105)(106), allied with the relatively low T_c's observed would appear to hinder development of superconducting applications for these materials. Even the critical current of 5×10^5 A/cm^2 observed for single crystalline thin films (39) is now considered low for a superconductor at 4.2K. However, when considering the applicability of a material to a task, several important factors must be weighed. These include, but are not limited to: ease of fabrication, energy gap, dimensionality, materials compatibility aspects, toxicity, transition temperature and critical parameters. It may be found that the bismuthates have properties which are hard to emulate with any other materials.

For example, a low critical current density has been shown useful in using $BaPb_{1-x}Bi_xO_3$ as a microwave switch. The transmission of 2.8 GHz microwaves through a polycrystalline thin film was switched on in less than 30 ns via a high current pulse (107), resulting in short microwave pulses.

Another example is the preparation of highly sensitive optical detectors using $BaPb_{1-x}Bi_xO_3$ (108)(109) suitable for IR and Visible light communications system applications. In this application, polycrystalline, rather than single crystalline films are required. The granularity introduces Josephson Junctions (JJ), the properties of which are strongly modified by optical signals. The low reflectivity of $BaPb_{1-x}Bi_xO_3$, due to its low carrier density, allows it to couple efficiently with the optical input, whereas most superconductors have high reflectivity. It should be noted that JJ's spontaneously form in oxide superconductors, and that materials modification and specification of these JJ's may well be requested of materials researchers in the near future. Such three dimensional, secondary structural engineering will present a tremendous challenge. The observed detector rise time of less than 10^{-10} s may make this challenge attractive.

The most exciting property of these materials is that thin polycrystalline films can behave as two dimensional Josephson Tunnel Junction Arrays (110) displaying remarkably good electromagnetic coupling (111) both as emitters and receivers. Alternating current to direct current conversion has been observed in bulk samples (112), but the thin film geometry allows use of one film as a radiation emitter, and another film as the receiver (111).

With the additional possibility of electrostatic coupling and charging of grain boundaries to modify the properties of the JJ's (48), it is attractive to consider the use of these JJ's in microelectronics, perhaps for on chip optical or microwave Josephson Junction communications.

8.0 CONCLUSION

As a prototypical perovskite superconductor system, the bismuthates have taught us much about the preparation and characterization of ceramic materials. These materials have inspired searches for means of experimentally identifying non-phonon mechanisms of superconductivity. The bismuthates have already matched the high T_c (28K) predicted for them in 1982 by Gabovich et al. (113). Certainly two dimensional analogs of these perovskites already exist (114), and superconducting ones will be synthesized. The greatest challenge these materials present at this time is whether T_c can become an appreciable fraction of the phase transformation temperature of $BaBiO_3$ (643 K). Perhaps this will lead us to truly high temperature superconductivity.

9.0 REFERENCES

1. Cox, D.E. and Sleight, A.W., Mixed - Valent $Ba_2Bi^{3+}Bi^{5+}O_6$: Structure and Properties vs Temperature. *Acta Cryst. B* 35:1 (1979).

2. Uchida, S., Kitazawa, K. and Tanaka, S., Superconductivity and Metal-Semiconductor Transition in $BaPb_{1-x}Bi_xO_3$. *Phase Transitions* 8: 95 (1987).

3. Balzarotti, A., Menushenkov, A.P., Motta, N. and Purans, J., EXAFS of the Superconducting Oxide $BaPb_{1-x}Bi_xO_3$. *Solid State Comm.* 49(9):887 (1984).

4. Schneemeyer, L.F., Thomas, J.K., Siegrist, T., Batlogg, B., Rupp, L.W., Opila, R.L., Cava, R.J. and Murphy, D.W., Growth and Structural Characterization of Superconducting $Ba_{1-x}K_xBiO_3$ Single Crystals. *Nature* 335:421 (1988).

5. Shuvayev, V.P. and Sazhin, B.I., Production of High- Temperature Polymer Superconductors. *Polymer Science U.S.S.R.* 24:229 (1982).

6. Batlogg, B., Remeika, J.P., Dynes, R.C., Barz, H., Cooper, A.S. and Garno, J.P., Structural Instabilities and Superconductivity in Single Crystal $Ba(Pb,Bi)O_3$. *Proceedings on the Symposia on Superconductivity in d- and f- Band Metals 1982*, Kernforschungszentrum Karlsruhe:401 (1982).

7. Sleight, A.W. and Cox, D.E., Symmetry of Superconducting Compositions in the $BaPb_{1-x}Bi_xO_3$ System. *Solid State Comm.* 58(6):347 (1986).

8. Oda, M., Hidaka, Y., Katsui, A. and Murakami, T., The Crystallographic Symmetries of Single $BaPb_{1-x}Bi_xO_3$ Crystals Grown from $BaCO_3$-PbO_2-Bi_2O_3 Solutions, *Solid State Comm.* 60(12):897 (1986).

9. Shebanov, L.A., Fritsberg, V. Ya. and Gaevskis, A.P., Crystallographic Properties and Superconductivity of Solid Solutions of the $BaBi_xPb_{1-x}O_3$ System. *Phys. Stat. Sol.(a)* 77:369 (1983).

10. Wignacourt, J.P., Swinnea, J.S., Stienfink, H. and Goodenough, J.B., Oxygen Atom Thermal Vibration Anisotropy in $Ba_{0.87}K_{0.13}BiO_3$. *Appl. Phys. Lett.* 53(18):1753 (1988).

11. Jones, N.L., Parise, J.B., Flippen, R.B. and Sleight, A.W., Superconductivity at 34 K in the K/Ba/Bi/O System. *J. Solid State Chem.* 78:319 (1989).

12. Hewat, E.A., Chaillout, C., Godinho, M., Gorius, M.F. and Marezio, M., Electron Beam Induced Superstructure in $Ba_{1-x}K_xBiO_{3-y}$. *Physica C* 157:228 (1989).

13. Sleight, A.W., Gillson, J.L. and Bierstedt, P.E., High-Temperature Superconductivity in the $BaPb_{1-x}Bi_xO_3$ System. *Solid State Comm.* 17:27 (1975).

14. Moiseev, D.P., Prikhotko, A.F. and Uvarova, S.K., Effect of Oxygen on the Superconductivity of Barium Lead Bismuth Oxide ($BaPb_{1-x}Bi_xO_3$) Ceramics. *Ukr. Fiz. Zh.* 27(9):1427 (1982).

15. Khan, Y., Nahm, K., Rosenberg, M. and Willner, H., Superconductivity and Semiconductor-Metal Phase Transition in the System $BaPb_{1-x}Bi_xO_3$. *Phys. Stat. Sol.(a)* 39:79 (1977).

16. Gilbert, L.R., Messier, R. and Roy, R., Bulk Crystalline $BaPb_{3/4}Bi_{1/4}O_3$: A Ceramic Superconductor. *Mat. Res. Bull.* 17:467 (1982).

17. Sleight, A.W., Superconductive Barium-Lead-Bismuth Oxides. US Patent 3,932,315 (1976).

18. Suzuki, M., Murakami, T. and Inamura, T., Superconductivity in $Ba_{1-x}Sr_xPb_{.75}Bi_{.25}O_3$. *Jpn. J. Appl. Phys.* 19(2):L72 (1980).

19. Sakudo, T., Uwe, H., Suzuki, T., Fujita, J., Shiozawa, J. and Isobe, M., Composition Effects on Properties of the Perovskite Superconductor $Ba(Pb,Bi)O_3$. *J. Phys. Soc. Jpn.* 55(1):314 (1986).

20. Mattheiss, L.R., Gyorgy, E.M. and Johnson, Jr., D.W., Superconductivity Above 20K in the Ba-K-Bi-O System. *Phys. Rev. B* 37(7):3745 (1988).

21. Cava, R.J., Batlogg, B., Krajewski, J.J., Farrow, R., Rupp,Jr., L.W., White, A.E., Short, K., Peck, W.F. and Kometani, T., Superconductivity Near 30K Without Copper: The $Ba_{0.6}K_{0.4}BiO_3$ Perovskite. *Nature* 332:814 (1988).

22. Hinks, D.G., Richards, D.R., Dabrowski, B., Mitchell, A.W., Jorgensen, J.D. and Marx, D.T., Oxygen Content and the Synthesis of $Ba_{1-x}Bi_xO_{3-y}$ *Physica C* 156:477 (1988).

23. Hinks, D.G., Mitchell, A.W., Zheng, Y., Richards, D.R. and Dabrowski, B., Synthesis of High-Density $Ba_{1-x}K_xBiO_3$ Superconducting Samples. *Appl. Phys. Lett.* 54(16):1585 (1989).

24. Hirano, S. and Takahashi, S., Hydrothermal Crystal Growth of $BaPb_{1-x}Bi_xO_3$ (0< x <0.30). *J. Cryst. Growth* 78:408 (1986).

25. Hirano, S. and Takahashi, S., Hydrothermal Synthesis and Properties of $BaPb_{1-x}Bi_xO_3$. *J. Cryst. Growth* 79:219 (1986).

26. Bogatko, V.V. and Venevtsev, Y.N., Dielectric Properties and Superconductivity of $BaPbO_3$ (I) - $BaBiO_3$ (II) Single Crystals. *Sov. Phys. Solid State* 25(5):859 (1983).

27. Katsui, A. and Suzuki, M., Single Crystal Growth of $Ba(Pb,Bi)O_3$ from Molten KCl Solvent. *Jpn. J. Appl. Phys.* 21(3):L157 (1982).

28. Katsui, A., Hidaka, Y. and Takagi, H., Growth of Superconducting $Ba(Pb,Bi)O_3$ Crystals. *J. Cryst. Growth* 66:228 (1984).

29. Norton, M.L., Electrodeposition of $Ba_{.6}K_{.4}BiO_3$. *Mat. Res. Bull.* 24(11):1391 (1989).

30. Gilbert, L.R., Messier, R. and Roy, R., Superconducting $BaPb_{1-x}Bi_xO_3$ Ceramic Films Prepared by R.F. Sputtering. *Thin Solid Films* 54:129 (1978).

31. Gilbert, L.R., Bulk and Sputtered Thin Film Ba(Pb,Bi)O$_3$: A Ceramic Superconductor. Ph.D. Thesis, Pennsylvania State U. (1979)

32. Gilbert, L.R., Messier, R. and Krishnaswamy, S.V., Resputtering Effects in Ba(Pb,Bi)O$_3$ Perovskites. *J. Vac. Sci. Technol.* 17(1):389 (1980).

33. Suzuki, M., Murakami, T. and Inamura, T., Preparation of Superconducting BaPb$_{1-x}$Bi$_x$O$_3$ Thin Films by RF Sputtering. *Jpn. J. Appl. Phys.* 19(5):L231 (1980).

34. Suzuki, M., Enomoto, Y. Murakami, T. and Inamura, T., Thin Film Preparation of Superconducting Perovskite-Type Oxides by rf Sputtering. *Jpn. J. Appl. Phys.* 20(s 20-4):13 (1981).

35. Suzuki, M., Enomoto, Y., Murakami, T. and Inamura, T., Preparation and Properties of Superconducting BaPb$_{1-x}$Bi$_x$O$_3$ Thin Films by Sputtering. *J. Appl. Phys.* 53(3):1622 (1982).

36. Hidaka, Y., Suzuki, M., Murakami, T. and Inamura, T., Effects of a Lead Oxide Annealing Atmosphere on the Superconducting Properties of BaPb$_{1-x}$Bi$_x$O$_3$ Sputtered Films. *Thin Solid Films* 106:311 (1983).

37. Suzuki, M. and Murakami, T., Superconducting BaPb$_{1-x}$Bi$_x$O$_3$ Thin Films by RF Magnetron Sputtering. *Jpn. J. Appl. Phys.* 22(12):1794 (1983).

38. Enomoto, Y., Murakami, T. and Suzuki, M., Infrared Optical Detector Using Superconducting Oxide Thin Film. *Physica C* 153-155:1592 (1988).

39. Suzuki, M. and Murakami, T., Epitaxial Growth of Superconducting BaPb$_{1-x}$Bi$_x$O$_3$ Thin Films. *J. Appl. Phys.* 56(8):2330 (1984).

40. Zaltsev, S.V., Martynyuk, A.N. and Protasov, E.A., Superconductivity of BaPb$_{1-x}$Bi$_x$O$_3$ Films Prepared by Laser Evaporation Method. *Sov. Phys. Solid State* 25(1):100 (1983).

41. Enomoto, Y., Murakami, T. and Moriwaki, K., Ba$_{1-x}$K$_x$BiO$_3$ Thin Film Preparation by ECR Ion Beam Oxidation, and Film Properties. *Jpn. J. Appl. Phys.* 28(8):L1355 (1989).

42. Hellman, E.S., Hartford, E.H. and Flemming, R.M., Molecular Beam Epitaxy of Superconducting (Rb,Ba)BiO$_3$. *Appl. Phys. Lett.* 55(20):2120 (1989).

43. Bogatko, V.V. and Venevtsev, Y.N., Spontaneously Polarized State and Superconduction in the System $BaPbO_3$ - $BaBiO_3$. *Izv. Adak. Nauk SSSR. Ser. Fiz.* 47(4): 637 (1983).

44. Horie, Y., Fukami, T. and Mase, S., First Order Structural Phase Transition in $BaPb_{1-x}Bi_xO_3$ and the Scaling Law. *Solid State Comm.* 62(7):471 (1987).

45. Hidaka, M., Fujii, H., Horie, Y. and Mase, S., Superconductor - Semiconductor Transition in $BaPb_{1-x}Bi_xO_3$. *Phys. Stat. Sol.(b)* 145:649 (1988).

46. Menushenkov, A.P., Protasov, E.A. and Chubunova, E.V., Influence of Oxygen Concentration on the Superconducting Properties of $BaPb_{1-x}Bi_xO_3$. *Sov. Phys. Solid State* 23(12): 2155 (1981).

47. Suzuki, M. and Murakami, T., Effect of Oxygen Vacancies on Carrier Localization in $BaPb_{1-x}Bi_xO_3$. *Solid State Comm.* 53(8):691 (1985).

48. Lin, T.H., Shao, X.Y., Wu, M.K., Hor, P.H., Jin, X.C., Chu, C.W., Evans, N. and Bayuzick, R., Observation of a Reentrant Superconducting Resistive Transition in Granular $BaPb_{0.75}Bi_{0.25}O_3$ Superconductor. *Phys. Rev. B* 29(3):1493 (1984).

49. Moiseev, D.P. and Uvarova, S.K., Microstructure and Superconductivity of Solid Solutions $BaPb_{1-x}Bi_xO_3$. *Izv. Akad. Nauk SSSR, Neorg. Mater.* 17(9):1685 (1981).

50. Tomeno, I. and Ando, K., Superconducting Properties of the $Ba_{1-x}Rb_xBiO_3$ System. *Phys. Rev. B* 40(4):2690 (1989).

51. Batlogg, B., Cava, R.J., Schneemeyer, L.F. and Espinosa, G.P., High-Tc Superconductivity in Bismuthates - How Many Roads Lead to High T_c? *IBM J. Res. Develop.* 33(3):208 (1989).

52. Bruder, C., Optical Absorption Line Shape at a CDW-Gap: Application to $BaBiO_3$. *Physica C* 153 - 155:693 (1988).

53. Tajima, S., Uchida, S., Masaki, A., Takagi, H., Kitazawa, K., Tanaka, S. and Sugai, S., Electronic States of $BaPb_{1-x}Bi_xO_3$ in the Semiconducting Phase Investigated by Optical Measurements. *Phys. Rev. B* 35(2):696 (1987).

54. Tajima, S., Kitazawa, K. and Tanaka, S., Effective Mass of Electrons in $BaPb_{1-x}Bi_xO_3$. *Solid State Comm.* 47(8):659 (1983).

55. Mattheiss, L.F., Plasma Energies for $BaPb_{1-x}Bi_xO_3$. *Phys. Rev. B* 28(12):6629 (1983).

56. Sofo, J.O., Aligia, A.A. and Nunez Regueiro, M.D., Electronic Structure of $BaPb_{1-x}Bi_xO_3$. *Phys. Rev. B* 39(13):9701 (1989).

57. Schlesinger, Z., Collins, R.T., Scott, B.A. and Calise, J.A., Superconducting Energy Gap of $BaPb_{1-x}Bi_xO_3$. *Phys. Rev. B* 38(13):9284 (1988).

58. McCarty, K.F., Radousky, H.B., Hinks, D.G., Zheng, Y., Mitchell, A.W., Folkerts, T.J. and Shelton, R.N., Electron-Phonon Coupling in Superconducting $Ba_{0.6}K_{0.4}BiO_3$: A Raman Scattering Study. *Phys. Rev. B* 40(4):2662 (1989).

59. Zasadzinski, J.F., Tralshawala, N., Hinks, D.G., Dabrowski, B., Mitchell, A.W. and Richards, D.R., Tunneling Spectroscopy in Superconducting $Ba_{1-x}K_xBiO_3$: Direct Evidence for Phonon-Mediated Coupling. *Physica C* 158:519 (1989).

60. Methfessel, C.E., Stewart, G.R., Matthias, B.T. and Patel, C.K.N., Why is there no Bulk Specific Heat Anomaly at the Superconducting Transition Temperature of $BaPb_{1-x}Bi_xO_3$? *Proc. Natl. Acad. Sci. USA* 77(11):6307 (1980).

61. Sato, M., Fujishita, H. and Hoshino, S., Specific Heat Anomaly of $BaPb_{1-x}Bi_xO_3$ at the Superconducting Transition. *J. Phys. C: Solid State Phys.* 16:L417 (1983).

62. Gabovich, A.M., Moiseev, D.P., Prokopovich, L.V., Uvarova, S.K. and Yachmenev, V.E., Experimental Demonstration of Bulk Superconductivity in the Perovskite System $BaPb_{1-x}Bi_xO_3$. *Sov. Phys. JETP* 59(5):1006 (1984).

63. Batlogg, B., Superconductivity in $Ba(Pb,Bi)O_3$. *Physica* 126B:275 (1984).

64. Graebner, J.E., Schneemeyer, L.F. and Thomas, J.K., Heat Capacity of Superconducting $Ba_{0.6}K_{0.4}BiO_3$ Near Tc. *Phys. Rev. B* 39(13):9682 (1989).

65. Kimball, C.W., Dwight, A.E., Farrah, S.K., Karlov, T.F., McDowell, D.J. and Taneja, S.P., Mossbauer Study of the Electronic and Vibrational Properties of Sn- and Sb- Substitutions in the Perovskite Superconductor $BaPb_{1-x}Bi_xO_3$. *Proceedings on the Symposia on Superconductivity in d- and f- Band Metals 1982*, Kernforschungszentrum Karlsruhe:409 (1982).

66. Clark, J.B., Dachille, F. and Roy, R., Resistance Measurements at High Pressure in the System $BaPb_{1-x}Bi_xO_3$. *Solid State Comm.* 19:989 (1976).

67. Sugiura, H. and Yamadaya, T., High Pressure Studies on the Perovskite-Type Compound $BaBiO_3$. *Physica* 139 & 140B:349 (1986).

68. Chu, C.W., Huang, S. and Sleight, A.W., Hydrostatic Pressure Effect on Tc of $Ba_{0.9}K_{0.1}Pb_{0.75}Bi_{0.25}O_3$. *Solid State Comm.* 18:977 (1976).

69. Wu, M.K., Meng, R.L., Huang, S.Z. and Chu, C.W., Superconductivity in $BaPb_{1-x}Bi_xO_3$ Near the Metal-Semiconductor Phase Boundary Under Pressure. *Phys. Rev. B* 24(7):4075 (1981).

70. Schirber, J.E., Morosin, B. and Ginley, D.S., Effect of Pressure on the Superconducting Transition Temperature of $Ba_{1-x}K_xBiO_3$. *Physica C* 157:237 (1989).

71. zur Loye, H-C., Leary, K.J., Keller, S.W., Ham, W.K., Faltens, T.A., Michaels, J.N. and Stacy, A.M., Oxygen Isotope Effect in High-Temperature Oxide Superconductors. *Science* 238:1558 (1987).

72. Gupta, H.C., Phonons in High-Tc $Ba_{0.6}K_{0.4}BiO_3$. *Physica C* 158:153 (1989).

73. Batlogg, B., Cava, R.J., Rupp Jr., L.W., Mujsce, A.M., Krajewski, J.J., Remeika, J.P., Peck Jr., W.F., Cooper, A.S. and Espinosa, G.P., Density of States and Isotope Effect in BiO Superconductors: Evidence for Nonphonon Mechanism. *Phys. Rev. Lett.* 61(14):1670 (1988).

74. Hinks, D.G., Richards, D.R., Dabrowski, B., Marx, D.T. and Mitchell, A.W., The Oxygen Isotope Effect in $Ba_{0.625}K_{0.375}BiO_3$. *Nature* 335:419 (1988).

75. Kondoh, S., Sera, M., Ando, Y. and Sato, M., Normal State Properties and Oxygen Isotope Effect of (Ba,K)BiO_3. *Physica C* 157:469 (1989).

76. Shirai, M., Suzuki, N. and Motizuki, K., Superconductivity in $BaPb_{1-x}Bi_xO_3$ and $Ba_xK_{1-x}BiO_3$. *J. Phys.: Condens. Matter* 1:2939 (1989).

77. Uemura, Y.J., Sternlieb, B.J., Cox, D.E., Brewer, J.H., Kadono, R., Kempton, J.R., Kiefl, R.F., Kreitzman, S.R., Luke, G.M., Mulhern, P., Riseman, T., Williams, D.L., Kossler, W.J., Yu, X.H., Stronach, C.E., Subramanian, M.A., Gopalakrishnan, J. and Sleight, A.W., Absence of Magnetic Order in (Ba,K)BiO$_3$. *Nature* 335:151 (1988).

78. Thanh, T.D., Koma, A. and Tanaka, S., Superconductivitiy in the BaPb$_{1-x}$Bi$_x$O$_3$ System. *Appl. Phys.* 22:205 (1980).

79. Gabovich, A.M. and Moiseev, D.P., Metal Oxide Superconductor BaPb$_{1-x}$Bi$_x$O$_3$: Unusual Properties and New Applications. *Sov. Phys. Usp.* 29(12):1135 (1986).

80. Uchida, S., Hasegawa, H., Kitazawa, K. and Tanaka, S., Magnetic Susceptibility of BaPb$_{1-x}$Bi$_x$O$_3$ System. *Physica C* 156:157 (1988).

81. Matsuyama, H., Takahashi, T., Katayama-Yoshida, H. Okabe, Y., Takagi, H. and Uchida, S., Ultraviolet Photoemission Study of Single - Crystal BaPb$_{1-x}$Bi$_x$O$_3$. *Phys. Rev. B* 40(4):2658 (1989).

82. Dzyaloshinskii, I.E., Chemical Nature of the Pairing of Holes in High-Temperature Suprconductors. *JETP Lett.* 49(2):142 (1989).

83. Lin, C.L., Qiu, S.L., Chen, J., Strongin, M., Cao, G., Jee, C-S. and Crow, J.E., Photoemission and Oxygen K-edge Absorption Studies of Ba(Pb,Bi)O$_3$. *Phys. Rev. B* 39(13):9607 (1989).

84. Sakamoto, H., Namatame, H., Mori, T., Kitazawa, K., Tanaka, S. and Suga, S., Photoemission Studies of Electronic Structures of BaPb$_{1-x}$Bi$_x$O$_3$. *J. Phys. Soc. Jpn.* 56(1):365 (1987).

85. Golovashkin, A.I., Superconductors with Unusual Properties and Possibilities of Increasing the Critical Temperature. *Sov. Phys. Usp.* 29(2):199 (1986).

86. Fleming, R.M., Marsh, P., Cava, R.J. and Krajewski, J.J., Temperature Dependence of the Lattice Parameters in the 30-K Superconductor Ba$_{0.6}$K$_{0.4}$BiO$_3$. *Phys. Rev. B* 38(10):7026 (1988).

87. Mattheiss, L.F. and Hamann, D.R., Electronic Structure of BaPb$_{1-x}$Bi$_x$O$_3$. *Phys. Rev. B* 28(8):4227 (1983).

88. Tajima, S., Ishii, H., Rittaporn, I., Uchida, S., Tanaka, S., Kitazawa, K., Seki, M. and Suga, S., Vacuum- Ultraviolet Spectra and Band Structure of BaPb$_{1-x}$Bi$_x$O$_3$. *Phys. Rev. B* 38(2):1143 (1988).

89. Takegahara, K. and Kasuya, T., APW Band Structure of Cubic BaPb$_{1-x}$Bi$_x$O$_3$. *J. Phys. Soc. Jpn.* 56(4):1478 (1987).

90. Mattheiss, L.F. and Hamann, D.R., Electronic Structure of the High-T$_c$ Superconductor Ba$_{1-x}$K$_x$BiO$_3$. *Phys. Rev. Lett.* 60(25):2681 (1988).

91. DeJongh, L.J., A Comparative Study of (Bi)Polaronic (Super)Conductivity in High- and Low T$_c$ Superconducting Oxides. *Physica C* 152:171 (1988).

92. Rice, M.J. and Wang, Y.R., Interpretation of Ba$_{1-x}$K$_x$BiO$_3$ as a Doped Peierls Insulator. *Physica C* 157:192 (1989).

93. Mattheiss, L.F. and Hamann, D.R., Electronic- and Crystal-Structure Effects on Superconductivity in the BaPb$_{1-x}$Bi$_x$O$_3$ System. *Phys. Rev. B* 26(5):2686 (1982).

94. Wenguo, W. and Benkun, M., The Theoretical Study of Superconductivity in Oxide BaPb$_{1-x}$Bi$_x$O$_3$. *Physica C* 153-155:1177 (1988).

95. Johnson, K.H., McHenry, M.E., Counterman, C., Collins, A., Donovan, M.M., O'Handley, R.C. and Kalonji, G., Quantum Chemistry and High-T$_c$ Superconductivity. *Physica C* 153-155:1165 (1988).

96. Shirai, M., Suzuki, N. and Motizuki, K., Microscopic Theory of Electron - Phonon Interaction and Superconductivity of BaPb$_{1-x}$Bi$_x$O$_3$. *Solid State Comm.* 60(6):489 (1986).

97. Rice, T.M. and Sneddon, L., Real-Space and k-Space Electron Pairing in BaPb$_{1-x}$Bi$_x$O$_3$. *Phys. Rev. Lett.* 47(9):689 (1981).

98. Kamaras, K., Porter, C.D., Doss, M.G., Herr, S.L., Tanner, D.B., Bonn, D.A., Greedan, J.E., O'Reilly, A.H., Stager, C.V. and Timusk, T., Excitonic Absorption and Superconductivity in YBa$_2$Cu$_3$O$_{7-y}$. *Phys. Rev. Lett.* 59(8):919 (1987).

99. Barbee III, T.W., Garcia, A. and Cohen, M.L., First-Principles Prediction of High-Temperature Superconductivity in Metallic Hydrogen. *Nature* 340:369 (1989).

100. Sleight, A.W., Chemistry of High-Temperature Superconductors. *Science* 242:1519 (1988).

101. Jurczek, E., Model Study of the Semiconductor-Metal Transition in $BaBi_{1-x}Pb_xO_3$ by Use of a Spatially Inhomogeneous Order-Parameter Approximation. *Phys. Rev. B* 35(13):6997 (1987).

102. Venevtsev, Yu.N., Bogatko, V.V., Popadeikin, I.A. and Tomashpolski, Yu.Ya., High Temperature Superconductivity and Ferroelectricity in Metal Oxides. *Ferroelectrics* 94:463 (1989).

103. Belous, N.A., Gabovich, A.M., Moiseev, D.P., Postnikov, V.M. and Chernyakhovskii, A.E., Coherence Properties and Current Transport in a Ceramic Josephson Medium, $BaPb_{1-x}Bi_xO_3$. *Sov. Phys. JETP* 64(1):159 (1986).

104. Belous, N.A., Gabovich, A.M., Moiseev D.P. and Postnikov, V.M., Detection of the Influence of Disorder and of Frustration of Weak Links on the Critical Current in a Three-Dimensional Granular Superconductor $BaPb_{0.75}Bi_{0.25}O_3$. *Sov. Phys. Solid State* 28(9):1615 (1986).

105. Welp, U., Kwok, W.K., Crabtree, G.W., Claus, H., Vandervoort, K.G., Dabrowski, B., Mitchell, A.W., Richards, D.R., Marx, D.T. and Hinks, D.G., The Upper Critical Field of $Ba_{1-x}K_xBiO_3$. *Physica C* 156:27 (1988).

106. Jin, S., Tiefel, T.H., Sherwood, R.C., Ramirez, A.P., Gyorgy, E.M., Kammlott, G.W. and Fastnacht, R.A., Transport Measurement of 32 K Superconductivity in the Ba-K-Bi-O System. *Appl. Phys. Lett.* 53(12):1116 (1988).

107. Minami, K. Saeki, K., Kubo, H., Ohtsuka, M. Awano, M. and Takai, H., Quick Extraction of Microwave Pulses from a Cavity by a Superconducting Polycrystalline $BaPb_{1-x}Bi_xO_3$ Thin Film. *J. Appl. Phys.* 62(5):1902 (1987).

108. Enomoto, Y., Suzuki, M. and Murakami, T., Highly Sensitive Optical Detector Using Superconducting Oxide $BaPb_{0.7}Bi_{0.3}O_3$ (BPB). *Jpn. J. Appl. Phys.* 23(5):L333 (1984).

109. Enomoto, Y. and Murakami, T., Optical Detector Using Superconducting $BaPb_{0.7}Bi_{0.3}O_3$ Thin Films. *J. Appl. Phys.* 59(11):3807 (1986).

110. Moriwaki, K., Suzuki, M. and Murakami, T., A Novel Discreet Current Behavior in the Current-Voltage Curve for Superconducting $BaPb_{0.7}Bi_{0.3}O_3$ Films. *Jpn. J. Appl. Phys.* 23(2):L115 (1984).

111. Moriwaki, K., Suzuki, M. and Murakami, T., Electromagnetic Coupling Effects in $BaPb_{0.7}Bi_{0.3}O_3$ Two-Dimensional Josephson Tunnel Junction Arrays. *Jpn. J. Appl. Phys.* 23(3):L181 (1984).

112. Ikegawa, S., Honda, T., Ikeda, H., Maeda, A., Takagi, H., Uchida, S-i., Uchinokura, K. and Tanaka, S., AC-DC Conversion Effect in Ceramic Superconductor $BaPb_{1-x}Bi_xO_3$. *J. Appl. Phys.* 64(10):5061 (1988).

113. Gabovich, A.M., Moiseev, D.P. and Shpigel', A.S., Anomalous Behavior of the Thermodynamic Properties of the Superconducting Ceramic $BaPb_{1-x}Bi_xO_3$. *Sov. Phys. Solid State* 24(6):1071 (1982).

114. Fu, W.T., Zandbergen, H.W., Xu, Q., van Ruitenbeek, J.M., de Jongh, L.J. and van Tendeloo, G., Structural and Transport Properties of the Triple-Layer Compounds $Ba_4(Pb_{1-x}Bi_x)_3O_{10}$ ($0 \leq x \leq 0.3$). *Solid State Comm.* 70(12):1117 (1989).

10

Crystal Chemistry of Superconducting Bismuth and Lead Oxide Based Perovskites

Robert J. Cava

1.0 INTRODUCTION

The discovery of high temperature superconductivity (1) in copper-oxide based compounds has changed the course of current research in solid state physics and materials science. Since the earliest 30-40K superconductors based on $La_{2-x}(Ca,Sr,Ba)_xCuO_4$, many new superconducting copper oxides have been discovered, with ever increasing chemical and structural complexity. Enough is currently known about the copper oxide based superconductors that certain structural, chemical and electronic generalizations can be made that apply simply to even the most complex known materials. These are: 1) The superconductivity occurs within two-dimensional copper-oxygen arrays based on the joining of CuO_4 squares at their oxygen corners to form infinite CuO_2 planes; 2) The superconducting CuO_2 planes are separated by charge reservoir layers which act to control the charge on the superconducting planes (through chemical doping) either through the transfer of holes, or, recently, electrons, (2) to some optimal concentration dependent in detail on the overall structure; 3) The electronic band at the Fermi level is one which is highly hybridized due to the similarity in energy between copper 3d and oxygen 2p states (3) and (4) the superconductivity occurs in all cases at the doping level where an originally antiferromagnetic insulator loses its local moments and becomes metallic.

With T_c's now having reached 120K, very few, if any, are arguing in favor of conventional electron-phonon coupling for the microscopic pairing mechanism in copper-based superconductors, especially in light of the surprisingly small changes in T_c with oxygen isotope substitution (4). Many have been drawn to the apparent relationship between antiferromagnetism and superconductivity and proposed magnetic origins for the pairing (5). Others point to the special hybridization of the Cu-O bond and the observed variable bond-charge (commonly referred to as variable formal copper valence) and propose electronic origins for the pairing (6). The questions are far from resolution at the present time.

Relatively few oxide superconductors were known before 1986. Most had low T_c's or otherwise unsurprising properties. One, however, $BaPb_{.75}Bi_{.25}O_3$, (7) (discovered in 1975) with a T_c of 12K, was recognized by some to be anomalous due to its surprisingly low density of states at the Fermi level. In 1988 we showed that the related compound $Ba_{.6}K_{.4}BiO_3$ displays a superconducting temperature near 30K, a T_c higher than that of conventional superconductors and surpassed only by copper containing compounds (8). As is true for the copper-oxide based superconductors, its T_c is especially high when compared with those of conventional superconductors with similar densities of states (9). These high T_c bismuth (lead)-based oxide compounds are different from the copper oxides for two reasons: 1) superconductivity occurs within the framework of a three dimensionally connected bismuth (lead)-oxygen array, and 2) there is no local or enhanced magnetism present in the chemical system, either in the superconductor itself or in the nonsuperconducting undoped (insulating) compounds. On the other hand, the two classes of materials are quite similar in that the Fermi level involves a strongly hybridized band (10) overlapping the Pb 6s, Bi 6s and O 2p levels. Further, the Bi-O bond charge is similar in variability to that of Cu-O. In analogy with the superconductivity-antiferromagnetism relationship in copper oxides, superconductivity in the Bi(Pb) based oxides occurs on doping away from an end member which is a "charge ordered" (charge density wave), not spin ordered, insulator. Recently, we have discovered another member of this family,

BaPb$_{.75}$Sb$_{.25}$O$_3$, to have a low (3.5K) T$_c$, extending the range of T$_c$'s within a single structure type to approximately one order of magnitude (11).

These materials are classical perovskites, a structure type which is the basis for many compounds of crystallographic and physical interest. The ideal undistorted perovskite has stoichiometry ABO$_3$, and consists of a three dimensional array of equally dimensioned "BO$_6$" octahedra sharing corner oxygens with each other. The "A" atom is at the center of the large resulting cavity and is coordinated to 12 oxygens. This ideal structure is shown in Figure 1. Small rotational relaxations of the octahedra about their shared corners are common. When this occurs, the ideally cubic symmetry becomes lower, and the elementary cell volume of approximately 4 x 4 x 4 angstroms increases in integer multiples (12). There are many well known derivatives of the basic perovskite in which different A site or B site atoms are ordered in cavities or octahedra of different sizes also yielding increased elementary cell sizes. For the superconducting Bi and Pb based perovskites, the correct carrier concentration for the occurrence of superconductivity is obtained by randomly mixing atoms of different charge on either the A or B type sites.

This chapter will describe primarily the crystal chemistry and crystallography of the superconducting compounds in this family. I begin with descriptions of the basic compounds BaBiO$_3$ and BaPbO$_3$. Next, the Ba(Pb,Bi)O$_3$ solid solutions, which have been studied in some detail, and the relatively new Ba(Pb,Sb)O$_3$ system are described. Finally, I describe what is known about (Ba,K)BiO$_3$, which, although it is a material of considerable importance, is not completely characterized as of yet because of the difficulties in materials preparation and crystal growth.

2.0 BaBiO$_3$

In many respects, this is the most interesting member of this family of compounds. For mixtures of the metal oxides with ratios between 1:3 and 3:1 Ba:Bi, it is the only ternary compound to form (13). This is in considerable contrast to the analogous BaO:PbO system where the compounds Ba$_4$Pb$_3$O$_{10}$ and Ba$_2$PbO$_4$

Superconducting Bismuth and Lead Oxide Based Perovskites 383

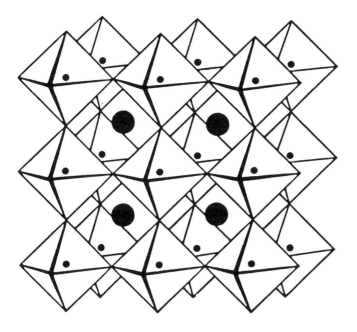

Figure 1: The ideal undistorted ABO_3 perovskite array. The A atoms are the large circles inside the cavity made of the BO_6 octahedra. The octahedra consist of a central B atom surrounded by 6 oxygens at the corners of an octahedron, generally represented by the geometric figure shown.

have been observed in addition to $BaPbO_3$. In fact, rather than forming the members of the Ruddesden - Popper series (compounds of stoichiometry $A_nB_{n-1}O_{3n-2}$) as is common in many perovskite based chemical systems, the ABO_3 type perovskite forms for Ba:Bi ratios between 1:1 and approximately 3:1. This occurs because Ba can apparently be accommodated in the B site resulting in solid solutions of the type $BaBi_{1-x}Ba_xO_{3-\delta}$ for $0 \le x < 0.5$. The ordered compounds with smaller divalent alkaline earths such as $Ba_2SrBiO_{2.5}$ are well known. This tendency for even the largest A type cations (excluding the large alkalis) to go into the octahedral sites severely limits the accessibility of solid solutions involving A site doping. The most obvious way to add electrons to $BaBiO_3$, for instance, by La substitution for Ba, does not work because La goes on the B site.

In metal oxides Bi occurs most commonly, and in fact almost exclusively, as Bi^{3+}. With the electronic configuration of the Bi atom $[Xe]4f^{14}5d^{10}6s^26p^3$, the trivalent ion leaves the 6p shell empty and the $6s^2$ lone pair. This lone pair is stereochemically active resulting in Bi^{3+}-O coordination geometries which are not regular. Even when in the presence of an oxygen environment where 6 neighbors would be possible, the coordination is strongly distorted to accommodate the lone pair.

There are relatively few oxides known where Bi is clearly 5^+, having been stripped of the $6s^2$ lone pair. These require a strongly chemically oxidizing environment, as is found for high ratios of electropositive elements (alkalis, alkaline earths) to Bi, e.g. in Na_3BiO_4 (14). The electronic configuration of Bi^{5+} is closed shell and thus both Bi^{3+} and Bi^{5+} are expected to be electronically stable. The same is certainly not true for the intermediate valence "Bi^{4+}" where the highly unfavorable $6s^1$ configuration would be expected.

In the context of other Bi oxides, then, the unusual nature of $BaBiO_3$ can be appreciated fully. In the conventional chemistry and valence counting of oxides, with oxygen and barium assuming the expected fixed valences of 2- (filling the oxygen 2p shell) and 2+ (emptying the Ba 6s shell) one would then expect $BaBiO_3$ to contain Bi in the 4+ valence state. Even looking at this compound from the point of view of the one-electron band structure calculations, the same result is expected. Thus $BaBiO_3$ is expected to be a metallic conductor with one delocalized electron. In fact $BaBiO_3$ is an insulator with a band gap of 2eV (15). The inherent instability of the $6s^1$ electronic state for Bi is the origin of the difficulty.

In fact the charge configuration in $BaBiO_3$ has been the matter of some controversy in the past mostly because some experimental probes appear to reflect the presence of a uniform Bi^{4+} valence state. Those results, however, have suffered from oversimplified interpretation. The most detailed information about the charge state of Bi in $BaBiO_3$ has come from analysis of the crystal structure. This analysis, described in some detail below, shows that there are indeed two types of Bi atoms in $BaBiO_3$, with different effective charges greater and smaller than +4. This picture of "charge disproportionation" is now generally accepted; that is that "$BaBiO_3$" is best represented as $Ba_2Bi^{4+\delta}Bi^{4-\delta}O_6$, with $\delta \approx 0.5-1$. This reflects the electronic instability of Bi^{4+}. In that way $BaBiO_3$ is analogous to other materials well known to display charge density waves (in which the electronic charge is modulated about some mean value in real space in the crystal) such as occurs for TaS_3 and $K_{.3}MoO_3$ (16). As in those materials, the charge modulation results in the localization of carriers and the subsequent result of electronically insulating rather than conductive properties. The case of $BaBiO_3$ appears to be relatively special, however, in that the coupling of the electronic charge to the atomic crystal lattice is relatively strong and a considerable structural distortion results.

The ideal simple cubic perovskite has a crystallographic cell of approximately 4 x 4 x 4 angstroms (a_p), with one ABO_3 formula unit per cell, crystallographic spacegroup Pm3m. At ambient temperature, $BaBiO_3$ is of monoclinic symmetry, space group I2/m, with approximate lattice parameters a=6.02, b=6.07, c=8.50, ($\sqrt{2}a_p$ x $\sqrt{2}a_p$ x $2 a_p$), β=90.2° and four $BaBiO_3$ units per cell (17)(18)(19). The cell axes are related to the simple perovskite by the matrix (a',b',c')=(1,1,0/ $\overline{1}$,1,0 / 0,0,2) (a,b,c). Inspection of the cell parameters shows that the distortions from cubic dimensionality are actually quite small, a situation that holds for all members of this family. The subtle nature of the dimensional distortions has led to considerable controversy concerning the "true" symmetry of $BaBiO_3$ and especially the solid solution series $BaPb_{1-x}Bi_xO_3$.

However, careful modern structural refinements, especially employing powder neutron diffraction, have yielded the most definitive results.

There have been several careful structural studies on $BaBiO_3$, but I will employ in this discussion the most recent set by Chaillout et al. (19) because the results illustrate the remarkable strength of the electronic-structural coupling in this compound. In spite of the relatively small dimensional distortions from cubic, the internal arrangements of the atoms differ from the ideal positions significantly. With respect to the simple cubic perovskite cell, the ideal atom positions in Pm3m are A:(1/2,1/2,1/2), B:(0,0,0), O:(0,0,1/2; 0,1/2,0; 1/2,0,0). Referring to the $\sqrt{2}a \times \sqrt{2}a \times 2a$ cell of a $BaBiO_3$ the expected ideal positions are: Ba:(1/2, 0, 1/4), Bi: (0,0,0; 0,0,1/2), and O:(0,0,1/4; 1/4,1/4,0). In the monoclinic spacegroup I2/m displacements from these ideal positions can occur and the two types of Bi and O sites become structurally and chemically distinct.

Chaillout et al. discovered that there are actually two polymorphs of $BaBiO_3$ at ambient temperature. Table 1 presents structural data and bondlengths from their paper illustrating the very subtle differences between the two polymorphs. Firstly, with the ideal undistorted perovskite atomic coordinates in mind, inspection of table 1 shows that the displacements of the atoms from those coordinates are small but highly significant compared to experimental precision. Comparison of the data for the two polymorphs shows extremely small but significant differences. These differences are most easily considered physically by inspection of the Bi-O bondlengths shown in the table. For the type 1 polymorph, produced when $BaBiO_3$ is synthesized or heat treated below 800°C, the two distinct Bi sites have significantly different Bi-O bondlengths. The Bi1 site is surrounded by 6 oxygens at an average distance of 2.258(2) angstroms, and the Bi2 site has 6 oxygen neighbors at 2.136(2) angstroms average. For the type 2 polymorph, synthesized or heat treated at 1125°C, the two distinct Bi sites have nearly identical Bi-O bondlengths, at 2.208(4) and 2.187(4) angstroms average for Bi1 and Bi2, respectively. Figure 2 presents the crystal structure of the type 1 polymorph.

There are several important points to appreciate about this result. Firstly, of course, one has to admire the skill of the

Table 1: Crystallographic Data for the Two Ambient Temperature Polymorphs of $BaBiO_3$, from Reference 19

Parameter	Type 1	Type 2
a	6.1796(1)	6.1804(2)
b	6.1337(1)	6.1347(2)
c	8.6632(2)	8.6650(3)
β	90.169(3)	90.095(6)
Ba(x,o,z)		
x	0.5043(5)	0.5023(8)
z	0.2483(7)	0.2461(14)
Bi1(0,0,0)		
Bi2(0,0,1/2)		
O1(x,o,z)		
x	0.0608(4)	0.0590(5)
z	0.2582(6)	0.2513(15)
O2(x,y,z)		
x	0.2615(4)	0.2612(8)
y	0.2515(10)	0.2411(12)
z	-0.0338(2)	-0.0332(2)
Bond Lengths		
Bi1-O1	2.267(5)	2.207(13)
Bi1-O2	2.254(4)	2.209(7)
Bi1-O Avg	2.258(2)	2.208(4)
Bi2-O1	2.129(5)	2.186(13)
Bi2-O2	2.140(4)	2.187(7)
Bi2-O Avg	2.136(2)	2.187(4)
"Valences"		
Bi1	3.49	3.85
Bi2	4.42	4.01

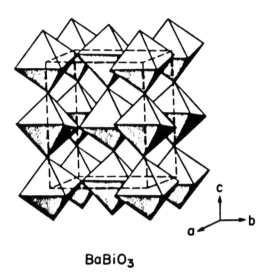

Figure 2: The crystal structure of $BaBiO_3$. The ordered charge-disproportionated larger and smaller BiO_6 octahedra are illustrated, as are the tilts of the octahedra about their shared corners.

experimentalists in detecting such a fine difference between the two polymorphs. Secondly, casual inspection of the difference in bondlengths of the Bi1-O and Bi2-O octahedra in the type 1 polymorph, 0.122(2) angstroms, might mislead one into thinking that the charge state at the two sites was not very different. From simple electrostatics, the more highly charged Bi atom will have a smaller octahedron than that of the more weakly charged Bi atom. How much different the charge states are can actually be put into relatively quantitative terms and expressed as a Bi "valence". The "valences" of the cations in oxides are commonly calculated following the work of Zachariesen and Brown, (20)(21) where the valence is merely taken as the sum of all the bond strengths between the cation and its surrounding oxygens. The bond strength for each individual bond is calculated from the bondlength and a bondlength bond-strength relationship of the type $v=(Ro/R)^n$ where Ro and n are determined by standardizing to compounds where valences and bondlengths are unambiguous. When applied to the type 1 polymorph of $BaBiO_3$ then, as shown in Table 1, one finds effective valences of approximately 3.5+ for Bi1 and 4.5+ for Bi2, quite a significant charge separation (see also ref. 22). The same is not true for polymorph 2, where the effective valences are near 4 and not significantly different from each other.

As Chaillout et al. pointed out, it is not appropriate to view the type 1 polymorph as containing charge disproportionated $Bi^{4\pm\delta}$ and the type 2 polymorph as containing a uniform Bi^{4+} valence state. Both polymorphs are in fact large band gap insulators. Rather, the type 1 polymorph (see Figure 2) is one where the more highly charged Bi and more weakly charged Bi are *ordered* in the Bi2 and Bi1 sites. The type 2 polymorph still contains more highly charged and more weakly charged Bi but they are *positionally disordered* so that on the average both Bi1 and Bi2 sites have a size which reflects the average 4+ valence. Similar kinds of order-disorder phenomena are of course common for perovskites with two chemically distinct atoms on the small cation sites, but $BaBiO_3$ is unusual in that the order-disorder phenomena are associated with different charge states of the *same* ion. The fact that this can occur suggests the presence of a very strong electronic-structural coupling.

There have been countless discussions by workers in the field of high T_c superconductivity concerning the "true" valence state of the copper, bismuth, and oxygen atoms in these materials. Whereas it is not appropriate to take a rigid view of valence as might pertain to completely ionic compounds, it is clear that the valences referred to in the above discussion do in fact have a physical meaning, even if somewhat more general than the simplest view might suggest. The problem, of course, is that there are both metal s and oxygen p states at the Fermi level, and therefore the bonds of interest are far from ionic. The holes in the oxygen p states which occur indicate that the oxygens are not strictly in the O^{2-} valence state (as is assumed in valence calculations), and therefore that the fractional valences assigned to the cations actually represent a fractional charge in a covalent bond shared by both the metal and oxygen atoms. Therefore calculated "valences" actually represent differences in the amount of charge involved in the Bi-O or Cu-O bonds. In the case of $BaBiO_3$ the difference in charge in the bonds between Bi1, Bi2, and their oxygen neighbors is calculated to be 1 electron/Bi, (4.5-3.5) a very significant charge ordering.

The existence of ordered and disordered polymorphs of $BaBiO_3$ implies the existence of an ordering transition temperature. This must be higher than 800°C where the type 1 polymorphs in the Chaillout et al. study were synthesized. The transition from disordered to ordered states must, however be extremely sluggish, as cooling at 6°/minute of the disordered polymorph from 1125°C through 800°C did *not* give rise to the ordered polymorph. Nonetheless Chaillout et al. attribute a DTA signal at 860°C on heating and 801°C on cooling (at 2°/min) in air to this transition. Many phase transitions have in fact been reported for $BaBiO_3$. A first order monoclinic to trigonal transition has been reported (23) to occur at 405K. Structure refinement of the trigonal phase indicated that the two distinct Bi sites are maintained, with exactly the same degree of charge ordering, and that the transition merely involves a change in the tilts of the BiO_6 octahedra. A later structural study found $BaBiO_3$ to be cubic by 900K (24). The trigonal to cubic transition has been reported to be in the range of temperature 750-800K (23). Dielectric anomalies have been reported to occur in the temperature regions 750-700K, 450K, and

250-300K (25)(26). (The dielectric constant of $BaBiO_3$ has been reported to be very large, implying that that material is a ferroelectric, results that are apparently not universally accepted (23).) It is not at all clear that a "cubic" average structure represents a truly uniform Bi^{4+} charge state for $BaBiO_3$ at high temperatures. It could of course merely reflect a completely disordered arrangement of highly charged and weakly charged BiO_6 octahedra with no tilting about their shared oxygens. Chaillout et al. in fact propose that the transition they observe at 800°C (1075K) is the one associated with the Bi^{4+} to $Bi^{4\pm\delta}$ charge disproportionation. The sequence of phase transitions is, apparently:

MELT	1075K	750K	405K	
CUBIC	CUBIC	TRIGONAL	MONOCLINIC	
Bi^{4+}	$Bi^{4\pm\delta}$	$Bi^{4\pm\delta}$	$Bi^{4\pm\delta}$	
		111 Tilt	110 Tilt	

However, considerably more research will be necessary to clarify the nature of these transitions.

As might be expected, $BaBiO_3$ can display a considerable range of oxygen stoichiometries. There does not appear, however, to be a continuous solid solution between $BaBiO_3$ and $BaBiO_{2.5}$, where all the Bi would be Bi^{3+}. Several carefully done studies have obtained essentially the same results (27)(28). For temperatures below 700°C, the phases occurring in equilibrium are $BaBiO_{3-\delta}$, $0 \leq \delta \leq 0.03$, $BaBiO_{2.8}$, and $BaBiO_{2.55}$, with two-phase regions for intermediate oxygen contents. The first phase is isostructural with $BaBiO_3$. The more oxygen deficient phases, although perovskites, differ significantly in structure from $BaBiO_3$. Their crystal structures have not yet been determined.

Finally, a few brief comments on the synthesis of $BaBiO_3$. Powdered materials are apparently easily prepared from stoichiometric mixtures of $Ba(NO_3)_2$, $BaCO_3$ or BaO_2 with Bi_2O_3 or $BiONO_3$ by heating in either air or oxygen at temperatures above 800°C. $BaBiO_{3-\delta}$ apparently melts congruently in air or oxygen near 1125°C, and single crystals of considerable size can be grown (19). Small single crystals have also been grown hydrothermally from BaO_2, $Ba(OH)_2 \cdot 8H_2O$ and Bi_2O_3 at 700°C under a pressure of 300 MPa (23). Oxygen deficient $BaBiO_3$ has been prepared either by heating at high temperature in reducing environments, (27)(28) or at low temperatures by the Zr-metal gettering technique (29).

3.0 $BaPbO_3$

Lead is a particularly interesting atom from the point of view of the crystal chemistry of perovskites because it can form compounds where it occupies either the large cation or the small cation site. This is because of the vast difference in the sizes and coordination geometries of the two stable valence ions Pb^{2+} and Pb^{4+} in oxides. Adjacent to Bi in the periodic table, elemental Pb has the electronic configuration [Xe] $4f^{14}5d^{10}6s^26p^2$. In the stable divalent state, the $6s^2$ lone pair is left and can be stereochemically active generally resulting in coordination environments with oxygen strongly distorted from perfect symmetry. The large Pb^{2+} ion forms in the A perovskite site when combined with small highly charged cations and is an integral part of a large number of perovskites of great practical interest for their ferroelectric properties such as the $PbZrO_3$ – $PbTiO_3$ family of compounds. In compounds which are strongly chemically oxidized, such as those which form with alkali or alkaline earth ions, lead assumes a 4+ valence state and is a stable closed shell ion. As with Bi, the intermediate valence state (for Pb it is 3+) where there might be an unpaired 6s electron, is highly energetically unfavorable. The 4+ state is assumed by Pb, for instance, in perovskite derived compounds like $BaPbO_3$, Ba_2PbO_4, $SrPbO_3$ and Sr_2PbO_4.

The $BaPbO_3$ perovskite has been studied with considerably less vigor than has $BaBiO_3$, probably because its crystal structure and electrical properties appear to be easily explained by conventional arguments. Although early crystallographic studies suggested that the symmetry of $BaPbO_3$ was cubic (30-33), a careful study by Shannon and Bierstedt established the true symmetry at ambient temperature to be orthorhombic (34). The distortion from cubic symmetry is very small, with the true unit cell containing 4 elementary perovskite units, with cell axes $\sqrt{2}a$ x $\sqrt{2}a$ x $2a$ that of the basic perovskite, as for $BaBiO_3$, with unit cell lengths reported by Shannon and Bierstedt as a=6.024(1), b=6.065-(1), and c=8.506(1) angstroms. The volume per ABO_3 unit in $BaPbO_3$ is 78 $Å^3$ compared to 82 $Å^3$ in $BaBiO_3$. Although a primitive unit cell was first assigned, subsequent neutron diffraction studies showed the orthorhombic cell to be body centered (35)(18). To date, there have been two refinements of the crystal structure of $BaPbO_3$ by powder neutron diffraction (35),(18). Neutron diffraction is necessary to determine the positions of light atoms such as oxygen in the presence of heavy atoms such as Ba and Pb, as the neutron scattering amplitudes are all roughly equal and not a function of atomic number. The resulting atomic positions are summarized in Table 2. Each of the references 18, 34 and 35 employs a different set of orientations for the crystallographic cell axes. The table adopts the assignments of reference 18 such that comparison to the structure of $BaBiO_3$ and the $Ba(Pb,Bi)O_3$ solid solutions is most easily made. To compare Shannon and Bierstedt's axial assignments to those employed in table 2, transform via (a,b,c)=(0,1,0/1,0,0/0,0,1) (a,b,c) $S.B.$. The transformation from the cell employed by Thornton and Jacobsen is (a,b,c)=(0,0,1/1,0,0/0,1,0)(a,b,c) $T.J.$. This transformation has been applied to the published data to make the entries in Table 2. The positions in the table can be compared to those of the ideal perovskite by recalling that in this type of cell the ideal coordinates would be: Ba:(1/2,0,1/4), Pb:(0,0,0), and O:(0,0,1/4;1/4,1/4,0). As for the case of $BaBiO_3$, the displacements from the ideal positions are small but significant.

As is expected, the crystal structure of $BaPbO_3$ displays one kind of Pb site and therefore there is a uniform Pb-O bonding charge state ("Pb^{4+}"). There are, however, two independent oxygen positions allowing the PbO_6 octahedra to relax from perfect symmetry if they care to. At 4.2K the distortion is extremely small, with 2 Pb-O1 bonds at 2.145(1) and 4 Pb-O2 bonds at 2.150(1) and intra-octahedral O-Pb-O bond angles differing from 90° by at most 1.4°. This is not a distortion which is significant from an electronic point of view. The data at room temperature do not display any distortion of the PbO_6 octahedra, but they appear to be somewhat less precise. The structural distortion in $BaPbO_3$ involves only a tilting of the octahedra around their shared oxygen corners. As in $BaBiO_3$, the tilt is about the [110] axis of

Table 2: Atomic Positional Parameters for $BaPbO_3$, space group Ibmm

Parameter	300K[18]	4.2K[35]
a*	6.0676(3)	6.063(1)
b	6.0343(4)	6.007(1)
c	8.5212(5)	8.476(1)
Ba(x,0,1/4)		
x	0.501(2)	0.5033(12)
Pb(0,0,0)		
O1(x,0,1/4)		
x	0.048(2)	0.0547(8)
O2(1/4,1/4,z)		
z	−0.025(2)	−0.0313(3)
Bond lengths		
Pb-O1	2.15	2.145(1)
Pb-O2	2.15	2.150(1)

* The cell dimensions obtained in reference 34 by X-ray diffraction at 300K, suitably transformed, are more accurate than those obtained by neutron diffraction. They are: a=6.056(1), b=6.024(1) and c=8.506(1).

the simple cubic perovskite. The tilt angle is apparently approximately 8° at 300K (10° in $BaBiO_3$) (18), and appears to grow to near 15° by 4.2K (35). Unlike the case for $BaBiO_3$, there appear to be no phase transitions reported for $BaPbO_3$, although a transition or transitions to higher symmetry structures at temperatures above 300K should certainly be expected.

$BaPbO_3$ has been prepared in polycrystalline and single crystal form by several techniques. Powders can be prepared in air at temperatures between 800 and 1000°C from mixtures of $BaCO_3$ and PbO heated in air over a period of 10 days (35). Considerable excess PbO was employed. We found that it could be prepared from stoichiometric mixtures of $BaCO_3$ and PbO_3O_4 in 24 hours of heating in O_2 at 825°C with intermediate grindings (8). There is one report of synthesis from $BaCO_3$ and PbO at 1200 and 1350°C (34). Those conditions would be in considerable disagreement with a published phase diagram (36), which claims that $BaPbO_3$ melts incongruently in air at approximately 1080°C. That diagram shows that solid $BaPbO_3$ is in equilibrium with PbO rich liquid between 1080°C where it first forms and the eutectic temperature of 850°C (Ba_2PbO_4 is found to melt *congruently* in air at approximately 1280°C).

Single crystals of $BaPbO_3$ can therefore be grown from the appropriate PbO rich melt at reasonable temperatures (36). Small crystals have been grown from mixtures of $Ba(OH)_2 \cdot 8H_2O$, $KClO_3$ (an oxidizing agent), and PbO heated at 700°C in a sealed tube at 3kbar external pressure for 24 hours (34). Crystal growth from a KF solvent has also been reported, but no details were given (37).

$BaPbO_3$ is a metallic conductor reported to have a resistivity of 290$\mu\Omega$-cm at 300K and a resistivity ratio of approximately 4 by 4.2K (34). There has been one report of a superconducting transition near 0.5K (37) but that transition was not seen in a later study (8). Simple valence arguments would suggest that $BaPbO_3$ should in fact be a normal valence compound and be an electronic insulator like the $BaSnO_3$ perovskite. One possible way in which carriers could be introduced would be through oxygen deficiency, but that has never been observed, and was specifically considered and rejected in reference 34. The commonly accepted reason for the observation of metallic conductivity is derived from band structure calculations (10) which show the presence of the strongly hybridized band at the fermi level involving the energetic overlap of Pb 6s and O 2p states. This broad band is only partly occupied by the available electrons and therefore metallic conductivity can occur. Thus $BaPbO_3$ is not really a "normal valence" compound like its ionic insulating neighbor $BaSnO_3$, where such cation-oxygen hybridization apparently does not occur.

4.0 $BaPb_{1-x}Bi_xO_3$

Because of the chemical similarity of Pb and Bi, and the fact that both form perovskite compounds with Ba, a continuous perovskite solid solution $BaPb_{1-x}Bi_xO_3$, $0 \le x \le 1$, exists. This means that Pb and Bi can be mixed in any proportion, on the perovskite B site. The term solid solution is employed because the basic structure is unchanged even though the crystallographic symmetry, reflecting dimensional and rotational distortions of the (BO_6) octahedral array, changes from orthorhombic for $BaPbO_3$, to monoclinic, for $BaBiO_3$.

The superconducting composition $BaPb_{.75}Bi_{.25}O_3$ has been studied extensively. In addition, the whole solid solution has been the subject of considerable research because the transition from the metallic, through superconducting, to semiconducting state as a function of the Pb to Bi ratio is of great interest to the theoretical explanation for the occurrence of superconductivity in this system. Bi has one more valence electron than does Pb, so the filling of the band at the fermi level changes significantly in the solid solution. Because the thrust of this review is crystal chemical, I will not describe physical properties in any detail. Several good reviews have been written in that regard. (See for instance reference 15.) Briefly, in the solid solution $BaPb_{1-x}Bi_xO_3$, metallic conductivity exists for $0 \le x \le 0.35$, and the materials are semiconducting for $0.35 \le x \le 1.0$. Superconductivity occurs in the metallic phase, with an optimal superconducting transition temperature of approximately 12K at x=0.25. Note three things about this composition: (1) It is a low rational ratio of 3Pb:1Bi, suggesting that some kind of simple Pb:Bi crystallographic ordering might be possible, (2) That T_c decreases on both sides of this value within a phase that is metallic, that is that the best superconductor is *not* right at the metal to semiconductor transition boundary and (3) That the optimal superconducting composition is very far in electron count from $BaBiO_3$. The latter fact has been a problem for theorists attempting to interpret the superconductivity in this system based on one electron band calculations (39). In addition to the fact that T_c is optimal at x=0.25, there is also a maximum in the observed Meissner effect, and a minimum in the superconducting transition width (9). This has suggested to some that there may be one unique superconducting composition at $BaPb_{.75}Bi_{.25}O_3$ perhaps due to some kind of Pb:Bi ordering, maybe short range, and that other compositions where degraded superconductivity occurs, $0.10 \leq x \leq 0.35$, are only superconducting due to microscopic Pb:Bi mixing inhomogeneities (9). This is an idea that has been much discussed with regard to superconductivity in the $La_{2-x}Sr_xCuO_4$ solid solution (40). At the present time there is not a consensus or strong evidence one way or the other for either chemical system. The difficulty in detecting such ordering in the $BaPb_{.75}Bi_{.25}O_3$ solid solution is that the usual diffraction techniques are not sensitive enough to see it. There has been one report (41) of anomalous behavior in the x dependence of observed intensities of certain X-ray diffraction lines which the authors interpreted as being due to partial Pb:Bi ordering. Further research into this question is certainly of considerable interest.

What is the manner in which the $BaPb_{1-x}Bi_xO_3$ solid solution evolves from a structure in which only tilts of the $(Pb,Bi)O_6$ octahedra are present to one in which a charge density wave (charge disproportionation) changes the sizes of the octahedra as well? This is a question that has been addressed by several studies, basically by measurement of crystallographic unit cell dimensions and symmetry by powder neutron or X-ray diffraction. Single crystal studies are not generally much use for answering questions of this sort due to the fact that materials such as these which are of higher symmetry at their growth temperature than at ambient temperature are often heavily twinned. There are at least five possible difficulties with the interpretation of symmetry and dimensional studies in $Ba(Pb,Bi)O_3$; they are: (1) The dimensional and angular distortions from cubic symmetry for the $BaPb_{1-x}Bi_xO_3$ solid solution are very small, meaning that high resolution diffraction techniques must be employed. (2) Diffraction peak widths may be broadened at high angles due to inhomogeneities in the Pb:Bi mixing in the sample, smearing out subtle diffraction line splitting, (3) Studies done at ambient temperature do not necessarily reflect what is happening near 10K where superconductivity occurs (perovskites are well known for their propensity toward having phase transitions) (4) Simple dimensional and symmetry measurements may miss something of importance, as for instance occurs for ordered and disordered $BaBiO_3$, and (5) There may well be quenchable phases occurring at high temperatures, with sluggish transitions to lower symmetry equilibrium states, resulting in the observation of apparently different unit cells at the same value of x dependent on the method of sample preparation.

With those caveats in mind, consider the data which exist for $BaPb_{1-x}Bi_xO_3$. The first of these, taken from reference 18, are re-plotted in Figure 3. These are the results most frequently quoted in the literature. They have apparently been obtained through interpretation of X-ray Guinier powder films. This technique, which has a focusing geometry, is capable of high resolution. The authors claim to observe four distinct distortions of the perovskite in the solid solution. For $BaPb_{1-x}Bi_xO_3$, they observe orthorhombic symmetry for $0 \leq x \leq 0.05$, tetragonal symmetry for $0.1 \leq x \leq 0.3$, orthorhombic symmetry again for $0.4 \leq x \leq 0.75$, and finally monoclinic symmetry for $BaBiO_3$. Although the cell volume is a continuous function of x, there is a significant discontinuity in the c axis length, which is anomalously large in the tetragonal phase. Of course there must be a corresponding discontinuity in the area of the a-b plane in the tetragonal phase to conserve the continuity in volume. As the tetragonal and orthorhombic symmetries reflect differences in the tilts of the $(Pb,Bi)O_6$ octahedra, the discontinuities in cell axis lengths imply discontinuous changes in tilt angles through the orthorhombic to tetragonal transition regions. One should view these results with caution; the canceling discontinuities which conserve cell volume continuity suggest that there may have been a misindexing of the powder pattern in the "tetragonal" phase. The most obvious error would be misassignment of 001 type reflections. If that is true then the new cell in the "tetragonal" region is actually orthorhombic, with $a'=(c/2) \cdot \sqrt{2}$, $b'=b$, $c'=(b/2) \cdot \sqrt{2}$. The data have been plotted for this cell assignment in Figure 4. Note that the continu-

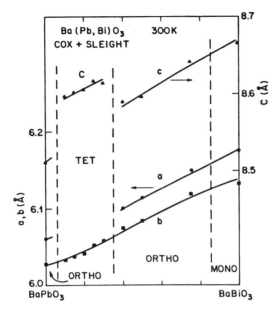

Figure 3: Crystallographic cell dimensions and symmetries for the $BaPb_{1-x}Bi_xO_3$ solid solution at 300K, from reference 18.

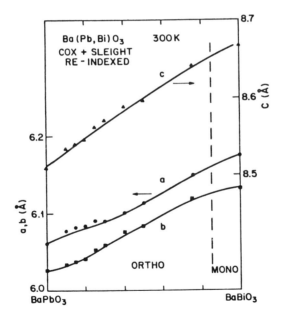

Figure 4: Crystallographic cell dimensions and symmetries for the $BaPb_{1-x}Bi_xO_3$ solid solution at 300K, data from reference 18 with cell re-assigned.

ity of tilt angle across the solid solution has been restored.

A second study of the complete $BaPb_{1-x}Bi_xO_3$ solid solution, published in 1985, does not report the same symmetry variations (42). Their results for the crystallographic cell parameters and symmetries at 25°C are re-plotted in Figure 5. The results have been obtained by powder diffraction on an X-ray diffractometer. These authors find that there are only two structural regions for $BaPb_{1-x}Bi_xO_3$. The material is orthorhombic with a $BaPbO_3$ related structure, for $0 \leq x < 0.9$, and monoclinic, related to $BaBiO_3$ for $0.9 \leq x \leq 1$. There is no change to tetragonal symmetry for compositions near the optimal superconducting composition ($x \approx 0.25$). Although the c axis length varies linearly in composition, there are subtle changes in slope for a and b vs. x. There are no axial discontinuities. Note that these results are virtually identical to the reassigned data from reference 18. The method of synthesis of the powder samples employed in the study was not described in any detail, but the final treatment was at 800-900°C in O_2 for 4 hours. Presumably this was not the whole synthetic procedure, as one would not expect single phase material to be obtained from starting oxides or carbonates in such a short anneal at such low temperatures. Thus, one cannot determine whether the difference between these results and those of Cox and Sleight is due to the temperature or procedure of synthesis or error in the determination of the crystallographic symmetry from the diffraction data.

This latter result has been corroborated by at least one other report (41) of the crystallographic symmetry of the $BaPb_{1-x}Bi_xO_3$ solid solution in the concentration range $0 \leq x \leq 0.4$. For polycrystalline samples prepared by an undisclosed route, the symmetry was found to be orthorhombic at 20C over the entire composition range $x \leq 0.4$ through the interpretation of data obtained on a powder diffractometer. (In this study and the previous diffractometer study, deconvolution of the $K\alpha_1$ and $K\alpha_2$ components of the Cu radiation employed was necessary to properly interpret the diffraction data). The authors make an additional very interesting observation, that is that the degree of orthorhombicity, i.e. how much a and b differ, has a distinct minimum in the solid solution $BaPb_{1-x}Bi_xO_3$ near x=0.2, where the orthorhombicity is very small. The identical effect can be seen in the data in Figure 5, with a and b most nearly equal near the

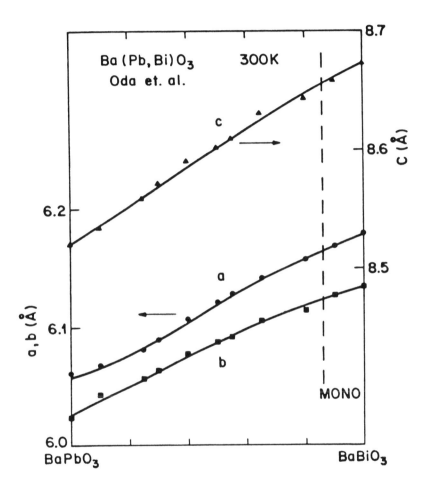

Figure 5: Crystallographic cell dimensions and symmetries for the $BaPb_{1-x}Bi_xO_3$ solid solution at 300K, from reference 42.

composition x=0.25. The proximity of this minimum in the structural distortion of the perovskite solid solution to the optimal superconducting composition is either a coincidence or a reflection of coupled origins for both phenomena. The minimum in orthorhombicity reflects a small minimum in the tilt angles of the $(Bi,Pb)O_6$ octahedra about their shared corner oxygens.

It is possible that both orthorhombic and tetragonal symmetry forms of $Ba(Pb,Bi)O_3$ exist in the superconducting composition region. The symmetry at room temperature in this composition region would then be sensitive to preparative conditions, which are not sufficiently described in any of the reports to come to an unambiguous understanding based on current evidence. Several groups, in fact, have claimed to have produced *both tetragonal and orthorhombic superconducting* $BaPb_{.75}Bi_{.25}O_3$ (43),(15). One of these groups (43) claims that the distinction is related to the incorporation of small amounts of Pb^{2+} (a few percent) on the Ba^{2+} site in the orthorhombic phase. More careful study of this issue might be of interest. It is not clear, however, that the symmetry is of much relevance to the superconductivity (both orthorhombic and tetragonal materials are superconducting) as the difference reflects merely a difference in the direction of the tilts of the octahedra, and not a significant difference in the magnitudes of the tilts. Even the magnitudes of the tilts may well be irrelevant to the superconductivity.

There have been a few reports of structural phase transitions above and below ambient temperature for compositions in the superconducting region. For samples which are orthorhombic at ambient temperature, a transition to cubic symmetry occurs between 360 and 410°C for $0 \le x \le 0.4$ (41). The transition is second order. There is another transition, at temperatures near 200°C, to a state of intermediate symmetry, perhaps tetragonal. This transition is apparently first order, and, if strongly hysteretic, could account for the existence of both tetragonal and orthorhombic ambient temperature phases. A second order orthorhombic to monoclinic transition has been reported to occur for superconducting $BaPb_{.75}Bi_{.25}O_3$ at 160K (42). The monoclinic cell is claimed to be different from that of $BaBiO_3$, in this case with cell parameters of a=6.041, b=6.074, c=8.528, γ=90.31° at 8K. This apparent transition to a low symmetry structure is of interest if it reflects something more than just a change in the tilts of the octahedra, and is worthy of detailed structural study. Dielectric anomalies at low temperatures observed (25) in $BaPb_{1-x}Bi_xO_3$ for $0.1 \leq x \leq 0.5$ correlate well to the proposed orthorhombic to monoclinic transition. The anomalies vary between approximately 150 and 250K, with a distinct minimum in temperature near x=0.3. Dielectric anomalies are also reported to occur for compositions in the nonsuperconducting region. Although the transitions are diffuse, the data suggest that there is a structural transition below ambient temperature for samples in which $x > 0.5$. The transition inferred is just below ambient temperature for $BaBiO_3$ and decreases to approximately 50K by x=0.5.

Keep in mind that $BaBiO_3$ is a charge disproportionated insulator which becomes superconducting through doping with Pb. By reference to the ambient temperature phase diagram of Cox and Sleight, most have assumed that the charge disproportionation in $BaPb_{1-x}Bi_xO_3$ disappears for $x < 0.9$, inferred from the monoclinic to orthorhombic transition composition. Other mechanisms, then, such as Pb:Bi ordering, or a very short range ordered CDW, have to be invoked to explain why superconductivity does not occur for x just barely lower than 0.9. If the dielectric anomalies below room temperature for $x > 0.5$ represent an orthorhombic to monoclinic (charge disproportionation) transition, then the picture changes: the CDW is not destroyed for low Pb doping. Further work on this apparent transition would be of interest to the superconductivity in this system. This is in contrast to most of the other symmetry changes, which may be relevant only from a geometric point of view.

The crystal structure of superconducting $BaPb_{.75}Bi_{.25}O_3$ has been determined for tetragonal symmetry material at 300K (18), and at high temperature (718K) for the cubic form (44) (Table 3). The latter is an ideal undistorted perovskite with unit cell a=4.31 angstroms. The Pb/Bi-O octahedron is regular with 2.16 angstrom bondlengths and 180° O-Bi/Pb-O inter-octahedra bond angles. The ambient temperature tetragonal material is distorted from the ideal structure by an 8° twist of the octahedra about one pseudo cubic [100] type axis (the tetragonal c axis). The Pb/Bi-O octahedra are essentially completely regular, with two bondlengths of 2.15 and four of 2.16 angstroms. There is relatively little unexpected about

Table 3: Atomic Positional Parameters for
BaPb$_{.75}$Bi$_{.25}$O$_3$

	300K[18]	718K[44]
Cell	a=6.0621(4) c=8.6090(8)	a=4.311(4)
Symmetry	I4/mcm	Pm3m
Ba	(1/2,0,1/4)	(1/2,1/2,1/2)
Bi/Pb	(0,0,0)	(0,0,0)
O1	(0,0,1/4)	(0,0,1/2)
O2	(1/2-x,x,0) x=0.2165(1)	
Bond Lengths		
Bi/Pb-O1	2.15	2.155(3)
Bi/Pb-O2	2.16	

these structures. It is interesting, however, that the Pb-, Pb/Bi-O bondlengths in $BaPbO_3$ and $BaPb_{.75}Bi_{.25}O_3$ are virtually identical and that only the direction (not the magnitudes) of the tilts of the octahedra has changed. The molecular volume for $BaPbO_3$, 78.00 $Å^3$, is slightly (1.4%) smaller than that of $BaPb_{.75}Bi_{.25}O_3$, 79.09 $Å^3$. It is not clear what the driving force for the change in tilt system is as Bi is introduced into $BaPbO_3$. Further detailed crystallographic study in the $BaPb_{1-x}Bi_xO_3$ solid solution, especially near 10K, would be of considerable interest.

As has been described earlier, $BaBiO_3$ is known to display a wide range of oxygen stoichiometry variation, and $BaPbO_3$ is relatively fixed in oxygen content. The superconducting composition $BaPb_{.75}Bi_{.25}O_3$ apparently displays an intermediate kind of behavior. It has been convincingly shown by several groups that oxygen deficient $BaPb_{.75}Bi_{.25}O_3$ has strongly degraded properties (43)(45)(46). The work was done on both thin films and bulk materials, generally by varying the pO_2 in the sputtering process or heating at high temperatures in a dynamic vacuum. Detailed chemical or structural work has not been performed. From a simpleminded point of view, based on the behavior of the endmembers, one might expect the maximum possible oxygen deficiency obtainable to be $BaPb_{1-x}Bi_xO_{3-x/2}$, which is in fact observed in bulk materials for $0 \le x \le 0.5$ (43).

The earliest physical property studies on $BaPb_{1-x}Bi_xO_3$ were performed on polycrystalline ceramics. The synthetic techniques employed were not in general described in detail. The materials can be synthesized, for example, from stoichiometric mixtures of $BaCO_3$, Bi_2O_3 and Pb_3O_4 heated in O_2 at temperatures greater than 850°C. Synthesis at temperatures in excess of 1000°C is not favorable due to volatilization of Pb oxides. A good technique for obtaining high quality polycrystalline samples is through hot-pressing prereacted powders (900°C, O_2) at 1000°C under 300 bars O_2 pressure (47).

Considerable effort has been expended in growing single crystals of compositions in the solid solution, with particular emphasis on the superconducting composition. A variety of methods have been employed, all flux type, as the solid solution members are not congruently melting. There have been a few detailed reports on growth of single crystals from melts in the

BaO-PbO$_2$-Bi$_2$O$_3$ ternary system (48)(49). Crystals of sizes greater than 2mm on an edge can be obtained, with T$_c$'s near 11K. BaPb$_{.7}$Bi$_{.3}$O$_3$, for instance (49), can be grown from a melt of initial composition 15 mol% BaO, 70 mol% PbO$_2$ and 15 mol% Bi$_2$O$_3$, in a platinum crucible. After soaking at 1050°C for 10 hours, the melt is slow cooled to a temperature near 825°C and held for 30 hours. The melt is decanted off. When the geometry is such that the top of the crucible is 10°C hotter than the bottom, relatively large crystals are obtained at the top surface of the melt. Alternately, superconducting Ba(Pb,Bi)O$_3$ can be obtained from a KCl melt (50). In this method, prereacted powder of composition BaPb$_{.9}$Bi$_{.1}$O$_3$ is mixed with KCl in a 1:9 to 1:19 weight ratio, heated to 1000-1150°C for several days in a tightly covered platinum crucible (to prevent volatilization of KCl) and then cooled to 800°C at 3°C/minute. Plate-like crystals are grown at the top and sides of the melt, with superconducting transitions near 11K.

5.0 BaPb$_{1-x}$Sb$_x$O$_3$

Antimony is directly above bismuth in the periodic table, with electronic configuration [Kr]4d^{10}5s^{2}5p^3. The valence electrons are therefore in the 5 sp shell as opposed to the 6 sp shell. Nonetheless antimony and bismuth display similar lone-pair stereochemistry. The 3+ and 5+ valence states are strongly preferred, with the 5s^2 lone pair configuration for 3+, and a highly unfavorable 4+ valence state. Amusingly, Sb$_2$O$_4$ exists but contains distinct Sb^{3+} and Sb^{5+} ions. The 5+ valence state of antimony is very stable in oxides in the presence of electropositive elements from the left side of the periodic table. In fact Sb^{5+} is virtually always found in such compounds, in relatively regular octahedral coordination with oxygen. This is in distinct contrast to the behavior of Bi, which is virtually always (with the notable exception of BaBiO$_3$) Bi^{3+} under the same chemical conditions. Thus the compound Ba$_2$BiSbO$_6$, an ordered perovskite isostructural with BaBiO$_3$, is claimed (17) to contain Bi^{3+} and Sb^{5+}. In that compound, the Sb^{5+}-O octahedra have bondlengths of 2.00(1) angstroms and the Bi^{3+}-O octahedra have bondlengths of 2.30(1) angstroms. Because Sb strongly favors the 5+ valence, the BaBiO$_3$

like perovskite analog $BaSbO_3$ is not known to exist.

It is interesting to consider the similarities and differences among ions in this part of the periodic table from the point of view of ionization energies. Table 4a compares the ionization energies (51) for removing electrons from Sn, Sb, Pb and Bi in comparison to the early transition metals of comparable valence Ti and V. There are several features to note in a comparison of this sort. Firstly, the ionization energies for Sb and Bi in the valence states of interest, 3+, 4+, and 5+, are remarkably similar to each other (and different from those of V), suggesting that their tendency to form 5+ and 3+ ions in oxides, respectively, may have to do with some other difference, maybe the relative differences in extent of the 5sp and 6sp orbitals. Sn and Pb are also somewhat similar (Sn has configuration $[Kr]4d^{10}5s^25p^2$) to each other and to Sb and Bi in the valence range of interest (2+,3+,4+). Since band structure calculations show that Bi, Pb and O form a broad strongly hybridized band at the Fermi level in the $Ba(Pb,Bi)O_3$ perovskites, the table suggests that the same should be true in the first approximation for Sn and Sb. Finally, the tendency toward instability of intermediate valence states, which clearly drives the charge disproportionation in $BaBiO_3$, can also be deduced from the ionization potentials, and is shown in Table 4b (6). For the early transition metal ions Ti^{3+} and V^{4+} there is no gain in energy in a disproportionation of the type $2M^n \rightarrow M^{n-1} + M^{n+1}$, and so one would not expect a "skipping" of the valence n. For the ions in this family of superconductors, the fact that the valence electrons are in 5 or 6 p and s states means that such disproportionation is energetically favorable. This tendency is very large for Bi and Sb, but is also large for Sn and Pb, as seen in the table.

There has been only one report to date on superconductivity in the $BaPb_{1-x}Sb_xO_3$ perovskite series (11). From $BaPbO_3$, where no superconductivity is observed, T_c rises to approximately 3.5K for x=0.25-0.3 and drops rapidly to below 1.5K for larger values of x. Because the $BaSbO_3$ perovskite does not exist, a complete solid solution is not expected. The limiting Sb concentration appears to be x=0.5. The electronic properties have not been studied in any detail, but the visual appearance of the materials suggests considerable changes within the solid solution

Table 4a: Ionization Potentials for Selected free Atoms[51]

Atom	Ionization Potential				
	I	II	III	IV	V
Ti	6.82	13.58	27.49	43.27	
V	6.74	16.50	30.96	49.10	69.3
Sn	7.34	14.63	30.50	40.713	
Sb	8.64	16.53	25.30	44.20	56.0
Pb	7.42	15.03	31.94	42.32	
Bi	7.29	16.69	25.56	45.30	56.0

Table 4b: Tendency Toward Valence Skipping[6], Expressed as $\Delta E = IP^{n+1} + IP^{n-1} - 2IP^n$

Ion	ΔE
Ti^{3+}	1.9
V^{4+}	2.1
Sn^{3+}	-5.6
Sb^{4+}	-7.1
Pb^{3+}	-6.5
Bi^{4+}	-9

series. Powdered $BaPbO_3$ is dark brown in color, with the addition of as little as 0.05 Sb the color changes to blue-black which deepens continuously as the Sb content increases. Between x=0.3 and 0.4 the materials become metallic purple in color and are dull black by x=0.5.

Preliminary measurements indicate that the symmetry of the crystallographic cell changes with Sb content. Single crystals of the superconducting composition $BaPb_{.75}Sb_{.25}O_3$ were found to be tetragonal at room temperature, with a = 6.028 and c = 8.511Å. The orthorhombic b axis of $BaPbO_3$ (a = 6.024, b = 6.065, c = 8.506) appears to decrease in a continuous manner and the c axis appears to increase continuously as x increases from 0 to 0.25. Thus the preliminary data suggest a continuous change in the tilt of the octahedra with Sb content. For Sb contents near x=0.4 the symmetry becomes cubic with a cell parameter a = 4.254Å. By x=0.5 the symmetry again becomes lower, with a pseudocubic cell parameter a = 4.248Å. As in the case of the $BaPb_{1-x}Bi_xO_3$, the symmetry changes are very subtle. Further higher resolution measurements are necessary to bring the understanding of the crystallography of this system to the same level that is presently available for $BaPb_{1-x}Bi_xO_3$. Detailed structural study of $BaPb_{1-x}Sb_xO_3$ has not yet been reported, and therefore metal-oxygen bondlengths are not available. However, for a "cubic" cell at x=0.4 the Pb/Sb-O bondlengths, (the octahedra are not tilted), would be $a_o/2$=2.13 angstroms, similar to the 2.16 angstrom bondlengths reported in superconducting $BaPb_{.75}Bi_{.25}O_3$.

What is the valence of Sb in these compounds? In the foregoing discussions, the similarities and differences in the chemical behavior of Bi and Sb in oxides have been described. Keep in mind that the best superconducting composition for *both* the $BaPb_{1-x}Bi_xO_3$ and $BaPb_{1-x}Sb_xO_3$ solid solutions occurs for x=0.25, suggesting a similar electronic charge donation to the energy band at the Fermi level for both Bi and Sb. For the $BaPbO_3$-$BaBiO_3$ solid solution, both endmembers contain the active cations in a 4+ average valence state. There are no discontinuities in crystallographic cell volume that would suggest a sudden change in the Bi valence in the $BaPb_{1-x}Bi_xO_3$ solid solution as a function of x, and therefore the average effective charge may be Bi^{4+} for all x. There has not, however, been a microscopic measurement of the effective charge state of Bi, and there could of course be some kind of gradual change in effective charge not apparent in the measurements thus far performed. The fact that both solid solutions are the best superconductors at x=0.25, the fact that the unit cell volume follows the behavior expected for a solution of "Sb^{4+}" in $BaPbO_3$, and the fact that the analogous solid solutions $BaPb_{1-x}M_xO_3$, M=Ta, Nb and V, could not be synthesized, led us to conclude (11) that Sb is acting in the same electronic fashion as is Bi when doped into $BaPbO_3$. None of these arguments are based on microscopic evidence, however. Fortunately Sb is an atom whose effective valence can be probed by Mossbauer measurements, and a more microscopic picture can be expected to emerge in the near future. Some inferences may also be made for the $BaPb_{1-x}Bi_xO_3$ superconductor through those results.

The $BaPb_{1-x}Sb_xO_3$ compounds were synthesized in polycrystalline form from starting materials $BaCO_3$, Pb_3O_4 and Sb_2O_3. Treatment was at 825 to 850°C for 20 hours with several intermediate grindings. Higher temperature synthesis or long time, low temperature annealing in O_2 degraded the samples. Single crystals were grown from a PbO based flux. Starting composition of the melt was 48.1 gm PbO, 13.5 gm $BaCO_3$ and 1.45 gm Sb_2O_3. These were mixed, placed in a covered Pt crucible and heated to 1150°C for a 3 hour soak after which the melt was cooled to 750°C at 4°/hour, annealed there for 24 hours and then pulled from the furnace.

6.0 $Ba_{1-x}K_xBiO_3$

The idea behind this solid solution is simple enough. Starting from $BaBiO_3$, the substitution of Pb for Bi removes electrons from the system, as Pb is one element to the left of Bi in the periodic table. Obviously, electrons can also be removed from the system by substitution of K^{+1} for Ba^{2+}. If we suppose that the key to the occurrence of superconductivity in $BaPb_{.75}Bi_{.25}O_3$ is related to the special charge fluctuations in Bi, then, in analogy to the copper oxides, a material with solely the active component on the electronically active sites should be a better superconductor. For the $Ba_{1-x}K_xBiO_3$ solid solution, Bi is formally oxidized as x increases. At the optimal superconducting composition x=0.4, T_c exceeds 30K, making this material the highest T_c compound known which is not based on copper oxide.

The chemistry of this material turns out to be highly nonconventional, a fact which apparently prevented its discovery in the early days of research on $Ba(Pb,Bi)O_3$. Superconductivity at 22K was first observed in the Ba-K-Bi-O system (38) in samples in which accidental melting resulted in trapped bubbles which acted to prevent the volatilization of K_2O in parts of the sample. Those results were apparently not reproduced elsewhere.

Conventional synthesis from nitrates, oxides or carbonates in air or oxygen does not result in the production of superconducting materials. Using an exotic synthetic technique, we were able to observe superconductivity reproducibly, isolate and characterize the superconducting phase, and increase its T_c to 30K. The superconducting compound $Ba_{.6}K_{.4}BiO_3$ was prepared from fresh BaO, KO_2 and Bi_2O_3 reacted in silver tubes which were sealed by mechanical crimping, and sealed in turn inside of evacuated quartz tubes. The reaction was carried out at 675°C for 3 days. Most satisfactory results were obtained when significant excess KO_2 was present in the starting mixture. The crimped silver tubes leak very slowly during the course of the reaction. This is not a very well controlled process and is therefore not completely satisfactory: perfectly sealed silver tubes made by welding did not work. The material after a successful 675°C synthesis is dark red or brown in color and is oxygen deficient $Ba_{.6}K_{.4}BiO_{3-x}$, with a powder diffraction pattern characteristic of a highly distorted perovskite structure such as is observed for oxygen deficient $BaBiO_{3-x}$. The correct oxygen content was obtained by annealing in flowing oxygen at 475°C for 45 minutes. The resulting powder is blue-black in color, with a simple cubic perovskite unit cell. Later neutron diffraction studies on materials prepared with a 475°C oxygen post anneal (53) showed the compound to have a full stoichiometric oxygen content (x=0). The Rb analog $Ba_{.6}Rb_{.4}BiO_3$ superconducts near 29K when prepared in the same manner. Superconductivity is not observed for $Ba_{1-x}Na_xOBiO_3$; it is not clear that the Na enters the compound.

A different synthetic route was later reported (53). Using the same starting materials and synthesis temperature the initial reaction was performed in flowing N_2 for 1 hour, followed by the same low temperature oxygen anneal which we had reported. Excess alkali was not employed. The success or failure of this method is very sensitive to the starting materials but because it does not involve reactions in sealed tubes (which can explode) it is considerably more straightforward than the one we initially employed. Neither of the methods yields perfect materials every time, although they are often quite good. In particular, sample homogeneity is a problem, especially with regard to the Ba/K distribution and oxygen deficiency.

Very small single crystals of the $Ba_{1-x}K_xBiO_3$ perovskite have now been grown from both hydrated and anhydrous KOH fluxes (54)(55). Both types of growth are very low temperature processes. For the hydrated fluxes, crystal growth occurs at 360°C in air out of a $Ba(OH)_2 \cdot 8H_2O$, Bi_2O_3, KOH mixture, with the crystals precipitating out of solution as the melt slowly dehydrates. The potassium content of the crystals is relatively low, and they are not superconducting. Growth from anhydrous KOH can also be performed in air by slow cooling through a higher (but still quite low) temperature range of 400-475°C. The crystals obtained by the anhydrous growth are at the present time larger than those obtained from hydrated fluxes, but neither technique as yet has yielded crystals of sufficient size for many desirable physical property measurements. As in the synthesis of bulk materials, careful attention must be paid to controlling the Ba:K ratio. Due to the low synthesis temperatures, the oxygen contents of the crystals appear to be satisfactory, and superconducting materials with T_c's in excess of 30K have been obtained. Both crystal growth techniques are not entirely reliable: often there can be no crystal growth at all.

The single crystals thus far prepared are too small for many physical property measurements. Also, the two techniques described for preparation of powders result in fine grained (and not sintered) materials which are also not appropriate for many measurements. A different technique, however, was recently developed which produces near theoretical density polycrystalline pellets (56). Stoichiometric mixtures of BaO, KO_2 and Bi_2O_3 are mixed and melted in N_2 gas and quickly quenched onto a copper block under the N_2 atmosphere. The process must be performed quickly to prevent excessive volatilization of K_2O. The solidified quenched button is then annealed in N_2 at 700°C and finally at 425°C in O_2. Superconducting transitions are generally in the neighborhood of 27K, but the high density of the resulting polycrystalline pellets lends these materials to certain physical property measurements which are not possible for the other materials. Note that for all synthetic methods requiring the use of BaO, great care must be taken to handle the BaO in an environment protected from air, as adverse reactions with air occur very quickly.

All of the bulk synthetic methods reported involve a low temperature (350-475°C) oxygen post synthesis anneal to obtain materials with optimized superconducting properties. This is due to the fact that, as in $BaBiO_3$, there is a wide range of oxygen deficiency possible for the $(Ba,K)BiO_3$ perovskites. As in the case of $BaBiO_{3-\delta}$, the oxygen deficient materials are highly distorted, low symmetry perovskites, but their crystal structures have not yet been studied. One report deals with the oxygen content of $Ba_{.6}K_{.4}BiO_{3-\delta}$ explicitly (57). The superconducting composition ($\delta=0$) begins to lose oxygen at 300°C when heated in N_2, attaining a relatively stable oxygen content $\delta=0.69$ by 700°C. This corresponds to the presence of Bi^{3+} only, and must have a very interesting highly distorted crystal structure due to the very large number of oxygen vacancies present (even more than the $\delta=0.5$ maximum in $BaBiO_{3-\delta}$). Oxidation of this material is reported to result in decomposition if temperatures in excess of 450°C are employed. The optimal oxidation temperature is probably right at 450°C; we reported initially that 475°C appears to work better without decomposition.

As for the B site substituted materials, the evolution from the charge density wave bearing insulator $Ba_{1-x}K_xBiO_3$ at $x=0.0$ to the superconductor at $x=0.4$ is of great relevance to understanding the mechanism for superconductivity. Some aspects of the physical properties of the solid solution are well established. Optical absorption measurements on thin films of $Ba_{1-x}K_xOBiO_3$ have shown that the materials are CDW insulators until a composition $x=0.35$, where there is an insulator-metal transition (58). The optical gap decreases with increasing K content within the insulating phase. The optimal superconducting composition, $x=0.4$, is just beyond the insulator-metal boundary. One report of a lower optimal doping concentration was incorrect (53), and has been retracted (59). Although the solid solution extends at least to x=0.5, the superconducting properties degrade for x > 0.4. There are several reports of a 34K T_c in a multiple phase materials, but those results have not yet been generally confirmed (60)(52).

The crystallographic properties of the $Ba_{1-x}K_xBiO_3$ solid solution perovskites have been studied by several groups. All now agree (8)(55)(59) that the superconducting composition $Ba_{.6}K_{.4}BiO_3$ is a simple cubic perovskite with one formula unit per cell, a_o=4.287Å at 300K (Ba and K disordered in the (1/2,1/2,1/2) sites, Bi in the (0,0,0) sites and O in the (0,0,1/2), (1/2,0,0) and (0,1/2,0) sites). This composition remains cubic even at temperatures below T_c, and does not undergo any structural transitions that increase the cell volume, e.g. through octahedral twists that would multiply cell constants by integer values (59)(61). The Bi-O bond distances are then simply $a_o/2$ = 2.14 angstroms at 300K, and shrink less than 0.01 angstrom by 10K. The octahedra are perfectly regular with O-Bi-O bond angles of 180°. These bond distances are very similar to those found in superconducting $BaPb_{.75}Bi_{.25}O_3$ and $BaPb_{.75}Sb_{.25}O_3$. The B site doped superconductors, as described, have O-Pb/M-O angles which differ from 180° by 8-10°.

There is at present, however, no agreement on the manner in which monoclinic $BaBiO_3$ evolves into cubic $Ba_{.6}K_{.4}BiO_3$. In this case, as for $BaPb_{1-x}Bi_xO_3$, the disagreements may be due to differences in the synthesis techniques for the materials studied, or experimental errors due to very subtle differences in the symmetries and atomic positions. The first descriptions of the structural evolution in the solid solution came from single crystal X-ray diffraction studies done on single crystals prepared in the fully oxidized form near 400°C. These studies found that $Ba_{1-x}K_xBiO_3$ evolves through orthorhombic symmetry at x=0.04 to cubic symmetry at the very low K content x=0.13 (54)(55). Thus the cubic phase region would extend from approximately x=0.1 to x >0.4. The x=0.13 sample is not superconducting. Based on these data, the insulator to metal transition at x=0.35 is not reflected in a change in symmetry at ambient temperature, into which one has to read that although the structural signature of the CDW in

$Ba_{1-x}K_xBiO_3$ is gone by x=0.13, the insulating properties are not gone until a much higher doping level is achieved. These results can be reconciled if the $BaBiO_3$ type CDW exists only on a very local scale (in a cluster of a few Bi atoms) or if a different kind of CDW with a different, smaller lattice distortion (with very weak diffraction effects) occurs for $0.1 \leq x \leq 0.35$. One group proposed that they had found such a distortion based on electron diffraction (62), but another study showed that the observed superlattice was actually induced by the electron beam (63) and was not due to a charge density wave. The superlattice was also not seen by neutron diffraction experiments which specifically looked for it (59). Thus if one considers only the data from single crystals, the electronic and structural signatures of the CDW cannot be reconciled.

An alternative picture has evolved from the study of polycrystalline materials prepared from the 700 C/N_2, 425 C/O_2 synthetic procedure (59). Note that since these materials are synthesized at higher temperatures and in a reduced oxygen content state they could in principle be different from the single crystals. The phase diagram was carefully studied by powder neutron diffraction for $Ba_{1-x}K_xBiO_3$ for temperatures between 10 and 473K, for $x \leq 0.5$, the K solubility limit. At ambient temperature, the phases for $Ba_{1-x}K_xBiO_3$ are as follows: monoclinic $BaBiO_3$ type for x<0.1; orthorhombic, for $0.1 \leq x \leq 0.25$; and cubic for $0.3 \leq x \leq 0.5$. Although the monoclinic -> orthorhombic -> cubic sequence is observed as for the single crystals, the compositions where the symmetry changes occur are at significantly higher K concentrations. In Figure 6 the data for the variation of cell parameters within the solid solution have been replotted for comparison to the data for $BaPb_{1-x}Bi_xO_3$. Several features are of note: there are no discontinuities in unit cell lengths across symmetry boundaries, implying smooth changes in tilt angles and bondlength readjustment as a function of K content, the cell volume changes continuously, as expected, and, finally, as for all the members of this family, the dimensional distortion from cubic symmetry is very small. The data for the evolution of the system at 10K are of considerable interest. The monoclinic phase exists for $x \leq 0.1$, the orthorhombic phase for $0.1 < x \leq 0.35$, and the cubic phase for $0.4 \leq x \leq 0.5$. At low temperatures, then, a crystallographic change from orthorhombic to

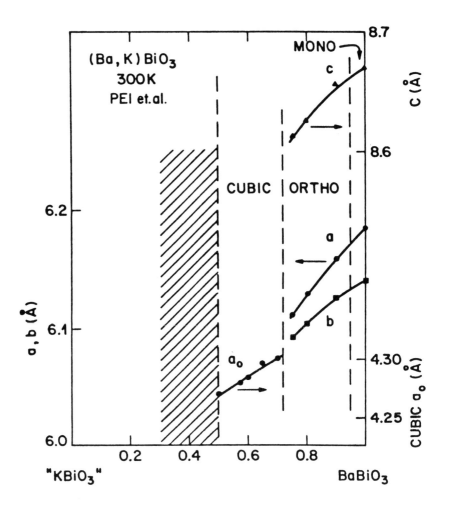

Figure 6: Crystallographic cell parameters and symmetries at 300K for the $Ba_{1-x}K_xBiO_3$ solid solution re-plotted from reference 59.

cubic symmetry is observed exactly where the physical properties indicate that the CDW disappears and superconductivity appears. Thus a detailed understanding of the crystal structure of the orthorhombic phase is important for understanding the evolution of the physical properties.

It is precisely in the structure of this orthorhombic phase where the published reports differ most significantly. The proposed crystal structures are presented in Table 5. Both groups report a body centered orthorhombic cell with $\sqrt{2}a_p \times \sqrt{2}a_p \times 2a_p$ lattice parameters. In the proposed spacegroups, however, the positional distortions from the expected ideal positions are very different. In the structure proposed by Schneemeyer et al. from single crystal X-ray data (55) there are two inequivalent Bi sites in which the Bi-O bond distances are modulated in an unusual manner. The structure would consist of alternating elongated and squat (BiO_6) octahedra with either 4 long and 2 short or 4 short and 2 long distances, with 180° O-Bi-O bond angles. There are several difficulties with the structure to be noted in Table 5 which suggest that confirmation of this structure by another group should be performed before it is generally accepted. They are: (1) *none* of the atom positions deviate from those of the ideal perovskite by more than 2 standard deviations, (2) the thermal parameters for the oxygen atoms are unusually large and (3) the agreement between observed and calculated structure factors ("R" in Table 5 = 6.2%) is considerably larger than the value the same group obtained for the structure of $Ba_{.6}K_{.4}BiO_3$ (R=2.2%) under the same experimental conditions.

Pei et al. looked at the structure of the orthorhombic phase of $Ba_{1-x}K_xBiO_3$ powders by time of flight powder neutron diffraction (59). Neutron diffraction intensities should be considerably more sensitive to oxygen atom positions than are X-ray diffraction intensities for compounds such as this. The structural parameters at 300K for x=0.1 and x=0.2 are presented in Table 5. The crystal structure is the same as that of $BaPbO_3$ (Table 2) with small differences in cell parameters and atomic coordinates. The refinements show that the atoms are displaced from the ideal positions by many standard deviations, and the thermal parameters for the oxygen atoms make physical sense. The crystal structure is one in which relatively regular Bi-O octahedra (Bi-O distances

Table 5: Proposed Structural Models for Orthorhombic $Ba_{1-x}K_xBiO_3$ at 300K

Parameter	Schneemeyer et. al.[55],*	Pei et. al.[59]	
x	0.04	0.1	0.2
a	6.1374(5)	6.1578(2)	6.1280(5)
b	6.1298(5)	6.1262(2)	6.1040(4)
c	8.6628(7)	8.6569(2)	8.6286(6)
Symmetry	Immm	Ibmm	Ibmm
Ba/K	(1/2,0,z) z=0.253(2)	(x,0,1/4) x=0.0504(1)	(x,0,1/4) x=0.0507(2)
Bi1	(0,0,0)	(0,0,0)	(0,0,0)
Bi2	(0,0,1/2)	-	-
O1	(0,0,z) z=0.244(4) B=7.3	(x,0,1/4) x=0.0527(6) B=1.3	(x,0,1/4) x=0.047(1) B=1.0
O2	(x,y,0) x=0.254(4) y=0.260(6) B=5.6	(1/4,1/4,z) z=-0.0265(3) B=2.0	(1/4,1/4,z) z=0.0116(7) B=2.0
R	6.2%	7.8	9.6
R_E		3.4	3.3
χ		2.3	2.9

* The cell of Schneemeyer et. al. has been transposed to the usual setting of axes a > b and the labels of the oxygen atoms have been switched to conform to the standard labels.

of 2.19 and 2.17 angstroms at x=0.1) are tilted such that the O-Bi-O bond angles differ from 180° by approximately 8-9°. The authors specifically tested the model of Schneemeyer et al. against the neutron data and found that the agreement was not satisfactory. Note, however, that there may be room for improvement of the Ibmm structural model, as the ratio of the agreement between model and data to that expected from statistics alone ($\chi=R/R_E$, Table 5) is larger, particularly at x=0.2, than is generally considered to be indicative of a perfectly refined structure ($\chi \leq 2$).

Looking at the crystal structure of the orthorhombic phase, then, we have the same difficulty in $Ba_{1-x}K_xBiO_3$ that we have in $BaPb_{1-x}Bi_xO_3$ in correlating crystal structures and the insulator to metal transition. If we accept the orthorhombic structure of Schneemeyer (55) et al. and the cubic x=0.13 structure of Wignacourt (54) et al., then the structural signature of the CDW has disappeared long before the appearance of superconductivity. If we accept the orthorhombic structure of Pei (59) et al., which appears to extend all the way to the insulator-superconductor composition, then again there is also no reconciliation between structural and physical properties, as the orthorhombic structure contains only *tilted* octahedra, and has no sign of maintaining the charge disproportionation found in $BaBiO_3$. Thus one has to invoke ideas such as the "local CDW" proposed for $Ba(Pb,Bi)_3$. Pei et al. have proposed (59) that the orthorhombic structure is actually an average over positionally disordered larger and smaller (charge proportionated) BiO_6 octahedra exactly as was proposed by Chaillout et al. for type 2 $BaBiO_3$, resulting in a neat picture which correlates structure and properties. It would be of considerable importance to confirm that picture through an alternative characterization technique. It would be truly remarkable if the great similarities in the structures of orthorhombic $BaPbO_3$ and $Ba_{1-x}K_xBiO_3$ were only skin deep, and the underlying basic structural principles (e.g. uniform charge vs. disordered charge disproportionated) were fundamentally different.

7.0 CONCLUSION

The perovskites $Ba_{.6}K_{.4}BiO_3$, $BaPb_{.75}Bi_{.25}O_3$ and $BaPb_{.75}$

$Sb_{.25}O_3$ are superconducting oxides with critical temperatures of 30, 12, and 3.5K. They are strongly related both structurally and chemically. Detailed study and comparison of their physical properties may help to understand the origin of their superconductivity. Even $BaPb_{.75}Sb_{.25}O_3$, with its relatively low T_c, can be considered as a high temperature superconductor, as its electronic density of states at the Fermi level is extremely low, similar to that of conventional superconductors with T_c's of only a fraction of a Kelvin (64). Figure 7 presents a composite T_c vs. electron count diagram for all 3 systems. Starting at the right with the end members $BaBiO_3$ and the imaginary "$BaSbO_3$", the partial substitution of Pb for Bi or Sb on the B site, or of K for Ba on the A site, results in the removal of electrons from the perovskite. The diagram illustrates the obvious fact that the B site substituted materials are optimal superconductors at the same electron count, but the Sb compound is superconducting at a significantly lower temperature. This count is considerably different from that of $Ba_{.6}K_{.4}BiO_3$ which has had significantly fewer electrons (about half that of the others) removed from the end member. Nonetheless, the large amount of doping required to induce superconductivity, 0.4 electrons removed per bismuth, is very large compared to what is expected from the band structure calculations.

Although the basic chemistry and crystallography of these compounds is relatively simple, there are many subtle but very important crystal chemical characteristics which make them a challenge to understand. Some of these characteristics have been unambiguously determined, and some have not, as has been described in the foregoing. There is considerable room for further work. In addition to more detailed characterization of the materials already known, it would be of interest if more members of this family could be discovered. The controversy over whether they are conventional superconductors, or operate via a new superconducting mechanism is bound to rage for many more years. One of the most pleasurable roles of the solid state chemist in this field is to come up with new materials which challenge the ideas of the theoretical physicists.

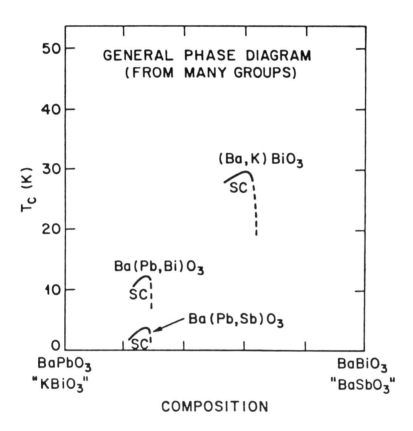

Figure 7: Composite T_c vs. electron count diagram for the $BaPb_{1-x}Bi_xO_3$, $BaPb_{1-x}Bi_xO_3$ and $Ba_{1-x}K_xBiO_3$ perovskites.

ACKNOWLEDGEMENT

I would like to acknowledge the collaboration of many colleagues at Bell Laboratories and elsewhere, and especially the very fruitful collaboration with B. Batlogg, J. Krajewski and W. F. Peck, Jr. I would also like to thank J. Jorgensen for providing the results of Pei et al. (Ref. 59) prior to publication.

8.0 REFERENCES

1. J. G. Bednorz and K. A. Muller, *Z. Phys.* B64:189 (1986).

2. Y. Tokura, H. Takagi and S. Uchida, *Nature* 337:345 (1988).

3. See, for instance, L. F. Mattheiss, *Phys. Rev. Lett.* 58:1029 (1987); Jaejun Yu, A. J. Freeman and J.-H. Xu, *Phys. Rev. Lett.* 58:103 (1987).

4. See for instance, B. Batlogg et. al., *Phys. Rev. Lett.* 58:2333 (1988); L. C. Bourne et. al., *Phys. Rev. Lett.* 58:2337 (1988).

5. See, for instance, P. W. Anderson, *Science* 235 (1987) 1196; J. R. Schrieffer, X. G. Wen and S. C. Zhang, *Phys. Rev. Lett.* 60:944 (1988).

6. See, for instance, C. M. Varma, *Phys. Rev. Lett.* 61:2713 (1988); W. A. Little, in *Novel Superconductivity*, eds. S. A. Wolf and Z. Kresin, Plenum, (1987) p. 341.

7. A. W. Sleight, J. L. Gillson and P. E. Bierstedt, *Solid State Commun.* 17:27 (1975).

8. R. J. Cava, B. Batlogg, J. J. Krajewski, R. Farrow, L. W. Rupp, Jr., A. E. White, K. Short, W. F. Peck, Jr. and T. Kometani, *Nature* 332:814 (1988).

9. See, for instance, B. Batlogg, *Physica*, 126B:275 (1984); K. Kitazawa, S. Uchida and S. Tanaka, *Physica* 135B:505 (1985).

10. See, for instance, L. F. Mattheiss and D. R. Hamann, *Phys. Rev.* B28:4227 (1983).

11. R. J. Cava, B. Batlogg, G. P. Espinosa, A. P. Ramirez, J. J. Krajewski, W. F. Peck, Jr., L. W. Rupp, Jr. and A. S. Cooper, *Nature* 339:291 (1989).

12. A. M. Glazer, *Acta Cryst.* B28:3384 (1972).

13. R. Scholder, K.-W. Ganter, H. Glaser and G. Merz, *Zeit. Anorg. Allg. Chem.* 319:375 (1963).

14. B. Schwedes and R. Hoppe, *Z. Anorg. Allg. Chem.* 393:136 (1972).

15. See for instance, S. Uchida, K. Kitazawa and S. Tanaka, *Phase Transitions*, 8:95 (1987).

16. See, for instance, "*Low Dimensional Conductors and Superconductors*", D. Jerome and L. G. Caron, eds., NATO ASI Series B, Physics Vol. 155, Plenum Publ. Co., NY (1989).

17. G. Thornton and A. J. Jacobson, *Acta Cryst.* B34:351 (1978).

18. D. E. Cox and A. W. Sleight, *Proceedings of the Conference on Neutron Scattering*, Gatlinberg Tennessee (1976), pp. 45-54, edited by R. M. Moon; available from National Technical Information Service, Springfield, VA 22161.

19. C. Chaillout, A. Santoro, J. P. Remeika, A. S. Cooper and G. P. Espinosa, *Sol. St. Comm.* 65:1363 (1988).

20. W. H. Zachariesen, *J. Less Common Metals*, 62:1 (1978).

21. I. D. Brown and Kang Kun Wu, *Acta Cryst.* B32:1957 (1976).

22. N. K. McGuire and M. O'Keefe, *Sol. St. Comm.* 52:433 (1984).

23. D. E. Cox and A. W. Sleight, *Acta Cryst.* B35:1 (1979)

24. H. Kusuhara, A. Yamanaka, H. Sakuma and H. Hashizume, *Jpn. J. Appl. Phys.*: (1987, 88?)

25. V. V. Bogatko and Yu. N. Venevtsev, *Sov. Phys. Sol. St.* 25:859 (1983).

26. E. G. Fesenko, E. T. Shuvaeva and Yu. I. Gol'tsov, *Sov. Phys. Cryst.* 17:362 (1972).

27. R. A. Beyerlein, A. J. Jacobson and L. N. Yacullo, *Mat. Res. Bull.* 20:877 (1985).

28. Y. Saito, T. Maruyama and A. Yamanaka, *Thermochem. Acta* 115:199 (1987).

29. C. Chaillout and J. P. Remeika, *Sol. St. Comm.* 56:833 (1985).

30. R. Hoppe and K. Blinne, *Z. Anorg. Allg. Chem.* 293:251 (1958).

31. G. Wagner and H. Binder, *Z. Anorg. Allg. Chem.* 297:328 (1958) and 298:12 (1959).

32. R. Weiss, *C. R. Acad. Sci.* 246:3073 (1958).

33. T. Nitta, K. Nagase, S. Hayakawa and Y. Iida, *J. Amer. Ceram. Soc.* 48:642 (1965).

34. R. D. Shannon and P. E. Bierstedt, *J. Amer. Ceram. Soc.* 53:635 (1970).

35. G. Thornton and A. J. Jacobson, *Mat. Res. Bull.* 11:837 (1976).

36. K. Oka and H. Unoki, *Jpn. Journal Appl. Phys.* 23:L770 (1984).

37. V. V. Bogatko and Yu. N. Venevtsev, *Sov. Phys. Sol. St.* 22:705 (1980).

38. L. F. Mattheiss, E. M. Gyorgy, and D. W. Johnson, Jr., *Phys. Rev.* B37:3745 (1988).

39. Werner Weber, *Jpn. J. Appl. Phys.*, 26: Supplement 26-3: 981 (1987).

40. See, for instance, D.R. Harshman, G. Appeli, B. Batlogg, R. J. Cava, A.S. Cooper, G.P. Espinosa, L.W. Rupp, E.J. Ansaldo, and D.L. Williams, *Phys. Lett. Comments*, 63:1187 (1989)

41. L. A. Shebanov, YV. Ya. Fritsberg and A. P. Gaevskis, *Phys. Stat. Sol.* A77:369 (1983).

42. M. Oda, Y. Hidaka, A. Katsui and T. Murakami, *Sol. St. Comm.* 55:423 (1985).

43. Y. Enomoto, M. Oda and T. Murakami, *Phase Transitions* 8:129 (1987).

44. G. Heger, G. Roth, C. Chaillout, B. Batlogg, To be published.

45. A. P. Menushenkov, E. A. Protasov and E. V. Chubunova, *Sov. Phys. Sol. St.* 23:2155 (1981).

46. Minoru Suzuki and Toshiaki Murakami, *Sol. St. Comm.* 53:691 (1985).

47. Troung D. Thanh, A. Koma and S. Tanaka, *Applied Physics*, 22:205 (1980).

48. Akinori Katsui, *Jpn. Jnl. Appl. Phys.* 21:L553 (1982).

49. A. Katsui, Y. Hidaka and H. Takagi, *J. Cryst. Growth* 66:228 (1984).

50. Akinori Katsui and Minoru Suzuki, *Jpn. Jnl. Appl. Phys.* 21:L157 (1982).

51. *Handbook of Chemistry and Physics*, Chemical Rubber Company Press, Inc., Boca Raton, Florida (1981)

52. S. Kondoh, M. Sera, F. Fukuda, Y. Ando and M. Sato, *Sol. St. Comm.* 67:879 (1988).

53. D. G. Hinks, B. Dabrowski, J. D. Jorgensen, A. W. Mitchell, D. R. Richards, Shiyou Pei and Donglu Shi, *Nature* 333:836 (1988).

54. J. P. Wignacourt, J. S. Swinnea, H. Steinfink, and J. B. Goodenough, *Appl. Phys. Lett.* 53:1753 (1988).

55. L. F. Schneemeyer, J. K. Thomas, T. Siegrist, B. Batlogg, L. W. Rupp, R. L. Opila, R. J. Cava and D. W. Murphy, *Nature* 335:421 (1988).

56. D. G. Hinks, A. W. Mitchell, Y. Zheng, D. R. Richards and B. Dabrowski, *Applied, Phys. Lett.* 54:1585 (1989).

57. D. G. Hinks, D. R. Richards, B. Dabrowski, A. W. Mitchell, J. D. Jorgensen and D. T. Marx, *Physica C* 156:477 (1988).

58. H. Sato, S. Tajima, H. Takagi and S. Uchida, *Nature* 338:241 (1989).

59. Shiyou Pei, J. D. Jorgensen, B. Dabrowski, D. G. Hinks, D. R. Richards, A. W. Mitchell, J. M. Newsam, S. K. Sinha, D. Vaknin and A. J. Jacobson, *Phys. Rev. B* 41:4126 (1990)

60. N. L. Jones, J. B. Parise, R. B. Flippen and A. W. Sleight, *J. Sol. St. Chem.* 78:319 (1989).

61. R. M. Fleming, P. Marsh, R. J. Cava and J. J. Krajewski, *Phys. Rev.* B38:7026 (1988).

62. Shiyou Pei, N. J. Zaluzec, J. D. Jorgensen, B. Dabrowski, D. G. Hinks, A. W. Mitchell and D. R. Richards, *Phys. Rev.* B39:811 (1989).

63. E. A. Hewat, C. Chaillout, M. Godinho, M. F. Gorius and M. Marezio, *Physica* C. 157:228 (1989).

64. B. Batlogg, R. J. Cava, L. W. Rupp, Jr., G. P. Espinosa, J. J. Krajewski, W. F. Peck, Jr., and A. S. Cooper, *Physica* C:1393 (1989).

11

Structure and Chemistry of the Electron-Doped Superconductors

Anthony C.W.P. James and Donald W. Murphy

1.0 INTRODUCTION

One of the most interesting recent developments in the field of high-temperature superconductivity has been the discovery of the electron-doped systems $Nd_{2-x}Ce_xCuO_4$ (0.15≤x≤0.18) (1), $Nd_{2-x}Th_xCuO_4$ (x~0.15) (2), $Nd_2CuO_{4-x}F_x$ (x~0.18) (3), and the corresponding Pr, Sm and Eu systems. These materials can all show magnetic transitions to bulk superconductivity above 20K; in some cases the onset of the resistive superconducting transition is as high as 27K. They differ from all other high-T_c cuprate superconductors in that superconductivity is induced by electron doping; i.e., by aliovalent substitutions in the T'-Nd_2CuO_4 structure (4) that lower the average copper valence below Cu^{2+}. This reduction of copper has been confirmed by thermogravimetric analysis and redox titration. It is especially important in view of the fact that there is only one crystallographic site available to copper in the T'-Nd_2CuO_4 structure; if the superconductors are single-phase and retain the T'-Nd_2CuO_4 structure, then *all* the Cu in these materials must be reduced below Cu^{2+} with far-reaching implications for theories of high-T_c superconductivity in cuprates. This situation would be quite different from that in, for example, $Pb_2Sr_2Y_{0.6}Ca_{0.4}Cu_3O_8$ (T_c=70K) (5), in which the average copper oxidation state is +1.8, but the copper ions are in two different crystallographic sites: linearly coordinated Cu^{1+} and sheets containing $Cu^{2.2+}$, with the latter actually responsible for the superconductivity.

If the superconductivity in $Nd_{1.85}Ce_{0.15}CuO_4$ and related systems is really due to overall reduction of copper by electron doping, then it would supply good evidence for theoretical models that predict an essential symmetry between the electronic properties induced by electron- and hole-doping in layered cuprates. Full evidence for such theories would still be lacking because, as we shall see, there is still no single layered cuprate that can be made superconducting by both hole- and electron-doping.

The purpose of this chapter is not to address the continuing controversy about the electronic nature of the electron-doped superconductors, but rather to review the crystal chemistry of the T'-Nd_2CuO_4 system and give practical details on how to prepare crystallographically pure electron-doped superconductors in ceramic or single-crystal form with high Meissner fractions.

2.0 THE T'-Nd_2CuO_4 STRUCTURE

The T'-Nd_2CuO_4 structure, on which all of the electron-doped superconductors are based, is shown in Figure 1 along with the related T-La_2CuO_4 (K_2NiF_4) and T* structures. The T and T' structures can be regarded as intergrowths of infinite, flat CuO_2^{2-} layers containing Cu in square planar coordination alternating with $Ln_2O_2^{2+}$ layers. In the T-La_2CuO_4 structure the $La_2O_2^{2+}$ layers have the rock salt structure, which results in La having a total oxygen coordination number of 9 and places oxygen atoms in the apical positions above and below Cu in the CuO_2^{2-} sheets, giving a tetragonally (Jahn-Teller) distorted octahedral Cu-O coordination sphere. In the case of T'-Nd_2CuO_4, the $Nd_2O_2^{2+}$ layers have the fluorite structure, which results in Nd having a total oxygen coordination number of 8 and places the O atoms in this layer well away from the Cu apical positions. The transition from T to T' structures on going from La to Pr, Nd, Eu, or Gd is driven by the lanthanide contraction; smaller lanthanide ions favour 8-fold over 9-fold coordination. The smallest lanthanide ions (Tb^{3+}-Lu^{3+}) and the lanthanide-like Y^{3+} do not form the T'-Nd_2CuO_4 structure.

Structure and Chemistry of the Electron-Doped Superconductors   429

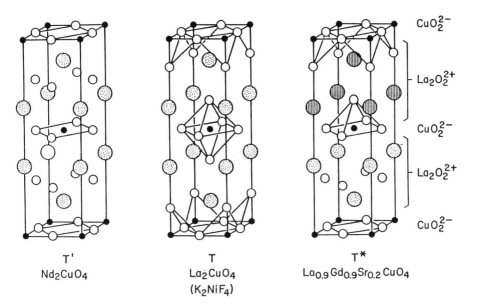

Figure 1: The T, T', and T* structures.

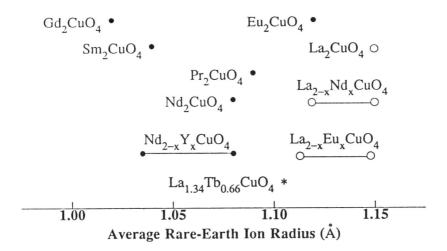

Figure 2: Stability fields of the T and T' structures in rare-earth cuprates as a function of average rare-earth ionic radius. Filled circles represent T' phases, open circles represent T phases. Asterisk represents T* phase.

Isovalent substitutions of rare-earth ions have enabled the systematics of the T-T' system to be mapped out. Figure 2 shows data from several sources (6)(7)(9) showing the transition from T to T' structures with decreasing average rare-earth ion radius. The recently discovered T* phase, an intergrowth of the T and T' structures (8)(9), occupies a very narrow region between the two in cases where the rare-earth ion site is occupied by two ions of very dissimilar radii; for example, $La_{1.33}Tb_{0.67}CuO_4$. The stability range of the T* structure can be greatly extended by substituting Sr for rare-earth ions, as in $La_{0.9}Gd_{0.9}Sr_{0.2}CuO_4$ and by applying high oxygen pressure. Sr substituted T* phases prepared under oxygen pressure are superconducting with $T_c \sim 25K$. Since these are clearly hole-doped superconductors they will not be discussed further here.

One very important consequence of the fact that the T and T' structures are layered intergrowths is the existence of a *lattice mismatch* between adjacent layers in both the T and the T' structures. In La_2CuO_4 CuO_2^{2-} layers are compressed by the adjacent $La_2O_2^{2+}$ layers, whereas in Nd_2CuO_4 the CuO_2^{2-} layers are in tension, hence the unusually short in-plane Cu-O bond in La_2CuO_4 (1.90 Å) and the unusually long in-plane Cu-O bond in Nd (1.973 Å) compared to the average Cu-O bond length in Cu(II) oxides of 1.93-1.94 Å. The compression of the CuO_2^{2-} layers in La_2CuO_4 results in the well known orthorhombic distortion of this compound at low temperatures (10), in which the layers buckle to relieve compressive stress. The relevance of the lattice mismatch to doping behaviour arises from the fact that in both $T-La_2CuO_4$ and $T'-Nd_2CuO_4$ the highest-energy electrons are in a band of primarily Cu-3d character that is antibonding with respect to in-plane Cu-O interactions (11). Removal of electrons from this band by hole doping will tend to shorten the in-plane Cu-O bond, whereas injection of electrons into this band by electron doping will tend to lengthen the in-plane Cu-O bond. It follows that hole doping will be favored in the T structure, where the Cu-O bonds are in compression, and electron doping will be favored in the T' structure, where the Cu-O bonds are in tension. Relief of compressive stress in the Cu-O plane provides the driving force for hole doping in La_2CuO_4 and explains why hole doping in this system is accompanied by a decrease in the transition temperature for the orthorhombic to tetragonal distortion. It also explains why electron doping in La_2CuO_4 is impossible. On the other hand, tension in the Cu-O layers means that electron doping is favored in the Nd_2CuO_4 system. Attempts have been made to hole dope Nd_2CuO_4 by substituting Sr^{2+} for Nd^{3+} (12) or Li^+ for Cu^{2+} (13). The limiting compositions were found to be $Nd_{1.9}Sr_{0.1}CuO_{4-\delta}$ and $Nd_2Cu_{0.85}Li_{0.15}O_{4-\delta}$. Both materials show no enhancement of conductivity over Nd_2CuO_4; oxygen loss has entirely compensated for the cationic substitution and there is no net hole doping. Thus, it appears that Cu-O bond length considerations make hole- and electron-doping mutually exclusive in layered cuprates. The absence of a driving force from bond tension also explains why other cuprates containing planar CuO_2^{2-} layers cannot be hole doped; an example is $Ca_{0.85}Sr_{0.15}CuO_2$ (Cu-O in-plane bond length 1.931 Å) (14).

3.0 SYSTEMATICS OF ELECTRON DOPING

Electron doping in $T'-Nd_2CuO_4$ can be achieved either by substituting a tetravalent ion (Ce^{4+}, Th^{4+}) for Nd^{3+} or by substituting a monovalent anion (F^-) for O^{2-}. In the case of Ce substitution there is the possibility that the cerium may be present in $Nd_{2-x}Ce_xCuO_4$ as Ce^{3+} or as some intermediate valence state between Ce^{3+} and Ce^{4+}. The fact that the $Nd_{2-x}Ce_xCuO_4$ unit cell contracts as x increases has been interpreted as evidence for Ce^{4+}, since the Ce^{4+} ion is smaller than Nd^{3+}; quantitative studies of this contraction assuming hard-sphere ionic radii and ignoring any change in the Cu-O bond length as a result of electron doping gave a Ce oxidation state close to, but slightly less than Ce^{4+} (15)(16). XPS studies of the cerium core levels indicate that all of the cerium is present as Ce^{4+} (16)(17). For thorium substitution, where there is no accessible Th^{3+} oxidation state, the question does not arise. The case of fluorine substitution for oxygen in $Nd_2CuO_{4-x}F_x$ is especially interesting; previous attempts to fluorinate superconducting cuprates had always been unsuccessful because the stability of SrF_2, BaF_2, and the lanthanide oxyfluorides resulted in these phases separating out. Fluorination of $Nd_2CuO_{4-x}F_x$ works (for $x \leq 0.18$) because the $Nd_2O_2^{2+}$ layers have a fluorite structure similar to the distorted fluorite structure of the very stable compound NdOF.

Of the electron-doped systems, the systematics of the Ce substituted phases $Nd_{2-x}Ce_xCuO_4$ have been studied in the most detail (18)(19)(20)(21). It appears that crystallographically pure ceramic samples of $Nd_{2-x}Ce_xCuO_4$ can be prepared for all $x \leq 0.20$. The variation in electronic properties and lattice parameters with doping level are shown in Figures 3 and 4; the corresponding properties of the hole-doped system $La_{2-x}Sr_xCuO_4$ are given in the chapter by Tarascon in this volume. Spin polarised neutron diffraction (23) and muon spin resonance (24) have shown that the parent compound, Nd_2CuO_4, is an antiferromagnetic insulator with localised electrons. Cerium doping results in a rapid increase in electrical conductivity and lowering of the magnetic ordering temperature until a superconducting region without long-range magnetic ordering is reached ($x \geq 0.14$). The superconducting region is very narrow; $Nd_{2-x}Ce_xCuO_4$ with $x \geq 0.18$ is metallic but nonsuperconducting. The highest T_c were seen for compositions right at the onset of superconductivity $x \simeq 0.15$. The general evolution of electronic and magnetic properties with doping level is very similar to that in $La_{2-x}Sr_xCuO_4$, with two major differences: first, several authors have reported a much broader superconducting compositional region ($0.05<x<0.2$) in $La_{2-x}Sr_xCuO_4$, with maximum T_c in the middle of this region at $x=0.15$ (10), and second, ceramic superconducting samples of $Nd_{2-x}Ce_xCuO_4$ are never metallic above T_c; their normal-state resistivity is generally high ($\sim 10^{-3}$ Ohm cm) and decreases with increasing temperature as shown in Figure 5 for $Nd_2CuO_{4-x}F_x$. This contrasts with superconducting single crystals of $Nd_{1.85}Ce_{0.15}CuO_4$, in which metallic conductivity has been seen in the normal state.

Less is known about the systematics of the other electron doped systems. It has been shown that the cuprates Pr_2CuO_4, Sm_2CuO_4, and Eu_2CuO_4 can all be made superconducting by substitution of cerium (7)(19)(21) or thorium (21)(22) for the rare earth and by substitution of fluorine for oxygen (3). Gd_2CuO_4 apparently cannot be made superconducting by electron doping (20). The superconducting transition temperatures appear to be lower than those for the corresponding Nd systems; for example, $Sm_{1.85}Ce_{0.15}CuO_4$ and $Eu_{1.85}Ce_{0.15}CuO_4$ show onset of the resistive

Structure and Chemistry of the Electron-Doped Superconductors 433

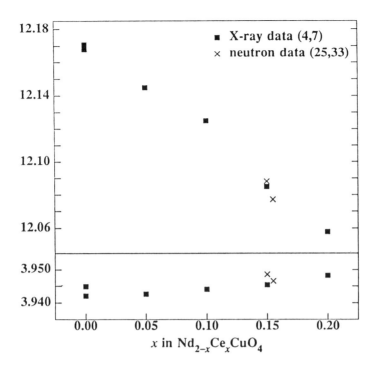

Figure 3: Lattice parameters of $Nd_{2-x}Ce_xCuO_4$ versus x.

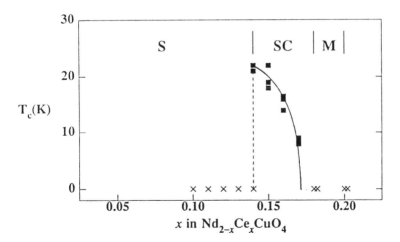

Figure 4: T_c versus x in $Nd_{2-x}Ce_xCuO_4$. Crosses represent samples that were not superconducting. Data from refs 18,19. M=metallic, S=semiconducting, SC=superconducting.

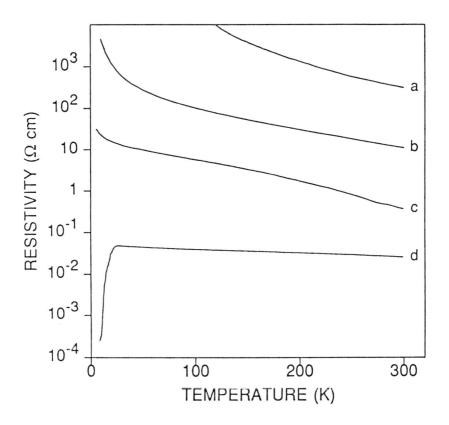

Figure 5: Resistivity of $Nd_2CuO_{4-x}F_x$ versus temperature. (a) x=0; sample made in air at 900°C. (b) x=0; sample reduced at 890°C under 60 ppm O_2N_2 (c) x≈0.1; fluorinated sample after initial firing in dry air at 900°C. (d) x=0.18; sample reduced 890°C in flowing N_2 containing 60 ppm O_2 and cooled at 300°C/hr (slow cooling replaces annealing step).

transition at 20K and 21K, respectively (19). Wang et al. have studied T_c as a function of rare-earth composition in the series $Nd_{1.85-x}La_xCe_{0.15}CuO_4$ (0≤x≤1) and $Eu_{1.85-x}La_xCe_{0.15}CuO_4$ (0 ≤ x ≤ 1.2). They find a clear maximum in T_c in these and other systems when rare-earth substitution results in a lattice parameter a=3.95 Å, corresponding to an in-plane Cu-O bond length of 1.95 Å (7). For comparison, the optimum Cu-O bond length for superconductivity in the hole-doped La_2CuO_4 based systems is 1.888 Å.

The thorium substituted system $Nd_{2-x}Th_xCuO_4$ resembles $Nd_{2-x}Ce_xCuO_4$ in its chemical and crystallographic behavior (2). The systematics of the fluorine substituted system $Nd_2CuO_{4-x}F_x$ are much harder to determine because of the possibility of fluorine loss by hydrolysis during synthesis and the difficulty of measuring fluorine content accurately. It is clear that $Nd_2CuO_{4-x}F_x$ has a smaller stability range than $Nd_{2-x}Ce_xCuO_4$, the limiting composition being approximately $Nd_2CuO_{3.82}F_{0.18}$ (3)(19). Hence, in this system overdoping into a metallic, nonsuperconducting regime is not possible. The superconducting region is quite narrow; no superconductors with x<0.15 have been made. The reasons for the upper limit for F substitution are discussed below.

The lattice parameters of $Nd_{2-x}Ce_xCuO_4$ show little variation with x. there is a small (~1%), apparently monotonic shortening of the c axis, which is believed to be due to the smaller ionic radius of Ce^{4+}. In the ab plane this effect is counteracted by the fact that electron-doping relieves tension in the CuO_2^{2-} layers, so that there is almost no change at all (~0.1%) in the a axis (Figure 3). Several X-ray (15) and neutron (25)(33) powder diffraction studies have concluded that the superconducting composition is single-phase with an undistorted T'-Nd_2CuO_4 structure down to 4.2K and few defects. X-ray diffraction data on the thorium and fluorine substituted superconductors also show an undistorted T' structure, with lattice parameters very little changed from Nd_2CuO_4.

However, at this point we should note that the electron-doped superconductors are *not* strictly single phase. Electron diffraction studies on both single crystals and ceramic superconducting samples have shown that they almost invariably contain two phases; one phase has the undistorted T'-Nd_2CuO_4 structure, the other exhibits a superlattice modulation of the T' structure. The superlattice reflections in electron diffraction were discovered independently by several groups, with differing interpretations of the nature of the supercell (25)(26). Measurements made in our laboratory have shown that the superlattice modulation is characterised by a wave vector $q=(1/4,1/4,0)$ parallel to the CuO_2^{2-} planes of the T' structure. The superlattice phase exists in four possible variants with its c axis parallel to the four [111] zone axes of the underlying tetragonal lattice (26). The nature and origin of the distortion giving rise to the superstructure is still unknown; it must be very small since the few superlattice reflections that have been seen in X-ray and neutron powder diffraction patterns are four orders of magnitude weaker that the principal reflections. Some clues are provided by the phenomenology of the superlattice, which can be summarised as follows (27):

1. Superconducting samples all contain both regions of superlattice and regions of undistorted T' structure, often within the same grain.
2. Samples quenched from high temperature (1100°C for $Nd_{2-x}Ce_xCuO_4$, 890°C for $Nd_2CuO_{4-x}F_x$) contain little or no superlattice. On annealing these samples at 400-700°C regions of superlattice form within minutes; this process can be observed directly in the electron microscope by means of a hot stage.
3. Identical regions of superlattice were seen in $Nd_{2-x}Ce_x CuO_4$, $Nd_2CuO_{4-x}F_x$, and undoped Nd_2CuO_4. We can conclude that superlattice formation is not due to substituent ordering. It could be due to ordering of oxygen defects or interstitials or to charge ordering on copper.

At present it is not known whether the superlattice or the undistorted T' phase or both are responsible for superconductivity, although there are reports that single crystals of $Nd_{1.85}Ce_{0.15}CuO_4$ with high Meissner fractions (>60%) do not contain any superlattice (28).

The two-phase nature of all ceramic samples of the electron-doped superconductors probably accounts for their less desirable properties, including low Meissner fractions in many samples, low and nonmetallic conductivity in the normal state, and the difficulty of making good single crystals and thin films. It must be taken into account when considering the best synthetic route to these materials. It may also explain why superconductivity in $Nd_{2-x}Ce_xCuO_4$ is only seen in the narrow composition range $0.15 \leq x \leq 0.18$, with an abrupt cutoff for $x<0.15$. The phase separation also means that data from physical measurements in ceramic samples of the electron-doped superconductors should be interpreted with caution; for example, two XPS studies have indicated that electron doping gives rise to localised Cu^+ ions (16)(17); this may be a result of charge ordering in the superlattice regions and not apply to the whole sample.

4.0 CHEMICAL SYNTHESIS AND ANALYSIS

The original report of superconductivity in $Nd_{2-x}Ce_xCuO_4$ by Tokura et al. (1) described the synthesis as follows: an initial reaction of Nd_2O_3, CeO_2, and CuO in the correct stoichiometric proportion at 1150°C in air followed by "reduction" at 900°C in flowing argon containing 100 ppm of O_2. Subsequent preparations reported in the literature are similar, but the initial reaction temperature is usually given as 1100°C with optimum conditions for the reduction step variously reported as 1050°C in argon containing 100 ppm of O_2 followed by 550°C for 10 hours in the same gas flow (18), 910°C in flowing CO_2 for 16 hours (32), 900°C in flowing argon for 12 hours (20), 910°C in flowing helium for 12-14 hours (21), or an unspecified temperature in a sealed, evacuated silica tube in the presence of zirconium metal to act as an oxygen getter (33). Reported Meissner fractions for ceramic samples are in the range 15-60%. Improved results have been reported when the CeO_2 starting material is freshly made from $Ce(CO_3)_2$ by calcining at 900°C (34), and when the initial reaction step is carried out over a period of days at 1100°C with intermittent grinding and pelletising. Both of these procedures result in a more uniform distribution of cerium in the product. From the published data and our own results the varying reaction conditions used to prepare $Nd_{2-x}Ce_xCuO_4$ can be rationalised as follows: the initial reaction step in air must be carried out at a temperature as near as possible to the onset of partial melting in $Nd_{2-x}Ce_x$

CuO_4, which takes place near 1150°C. The high temperature is needed because of the refractory nature of CeO_2 and in order to drive the equilibrium

$$(1-x/2)\ Nd_2CuO_4 + x\ CeO_2 + x/2\ CuO \rightleftharpoons Nd_{2-x}Ce_xCuO_4 + x/2\ O_2$$

all the way to the right-hand side. The reversible nature of this reaction requires that samples be quenched after this step to prevent reoxidation on cooling. The quenched samples are single-phase T' material by X-ray diffraction. However, samples of $Nd_{1.85}Ce_{0.15}CuO_4$ quenched from 1100°C show little or no superconductivity. The "reduction" step at lower temperature is needed to make strongly superconducting material. It is questionable whether much actual reduction takes place in this step; it is probably better described as an annealing. Some groups have found that the annealing step results in loss of a small amount of oxygen (approx. 0.03 oxygens per formula unit), based on iodometric titration and thermogravimetric reduction data (7)(18). Others report no change in oxygen content (21)(32), and two high-resolution neutron powder-diffraction studies have failed to find interstitial oxygen in the quenched samples or any change in lattice parameters or oxygen content in the annealing step (25)(33). In fact, we have found that the reducing conditions can be dispensed with completely and replaced by annealing the samples quenched from high temperature at 900°C overnight in a sealed, evacuated silica tube. Electron diffraction shows that this annealing results in a microscopic disproportionation of the material into regions with and without the superlattice modulation of the T' structure described above. Clearly, it is one of the phases formed in this disproportionation that is responsible for superconductivity, hence the need for a final reaction step at temperatures low enough for disproportionation to take place, under conditions that prevent reoxidation of the Ce-doped material. Since the superlattice modulation could be due to ordering of oxygen defects/interstitials it may also be possible to improve superconducting properties of ceramic $Nd_{1.85}Ce_{0.15}CuO_4$ by reduction or cycling of oxygen. However, as we shall see, the deviations from the ideal oxygen stoichiometry of 4.00 in the superconductors are so small that systematic studies are difficult.

The conditions for the synthesis of the thorium-doped superconductor $Nd_{1.85}Th_{0.15}CuO_4$ are reported to be the same as those for the synthesis of $Nd_{1.85}Ce_{0.15}CuO_4$. The final annealing step was carried out at 910°C in flowing He. The synthesis of the fluorine-doped superconductor $Nd_2CuO_{3.82}F_{0.18}$ is somewhat different because the higher mobility of fluoride ions makes high temperatures unnecessary for the initial reaction step. In fact, high temperatures should be avoided because of the volatility of CuF. The starting materials are Nd_2O_3, NdF_3, and CuO with the NdF_3 present in slight excess. The best results were obtained with NdF_3 freshly prepared by precipitation, probably because its extremely small particle size improves product homogeneity. The initial reaction step is carried out at 900°C in dry air for 14 hours and results in a mixture containing $Nd_2CuO_{3.9}F_{0.1}$ with small quantities of NdOF and CuO. More fluorine substitution is needed to make $Nd_2CuO_{4-x}F_x$ superconducting; this is achieved by reducing the mixture under carefully controlled conditions; typically 890°C for 14 hours under flowing N_2 containing 60 ppm of O_2. Figure 6 shows the results of a study of the stability of $Nd_2CuO_{4-x}F_x$ as a function of oxygen partial pressure (pO_2) at 890°C. The pressure was controlled by means of an yttria-stabilized zirconia cell that could supply potentiostatically controlled pO_2's down to 10^{-9} atm. The figure shows that $Nd_2CuO_{4-x}F_x$ with x>0.1 is only stable at 890°C when the $pO_2 < 2 \times 10^{-4}$ atm and that a slow reductive decomposition to Nd_2O_3, NdOF, and Cu_2O takes place for $pO_2 < 5 \times 10^{-5}$ atm. Hence, superconducting samples can only be made in the narrow pO_2 range indicated. Data obtained at other temperatures showed that $Nd_2CuO_{4-x}F_x$ with x>0.1 was stable in a narrow but higher pO_2 range at higher temperature and at lower pO_2 at lower temperature. For this reason, there is no single pO_2 for which the superconducting material is stable at both high and low temperature and it is essential to quench samples after the reduction step to prevent reoxidation on cooling (29). The relative ease with which reductive decomposition takes place in $Nd_2CuO_{4-x}F_x$ limits the maximum level of fluorine doping to x~0.18. Reductive decomposition takes place even more readily in $Ln_2CuO_{4-x}F_x$, where Ln=Sm, Eu, or Gd, with the result that pure, superconducting ceramics could not be made in these systems, although there are indications of superconductivity in the Sm and

440 Chemistry of Superconductor Materials

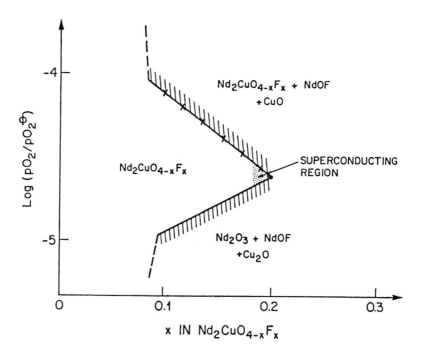

Figure 6: Stability versus oxygen partial pressure of $Nd_2CuO_{4-x}F_x$ at 890°C.

Eu systems. On the other hand, superconducting $Pr_2CuO_{4-x}F_x$ can be made; it appears to have a lower T_c than the Nd compound.

Samples of $Nd_2CuO_{4-x}F_x$ quenched from high temperature (890°C) are never superconducting (Figure 5). Suitably doped samples must be annealed in sealed, evacuated silica tubes at a temperature between 300 and 700°C for 2-6 hours to make them superconducting; in some cases samples prepared in this way have Meissner fractions of over 60%. There is no change in the average oxygen content, average copper oxidation state, or X-ray diffraction pattern in the final annealing step. As discussed above, the annealing induces a microscopic disproportionation of the $Nd_2CuO_{4-x}F_x$ into two phases, one of which has a superlattice modulation of the T'-Nd_2CuO_4 structure that is nearly invisible by X-ray diffraction but clearly seen in electron diffraction. It appears that the superlattice in $Nd_{2-x}Ce_xCuO_4$ is thermally more stable than that in $Nd_2CuO_{4-x}F_x$, since $Nd_2CuO_{4-x}F_x$ samples quenched from 890°C contain no superlattice, whereas $Nd_{2-x}Ce_xCuO_4$ samples annealed at 900°C have fully formed superlattice regions.

Determination of Ce or Th concentrations in the $Nd_{2-x}Ce_xCuO_4$ and $Nd_{2-x}Th_xCuO_4$ systems has been done by XPS, flame photometry, and refinement of neutron powder diffraction profiles. The last method gives a direct measurement of the amount of Ce actually substituted for Nd in the structure. However, determination of fluorine content in $Nd_2CuO_{4-x}F_x$ is a more difficult problem. Fluorine substitution has virtually no effect on crystallographic lattice parameters or X-ray and neutron diffraction peak intensities. Fluorine content cannot be determined by the standard method of dissolution in acid followed by measurement of F^- activity with an ion-selective electrode because of interference from Nd^{3+}. The only analytical technique to have been employed is reaction with steam at 900°C in a platinum tube followed by determination of the liberated HF, a difficult and inevitably inaccurate procedure. The fluorine content has also been measured by comparison of O and F 2p intensities in XPS (19). Fluorine content can be determined indirectly from the average copper oxidation state obtained by iodometric titration or thermogravimetric reduction in hydrogen, provided that there are no oxygen defects/interstitials present in the structure (see below) and that the sample is completely free from CuO and Cu_2O. The latter condition is seldom satisfied in real samples.

The other compositional variable in the electron-doped superconductors is the oxygen content. It is well known that low-temperature (500°C) oxygenation of La_2CuO_4 results in a material of composition $La_2CuO_{4+\delta}$, where $\delta \approx 0.13$ and the extra oxygen occupies interstitial sites in the La_2CuO_4, giving rise to a complex distortion of the structure (30). It is important to know the exact oxygen stoichiometry in the electron doped superconductors both because oxygen defects and interstitials have an effect on the average copper oxidation state and because ordering of oxygen defects or interstitials could be responsible for superlattice formation in these materials. There is some disagreement in the literature on the subject of oxygen content in the superconducting phases, although it appears that any deviation from the ideal oxygen stoichiometry of 4.00 is small. One report based on iodometric double titration claims that there is interstitial oxygen present and that the superconducting phase in the $Nd_{2-x}Ce_xCuO_4$ system is $Nd_{1.85}Ce_{0.15}CuO_4$, where $\delta=0.05$ (31). Other reports based on iodometric titration and thermogravimetric reduction give $\delta=-0.01$ (7)(32), $\delta=-0.02$ (21), and $\delta=-0.04$ (18). Recent high-resolution powder neutron diffraction studies on superconducting $Nd_{1.85}Ce_{0.15}CuO_4$ have given δ values of -0.06 (25) and 0.00 (33). The fact that the overall oxygen content is near stoichiometric does not rule out the possibility that segregation into superlattice/non-superlattice regions in these materials is due to oxygen disproportionation within the sample, although such a disproportionation would be expected to show up in the neutron powder diffraction.

5.0 SINGLE CRYSTALS AND THIN FILMS

Very soon after the first report of superconductivity in $Nd_{1.85}Ce_{0.15}CuO_4$, Hidaka et al. succeeded in making superconducting single crystals of this composition (35). This achievement was soon repeated by others using basically the same technique (36)(37). The crystals are grown in air from a CuO flux containing excess cerium at high temperature. Typical charges are composed of simple oxides with compositions described by Nd:Ce:Cu atomic

Table 1. Structural Parameters and Calculated X-ray Powder Diffraction Pattern of $Nd_{1.85}Ce_{0.15}CuO_4$ (Structural Parameter From ref. 25)

Tetragonal - Space Group I4/mmm $a = 3.9469$ Å $c = 12.0776$ Å

Atom	Site	x/a	y/b	z/c	B_{iso} (Å²)	Fractional Occupancy
Nd	4e	0	0	0.3525	0.32	0.925
Ce	4e	0	0	0.3525	0.32	0.075
Cu	2a	0	0	0	0.48	1
O (1)	4c	0	0	1/2	0.71	0.99
O (2)	4d	0	1/2	1/4	0.49	0.97

Calculated pattern for CuK_α ($\lambda = 1.5418$Å) diffractometer as follows:

h k l	2Θ	d/Å	Relative Intensity
002	14.67	6.039	0
101	23.72	3.75	19
004	29.58	3.0194	3
103	31.75	2.8184	100
110	32.07	2.7909	36
112	35.43	2.5334	0
105	43.95	2.0603	3
114	44.19	2.9495	19
006	45.04	2.0129	7
200	45.99	1.9734	27
202	48.53	1.8758	0
211	52.38	1.7466	5
204	55.64	1.6519	3
116	56.35	1.6326	13
213	56.96	1.6166	37
107	58.37	1.5809	7
008	61.41	1.5097	2
215	65.49	1.4252	2
206	66.33	1.4092	8
220	67.07	1.3954	8

percentage ratios such as 35:10:55 or 26:2:72. The charge is heated to 1300°C in an alumina crucible, held at this temperature for 2-4 hours and cooled at 2-6°C/hour to 1000°C, followed by rapid cooling. Since the molten CuO has a strong tendency to creep out of the crucible at 1300°C the crystal growth is best carried out in a tall, narrow crucible which is then placed in a wider and shorter crucible; the resulting temperature gradient in the inner crucible helps minimise loss of CuO. After this procedure crystals of $Nd_{2-x}Ce_xCuO_4$ can be separated from the residual charge. The crystals do not have the same composition as the bulk charge and are frequently contaminated by CuO inclusions. Those that contain enough cerium still need to be annealed at 900°C in flowing argon to induce superconductivity. Superconducting single crystals prepared in this way show sharp resistive transitions and metallic conductivity above T_c. Superconducting single crystals have so far been made only in the cerium-doped system.

There have been some reports of superconducting thin films of $Nd_{2-x}Ce_xCuO_4$ prepared by laser ablation (38)(39). The films are prepared from sintered $Nd_{1.85}Ce_{0.15}CuO_4$ targets on heated (100)-$SrTiO_3$ substrates. Typical conditions are 500°C at an ambient oxygen pressure of 10^{-4} atm, followed by annealing at 900°C.

6.0 SUMMARY

This chapter has reviewed the state of knowledge about the electron-doped superconductors approximately one year after the discovery of superconductivity in $Nd_{1.85}Ce_{0.15}CuO_4$. Since then, it has been shown that superconductivity exists in the Pr, Sm, and Eu (but not Gd) analogs of $Nd_{1.85}Ce_{0.15}CuO_4$ and in solid solutions of these. It has also been shown that electron doping can be achieved in these systems by substituting Th^{4+} for Ln^{3+} (Ln=Nd, Sm,Eu) and by substituting F^- for O^{2-}. The systematics of $Nd_{2-x}Ce_xCuO_4$ have been worked out and single crystals in this system grown. According to X-ray and neutron diffraction studies, the electron doped superconductors have an undistorted T'-Nd_2CuO_4 structure with oxygen stoichiometry very close to 4.00. However, electron diffraction has revealed that all ceramic superconducting samples are biphasic: they contain regions with the T'-Nd_2CuO_4 structure and regions with a superlattice modulation of this structure, often within the same grain. It has been shown that the final low-temperature annealing step that is essential for the synthesis of all electron-doped superconductors has very little effect on the overall chemical composition of the materials, but is necessary to form the regions of superlattice. The two-phase nature of the superconducting ceramics could account for their low and non-metallic conductivity in the normal state and for the very narrow ($0.14 \leq x \leq 0.17$) and asymmetric composition range of superconducting $Nd_{2-x}Ce_xCuO_4$. The two-phase nature of the superconducting ceramics also means that the true composition of the superconducting phase in this system is unknown, so that it still cannot be stated with absolute certainty that reduced copper is responsible for superconductivity. Current research in high-resolution neutron diffraction and the preparation of single crystals free from superlattice should answer some of these questions.

7.0   REFERENCES

1. T. Tokura, H. Takagi and S. Uchida *Nature* 337:345-347 (1989).

2. J.T. Markert and M.B. Maple, *Solid State Commun.* 70:145-147 (1989).

3. A.C.W.P. James, S.M. Zahurak and D.W. Murphy, *Nature* 338:240-241 (1989).

4. H.-K. Muller-Buschbaum and W. Wollschlager, *Z. Anorg. Allg. Chem.* 414:76-80 (1975).

5. R.J. Cava, B. Batlogg, J.J. Krajewski, L.W. Rupp, L.F. Schneemeyer, T. Siegrist, R.B. van Dover, P. Marsh, W.F. Peck, Jr., P.K. Gallagher, S.H. Glarum, J.H. Marshall, R.C. Farrow, J.V. Waszczak, R. Hull and P. Trevor, *Nature* B 336:211-217 (1988).

6. K.K. Singh, P. Ganguly and C.N.R. Rao, *Mater. Res. Bull.* B 17:493-500 (1982).

7. E. Wang, J.-M. Tarascon, L.H. Greene, G.W. Hull and W.R. McKinnon, *Phys. Rev.* B, (In press).

8. H. Sawa, S. Suzuki, M. Watanabe, J. Akimutsu, H. Matsubara, H. Watabe, S. Uchida, K. Kokusho, H. Asano, F. Izumi and E. Takayama-Muromachi, *Nature* 337:347-348 (1989).

9. S.-W. Cheong, Z. Fisk, J.D. Thompson and R.B. Schwartz, *Physica C* (In press).

10. S.-W. Cheong, J.D. Thompson and Z. Fisk, *Physica C* 158:109-126 (1989).

11. M.-H. Whangbo, M. Evain, M. Beno and J.M. Williams, *Inorg. Chem.* 26:1829-1833 (1987).

12. J. Gopalakrishnan, M.A. Subramanian, C.C. Torardi, J.P. Attfield and A.W. Sleight, *Mat. Res. Bull.* 24:321-330 (1989).

13. D.B. Currie and M.T. Weller, *Mat. Res. Bull.* 24:1155-1162 (1989).

14. T. Siegrist, S.M. Zahurak, D.W. Murphy and R.S. Roth, *Nature* 334:231-234 (1988).

15. T.C. Huang, E. Moran, A.I. Nazzal, J.B. Torrance and P.W. Wang, *Physica C* 159:625-628 (1989).

16. P.H. Hor, Y.Y. Xue, Y.Y. Sun, Y.C. Tao, Z.J. Huang, W. Rabalais and C.W. Chu, *Physica C* 159:629-633 (1989).

17. A. Grassmann, J. Strobel, M. Klauda, J. Schlotterer and G. Saemann-Ischenko, *Europhys. Lett.* 9:827-832 (1989).

18. H. Takagi, S. Uchida and Y. Tokura, *Phys. Rev. Lett.* 62:1197-1200 (1989).

19. J.P. Strobel, M. Klauda, M. Lippert, B. Hensel, G. Saemann-Ischenko, W. Gerhauser, H.-W. Neumuller, W. Ose and K.F. Renk, *Physica C, Proceedings of the International HTSC Conference*, Stanford, California July 23-28 1989.

20. T.C. Huang, E. Moran, A.I. Nazzal and J.B. Torrance, *Physica C* 158:148-152 (1989).

21. J.T. Markert, E.A. Early, T. Bjornholm, S. Ghamaty, B.W. Lee, J.J. Neumeier, R.D. Price, C.L. Seaman and M.B. Maple, *Physica C* 158:178-182 (1989).

22. E.A. Early, N.Y. Ayoub, J. Beille, J.T. Markert and M.B. Maple, *Physica C* 160:320-322 (1989).

23. S. Skanthakumar, H. Zhang, T.W. Clinton, W.-H. Li, J.W. Lynn, Z. Fisk and S.-W. Cheong, *Physica C* 160:124-128 (1989).

24. G.M. Luke, B.J. Sternlieb, Y.J. Uemura, J.H. Brewer, K. Kadono, R.F. Kiefl, S.R. Kreitzman, T.M. Riseman, J. Gopalakrishnan, A.W. Sleight, M.A. Subramanian, S. Uchida, H. Takagi and Y. Tokura, *Nature* 338:49-51 (1989).

25. F. Izumi, Y. Matsui, H. Takagi, S. Uchida, Y. Tokura and H. Asano, *Physica C* 158:433-439 (1989).

26. C.H. Chen, D.J. Werder, A.C.W.P. James, D.W. Murphy, S.M. Zahurak, R.M. Fleming, B. Batlogg and L.F. Schneemeyer, *Physica C* 160:375-380 (1989).

27. A.C.W.P. James, D.W. Murphy, C.H. Chen, D.J. Werder, J. Chiang, S.M. Zahurak, R.M. Fleming, B. Batlogg and L.F. Schneemeyer, to be published in the Proceedings of the *MRS Symposium on HTSC*, Boston, Mass. Nov 27-Dec 1, 1989.

28. J.-M. Tarascon Private Communication.

29. M.E. Lopez-Morales and P.M. Grant, *J. Solid State Chem.* (In press).

30. J.D. Jorgensen, B. Dabrowski, S. Pei, D.G. Hinks, L. Soderholm, B. Morosin, J.E. Schirber, E.L. Venturini and D.S. Ginley, *Phys. Rev. B* 38:11337-11350 (1988).

31. E. Moran, A.I. Nazzal, T.C. Huang and J.B. Torrance, *Physica C* 160:30-34 (1989).

32. E. Takayama-Muromachi, F. Izumi, Y. Uchida, K. Kato and H. Asano, *Physica C* 159:634-638 (1989).

33. G.H. Kwei, S.-W. Cheong, Z. Fisk, F.H. Garzon, J.A. Goldstone and J.D. Thompson, *Phys. Rev. B* (In press).

34. T.A. Vanderah Private Communication.

35. Y. Hidaka and M. Suzuki, *Nature* 338:635-637 (1989).

36. J.-M. Tarascon, E. Wang, L.H. Greene, B.G. Bagley, G.W. Hull, S.M. D'Egidio, P.F. Miceli, Z.Z. Wang, T.W. Jing, J. Clayhold, D. Brawner and N.P. Ong, *Phys. Rev. B* 40:4494-4502 (1989).

37. J.-M. Tarascon, E. Wang, L.H. Greene, R. Ramesh, B.G. Bagley, G.W. Hull, P.F. Miceli, Z.Z. Wang, D. Brawner and N.P. Ong, *Physica C, Proceedings of the International Conference on HTSC*, Stanford, July 23-29 (1989).

38. H. Adachi, S. Hayashi, K. Setsune, S. Hatta, T. Mitsuyu and K. Wasa, *Appl. Phys. Lett.* 54:2713-2717 (1989).

39. A. Gupta, G. Koren, C.C. Tsuei, A. Segmuller and T.R. McGuire. To be published in the Proceedings of the *MRS Symposium on HTSC*, Boston, Mass. Nov 27-Dec 1 (1989).

Part III
Sample Characterization

12

X-Ray Identification and Characterization of Components in Phase Diagram Studies

J. Steven Swinnea and Hugo Steinfink

The discovery of unusual physical or chemical properties in multicomponent systems demands the isolation and chemical and physical characterization of the single component which is responsible for the observed effect. In many instances the resulting search is less than systematic and depends more on serendipity than on careful experimentation. As an example, many of the early attempts to discover the compound responsible for superconducting transition temperatures in the 90K range were sometimes haphazard when viewed in terms of synthetic techniques.

While certainly an exciting event, the discovery of truly novel effects in solid state materials is rare. The design of new materials that will incorporate desired properties requires careful investigation of compounds in a given chemical system. Generally, the approach most often used for solid state materials is to mix various ratios of starting components, process this mixture in some way, usually by heating, to effect reactions among the components, and then to analyze the results of the processing step. The characterization step in this approach may include, but not be limited to, optical and electron microscopy, X-ray diffraction, or separation of reaction products on the basis of physical properties. The results of the characterization steps then serve as a guide towards the synthesis of the desired phase. If starting materials appear in the reaction product, the mixture might be reprocessed to insure that the reaction has reached equilibrium. If the characterization indicates that a new phase or phases are present in the reaction product, an attempt is made to obtain them in pure form. The information gleaned from the reaction is used to prepare new reaction mixtures or the suspected new phases are culled directly from the multicomponent product. This is a laborious, time consuming process and the effort can be shortened considerably by a systematic approach to synthesis of new materials. The basis for this type of approach is the phase diagram and for most solid materials X-ray diffraction, supplemented by a variety of other techniques, is the tool of choice for identifying components in multicomponent mixtures. Recognizing the importance of these two subjects, the phase diagram and X-ray diffraction techniques, it seems appropriate to review the applicable theories and practices before turning to a practical example, the isolation of the high-T_c superconductor $YBa_2Cu_3O_7$.

1.0 THE GIBBS PHASE RULE

In discussing chemical systems, one must be aware of the rules which determine the chemical species that are permitted to occur for a given set of conditions. The basic rule governing systems which are considered to be in thermodynamic equilibrium was first stated by J. Willard Gibbs as early as 1876. The Gibbs phase rule relates the physical state of a mixture with the chemical species of which it is composed and is given in its simplest form as

$$P + F = C + 2.$$

This relationship holds for any chemical system which is subject to variations in temperature, pressure, and proportions of its basic components and describes the number of phases P present in terms of the system's degrees of freedom F and the number of component species C. Even though the phase rule is simple in form, it is not limited in its ability to describe very complex systems. Equilibrium effects arising from the presence of surface tension, stress, magnetic fields, *etc.* can be accounted for by the incorporation of additional degrees of freedom into the phase rule. Such effects, however, will not be considered in this discussion.

We normally think of gases, liquids, and solids as phases. In a stricter sense the phases in a chemical system are the chemically and physically homogeneous states of aggregation of the system which could conceivably be segregated by mechanical means. A single phase may be one unit or may consist of finely divided subunits. Thus ice is a single solid phase whether it exists as a block or is divided into fine chips. Only one gaseous phase may exist in a given system as gases are considered to be completely soluble in one another. While the liquid phase is usually a single phase, some liquids are insoluble or have limited solubility in one another, and thus multiple liquid phases are possible in a system. Solids with different compositions as well as solids with the same composition and different crystal structure are normally considered to be separate phases because they can be physically separated from one another.

The degrees of freedom F in the phase rule refer to the number of externally controllable conditions of the system which must be specified to define uniquely the state of the system at equilibrium. In chemical systems the controllable variables are the temperature, pressure, and the proportions of the components of the system. The degree of freedom has a direct parallel in algebra where the "phase rule" is

$$F = U - J$$

where U is the total number of variables (components) and J is the number of independent equations (phases). Thus a system of equations such as

$$x = 2y + 3z$$
$$4x = 7y + 28$$

has one degree of freedom even if a variable (and an equation) is eliminated. This single degree of freedom indicates that the algebraic system is completely determined by arbitrarily selecting the value of one variable.

The number of components in a chemical system is often more difficult to determine than the number of variables in an algebraic system of equations. Ideally, the number of components is defined as the smallest number of compositional terms capable of independent variation in their proportion in the system and necessary to describe the composition of each and every phase in the system. In elemental systems the number of components is obvious. Iron is a one component or unary system, two metals comprise a two component or binary system and three metals comprise a three component or ternary system. If the substances in the system are compounds rather than simple elements, a more careful consideration of possible species is necessary to determine the number of components. NH_4Cl for instance is considered to be a unary system even though the gas phase dissociates into discrete NH_3 and HCl particles. In this system, the gas phase always has the overall composition NH_4Cl and upon condensation will always give the single component NH_4Cl. Because $CaCO_3$ dissociates into solid CaO and gaseous CO_2, it is part of a binary system if a gas phase is present. The single component $CaCO_3$ cannot properly describe the gas and solid phases in this system and the components CaO and CO_2 must be chosen.

In most systems the components are chosen so that all the phases in the system can be represented by additive proportions of the components. The system Na_2SO_4-SO_3-H_2O is ternary, all known phases being reaction products of additive mixtures of the three components. In addition to the simpler additive system is the metathetical or reciprocal system in which negative proportions must be introduced. An example of this is the system consisting of the four solids, $BaCO_3$, Na_2SO_4, $BaSO_4$, and Na_2CO_3. Although there are four interacting ions in this system, their proportions are not independently variable and thus this is actually a ternary system in which any phase may be represented in terms of three of the four salts, *e.g.* pure $BaCO_3$ = $BaSO_4$ + Na_2CO_3 - Na_2SO_4.

The previous examples help us to understand the nature of a component. While the ions in a system are instrumental in determining the chemical species possible in a system, they cannot be considered components of the system because they are not substances which are capable of separate existence in the system. Thus it becomes apparent that only true chemical substances such as $BaSO_4$ above are capable of independent existence and, hence, independent variation in proportion and therefore may serve as components.

Before moving on, it is wise here to note two important limitations of the phase rule. The criteria for components only prescribe that they be able to represent each phase in the system. The phase rule says nothing about how these components may combine to give other species and, thus, does not define the number or nature of other species in the system. That is, given the components CaO and CO_2, the phase rule cannot predict the existence of the intermediate compound $CaCO_3$.

In a like manner, the phase rule cannot differentiate between global and local equilibrium states. For a given temperature, pressure, and composition, there is a minimum value of the free energy, *the equilibrium state*. However, local minima, lying at higher values of the free energy, may exist. As the system approaches equilibrium, it may become trapped in one of these higher lying minima. Such a state is considered to be *metastable*. To reach the true energy minimum an energy barrier must be surmounted. This may not be possible for the given conditions of pressure and temperature and the metastable state appears to be the equilibrium state. A well known example is the diamond-graphite system. At ambient conditions diamond is thermodynamically metastable, yet kinetically stable, relative to graphite. It may be difficult or even impossible to determine whether a new system is in a metastable or equilibrium state.

2.0 PHASE DIAGRAMS

While the phase rule itself is a powerful tool in the study of phase equilibria, it becomes even more elegant in its graphical manifestation, the phase diagram. Phase diagrams provide a convenient means for recording exactly which phases of a given system are present for a given set of conditions. Because of phase rule restrictions, the phase diagrams for unary, binary, and ternary systems vary in complexity and are constructed in different manners.

The unary phase diagram is seldom used in solid state syntheses. However, the unary diagram forms the basis for the phase diagrams of multicomponent systems. Since there are no composition variables, the only externally controllable variables in a unary system are simply the temperature and pressure. For this reason it is common practice to express the phase diagram as a two dimensional plot of pressure versus temperature as shown in Figure 1. The phase rule for a unary system allows for bivariant, univariant, and invariant behavior, and these behaviors may all be represented on the diagram.

Bivariant behavior, for which both temperature and pressure are allowed to vary independently, fixes two degrees of freedom and restricts the system to a single phase. This is represented on the phase diagram as an open area. Since there is only one component, there may only be single areas for the liquid and vapor phases, but several areas corresponding to solids of different crystal structure may exist.

Univariant equilibrium for which there is one degree of freedom, represents the equilibrium between two co-existing phases. Since there is only one degree of freedom, choosing a value for one external variable, *e.g.* temperature, determines the remaining variable in a dependent manner, and the locus of points represented on the phase diagram for univariant behavior must lie on a line or curve. Thus the curves on the unary phase diagram represent solid-liquid, solid-vapor, solid-solid, and liquid-vapor equilibrium.

Invariant behavior occurs at the intersection of three univariant curves. This intersection defines a point at which three phases are in equilibrium. At these so called triple or invariant points, there are no degrees of freedom and both temperature and pressure assume fixed values.

The unary diagram is used to predict the phase behavior of a pure substance undergoing a change in temperature or pressure. The effect of heating or cooling a material at a fixed pressure (an isobar) is studied by traversing the diagram of Figure 1 horizontally. At any given temperature the vapor pressure of the substance can be read from the solid-vapor or liquid-vapor curves. In a like manner one can study the effect of changes in pressure at constant temperature by following a vertical line, an isotherm.

Because there is an added term, the composition, binary systems are inherently more complex than unary systems. In order to completely represent the phase diagram of a binary system a three dimensional pressure-temperature-composition (P-T-x) diagram can be constructed. However, it is a more common

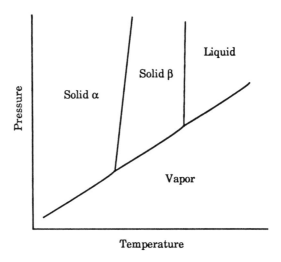

Figure 1. Phase diagram for a one-component system with two solid phases.

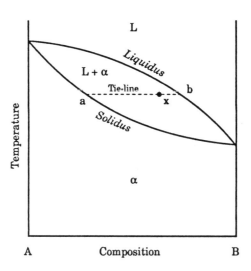

Figure 2. Phase diagram for a condensed two-component isomorphous system.

practice to present the phase diagram as a two dimensional T-x diagram with fixed pressure.

The appearance of a binary T-x diagram depends on the mutual solubility between the two components. An isomorphous system is one in which there is complete solubility and can be represented by a diagram such as Figure 2. In a condensed, *i.e.* no vapor phase, isomorphous system, there are two single phase regions, a liquid L and a solid α and a two phase region L+α. The single phase regions are trivariant so that all three external variables must be fixed to satisfy the phase rule. The two phase region must be bivariant according to the phase rule and since the two phases must be in physical contact, these are subjected to identical temperature and pressure. This implies that once the temperature is chosen at the fixed pressure, the composition of the solid and liquid phase in equilibrium must also be fixed. This is the case and the graphical representation of this equilibrium is the tie-line.

The dotted horizontal line (an isotherm) in Figure 2 is a tie-line which connects a point (a) on the solidus line with a point (b) on the liquidus line and passes through the overall system composition at x. The compositions at a and b are respectively the compositions of the solid and liquid in equilibrium at the given conditions. Given the overall composition x one can calculate the relative amounts of the solid and liquid phases using the "lever principle." If the total length of the tie-line is taken to represent 100%, then the length xb divided by the length ab represents the fraction of solid present and the length ax divided by the length ab represents the fraction of liquid present. While the concepts of the tie-line and lever principle are rather simple, they are powerful tools in the study of phase diagrams and are used extensively in the study of equilibrium systems.

In binary systems in which there is no mutual solubility of the two components, equilibrium between two solid phases can occur. A region such as this is indicated as α + β on the condensed phase T-x diagram depicted in Figure 3. There are three single phase regions in this diagram. The solid phase regions α and β are respectively a region in which a limited amount of component B is allowed to dissolve in crystal structure α and a region in which a limited amount of A is allowed to dissolve in β.

458 Chemistry of Superconductor Materials

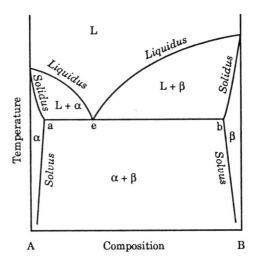

Figure 3. Phase diagram for a two-component system with no intermediate compounds displaying limited solid solubility for α and β.

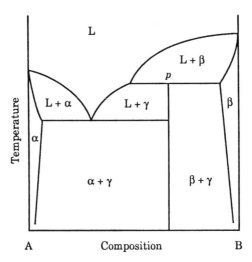

Figure 4. Phase diagram from a two component system with one intermediate compound, γ, which exhibits peritectic melting and displaying limited solid solubility for α and β.

The other single phase region is the liquid L. In addition to the two phase $\alpha+\beta$ region, there are two other two phase regions L + α and L + β. Just as in the isomorphous diagram the solidus and liquidus lines are connected by tie-lines of constant temperature. In a like manner, the $\alpha + \beta$ region is also considered to possess tie-lines joining the two solid-solution or solvus curves.

In Figure 3 there is a special point at which a liquid of composition e is in equilibrium with two solids of composition a and b. This is called a eutectic point and is the only liquid that can coexist with two solid phases. Notice that the phase rule allows one degree of freedom at the eutectic and since the pressure has been previously selected, there are no other choices for conditions. In other words, three phase equilibrium in binary systems occurs at fixed temperature and composition at an isobaric section through phase space. The phase diagram indicates that a liquid with the eutectic composition will decompose into two solid phases upon cooling. This decomposition is called a eutectic reaction. Other three phase binary reactions which can occur upon cooling are the monotectic in which a liquid decomposes into a solid and a second liquid and the eutectoid in which a solid decomposes into two new solids. Likewise, decomposition reactions may occur upon heating. The possible reactions in this case are the peritectic in which a solid decomposes into a solid and a liquid, point p, Figure 4, the syntectic in which a solid decomposes into two insoluble liquids, and the peritectoid in which a solid decomposes into two new solids.

The presence of intermediate compounds in a binary system alters the appearance of the phase diagram. Figure 4 illustrates a simple system in which compound γ forms and decomposes at p at a temperature above the eutectic, *i.e.* a peritectic reaction. Because the composition of γ is fixed, it is represented by a vertical line on the diagram and is designated as a line compound. The effect of this compound is to divide the solid region $\alpha + \beta$ into two new regions, $\alpha + \gamma$ and $\beta + \gamma$, and to add another solid-liquid equilibrium region L + γ. Similar constructions are possible for compounds which decompose below the eutectic and for compounds which are stable at their melting points.

The construction of a phase diagram in a ternary system poses an added difficulty. In this system there are four independent variables, temperature, pressure, and two compositional terms, requiring a four-dimensional diagram. Once the pressure is fixed, there are still three variables, and the complete phase diagram must be represented by a three dimensional figure. This is normally done by assigning the two compositional variables to a plane and taking temperature as a perpendicular axis to this plane. The composition plane is normally taken to be an equilateral triangle so that the proportions of all three components can be read directly. The three dimensional figure is then a triangular prism with a horizontal section being an isothermal, isobaric composition triangle.

Figure 5 is such a composition triangle. The three corners A, B, and C represent the pure components of the ternary. The sides of the triangle represent binary combinations of the three components. In a full three dimensional representation of the ternary diagram the sides become binary T-x diagrams. Interior points on the composition triangle represent ternary mixtures or compounds. Since an apex of the triangle represents 100% of a component and the opposite side 0% of that component, a set of equally spaced lines drawn parallel to the opposite side divides the triangle into lines of constant composition for the apical component. Constructing parallels for all sides then allows the proportions of all three components to be read directly from the diagram.

A more useful method for determining composition allows the fractions to be read from the sides of the triangle. Given a point on the triangle, one of the components is chosen and lines are drawn from the point of interest parallel to the sides adjacent to the chosen component. In Figure 5 the point is P, the chosen component is B, and the lines FP and GP are constructed parallel to the sides BA and BC. This divides the side AC into three line segments AF, FG, and GC. The lengths of these line segments are then proportional to the amounts of A, B, and C in the following manner, x_A = GC/AC, x_B = FG/AC, and x_C = AF/AC. Note that the central line segment FG is always a measure of B, the component not located on the line. The two outer line segments, GC and AF, just as in the case of the two component lever principle, always determine the proportion of the components at the opposite end of the line, A and C respectively. This construction also applies to composition triangles which are not equilateral.

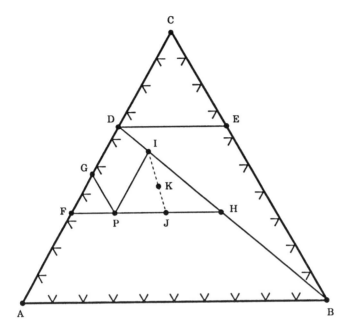

Figure 5. The three component composition triangle.

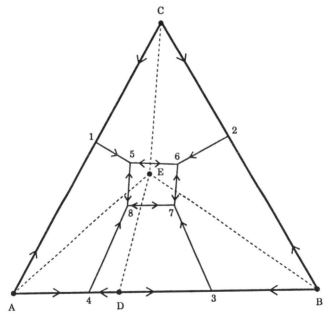

Figure 6. Phase diagram of a three-component system consisting of one binary and one ternary compound. Boundary curves are solid and composition lines are dashed. The arrows indicate the direction of falling temperature.

Consider the triangle ABD and a mixture at P composed of A, B, and D. In this case D is a binary compound composed of A and C. The component A is chosen and the lines IP and HP are constructed parallel to AD and AB. This divides the side DB into segments DI, IH, and HB and the proportions are then x_A = IH/DB, x_B = DI/DB, and x_D = HB/DB. This extension to non-equilateral triangles is useful when working in systems in which equilibrium is established among compounds in the ternary system which do not lie at the apices of an equilateral triangle.

There are a few other useful properties of the composition triangle which should be noted:

1. The proportion of a component along a line parallel to the opposite side is constant. Thus the fraction of C in any composition along the line DE is constant. Such a line is called an isopleth.

2. The ratio of two components is constant along a line through the third. The ratio A/C is then constant along the line DB and is equal to DC/AD. This is useful in situations in which a diluent is added to a mixture. B can be added to the composition I and the resulting mixture still lies on the line DB closer to B.

3. The composition of any point, say I, can then be represented not only as %A and %B, but also in terms of %B, and the ratio A/C.

4. The above holds true for any subtriangle of the composition triangle. Thus in any arbitrary triangle, the ratio of two of the components is constant along a line passing through the third. In the triangle IHP the ratio H/P is constant on the line IJ.

5. If two compositions on the composition triangle are mixed, the resulting composition lies on a line joining them. In this case the starting compositions lie at the endpoint of a binary tie-line and the mixture composition can be located using the lever principle.

6. If a complex such as K exists as a mixture of three substances I, H, and P, the relative amounts are found as follows

$$\%I = \frac{JK}{IJ} \times 100$$

$$\%P = \frac{JH}{HP} \frac{IK}{IJ} \times 100$$

$$\%H = \frac{JP}{HP} \frac{IK}{IJ} \times 100$$

This is just a special case of the concept of center of gravity.

The analysis of ternary phase diagrams is quite involved. The behavior of condensed systems is often displayed on composition triangles onto which the temperature information has been projected so that all information about solid-liquid equilibria may be obtained. A somewhat simple system containing a ternary compound E and a binary compound D is depicted in Figure 6. The composition triangle is broken into fields by the solid lines called boundary curves. The field 1-5-6-2-C is one in which solid C is the major phase and is the first solid to appear when liquids with compositions in this field are cooled. There is a temperature, a melting point, associated with each labeled point on the diagram and the arrows along the boundary curves indicate the direction of falling temperature, *i.e.* the melting temperature at C is higher than that at point 1. Along the boundary curves a three-phase equilibrium exists between two solid phases and a liquid. For example, the two solid phases in equilibrium along 5-6 are C and E and along 4-8 the solids are A and D. In this situation the phase rule allows two degrees of freedom. One degree of freedom is used in fixing the system pressure. The remaining degree of freedom demands that the two remaining variables, temperature and composition, vary dependently defining the boundary curve. Points 1, 2, 3, and 4 represent binary eutectics. Points such as 5, 6, 7, and 8 are ternary eutectics. At point 7, then, there is a four-phase equilibrium between three solids (B, D, and E) and a liquid. The phase rule allows only one degree of freedom, so fixing the pressure determines all the other external variables and the equilibrium appears as a point on the diagram. Upon cooling a ternary, only certain solid phases remain. In Figure 6 these are the three primary components A, B, and C, the binary compound D, and the ternary compound E. The dashed lines joining these phases are called Alkemade lines and divide the composition triangle into four smaller triangles. The final product of crystallization from any composition in one of these triangles will always consist of the three phases found at the apices of that triangle. Generally, it is this sub-solidus composition triangle that is useful in the study of phase relations in solid state systems.

The study of phase equilibria and the construction of phase diagrams is a complex and difficult undertaking. This review has only been offered as a means to introduce concepts and situations that aid in the study of condensed materials systems. Nonetheless, it does provide the foundation for understanding practical problems in materials synthesis. For those who wish to study further, *Phase Diagrams for Ceramists* (1) has an excellent introductory explanation of phase equilibria which includes an extensive bibliography.

3.0 PHASE DIAGRAM STUDIES

Given a system of interest, where does one begin? The first step in any phase diagram study is to determine which, if any, phases in the chosen system are already known. There are several sources for phase information and not all of them are obvious. Of course the first place to check is in a compilation of phase diagrams. For chemical nonalloy systems there are two good references: *Phase Diagrams for Ceramists* (1), a compilation of phase diagrams for condensed systems, and *Phase Diagrams: A Literature Source Book* (2), a bibliographic compilation of general systems. In searching for information on a given system it is important to also gather information on any subsystems. For instance when investigating a ternary, there is often information available on at least one of the binaries involved.

Lacking complete information on a given system, there are other avenues available for the preliminary investigation of the phase diagram. A thorough search of the literature may turn up individual phases in the system. While we have not yet reviewed X-ray crystallography, there are several crystallographic databases that provide valuable information. The first of these is the

Powder Diffraction File maintained and distributed by JCPDS-International Centre for Diffraction Data (3). While this database is normally used to identify materials via X-ray diffraction techniques, its 50,000 entries can also be searched manually and electronically for the existence of substances with given elemental composition. In addition to X-ray diffraction data and elemental information, each entry in this database also contains physical data and bibliographic information. CRYSTDAT and CRYSTIN are two programs administered by the Canadian Institute for Scientific and Technical Information (4). CRYSTDAT is a search tool designed to access the NIST Crystal Data File (5) which contains information on some 120,000 compounds which have been characterized by diffraction techniques. In addition to crystallographic information, the file contains empirical and chemical formula information as well as bibliographic information. The Crystal Data File overlaps somewhat with the Inorganic Crystal Structure Database (6) which is searched by CRYSTIN and contains crystallographic data compiled by scientists at the University of Bonn, West Germany and McMaster University, Hamilton, Canada. There are approximately 30,000 compounds in the database for which structural, elemental, and bibliographic information is available.

Once the literature is searched for existing information, the process of synthesis and characterization can begin. As mentioned, the synthesis step normally consists of mixing fixed proportions of the components and then processing them in some manner. Following processing, the reaction product must be characterized by some means so that the individual phases are identified. This characterization step may employ several techniques, but, because it is sensitive to crystalline atomic structure, the technique of X-ray diffraction is often the most powerful tool.

4.0 X-RAY DIFFRACTION

Most solid materials are crystalline. The atoms that are contained in the solid are arranged so that a specific structural motif is repeated in a periodic manner in all directions. This property is remarkable in that it permits the description of the location of every atom in a macroscopic crystalline substance by only describing the location of a few atoms with respect to a three dimensional crystal lattice. A lattice is an array of points such that the environment about each point in the lattice is identical under translations of the type i**a** + j**b** + k**c** where **a**, **b**, and **c** are non-coplanar, non-colinear unit vectors and i, j, and k are integers. Once the basis vectors are chosen, they can then be considered to define an enclosure of the basic repeating unit of the crystal. This parallelepiped, known as the unit cell, is described by the crystallographic axes a, b, and c and their interaxial angles α, β and γ. The choice of a particular unit cell for a given crystal structure is not necessarily unique but it is normally dictated by the symmetry elements present in the crystal structure. The unit cell can contain either a single lattice point, when it is designated as a primitive cell, or multiple lattice points, when it is labeled as a centered cell. Symmetry constraints allow for the classification of all crystals and their lattices into 14 conventional lattice types. These so called Bravais lattices are distributed among the seven familiar crystal systems listed in Table 1.

When X-rays illuminate a crystalline material, the atoms in the crystal act as scattering centers. Because of the periodic nature of crystals, the scatterers can be considered to be associated with periodically spaced parallel planes a distance d apart. For certain angles of incidence to these planes the X-rays are scattered coherently and in phase. The coherent scattering is known as X-ray diffraction and the geometric condition required for diffraction, the Bragg equation, is given by

$$n\lambda = 2d\sin\theta$$

where n is an integer, λ is the wavelength of the incident X-rays, d is the interplanar spacing, and θ is the angle between the plane and incident X-ray beam. If the X-rays illuminating the sample are of a fixed wavelength, *i.e.* monochromatic, then there is a simple relationship between diffraction angle and interplanar spacing.

In X-ray diffraction one is interested in exploring the intensity of X-rays diffracted from the crystal planes. Note that the Bragg equation does not contain information about the scattered intensity from a given plane. It only provides the

Table 1: The 14 Bravais Lattices

Crystal System	Lattice Types	hkl Restrictions	Unit Cell Constraints
Cubic	P	no restrictions	$a = b = c; \alpha = \beta = \gamma = 90°$
	F	h, k, l all even or all odd	
	I	$h + k + l$ even	
Hexagonal	P	no restrictions	$a = b; \gamma = 120°$
	R	$-h + k + l = 3n$	
Tetragonal	P	no restrictions	$a = b; \alpha = \beta = \gamma = 90°$
	I	$h + k + l$ even	
Orthorhombic	P	no restrictions	$\alpha = \beta = \gamma = 90°$
	C	$h + k$ even	
	F	h, k, l all even or all odd	
	I	$h + k + l$ even	
Monoclinic	P	no restrictions	$\alpha = \gamma = 90°$
	C	$h + k$ even	
Triclinic	P	no restrictions	no restrictions

geometric condition for scattering to be observed. It then becomes necessary to label these planes in some systematic way. This is best done by describing the planes in terms related to their intercepts on the crystallographic axes, a method popularized by W. H. Miller. In this construction the fractional intercepts of the plane on each axis are determined, the reciprocals of these intercepts are calculated, and finally any fractions are cleared to leave three integers h, k and l. In this way the equation of the plane can be written in intercept form as $(hx/a) + (ky/b) + (lz/c) = n$, where n is integral. This form is consistent with the concept of periodicity in crystals as consideration of all possible values of n leads to the description of an infinite set of equally spaced parallel planes for a given set of Miller indices hkl. The concept of Miller indices also provides a convenient method for determining interplanar spacing, d_{hkl}, which replaces d/n in the Bragg equation. These relationships which relate d_{hkl} to the unit cell parameters are listed in Table 2.

It is evident from the above that the relative dimensions of the crystal lattice determine the diffraction angles in an X-ray diffraction experiment. What then determines the diffracted intensity? It has been mentioned that the atoms in a crystal scatter the incoming X-ray beam. This is not entirely correct, though, because the X-rays are actually scattered by the electrons associated with the individual atoms. In general a single electron has a fixed scattering power and thus the Z electrons surrounding an atom of given atomic number have a combined scattering power proportional to that number. This implies that the observed intensity is due at least in part to the *types* of atoms in the unit cell. In general the atoms in a crystal do not lie conveniently on the lattice planes but instead at some position with coordinates xyz in the unit cell. These atoms with different positions in the unit cell then impart different phases to the scattered X-rays affecting the scattered intensity. Thus there is also an atom *position* effect on observed intensity. When both of these effects, atom type and atom position, are considered together, it becomes obvious that the scattered intensity in an X-ray diffraction experiment is determined by *crystal structure*.

It is the goal of any X-ray diffraction experiment to exploit both the geometric and structural information available in the d_{hkl}-

Table 2: $\dfrac{1}{d_{hkl}^2}$ as a Function of Cell Parameters for the Six Crystal Systems

Cubic	$\dfrac{1}{a^2}(h^2 + k^2 + l^2)$
Hexagonal	$\dfrac{4}{3a^2}(h^2 + hk + k^2) + \dfrac{l^2}{c^2}$
Tetragonal	$\dfrac{h^2 + k^2}{a^2} + \dfrac{l^2}{c^2}$
Orthorhombic	$\dfrac{h^2}{a^2} + \dfrac{k^2}{b^2} + \dfrac{l^2}{c^2}$
Monoclinic	$\dfrac{\dfrac{h^2}{a^2} + \dfrac{l^2}{c^2} - \dfrac{2hl\cos\beta}{ac}}{\sin^2\beta} + \dfrac{k^2}{b^2}$
Triclinic	$\dfrac{\dfrac{h^2}{a^2}\sin^2\alpha + \dfrac{k^2}{b^2}\sin^2\beta + \dfrac{l^2}{c^2}\sin^2\gamma + \dfrac{2hk}{ab}(\cos\alpha\cos\beta - \cos\gamma) + \dfrac{2kl}{bc}(\cos\beta\cos\gamma - \cos\alpha) + \dfrac{2lh}{ca}(\cos\gamma\cos\alpha - \cos\beta)}{1 - \cos^2\alpha - \cos^2\beta - \cos^2\gamma + 2\cos\alpha\cos\beta\cos\gamma}$

I_{hkl} data. X-ray diffraction can be broadly divided into two major techniques, single crystal and powder diffraction.

The technique of single crystal X-ray diffraction is quite powerful. In this technique an individual crystal is oriented so that each hkl plane may be examined separately. In this manner it becomes a simple matter to determine the unit cell parameters and symmetry elements associated with the crystal structure. Furthermore, it is also possible to record the intensity for each reflection from a given hkl plane and from this determine the location of atoms in the crystal, i.e. the crystal structure. While the data derived from single crystal X-ray diffraction are very valuable, the experiments are sometimes quite time consuming and so the technique is limited in its appeal as a day to day analytical tool.

The data from an X-ray powder diffraction experiment, on the other hand, can be obtained quickly and are useful in the identification of phases in mixtures of crystalline substances. In this technique the sample is ground so that it is composed of micron sized crystallites and the X-ray diffraction pattern of this powdered sample is then recorded. Because the sample consists of *randomly* oriented crystallites, each plane, (hkl), has a definite probability of being positioned at its proper Bragg angle. The powder pattern can then be either recorded on a photographic film or with an electronic X-ray detector. In general the photographic method is employed only when small amounts of material, i.e. micrograms are available. The powder diffraction method reduces the three dimensional hkl information to a single dimension, d_{hkl}. This, of course, makes detailed structural determinations difficult, but the d_{hkl}-I_{hkl} information can be considered to be a "fingerprint" of a given crystalline substance and the powder method is thus a useful tool for the identification of unknown phases.

If the powder diffraction pattern is indeed a fingerprint for a given material, then it would seem desirable to have a library of the patterns for all known substances. In this way, the powder pattern of an unknown material could be compared to the known patterns in the library until a match is made. Such a file does exist. It is the JCPDS Powder Diffraction File (PDF) mentioned earlier (3). The PDF is published annually as a set of consecutively numbered "cards" (the idea of a card becoming somewhat obsolete as the PDF becomes more available on electronic media). Each card consists of the following information:

- A PDF card number.
- The compound name and the chemical formula.
- The three most intense d-I pairs.
- The largest observed interplanar spacing.
- An area containing unit cell data, physical properties, experimental conditions, and references.
- For each observed reflection a listing of d, I, and hkl; the latter when known.

Because of the number of standards in the PDF, it is not possible, in general, to search each entry individually. However, in many situations the analyst either knows or can make an educated guess as to the probable identity of the unknown. In this intuitive case, the PDF can be consulted directly and the cards for the suspected compounds can be consulted quickly. In fact, when a specific phase diagram is being investigated it is common practice to obtain powder diffraction patterns of all known phases in the system. If the 2θ axis of these "reference" patterns is the same as that for unknown patterns, an unknown pattern can be compared directly to these relatively few standards on an illuminated viewing table and the diffraction lines from the known phases can be identified on the unknown pattern. As new phases are discovered and identified, their patterns may then be added to this reference file.

In some systems for which there are known phases and even single crystal studies, the powder patterns for the known phases have not been deposited in the PDF. In this case there are at least three methods for obtaining the necessary patterns. The first is obvious; the known phase is prepared and the powder diffraction pattern is recorded. However, it is not always possible to prepare single phase material of the known phase and so other methods must be employed. These other methods depend on the availability of crystal structure data for the desired phase. If the crystal structure is known, the complete powder pattern can be calculated. If only unit cell data are known, the d-spacings of all possible lines for the phase can be determined from the relationships in Table 2. In this way Miller indices can be assigned to each line arising from the phase in any unknown pattern. This process is known as indexing and in principle can be carried out in reverse, determining the unit cell parameters by assigning Miller indices to lines in an unindexed pattern.

The last method for producing standard patterns for phases not in the PDF is more involved. In many instances single crystals of unknown phases can be removed from reaction mixtures. If this is the case, a full three dimensional crystal structure analysis will yield the positions of all atoms in the structure. Once the crystal structure is known, it can be used to calculate the X-ray powder diffraction pattern for the phase. This powder diffraction information can then be used with confidence as a standard powder pattern.

In many instances the constituents in an unknown can not be guessed and the entire file must be searched. The goal of the search is to reduce the number of possibilities to a reasonable number of PDF cards which can then be checked line by line for a match. For this type of search JCPDS publishes the Hanawalt search manual. This manual is organized into groups and subgroups based on d-spacing and intensity. The d-spacing of the most intense line determines the group into which the pattern falls. The groups are then ordered according to the d-spacing of the second most intense line in the pattern. In total the eight most intense lines of the pattern are listed in order of decreasing intensity for each entry.

To identify an unknown using the Hanawalt manual, the first and second most intense lines of the diffraction pattern are chosen as a line-pair and the manual is checked for a matching entry. If all eight lines in the entry match the unknown pattern, the match is confirmed by checking the complete pattern in the PDF. Consider the diffraction data from a black powder given in Table 3. The strongest line at 2.44 Å and the next strongest at 1.43 Å are chosen as line-pairs with which to enter the manual. The line 2.44 Å is found in the main group 2.50 - 2.44 Å (±0.01 Å to account for error) and the group is searched for the occurrence of 1.43 Å. When this is done some six entries are found which match most of the lines in the unknown pattern and all are black solids. Upon close inspection it is noted that all of these

Table 3: X-ray Powder Diffraction Pattern for Unknown Black Powder

d_{obsd} (Å)	I_{obsd}	Co_3O_4 9-418
4.68	15	4.67_{20}
2.87	32	2.86_{40}
2.44	100	2.43_{100}
2.34	9	2.33_{12}
2.13	5	
2.024	21	2.021_{25}
1.652	8	1.651_{12}
1.558	29	1.556_{35}
1.431	38	1.429_{45}

Table 4: X-ray Powder Diffraction Pattern for Unknown White Powder

d_{obsd} (Å)	I_{obsd}	CaO 4-777	$CaCO_3$ 5-586	$Ca(OH)_2$ 4-733
4.92	15			4.90_{74}
3.85	5		3.86_{12}	
3.11	5			3.11_{23}
3.04	95		3.035_{100}	
2.85	5		2.845_{5}	
2.78	50	2.78_{34}		
2.62	20			2.628_{100}
2.50	20		2.495_{14}	
2.41	100	2.41_{100}		
2.29	30		2.285_{18}	
2.096	25		2.095_{18}	
1.928	10		1.927_{5}	1.927_{42}
1.916	20		1.913_{17}	
1.876	25		1.875_{17}	
1.795	5			1.796_{36}
1.702	60	1.701_{45}		
1.686	5			1.687_{21}
1.627	5		1.626_{4}	
1.605	8		1.604_{8}	
1.524	6		1.525_{5}	
1.484	2			1.484_{13}
1.451	20	1.451_{10}		

compounds possess similar chemical formulae, i.e. AB_2O_4. Since X-ray diffraction is sensitive to crystal structure and types of atomic constituents, it must be concluded that these entries must possess identical crystal structures and similar composition. In this case all the listed compounds have the spinel type structure and the constituent atoms are all transition elements that have similar atomic number. This observation presents a problem in the present example, yet leads to an important caveat when relying exclusively on X-ray diffraction patterns, namely, that chemically different compounds with the same crystal structure may have strikingly similar X-ray diffraction patterns.

How then can an exact match be made? In this case there must be some outside knowledge supplied and an examination of the X-ray fluorescence spectra would indicate the presence of only Co and O in the unknown powder. When this information is used, the unknown is identified as Co_3O_4, card 9-418. However, when the card is examined the weak line 2.13 Å remains unidentified in the unknown pattern. X-ray intensity is proportional to the amount of the diffracting material present, so it is natural to suspect that a contaminant phase will only display its most intense lines. The problem in this case is that there is only one line remaining so that the Hanawalt manual is useless to help in the identification. The intuitive method must be used. The key question to be answered is, what contaminants might be found in a sample containing only Co and O? The two possibilities that come to mind are elemental Co or another oxide of Co. When these two possibilities are checked, it is quickly found that the most intense line of CoO, card 9-402, is indeed 2.13 Å, and the identification is complete.

The Hanawalt method is relatively straightforward for unknowns which are nearly single phase. For multiphase mixtures the identification becomes tedious. Consider the data for a white powder given in Table 4 (7). Again the search is begun with the two most intense lines, 2.41 Å and 3.04 Å, but no match is found for this line-pair. Now a new line-pair must be chosen and the first and third most intense lines seem likely. Searching the Hanawalt manual with the pair 2.41 Å and 1.70 Å yields the possibilities γ-TaH, CaO, and $NaYO_2$. Only the four lines of the CaO standard pattern lines match the lines in the unknown pattern.

Now it must be determined whether or not the intensities on the CaO card account for the observed intensity of these four matched lines. At first there appears to be a discrepancy, that is, three of the four matched CaO lines are too intense in the unknown pattern. If the intensities of these lines are divided by a common factor of about 1.5, however, their intensities match those on the PDF card reasonably well. This effect is not unusual as it is not uncommon for a detection system, whether photographic or electronic, to saturate at high count rates so that weaker reflections are artificially enhanced. It is reasonably certain then that the four lines 2.78 Å, 2.41 Å, 1.702 Å, and 1.451 Å in the unknown can be completely ascribed to the presence of CaO. A check of the remaining lines suggests the use of 3.04 Å and 2.29 Å as a possible line-pair. The Hanawalt manual gives at least three possibilities, $Rb_2Mg(NH)_2$, LuSI, and $CaCO_3$, with $CaCO_3$, card 5-586, giving the best match. Again the lines in the unknown are checked against the lines on the card and 12 lines match $CaCO_3$. The final line-pair 2.62 Å and 4.92 Å is chosen and a search of the manual is made. Again there are several possibilities, with $CaRuO_4 \cdot 2H_2O$ and $Ca(OH)_2$ giving equally good matches. Intuition favors the presence of the latter and the remaining lines in the pattern can be ascribed to $Ca(OH)_2$. Knowledge of the elemental composition, *e.g.* from X-ray fluorescence data, would rule out $CaRuO_4 \cdot 2H_2O$ if no Ru spectral lines are observed.

It is obvious that the Hanawalt search method is powerful, but in the general case of multiphase unknowns with overlapped lines, the search can become onerous. The use of outside information such as color or chemical composition, may not always provide the answer. The availability of the PDF on electronic media has made computer search/match procedures practical. These programs allow an unknown pattern to be tested for the presence of each and every entry in the PDF. This is quite different from the Hanawalt and intuitive methods for which only a few patterns are actually searched or matched. Most of these programs assign a "figure of merit" or "goodness of fit" to each PDF entry during the search phase and report the highest ranked FOM's for matching. The match phase of the program usually involves comparing phases with high FOM to the unknown on a graphics monitor. In most cases the lines from an identified phase can be normalized to the PDF entry and subtracted from the unknown pattern. In this manner the residual pattern may be searched until all lines in the unknown have been assigned to known phases.

This discussion of X-ray diffraction has merely been introductory, of course. At most we have tried to introduce concepts which will be useful to the solid state chemist who is unfamiliar with the subject. More detailed discussions can be found in X-ray crystallography texts such as *Elements of X-Ray Diffraction* by Cullity (8).

5.0 TYING IT ALL TOGETHER

The only remaining task is to study a specific example that uses the ideas that have been presented here. Quite possibly one of the most interesting recent examples of unknown phase identification is the discovery of the high-T_c superconducting material $YBa_2Cu_3O_7$. When initial reports of 90 K superconductivity in multiphase reaction mixtures in the Y-Ba-Cu-O system surfaced, there was a rush in the scientific community to identify the active phase. The immediacy of the problem led to numerous methods being employed in the search. The following shows how the phase diagram and X-ray diffraction can combine to solve problems like this.

The initial reports of 90 K superconductivity all indicated that this property was found in reaction mixtures containing oxides of the elements Y, Ba, and Cu. Most of the investigators centered their attention on mixtures corresponding to compositions mimicking the $La_{2-x}M_xCuO_4$ (M = Ba, Sr, or Ca) 30 K superconductors. However, some investigators were more adventuresome and departed from this composition. Nonetheless, no one seemed able to produce a single phase superconductor. This immediately suggests the use of the phase diagram to determine the proper composition.

The first step, as always, is to choose the components of the system and determine which phases are already known. The components chosen in this instance are the simple oxides Y_2O_3, BaO, and CuO. For the phase diagram we will use $YO_{1.5}$ instead of Y_2O_3 as this will make plotting compositions on the diagram easier. Now, which phases are known? If *Phase Diagrams for Ceramists* (1) is consulted the binary system Y_2O_3 - BaO is noted with phases Y_2BaO_4, $Y_4Ba_3O_9$, $Y_2Ba_2O_5$, and $Y_2Ba_4O_7$. Of these Y_2BaO_4, $Y_2Ba_4O_7$, $Y_2Ba_2O_5$ form at the temperature of interest, 950°C. The PDF is consulted next and the compounds $Y_2Cu_2O_5$ and $BaCuO_2$ are found. Finally the NIST crystal data file may be consulted for occurrences of phases in this system. CRYSTDAT is used to search this database for all compounds containing Y, Ba, Cu, and O and the compound Y_2BaCuO_5, an orthorhombic structure reported by Michel and Raveau (9), is discovered. The composition triangle for these known phases is given in Figure 7.

After the known phases are identified, the process of synthesis and characterization of reaction products can begin. But where does one begin? In this case we can consult the literature for probable compositions. Let us begin with the X-ray powder diffraction data given by Tarascon *et al* (10). They found superconductivity in a sample of nominal composition $Y_{0.583}Ba_{0.083}$-$Cu_{0.333}O_x$ (point 1, Figure 7) and reported a powder pattern from a mixture of several phases. They were able to identify the components Y_2O_3, and $Y_2Cu_2O_5$, but several lines remained unidentified in the pattern. They fail to mention the compound Y_2BaCuO_5 in their paper and five of the extra lines can be attributed to this compound. Three of the remaining lines match CuO leaving only the lines 2.745 Å and 2.721 Å from what must be the superconducting compound. There is a problem with this result, however. Our study of the phase diagram tells us that ternary systems with intermediate phases crystallize into at most three crystalline phases upon equilibrium solidification. Clearly, the appearance of *five* phases indicates that the reaction did not produce equilibrium products. In general the approach to equilibrium for a given solid state reaction can be tested by grinding the reaction product and reheating it. Any changes in the powder diffraction pattern indicate that the original product was not an equilibrium product. In this case we know that the initial reaction product can not be due to an equilibrium reaction and subsequent investigation would indicate that the proper equilibrium products are Y_2O_3, $Y_2Cu_2O_5$, and Y_2BaCuO_5.

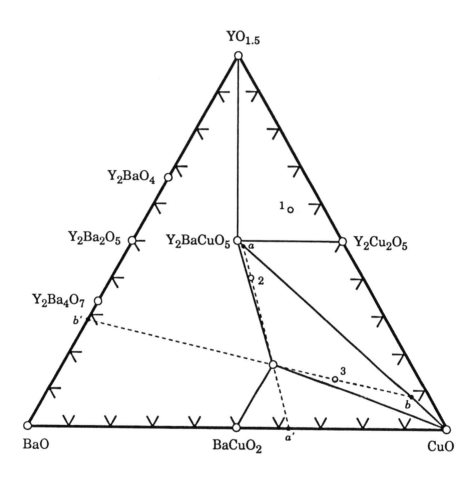

Figure 7. Partial phase relations at 950°C and 0.21 atm O_2 in the Y_2O_3-BaO-CuO system.

Now we will examine two more interesting compositions. Chu et al. (11) reported superconductivity in a mixture of overall composition $Y_{1.2}Ba_{0.8}CuO_x$ and Stacy and coworkers (12) prepared copper rich compositions near $Y_{1-x}Ba_xCu_2O_y$. As part of our investigation of this system (13), we prepared Chu's composition and also a mixture corresponding to x = 0.6 in Stacy's work, points 2 and 3, Figure 7. The X-ray powder diffraction patterns for these two preparations are listed in Table 5. In both of these patterns the phases Y_2BaCuO_5 and CuO can be identified. The remaining lines are common to both patterns and, since only three phases can exist at equilibrium upon solidification, they must represent the powder pattern of a third component, the superconducting phase.

How can these patterns be used to determine an approximate composition for the third component? The powder pattern indicates that the mixtures 2 and 3 must lie in a three-phase region bounded by Y_2BaCuO_5, CuO, and a third component of unknown stoichiometry. The ratio of the relative amounts of the two known compounds can be determined at least semi-quantitatively by using the ratios of the intensities of a non-overlapped, intense reflection from each compound. This property arises from the fact that X-ray powder diffraction patterns are additive and the presence of one component does not affect the powder pattern of the other components. The technique may be made more accurate by preparing standard mixtures of the components and preparing a working curve of intensity ratio versus composition. Once this ratio is determined it is used to define a point on the tie-line joining the two known components by applying the lever principle (property five in the discussion of ternary systems). For the present example an examination of the ratio of the 3.00 Å reflection from Y_2BaCuO_5 and the 2.52 Å line from CuO determines points *a* and *b* on the tie-line between Y_2BaCuO_5 and CuO. Now, the ratio of Y_2BaCuO_5 and CuO in the starting mixture and reaction product must be the same. Since the ratio of two components in a ternary is constant along a line through the third (property four), the composition of the third component must lie on a line extended from point *a* through point 2, *i.e.* a-a'. Likewise, the composition of the unknown must lie along the line determined by reaction 3, b-b'. The intersection of these two lines

Table 5: X-ray Powder Diffraction Patterns for Two Reaction Products in the Study of the Y_2O_3–BaO–CuO System

Composition 2		Composition 3		
d_{obsd} (Å)	I_{obsd}	d_{obsd} (Å)	I_{obsd}	Identification
		11.80	2	U
6.16	2			G
		5.87	3	U
4.16	2			G
3.89	4	3.90	10	U
3.84	2	3.84	4	U
3.57	2			G
3.43	6			G
		3.23	4	U
		3.20	5	U
3.08	5			G
3.05	7			G
3.00	100	3.00	16	G
2.93	67	2.93	12	G
2.83	46	2.83	9	G
2.80	19	2.80	4	G
2.75	18	2.75	60	U
2.73	31	2.73	100	U
2.71	19			G
2.57	5			G
2.52	4	2.52	39	C
2.51	11			G
		2.47	3	U
2.42	9			G
2.34	8	2.34	10	U
2.32	14	2.32	46	C
2.24	6	2.24	18	U
2.22	23	2.22	4	G
2.183	11			G
2.159	7			G
2.137	3			G
2.074	7			G
2.031	8			G
1.992	33	1.992	6	G
1.946	11	1.946	17	G,U
1.914	3	1.913	11	U
1.875	9			G
		1.868	11	C

G = Y_2BaCuO_5 C = CuO   U = Unknown

must approximate the composition of the unknown and provides a starting composition for the next synthesis.

The intersection located by the above method gives an approximation of the desired composition. Suppose instead of $Y_{0.167}Ba_{0.333}Cu_{0.500}O_{1.166}$, the composition $Y_{0.15}Ba_{0.35}Cu_{0.50}O_{1.15}$ is estimated. Can the proper composition be discerned from this information? In general, this depends on the stoichiometry of the unknown compound. In the present case the ratio of Ba and Cu to Y appears to be about 2:1 and 3:1 respectively. Multiplying by 6 then yields the stoichiometry $Y_{0.9}Ba_{2.1}Cu_3O_{6.45}$. Now the question arises, is integral stoichiometry, i.e. $YBa_2Cu_3O_{6.5}$, correct or might there be a solid solution between Y and Ba in this compound? If there is solid solubility of the type $Y_{1-x}Ba_{2+x}Cu_3O_{6.5-0.5x}$, it must be confirmed by reacting mixtures of varying composition to test the hypothesis. For this material the answer is that for any composition other than $YBa_2Cu_3O_{6.5}$, impurity phases appear and thus, within the limits of detectability of the impurity phases by the powder diffraction method, the compound has integral stoichiometry for the cations (13). Subsequent work established, however, that there is a solubility region for O and the proper stoichiometry is $YBa_2Cu_3O_{7-\delta}$ (14).

Once a material has been prepared as a single phase, it is imperative that the powder pattern be indexed, that is Miller indices must be assigned to each line in the diffraction pattern. Indexing provides proof that the observed powder pattern can be attributed to a single phase material and provides unit cell constants for the material. This is not always a simple task and one can try a variety of methods for indexing. Generally, indexing is attempted in higher, uniaxial symmetry systems (cubic, tetragonal, hexagonal) first and lower symmetry systems last. In many instances manual methods can be used for the higher symmetry crystal systems, but automated computer algorithms are preferred for lower symmetry systems where the variables are h, k, l for each reflection and the lattice constants that may consist of as many as six unknown parameters for the triclinic system.

The apparent difficulty of indexing can be overcome if single crystals of the desired phase exist and this brings to mind a characterization step that should never be omitted. Anytime a reaction product is obtained, it should be examined carefully under an optical microscope. The magnification need not be high, 30-50× is quite sufficient. The first thing to be noted is the overall appearance of the material. Are there phases of different habit or color present? Has melting occurred or is the material still a powder sample? Such observations are invaluable and can be just as important as a powder diffraction pattern. If any material is present that might consist of single crystallites, it is always advisable to harvest a few for immediate or future X-ray analysis. It is also good practice to remove samples of any identifiable phases for x-ray fluorescence analysis. In this manner compositional information can be obtained for even unknown phases.

The optical examination of a reaction product containing the superconducting phase $YBa_2Cu_3O_{7-\delta}$ yielded crystals of about 50 μm. The single crystal analysis of this phase is complicated by the fact that twinning occurs, but it does indicate that the phase is orthorhombic with lattice constants a = 3.81 Å, b = 3.87 Å, c = 11.68 Å. This knowledge of the lattice constants makes the indexing of the powder pattern trivial. The lattice constants of $YBa_2Cu_3O_{7-\delta}$ suggest that the structure might be approximated by a tetragonal unit cell. A search of the NIST Crystal Data File using CRYSTDAT for all rare-earth, scandium, or yttrium oxides with tetragonal symmetry yields the compound $La_3Ba_3Cu_6O_{14.1}$ (15). This compound exists in a solid solution series in which there is mixing of La and Ba in the structure. An examination of this structure and its similarity to $YBa_2Cu_3O_{7-\delta}$ determined the gross structural features of $YBa_2Cu_3O_{7-\delta}$, i.e. an ordered phase with yttrium located between coplanar CuO_2 layers with no oxygen at the yttrium level and a second crystallographically independent copper atom located between barium layers.

The identification of the superconducting phase $YBa_2Cu_3O_{7-\delta}$ provides an example in which knowledge of thermodynamics, i.e. the Gibbs phase rule and the theory of equilibrium phase diagrams coupled with X-ray diffraction techniques led to success. Further, the use of databases that can now be easily accessed and searched on-line provided leads to a preliminary structure determination. The procedures outlined here are among the basic approaches used in solid state chemistry research, but by no means are they the only ones. Clearly the results from other analytical techniques such as electron microscopy and diffraction, thermal analysis, *etc.* provide valuable data. The thrill of synthesizing and identifying new materials and the hope that they will possess properties to advance technology for the benefit of mankind are the driving motivations of scientists and engineers. Solid-state chemists have been the participants in an enormously exciting chapter—the discovery of high-T_c superconducting compounds that hold great promise for the future.

ACKNOWLEDGMENT

The financial support of the Robert A. Welch Foundation, Houston, Texas, and of the Microelectronics and Computer Technology Corporation (MCC) of Austin, Texas, is gratefully acknowledged.

6.0 REFERENCES

1. Levin, E.M., Robbins, C.R, and McMurdie, H.R., *Phase Diagrams for Ceramists*, American Ceramic Society (1987).

2. Wisniak, J., *Phase Diagrams: A Literature Source Book*, Elsevier Scientific Publishing Company, NY (1981),

3. McClune, W.F., editor-in-chief, *Powder Diffraction File*, JCPDS-International Centre for Diffraction Data, Swarthmore, PA.

4. Wood, G.H., manager, *CAN/SAN Scientific Numeric Databases*, Canada Institute for Scientific and Technical Information, Ottawa, Canada.

5. Stalynick, J.K. and Mighell, A.D., *Crystal Data File*, National Institute for Standards and Technology, Gaithersburg, MD (1982).

6. Bergerhoff, G. and Brown I.D., *The Inorganic Crystal Structure Database*, Fachinformationszentrum, Energie, Physik, Mathematik, Karlsruhe (1981).

7. Hubbard, C.R., McCarthy, G.J., and Foris, C.M., *PDF Workbook*, International Centre for Diffraction Data (1980).

8. Cullity, B.D., *Elements of X-Ray Diffraction, 2nd ed.*, Addison-Wesley, Menlo, Park, California (1978).

9. Michel, C. and Raveau, B., *J. Solid State Chem.* 43:73 (1982).

10. Tarascon, J.M., Greene, L.H., McKinnon, W.R., and Hull, G.W., *Physical Review B* 35(13):7115 (1987).

11. Wu, M.K., Ashburn, J.R., Torng, C.J., Hor, P.H., Meng, R.L., Gao, L., Huang, Z.J., Wang, Y.Q., and Chu, C.W., *Physical Review Letters* 58(9):908 (1987).

12. Stacy, A.M., Badding, J.V., Geselbracht, M.J., Ham, W.K., Holland, G.F., Hoskins, R.L., Keller, S.W., Millikan, C.F., and zur Loye, H., *J. Am. Chem. Soc.* 109:2528 (1987).

13. Steinfink, H., Swinnea, J.S., Sui, Z.T., Hsu, H.M., Goodenough, J.B., *J. Am. Chem. Soc.* 109:3348 (1987).

14. Manthiram, A., Swinnea, J.S., Sui, Z.T., Steinfink, H., and Goodenough, J.B., *J. Am. Chem. Soc.* 109:6667 (1987).

15. Er-Rakho, L., Michel, C. Provost, J., and Raveau, B., *J. Solid State Chem.* 37:151 (1981).

13

Structural Details of the High T_c Copper-Based Superconductors

Charles C. Torardi

1.0 INTRODUCTION

The structures of the new high-temperature copper-based superconductors are related to that of perovskite, $CaTiO_3$, which can be represented by the general formula ABO_3. The so-called 1-2-3 superconductor, $Y_1Ba_2Cu_3O_7$, is most closely related to perovskite. This is so because $YBa_2Cu_3O_7$ is an oxygen "deficient" perovskite ABO_{3-x} with $A = (Y_{0.33}Ba_{0.67})$, $B = Cu$, and $x = 0.67$ (see Figure 1[1]). In comparison, the structures of other copper-containing superconductors such as La_2CuO_4, $Bi_2Sr_2CaCu_2O_8$, and $Tl_2Ba_2Ca_2Cu_3O_{10}$ may be viewed as intergrowths between the perovskite structure (the part containing the CuO_2 layers and the cations above and below the layers) and the rock salt structure (the part containing the LaO, BiO, and TlO sheets) as shown in Figure 2. These structures are all discussed in more detail in the following sections.

---

[1] Structural figures were drawn with the assistance of the ORTEP program, C. K. Johnson (1976).

486   Chemistry of Superconductor Materials

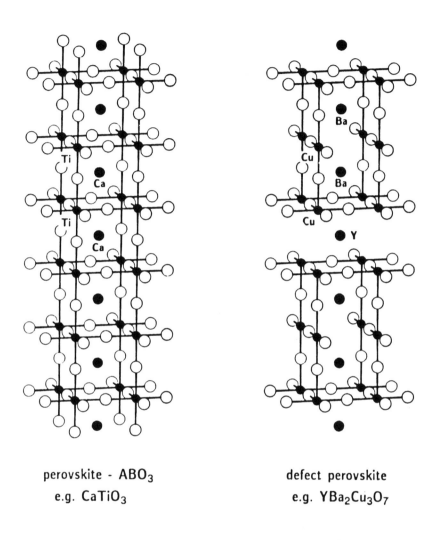

perovskite - ABO_3
e.g. $CaTiO_3$

defect perovskite
e.g. $YBa_2Cu_3O_7$

Figure 1: Relationship between the perovskite structure ABO_3 (left) and the defect-perovskite superconductor $YBa_2Cu_3O_7$ (right). Metal atoms are shaded. Note the missing oxygen atoms in the latter drawing that result in formation of copper-oxygen sheets (above and below the Y atoms), and copper-oxygen chains (between the Ba atoms).

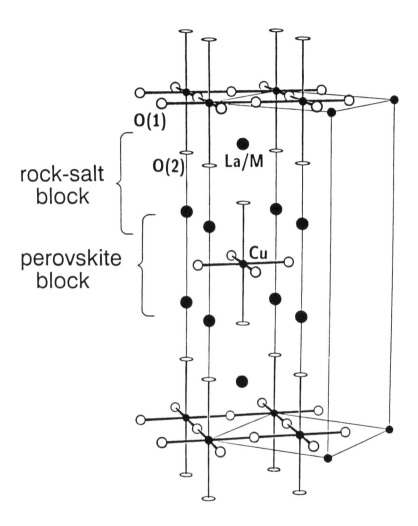

Figure 2: Tetragonal and orthorhombic unit cells of $La_{2-x}M_xCuO_4$ (e.g., M = Sr or Na) and La_2CuO_4, respectively. Metal atoms are shaded and Cu-O bonds are shown. The structure may be viewed as an intergrowth of perovskite and rock salt blocks.

The most obvious structural feature common to all of the Cu-containing superconductors is the CuO_2 sheet. Superconductivity of all known copper-oxide superconductors, with the possible exception of electron-superconductors such as $Nd_{2-x}Ce_xCuO_4$ and $Nd_2CuO_{4-x}F_x$ (1)(2), is believed to originate from the presence of holes in these CuO_2 layers (3). Holes are created when the Cu^{2+} ions lose additional electrons. Mechanisms for hole-doping are several and include cation substitutions, cation vacancies, interstitial or extra oxygen, and overlapping of energy bands at the Fermi level (i.e., internal redox mechanisms) (4)(5)(6). In this chapter, the concept of incorporating holes into the CuO_2 sheets of these high-T_c superconductors via the above mechanisms is also highlighted.

The intent of this chapter is twofold. One is to give brief structural descriptions of many of the copper-oxide superconductors. For in-depth information, the reader is referred to the original publications. The second is to provide detailed crystallographic information (lattice constants, positional and thermal parameters, space groups, etc.), and compositional data on many of the superconductors discussed. Also, calculated x-ray powder diffraction patterns for these same compounds are tabulated. It is hoped that such information will prove useful to the superconductivity researcher.

2.0 STRUCTURES OF THE PEROVSKITE-RELATED $YBa_2Cu_3O_7$, $YBa_2Cu_4O_8$, AND $Y_2Ba_4Cu_7O_{15}$ SUPERCONDUCTORS

2.1 123 Superconductor

The structure of superconducting, orthorhombic $YBa_2Cu_3O_{6+x}$ (x = 0.5-1.0) (7) contains double CuO_2 layers oriented in the (001) plane. Layers are composed of corner-sharing approximately planar CuO_4 moieties (see Figure 3). However, each copper atom is slightly displaced out of the sheet towards a fifth oxygen atom. This completes a square pyramidal oxygen environment around copper with the apices directed above and below the CuO_2 double layer. Yttrium cations reside between the copper-oxygen sheets of the double layer in 8-fold coordination with oxygen. Many of the other rare-earth cations can replace yttrium. Barium ions are found above and below the double Cu-O sheets. The $BaO/CuO_2/Y/CuO_2/BaO$ slabs are

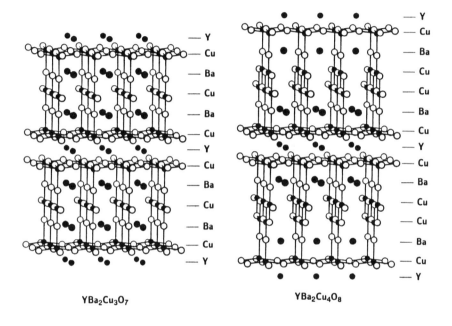

Figure 3: Orthorhombic structures of $YBa_2Cu_3O_7$ (left) containing single Cu-O chains between the copper-oxygen sheets, and $YBa_2Cu_4O_8$ (right) with double Cu-O chains between the copper-oxygen sheets. Metal atoms are shaded and Cu-O bonds are shown. The c axis is vertical, and the Cu-O chains run along the b axis direction which is oriented in and out of the plane of the drawing.

interconnected by a sheet of Cu and O atoms with variable composition CuO_x. These Cu atoms bond to the apical oxygen atoms of the square pyramidal CuO_5 units. The oxygen atoms of this central Cu-O sheet are ordered in such a way as to give strings of Cu and O atoms, with a variable oxygen stoichiometry, running along the b axis of the unit cell. T_c's are greater than 90 K when x is close to 1. Initially, it was believed that the Cu-O strings were necessary for superconductivity, but it is now known that they serve only as a means of oxidizing the CuO_2 sheets by allowing "extra" oxygen into the structure (8).

2.2 124 and 247 Superconductors

The CuO_2 sheets in orthorhombic $YBa_2Cu_4O_8$ are interconnected by a double layer of Cu-O chains (Figure 3) in comparison to the single Cu-O chains found in $YBa_2Cu_3O_7$. The double strings of the 124 superconductors are formed from CuO_4 units that share edges and corners along the b axis (9)(10). Yttrium can be replaced by a variety of rare earth cations and the T_c's range from 60 to 80 K. The variable oxygen stoichiometry of the single Cu-O chains in the 123 materials is not observed in the double chains of the 124 compounds.

$Y_2Ba_4Cu_7O_{15}$ can be viewed as an ordered 1:1 intergrowth of the 123 and 124 compounds ($YBa_2Cu_3O_7$ + $YBa_2Cu_4O_8$ = $Y_2Ba_4Cu_7O_{15}$) (11). Of course, the 123 portion of the structure can exhibit variable oxygen content in the single Cu-O chains, and T_c can therefore be changed. Onset of superconductivity can be as high as 90 K (12).

3.0 STRUCTURES OF THE PEROVSKITE/ROCK SALT SUPERCONDUCTORS

3.1 Lanthanum-Containing Superconductors

The body-centered tetragonal cell of $La_{2-x}M_xCuO_4$ (M = Sr, Ba, Na) and the related A-centered orthorhombic cell of unsubstituted La_2CuO_4 are given in Figure 2. For clarity, only the atoms of the tetragonal cell are shown along with the Cu atoms at the corners of the orthorhombic cell. This K_2NiF_4 structure-type contains sheets of corner-sharing CuO_6 octahedra that are oriented in the ab plane. The octahedra are axially elongated (along the c axis) to give four intralayer Cu-O bond lengths of ~1.9 Å, and two longer bonds, ~2.5 Å, perpendicular to the sheets. In the orthorhombic structure, the Cu-O sheets are buckled in a manner that gives CuO_6 octahedra that are alternately tilted (13). Lanthanum and M cations are in nine-coordination sites directly above and below the CuO_2 sheets. This $LaO/CuO_2/LaO$ slab is the "perovskite" part of the structure. The LaO layers occur in pairs to form La_2O_2 slabs separating the CuO_2 sheets. The La_2O_2 slabs can be viewed as the intergrown "rock salt" portion of the structure (see Figure 2). However, there are displacements of the La and O atoms from the ideal rock salt positions (discussed below).

Holes are created in the CuO_2 sheets by the substitution of lower-valent cations such as Sr^{2+} and Na^+ for La^{3+}. A shortening of the in-plane Cu-O bond is observed with increasing substitution because electrons are removed from antibonding bands upon oxidation. $La_{1.85}Sr_{0.15}CuO_4$ has the highest T_c in this family, 40 K. It is also possible to oxidize the CuO_2 sheets by preparing an oxygen-rich material, $La_2CuO_{4+\delta}$, under high oxygen pressure (14).

3.2 Bismuth-Containing Superconductors

The subcell structures of orthorhombic $Bi_2Sr_2CuO_6$ (15) and $Bi_2Sr_2CaCu_2O_8$ (16) are shown in Figure 4. They contain single or double sheets, respectively, of corner-sharing CuO_4 units oriented in the (001) plane, and each copper atom has one or two additional oxygen atoms positioned above or below the CuO_2 sheet to form axially elongated octahedra in $Bi_2Sr_2CuO_6$ and square pyramids in $Bi_2Sr_2CaCu_2O_8$. The latter compound contains calcium (and some strontium and bismuth) between the CuO_2 sheets in eight-fold coordination with oxygen. (As described above, this structural feature is found in $YBa_2Cu_3O_7$ where the copper-oxygen sheets are separated by yttrium cations.) Strontium cations reside just above and below the single and double Cu-O sheets. These Sr-Cu-(Ca)-O perovskite slabs are interconnected by a rock salt related double bismuth-oxygen layer. Until recently, the detailed atomic arrangement in the double Bi-O layers was not known. Analysis of the substructure (17) and of the incommensurate superstructure modulation (18) that exists in these compounds has given a much better understanding of the Bi-O layer structure. Also, a new compound, $Bi_2Sr_3Fe_2O_{9.2}$, isostructural with

492 Chemistry of Superconductor Materials

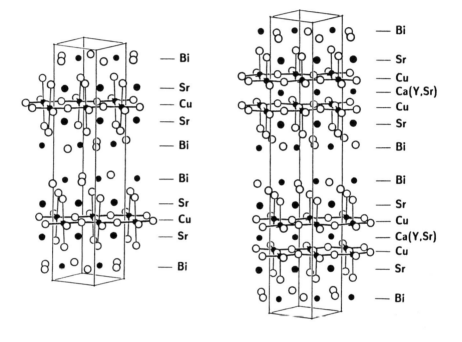

Figure 4: Subcell structures of $Bi_2Sr_2CuO_6$ (left) and $Bi_2Sr_2CaCu_2O_8$ (right). Metal atoms are shaded and Cu-O bonds are shown.

$Bi_2Sr_2CaCu_2O_8$ but containing a commensurate modulated superstructure has been prepared and characterized (19) and gives further insight into the structures of the Bi-O layers. This is discussed in further detail below. A material containing three copper-oxygen sheets, with ideal formula $Bi_2Sr_2Ca_2Cu_3O_{10}$, can also be made. The T_c's of the one, two, and three copper-sheet compounds are 10, 85, and 110 K, respectively.

If these phases were truly stoichiometric, the formal oxidation state of copper would be 2+ and they would not be superconductors. Because these compounds are hole-superconductors, there has to be some way of introducing holes into the copper-oxygen sheets. Two mechanisms for oxidizing the CuO_2 sheets which have been studied are the insertion of interstitial oxygen into the Bi-O sheets (17) and cation deficiencies on the Bi and Sr sites (4)(20). In addition, substitutional defects, such as partial exchange of the Ca^{2+} with Sr^{2+} or Bi^{3+} between the adjacent CuO_2 sheets has been established. Substitution of Sr^{2+} for Ca^{2+} does not affect the oxidation state of copper, but does push apart the CuO_2 sheets and this, in turn, could affect the superconducting transition temperature.

3.3 Thallium-Containing Superconductors

The homologous series $Tl_2Ba_2Ca_{n-1}Cu_nO_{2n+4}$ with n = 1, 2, 3 and 4 are known (5)(21). The structures of the first three members are shown in Figure 5. For n = 1 and 2, the structures are very similar to those of the analogous bismuth-strontium compounds discussed above, but possess higher crystallographic symmetry (tetragonal) and no obvious superstructures. For simplicity, the Tl ions in Figure 5 are shown in octahedral coordination with oxygen. There are, however, atomic displacements that occur in the thallium-oxygen double layers, but only the average atomic arrangement with tetragonal symmetry is observed (5)(15)(17)(22). This is discussed below in Section 5 on distortions. For the cases where n = 2 and 3, it has been shown that thallium deficiencies and/or Ca^{2+} substitution for Tl^{3+} exist (5)(22), and that thallium substitution for calcium also occurs. These defects provide a way of incorporating holes or changing the carrier concentration in the CuO_2 sheets. For n = 1, it is clear that all the metal sites can be fully occupied and a high T_c (~ 90 K) achieved (15)(23). There is evidence for interstitial oxygen in the Tl-O double layers of

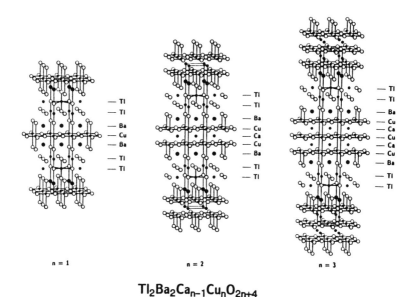

Figure 5: Structures of $Tl_2Ba_2Ca_{n-1}Cu_nO_{2n+4}$ (n = 1,2,3). Thallium atoms are shown on ideal (octahedral-type) sites for clarity (see text). Metal atoms are shaded and Cu-O bonds are shown.

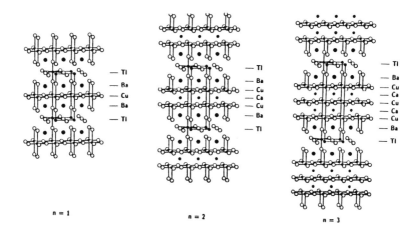

Figure 6: Structures of $TlBa_2Ca_{n-1}Cu_nO_{2n+3}$ (n = 1,2,3). Thallium atoms are shown on ideal (octahedral-type) sites for clarity (see text). Metal atoms are shaded and Cu-O bonds are shown.

$Tl_2Ba_2CuO_6$ (23) yielding a means of oxidizing the CuO_2 sheets. Also, recent band structure calculations (discussed below) show that there is considerable overlap of the Tl 6s-block bands with the Cu 3d x^2-y^2 bands providing a mechanism to transfer electrons from the CuO_2 sheets to the Tl-O layers.

Another closely related family of superconductors is represented by the formula $TlBa_2Ca_{n-1}Cu_nO_{2n+3}$ (n = 1, 2, 3, 4, 5). They contain single layers of Tl and O atoms that separate the perovskite-like Ba-Cu-Ca-O slabs (24)(25)(26)(27) (Figure 6). Distortions in the Tl-O sheets are also found in these compounds (26)(27). Note that if these phases were stoichiometric, copper would always have a formal oxidation state of greater than two. Therefore, the chemical composition of this homologous series allows the existence of holes in the copper-oxygen sheet.

3.4 Thallium-Lead Containing Superconductors

In the $TlBa_2Ca_{n-1}Cu_nO_{2n+3}$ family, it is possible to replace half the thallium with lead and all of the barium with strontium to give compounds of composition $(Tl_{0.5}Pb_{0.5})Sr_2Ca_{n-1}Cu_nO_{2n+3}$ with n = 1, 2, and 3 (28)(29)(30)(31). It appears that the Tl/Pb ratio deviates only slightly from the 1:1 stoichiometry (30)(31). The T_c's range from no superconductivity for the n = 1 compound (29) to 120 K for the n = 3 compound (30). The structures are essentially the same as those shown in Figure 6. As expected, atomic displacements of the Tl/Pb and O atoms of the Tl/PbO sheets are observed, and calcium is substituted by Tl and/or Pb and/or Sr. It is not possible to distinguish which heavy cations replace calcium at the substitution levels observed using conventional single crystal x-ray diffraction. (For example, the scattering factor curves for Tl and Pb are too similar to easily distinguish between them.)

4.0 STRUCTURES OF Pb-CONTAINING COPPER-BASED SUPERCONDUCTORS

Another family of superconductors is exemplified by $Pb_2Sr_2(Y_{0.5}Ca_{0.5})Cu_3O_8$. Yttrium can be replaced by most of the rare earths, and the rare-earth/calcium ratio can be varied (32)(33). The structure, shown in Figure 7, possesses double Cu-O sheets separated by Y and Ca ions analogous to the double sheets seen in $YBa_2Cu_3O_7$, $Bi_2Sr_{3-x}M_xCu_2O_8$ (M = Ca, Y), and $Tl_2Ba_2CaCu_2O_8$. Strontium ions reside directly above and below the double Cu-O sheets. A double Pb-O layer separates the Cu-O double layers. However, the Pb-O layers themselves sandwich a layer of copper atoms. These copper atoms, which are in approximately two-fold linear coordination with oxygen, are very similar to the Cu^{1+} ions in $YBa_2Cu_3O_6$ (34). Variable oxygen stoichiometry exists with oxygen going into the single copper sheet forming Cu-O chains as in the 123 materials. Lead appears to be present as Pb^{2+}, and, not unexpectedly, distortions in the Pb-O layers are found (33)(35). The highest T_c recorded so far in this family is 77 K.

The structure of another lead-containing compound related to $YBa_2Cu_3O_{6+x}$ was determined from crystals found as an "impurity" phase in the preparations of $Pb_2Sr_2(Y,Ca)Cu_3O_8$ (33). The composition was determined to be $(Y_{0.85}Ca_{0.15})Sr_2(Cu_{2.3}Pb_{0.7})O_{6.8}$. It has the 123 structure with Y and Ca between the CuO_2 sheets, strontium replacing barium, and Pb substituting for some of the copper in the "chain" sites. Atomic displacements of the atoms in the Pb-Cu-O layer are apparent (33). This phase does not superconduct.

5.0 DISTORTIONS IN THE ROCK SALT LAYERS AND THEIR EFFECT ON ELECTRONIC PROPERTIES

The most obvious common structural feature of all the high T_c copper-based superconductors is, of course, the CuO_2 sheets. The sheets form a fairly rigid, covalently-bonded two-dimensional network with short Cu-O bonds (~1.9 Å). The structures are constrained by the rigid Cu-O sheets, so a mismatch occurs between the Cu-O sheets and the M-O sheets (M = La, Bi, Tl, or Pb) for La_2CuO_4, the Tl/Ba, TlPb/Sr, Bi/Sr, and Pb/Sr materials (17). As drawn in Figures 5 and 6, the Tl-O coordination appears octahedral, but the Tl and O atoms in the sheets are actually displaced from the ideal rock salt positions (which are shown in the figures) to form a more favorable bonding environment. If the atoms were actually located on the ideal sites, then Tl-O bond lengths of ~2.7 Å would result. These

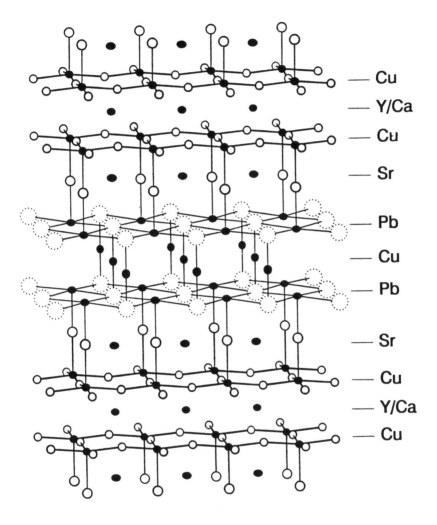

Figure 7: Structure of $Pb_2Sr_2(Y,Ca)Cu_3O_8$. Metal atoms are shaded, Cu-O and Pb-O bonds are shown. Oxygen atoms in the Pb sheets are dotted and shown on the "ideal" sites for clarity.

are too long for normal Tl-O bonds that should be ~ 2.2-2.3 Å in length. In the tetragonal unit cells of these materials, the displacements are seen as large thermal parameters for the Tl and O atoms in the sheets. Anisotropic refinements show that the large motion is always in the plane of the sheet. However, what we are seeing is an average of the true Tl-O layer structure in the tetragonal symmetry. The metal atoms are actually displaced to form two short and two long Tl-O bonds within the sheet. This is consistent with radial distribution function studies (36) showing that the Tl-O distances are actually shorter than those obtained when the atoms are placed on the ideal positions - 2.3 vs 2.7 Å. A similar situation exists for the Bi-Sr and Pb-Sr materials.

In the orthorhombic structures of the Bi-Sr compounds, or in orthorhombic $Tl_2Ba_2CuO_6$ (37), or even in orthorhombic La_2CuO_4 (17), we begin to see a clear displacement of the atoms in the M-O layers. Figure 8 shows how the atoms have moved in the Bi-O layer of the Bi-Sr compounds, but the same situation is seen in all of the orthorhombic superconductors. Figure 8A shows "ladder-like" structures of Bi and O atoms with Bi-O bonds of ~2.2 Å across the ladder as expected. At the same time, the Bi-O distance between the ladders is now ~3.2 Å creating a non-bonded gap. In the ladder structure, there are still Bi-O bonds that are too long, ~2.7 Å, along the ladder. In the structural analyses of the Bi-Sr phases, it is clear from the thermal parameters and from the electron density maps that the atoms do not actually reside on the ideal positions of the orthorhombic space group, but are displaced to one side or the other side of the mirror planes shown in Figure 8A. From the ladder structure, therefore, two models can be constructed which contain M-O bond lengths of ~2.1-2.3 Å. Figure 8B shows how island-like structures of Bi and O atoms can be made, while Figure 8C shows how chain-like structures can be constructed. Each M atom forms two short and two long bonds to oxygen in the plane of the sheet. These pictures are still idealized because they do not take into account any superstructure. However, the overall structure is similar to that proposed for the Tl-O layers in $Tl_2Ba_2CaCu_2O_8$ (36), and the Bi-O layers in $Bi_2Sr_3Fe_2O_{9.2}$ (19) and $Bi_2Sr_2CaCu_2O_8$ (18). When the atoms are clustered to form islands or chains, there are then sites available for extra oxygen as shown by the dotted circles (17).

Tight-binding band electronic structure calculations have been

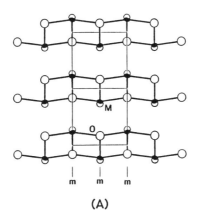

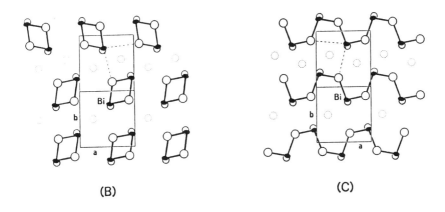

Figure 8: Atomic displacements in the M-O (M = La, Bi, Tl, Pb) rock salt-like layers of the copper-based superconductors. The view is perpendicular to the MO sheet. (A): Ladder-like arrangement of M and O atoms observed in orthorhombic La_2CuO_4, $Bi_2Sr_2Ca_{n-1}Cu_nO_{2n+4}$, and $Tl_2Ba_2CuO_6$. (B) and (C): Atomic arrangement of Bi and O atoms as islands and chains that form when the atoms are displaced from the mirror positions shown in (A) (see text).

performed on the various structures of the Bi-O (38) and Tl-O (6) double layers discussed above. For the Bi-O layer materials, previous band structure calculations (39) using the ideal rock salt-type positions showed the Bi 6p bands overlapping the Cu 3d bands implying that Bi^{3+} oxidizes Cu^{2+}. However, the position of the Bi 6p-block bands, with respect to the Fermi level, is strongly affected by distortion around the Bi atoms. Results show that the bottom of the Bi 6p-block bands actually lies more than 1.0 eV above the Fermi level, so that the formal oxidation state appropriate for Bi is no lower than +3 as expected.

When the distortions in the Tl-O sheets are taken into account, the *double* layers are found to have the bottom of the Tl 6s-block bands significantly below the Fermi level, forming a large electron pocket to remove electrons from the x^2-y^2 bands of the CuO_2 layers (6). Thus, in contrast to the Bi-system discussed above, Tl^{3+} can oxidize the CuO_2 layers resulting in an intermediate oxidation state of Tl between +1 and +3. In sharp contrast, the Tl 6s-block bands of the Tl-O *single* layers are found to lie well above the Fermi level (Tl^{3+}), regardless of whether or not the distortions in the sheets are taken into consideration (6). Therefore, the single layers do not remove electrons from the x^2-y^2 bands of the CuO_2 layers. This is consistent with the observation that stoichiometric $TlBa_2Ca_{n-1}Cu_nO_{2n+3}$ already has a high formal oxidation state for copper, so that further hole generation to produce superconductivity is not necessary.

6.0 CORRELATIONS OF T_c WITH IN-PLANE Cu-O BOND LENGTH

Oxidation of the CuO_2 layer removes electrons from the x^2-y^2 bands which have antibonding character in the in-plane Cu-O bonds. Therefore, as the number of holes (n_H) increases, the in-plane Cu-O bond length (r_{Cu-O}) is shortened. In addition to this electronic factor, the in-plane r_{Cu-O} is also controlled by a nonelectronic factor (i.e., steric strain) associated with the cations located at the 9-coordinate sites adjacent to the CuO_2 layers (e.g., La, Sr, Ba) (40). With the increasing size of the 9-coordinate site cations, the in-plane Cu-O bond is lengthened to reduce the extent of the resulting steric strain.

Plots of T_c vs in-plane r_{Cu-O} for the series of p-type cuprate superconductors are grouped into three classes distinguished by the size of the 9-coordinate site cations (that is, La-, Sr- and Ba-classes) because of the combined electronic and nonelectronic effects. Every class of the T_c vs in-plane r_{Cu-O} plot shows a maximum, so that every class of the p-type cuprate superconductors possesses an optimum hole density for which the T_c is maximum (40).

The bond valence s_i of a bond i is identified as $s_i = \exp[(r_o - r_i)/0.37]$ (41) where r_i is the length of the bond i, and r_o is the constant that depends upon the atoms constituting the bond. For a metal atom surrounded by several identical ligands with bond lengths r_i, its bond valence sum (BVS) is given by the sum of all the bond valences Σs_i. The BVS of the metal atom is a measure of the total amount of electrons it loses, that is, the formal oxidation state of the atom. The lengths of the chemical bonds in crystalline materials are determined by the electronic factor, which reflects the amounts of electrons in the bonds, and also by the nonelectronic factor which alters the bond lengths without changing the amounts of electrons in the bonds. Thus, BVS values cannot be used as a measure of formal oxidation states unless the steric factor is constant (42).

By definition, the bond valence of any given bond should increase with the shortening of its length. Since the in-plane r_{Cu-O} decreases with increasing n_H, the BVS of an in-plane Cu atom obtained only from its in-plane Cu-O bonds (referred to as the in-plane BVS) increases with increasing n_H. The T_c vs in-plane BVS plots (42) are grouped into La-, Sr- and Ba-classes just as in the case of the T_c vs in-plane r_{Cu-O} correlation (described above) (40). With the larger 9-coordination site cation, the in-plane Cu-O bond is more stretched out so that the maximum of the corresponding T_c vs in-plane BVS plot is shifted toward the direction of smaller BVS values. Within each class, the steric effect of the 9-coordinate site cation is fairly constant so that a change in the in-plane BVS value becomes a reliable measure of the change in n_H (42). For details of these correlations, the reader should see References 40 and 42.

7.0 TABLES OF CRYSTALLOGRAPHIC INFORMATION

At the end of the references section can be found reference

Tables 1 through 7 giving positional and isotropic thermal parameters for most of the compounds discussed in this chapter. These data are taken, for the most part, from the literature, but, for a few materials that have not been structurally characterized, calculated positions are given. The tables also include lattice constants, space groups, and compositional data. Table 8 gives calculated x-ray powder diffraction information (2θ (Cu), d-spacing, hkl, and intensity) for the same oxide compounds. In regard to the diffraction patterns, it should be remembered that preferred orientation and absorption effects, and cation substitutions will make the experimentally observed intensities differ from the calculated ones. Additional information on the crystal structures of high-T_c oxides can be found in a recent review (48).

Table 1. Positional and Isotropic Thermal ($Å^2$) Parameters for $YBa_2Cu_3O_6$, $YBa_2Cu_3O_7$, $DyBa_2Cu_4O_8$, and $Y_2Ba_4Cu_7O_{14+x}$

		$YBa_2Cu_3O_6$[a]	$YBa_2Cu_3O_7$[b]	$DyBa_2Cu_4O_8$[c]	$Y_2Ba_4Cu_7O_{14+x}$[d]
Ba(1)	site	2h	2t	4j	4j
	x	0.50	0.50	0.50	0.50
	y	0.50	0.50	0.50	0.50
	z	0.1952(4)	0.1843(3)	0.13496(3)	0.4310(3)
	B	0.9(1)	0.54(5)	0.61(2)	1.00(4)
Ba(2)	site	- -	- -	- -	4j
	x	- -	- -	- -	0.50
	y	- -	- -	- -	0.50
	z	- -	- -	- -	0.18797(2)
	b	- -	- -	- -	0.51(3)
Y(Dy)	site	1d	1h	2c	4j
	x	0.50	0.50	0.50	0.50
	y	0.50	0.50	0.50	0.50
	z	0.50	0.50	0.00	0.11545(4)
	B	0.6(1)	0.46(4)	0.46(2)	0.06(5)
Cu(1)	site	1a	1a	4i	2a
	x	0.00	0.00	0.00	0.00
	y	0.00	0.00	0.00	0.00
	z	0.00	0.00	0.21312(5)	0.00
	B	1.0(1)	0.50(5)	0.66(3)	2.9(2)
Cu(2)	site	2g	2q	4i	4i
	x	0.00	0.00	0.00	0.00
	y	0.00	0.00	0.00	0.00
	z	0.3605(3)	0.3556(1)	0.06179(6)	0.08293(5)
	B	0.3(1)	0.29(4)	0.52(3)	0.45(7)
Cu(3)	site	- -	- -	- -	4i
	x	- -	- -	- -	0.00
	y	- -	- -	- -	0.00
	z	- -	- -	- -	0.14831(5)
	B	- -	- -	- -	0.38(7)
Cu(4)	site	- -	- -	- -	4i
	x	- -	- -	- -	0.00
	y	- -	- -	- -	0.00
	z	- -	- -	- -	0.23012(5)
	B	- -	- -	- -	0.46(7)
O(1)	site	2g	1e	4i	4i
	x	0.00	0.00	0.00	0.00
	y	0.00	0.50	0.00	0.00
	z	0.1518(4)	0.00	0.1460(3)	0.0353(8)
	B	1.5(1)	1.35(5)	1.2(2)	5.3(18)

(continued)

Table 1. (Continued)

		YBa$_2$Cu$_3$O$_6$[a]	YBa$_2$Cu$_3$O$_7$[b]	DyBa$_2$Cu$_4$O$_8$[c]	Y$_2$Ba$_4$Cu$_7$O$_{14+x}$[d]
O(2)	site	4 i	2 s	4 j	4 j
	x	0.00	0.50	0.50	0.50
	y	0.50	0.00	0.00	0.00
	z	0.3794(2)	0.3773(2)	0.0526(3)	0.0871(3)
	B	0.7(1)	0.56(5)	0.7(2)	0.5(2)
O(3)	site	- -	2 r	4 i	4 i
	x	- -	0.00	0.00	0.00
	y	- -	0.50	0.50	0.50
	z	- -	0.3789(3)	0.0532(3)	0.0865(3)
	B	- -	0.37(5)	0.8(2)	0.7(2)
O(4)	site	- -	2q	4 i	4 j
	x	- -	0.00	0.00	0.50
	y	- -	0.00	0.50	0.00
	z	- -	0.1584(2)	0.2186(3)	0.1430(3)
	B	- -	0.67(5)	1.2(2)	0.7(2)
O(5)	site	- -	- -	- -	4 i
	x	- -	- -	- -	0.00
	y	- -	- -	- -	0.50
	z	- -	- -	- -	0.1432(3)
	B	- -	- -	- -	0.8(2)
O(6)	site	- -	- -	- -	4 i
	x	- -	- -	- -	0.00
	y	- -	- -	- -	0.00
	z	- -	- -	- -	0.1937
	B	- -	- -	- -	0.4(2)
O(7)	site	- -	- -	- -	4 i
	x	- -	- -	- -	0.00
	y	- -	- -	- -	0.50
	z	- -	- -	- -	0.2328(3)
	B	- -	- -	- -	0.8(2)
a(Å)		3.8519(1)	3.8231(1)	3.8463(3)	3.851(1)
b(Å)		- -	3.8864(1)	3.8726(3)	3.869(1)
c(Å)		11.8037(4)	11.6807(2)	27.237(2)	50.29(2)
Space group		P4/mmm	Pmmm	Ammm	Ammm

[a]Reference 34.
[b]Reference 7.
[c]Reference 10.
[d]Reference 11.

Table 2. Positional and Isotropic Thermal ($Å^2$) Parameters for La_2CuO_4, $La_{2-x}Sr_xCuO_4$, and $La_{1.8}Na_{0.2}CuO_4$.

		La_2CuO_4[a]	$La_{1.92}Sr_{0.08}CuO_4$[b]	$La_{1.8}Na_{0.2}CuO_4$[c]
La/M	site	8f	4e	4e
	x	0.007	0.00	0.00
	y	0.00	0.00	0.00
	z	0.362	0.36094(2)	0.3608(2)
	B	--	0.58(1)	0.6(1)
Cu(1)	site	4a	2a	2a
	x	0.00	0.00	0.00
	y	0.00	0.00	0.00
	z	0.00	0.00	0.00
	B	--	0.32(1)	0.3(1)
O(1)	site	8e	4c	4c
	x	0.25	0.00	0.00
	y	0.25	0.50	0.50
	z	0.007	0.00	0.00
	B	--	0.99(7)	0.8(1)
O(2)	site	8f	4e	4e
	x	0.969	0.00	0.00
	y	0.00	0.00	0.00
	z	0.187	0.1825(3)	0.1830(2)
	B	--	1.9(1)	1.3(1)
a(Å)		5.406	3.7966(7)	3.7796(1)
b(Å)		5.370	--	--
c(Å)		13.15	13.186(4)	13.187(4)
Space Group		Abma[d]	I4/mmm	I4/mmm

[a]Reference 13.
[b]Reference 43.
[c]Reference 44.
[d]Nonconventional setting of Cmca (No. 64): (0,0,0; 0,1/2,1/2) + [x,y,z; 1/2+x,-y,1/2-z; 1/2-x,-y,1/2+z; -x,y,-z].

Table 3. Positional[a] and Isotropic Thermal (Å^2) Parameters for the Subcells of $Bi_2Sr_2Ca_{n-1}Cu_nO_{2n+4}$ (n = 1,2,3).

		$Bi_2Sr_2CuO_6$[b]	$Bi_2Sr_2CaCu_2O_8$[c]	$Bi_2Sr_2Ca_2Cu_3O_{10}$[d]
Bi(1)	site	8l	16m(8l)[e]	8l
	x	0.50	0.4491(7)	0.50
	y	0.2244(4)	0.2276(3)	0.225
	z	0.0659(2)	0.0522(1)	0.0435
	B	5.7(2)	2.7(1)	- -
Sr(1)	site	8l	8l	8l
	x	0.00	0.00	0.00
	y	0.2518(7)	0.2528(7)	0.252
	z	0.1790(4)	0.1409(1)	0.1052
	B	2.7(2)	3.2(1)	- -
Cu(1)	site	4e	8l	4f
	x	0.00	0.50	0.50
	y	0.75	0.2511(8)	0.25
	z	0.25	0.1972(2)	0.25
	B	4.2(5)	2.5(1)	- -
Cu(2)	site	- -	- -	8l
	x	- -	- -	0.50
	y	- -	- -	0.25
	z	- -	- -	0.1620
	B	- -	- -	- -
Ca(1)	site	- -	4e[f]	8l
	x	- -	0.00	0.0
	y	- -	0.25	0.25
	z	- -	0.25	0.2037
	B	- -	2.3(2)	- -
O(1)	site	8g	8g	8g
	x	0.25	0.75	0.75
	y	0.50	0.00	0.50
	z	0.252(2)	0.1971(9)	0.25
	B	0.8(5)	1.5(6)	- -
O(2)	site	16m(8l)[e]	8g	8g
	x	0.59(1)	0.25	0.75
	y	0.275(8)	0.50	0.00
	z	0.149(3)	0.203(1)	0.1625
	B	3(1)	3.0(8)	- -
O(3)	site	16m(8l)[e]	16m(8l)[e]	8g
	x	0.03(4)	0.435(8)	0.25
	y	0.17(1)	0.257(8)	0.50
	z	0.065(4)	0.119(1)	0.1625
	B	9(3)	3.7(9)	- -

(continued)

Table 3. (Continued)

		$Bi_2Sr_2CuO_6$[b]	$Bi_2Sr_2CaCu_2O_8$[c]	$Bi_2Sr_2Ca_2Cu_3O_{10}$[d]
O(4)	site	- -	16m(8l)[e]	8l
	x	- -	0.13(1)	0.50
	y	- -	0.20(2)	0.25
	z	- -	0.067(3)	0.0945
	B	- -	13(3)	- -
O(5)	site	- -	- -	16m(8l)[e]
	x	- -	- -	0.10
	y	- -	- -	0.20
	z	- -	- -	0.0435
	B	- -	- -	- -
a(Å)		5.362(3)	5.399(1)	5.413(3)
b(Å)		5.374(1)	5.414(1)	5.413(3)
c(Å)		24.622(6)	30.90(2)	37.10(1)

[a] Space group Amaa, nonconventional setting of Cccm (No. 66): (0,0,0; 0,1/2,1/2) + (x,y,z; 1/2+x, -y, z; -x,y,z; 1/2+x, y, -z).
[b] Reference 15.
[c] References 16 and 28.
[d] Calculated atomic positions; lattice parameters taken from Reference 45.
[e] Atom displaced from mirror plane (8l sites) into two sites (16m) each with half occupancy.
[f] This Ca site contains some Sr and/or Bi. It was modeled as 0.67 Ca + 0.33 Sr.

Table 4. Positional[a] and Isotropic Thermal ($Å^2$) Parameters for $Tl_2Ba_2Ca_{n-1}Cu_nO_{2n+4}$ (n = 1,2,3,4).

		$Tl_2Ba_2CuO_6$[b]	$Tl_2Ba_2CaCu_2O_8$[c]	$Tl_2Ba_2Ca_2Cu_3O_{10}$[d]	$Tl_2Ba_2Ca_3Cu_4O_{12}$[e]
Tl(1)	site	4e	4e[f]	4e[g]	4e
	x	0.50	0.50	0.50	0.50
	y	0.50	0.50	0.50	0.50
	z	0.20265(2)	0.21359(2)	0.2201(1)	0.2242
	B	1.3(1)	1.7(1)	2.1(1)	- -
Ba(1)	site	4e	4e	4e	4e
	x	0.00	0.00	0.00	0.00
	y	0.00	0.00	0.00	0.00
	z	0.08301(3)	0.12179(3)	0.1448(1)	0.1593
	B	0.5(1)	0.8(1)	0.6(1)	- -
Cu(1)	site	2b	4e	2b	4e
	x	0.50	0.50	0.50	0.50
	y	0.50	0.50	0.50	0.50
	z	0.00	0.0540(1)	0.00	0.0362
	B	0.4(1)	0.4(1)	0.3(1)	- -
Cu(2)	site	- -	- -	4e	4e
	x	- -	- -	0.50	0.50
	y	- -	- -	0.50	0.50
	z	- -	- -	0.0896(2)	0.1124
	B	- -	- -	0.7(1)	- -
Ca(1)	site	- -	2a[h]	4e[i]	2a
	x	- -	0.00	0.00	0.00
	y	- -	0.00	0.00	0.00
	z	- -	0.00	0.0463(2)	0.00
	B	- -	0.6(1)	1.2(1)	- -
Ca(2)	site	- -	- -	- -	4e
	x	- -	- -	- -	0.50
	y	- -	- -	- -	0.50
	z	- -	- -	- -	0.5755
	B	- -	- -	- -	- -
O(1)	site	4c	8g	4c	8g
	x	0.00	0.00	0.00	0.00
	y	0.50	0.50	0.50	0.50
	z	0.00	0.0531(2)	0.00	0.5362
	B	0.7(1)	0.8(1)	0.9(9)	- -
O(2)	site	4e	4e	8g	8g
	x	0.50	0.50	0.50	0.50
	y	0.50	0.50	0.00	0.00
	z	0.1168(4)	0.1461(3)	0.0875(6)	0.6116
	B	1.1(1)	1.5(2)	1.2(6)	- -

(continued)

Table 4. (Continued)

		$Tl_2Ba_2CuO_6$[b]	$Tl_2Ba_2CaCu_2O_8$[c]	$Tl_2Ba_2Ca_2Cu_3O_{10}$[d]	$Tl_2Ba_2Ca_3Cu_4O_{12}$[e]
O(3)	site	16n(4e)[j]	16n(4e)[j]	4e	4e
	x	0.595(5)	0.604(9)	0.50	0.50
	y	0.50	0.50	0.50	0.50
	z	0.2889(5)	0.2815(7)	0.159(1)	0.1752
	B	0.4(3)	3(1)	2.3(9)	- -
O(4)	site	- -	- -	4e	4e
	x	- -	- -	0.50	0.50
	y	- -	- -	0.50	0.50
	z	- -	- -	0.272(1)	0.7291
	B	- -	- -	2.1(1)	- -
a(Å)		3.866(1)	3.8550(6)	3.8503(6)	3.856(1)
c(Å)		23.239(6)	29.318(4)	35.88(3)	42.051(4)

[a]Space group I4/mmm.
[b]Reference 15.
[c]Reference 22.
[d]Reference 5.
[e]Calculated atomic positions; lattice parameters taken from reference 28.
[f]Site composition fixed at 0.90 Tl + 0.10 Ca. However, site can also be modeled as containing only thallium with 7.5% vacancies.
[g]Site composition fixed at 0.85 Tl + 0.15 Ca. However, site can also be modeled as containing only thallium with 11.3% vacancies.
[h]Site composition fixed at 0.90 Ca + 0.10 Tl.
[i]Site composition fixed at 0.88 Ca + 0.12 Tl.
[j]Atom displaced from 4e sites to 16n sites with quarter occupancy.

Table 5. Positional[a] and Isotropic Thermal ($Å^2$) Parameters for $TlBa_2Ca_{n-1}Cu_nO_{2n+3}$ (n = 1,2,3,4).

		$TlBa_2CuO_5$[b]	$TlBa_2CaCu_2O_7$[c]	$TlBa_2Ca_2Cu_3O_9$[d]	$TlBa_2Ca_3Cu_4O_{11}$[e]
Tl(1)	site	1a	1a	4l(1a)[f]	1a
	x	0.00	0.00	0.085(2)	0.00
	y	0.00	0.00	0.00	0.00
	z	0.00	0.00	0.00	0.00
	B	- -	1.1(2)	1.7(1)	- -
Ba(1)	site	2h	2h	2h[g]	2h
	x	0.50	0.50	0.50	0.50
	y	0.50	0.50	0.50	0.50
	z	0.2868	0.2158(1)	0.1729	0.1447
	B	- -	0.8(1)	0.6(1)	- -
Cu(1)	site	1b	2g	1b	2g
	x	0.00	0.00	0.00	0.00
	y	0.00	0.00	0.00	0.00
	z	0.50	0.3745(2)	0.50	0.2504
	B	- -	0.5(1)	0.4(1)	- -
Cu(2)	site	- -	- -	2g	2g
	x	- -	- -	0.00	0.00
	y	- -	- -	0.00	0.00
	z	- -	- -	0.2991(2)	0.4187
	B	- -	- -	0.4(1)	- -
Ca(1)	site	- -	1d[h]	2h[i]	1d
	x	- -	0.50	0.50	0.50
	y	- -	0.50	0.50	0.50
	z	- -	0.50	0.3953(3)	0.50
	B	- -	0.7(2)	0.7(1)	- -
Ca(2)	site	- -	- -	- -	2h
	x	- -	- -	- -	0.50
	y	- -	- -	- -	0.50
	z	- -	- -	- -	0.3346
	B	- -	- -	- -	- -
O(1)	site	2e	4i	2e	4i
	x	0.00	0.00	0.00	0.00
	y	0.50	0.50	0.50	0.50
	z	0.50	0.3790(8)	0.50	0.2525
	B	- -	0.7(2)	0.9(3)	- -
O(2)	site	2g	2g	4i	4i
	x	0.00	0.00	0.00	0.00
	y	0.00	0.00	0.50	0.50
	z	0.2162	0.158(1)	0.3019(7)	0.4187
	B	- -	0.9(2)	1.1(2)	- -

(continued)

Structural Details of the High T_c Copper-Based Superconductors 511

Table 5. (Continued)

		$TlBa_2CuO_5^b$	$TlBa_2CaCu_2O_7^c$	$TlBa_2Ca_2Cu_3O_9^d$	$TlBa_2Ca_3Cu_4O_{11}^e$
O(3)	site	1c	1c	2g	2g
	x	0.50	0.50	0.00	0.00
	y	0.50	0.50	0.00	0.00
	z	0.00	0.00	0.1277(11)	0.1069
	B	- -	2.6(6)	0.6(3)	- -
O(4)	site	- -	- -	1c	1c
	x	- -	- -	0.50	0.50
	y	- -	- -	0.50	0.50
	z	- -	- -	0.00	0.00
	B	- -	- -	2.6(7)	- -
a(Å)		3.85	3.8566(4)	3.853(1)	3.850
c(Å)		9.59	12.754(2)	15.913(4)	19.01

[a]Space group P4/mmm.
[b]Calculated atomic positions; lattice parameters taken from reference 29.
[c]Reference 26.
[d]Reference 27.
[e]Calculated atomic positions; lattice parameters taken from reference 25.
[f]Atom displaced from 1a sites to 4l sites with quarter occupancy.
[g]Occupancy factor refined to 0.94(1) atom/site.
[h]Site composition fixed at 0.83 Ca + 0.17 Tl.
[i]Site composition fixed at 0.95 Ca + 0.05 Tl.

Table 6. Positional[a] and Isotropic Thermal ($Å^2$) Parameters for $(Tl_{0.5}Pb_{0.5})Sr_2Ca_{n-1}Cu_nO_{2n+3}$ (n = 1,2,3).

		$(TlPb)Sr_2CuO_5$[b]	$(TlPb)Sr_2CaCu_2O_7$[c]	$(TlPb)Sr_2Ca_2Cu_3O_9$[d]
Tl/Pb	site	1a	4l(1a)[e]	4l(1a)[e]
	x	0.00	0.0737(6)	0.067(3)
	y	0.00	0.00	0.00
	z	0.00	0.00	0.00
	B	- -	0.4(1)	1.8(2)
Sr(1)	site	2h	2h	2h
	x	0.50	0.50	0.50
	y	0.50	0.50	0.50
	z	0.2950	0.2162(1)	0.1709(2)
	B	- -	0.7(1)	1.6(1)
Cu(1)	site	1b	2g	1b
	x	0.00	0.00	0.00
	y	0.00	0.00	0.00
	z	0.50	0.3638(2)	0.50
	B	- -	0.5(1)	1.2(1)
Cu(2)	site	- -	- -	2g
	x	- -	- -	0.00
	y	- -	- -	0.00
	z	- -	- -	0.2868(2)
	B	- -	- -	1.1(1)
Ca(1)	site	- -	1d[f]	2h[g]
	x	- -	0.50	0.50
	y	- -	0.50	0.50
	z	- -	0.50	0.3928(3)
	B	- -	0.5(1)	1.5(1)
O(1)	site	2e	4i	2e
	x	0.00	0.00	0.00
	y	0.50	0.50	0.50
	z	0.50	0.3696(6)	0.50
	B	- -	1.0(2)	1.2(3)
O(2)	site	2g	2g	4i
	x	0.00	0.00	0.00
	y	0.00	0.00	0.50
	z	0.2236	0.1655(9)	0.2924(7)
	B	- -	0.8(2)	1.5(2)
O(3)	site	1c	4n(1c)[h]	2g
	x	0.50	0.59(1)	0.00
	y	0.50	0.50	0.00
	z	0.00	0.00	0.131(1)
	B	- -	1.2(7)	1.6(3)

(continued)

Table 6. (Continued)

		(TlPb)Sr$_2$CuO$_5$[b]	(TlPb)Sr$_2$CaCu$_2$O$_7$[c]	(TlPb)Sr$_2$Ca$_2$Cu$_3$O$_9$[d]
O(4)	site	- -	- -	4n(1c)[h]
	x	- -	- -	0.60(2)
	y	- -	- -	0.50
	z	- -	- -	0.00
	B	- -	- -	3(1)
a(Å)		3.731	3.791(2)	3.808(1)
c(Å)		8.98	12.06(1)	15.232(7)

[a] Space group P4/mmm.
[b] Lattice parameters from Reference 29. Positional parameters are for TlSr$_2$CuO$_5$ taken from Reference 46.
[c] Reference 28.
[d] Reference 30.
[e] Atom displaced from 1a sites to 4l sites with quarter occupancy.
[f] Site composition fixed at 0.92 Ca + 0.08 Tl/Pb. However, site can also be modeled as 0.73 Ca + 0.27 Sr.
[g] Site composition fixed at 0.94 Ca + 0.06 Tl/Pb. However, site can also be modeled as 0.80 Ca + 0.20 Sr.
[h] Atom displaced from 1c sites to 4n sites with quarter occupancy.

Table 7. Positional[a] and Isotropic Thermal ($Å^2$) Parameters for $Pb_2Sr_2(Y,Ca)Cu_3O_8$ and $(Y,Ca)Sr_2(Cu,Pb)_3O_{6.8}$.

		$Pb_2Sr_2(Y,Ca)Cu_3O_8$[b]	$(Y,Ca)Sr_2(Cu,Pb)_3O_{6.8}$[b]
Pb(1)/Cu(1)	site	2h[c]	1a[d]
	x	0.50	0.00
	y	0.50	0.00
	z	0.3874(1)	0.00
	B	0.9(1)	2.2(1)
Y(1)	site	1a[e]	1d[f]
	x	0.00	0.50
	y	0.00	0.50
	z	0.00	0.50
	B	0.9(3)	0.8(1)
Sr(1)	site	2g	2h
	x	0.00	0.50
	y	0.00	0.50
	z	0.2206(3)	0.2072(3)
	B	1.1(2)	1.2(1)
Cu(1)	site	2h	--
	x	0.50	--
	y	0.50	--
	z	0.1070(4)	--
	B	0.9(2)	--
Cu(2)	site	1b	2g[g]
	x	0.00	0.00
	y	0.00	0.00
	z	0.50	0.3612(4)
	B	0.9(3)	0.8(1)
O(1)	site	4i	4i
	x	0.00	0.00
	y	0.50	0.50
	z	0.096(1)	0.376(1)
	B	1.1(4)	1.2(3)
O(2)	site	2h	2g
	x	0.50	0.00
	y	0.50	0.00
	z	0.249(2)	0.150(3)
	B	1.7(6)	4.6(9)
O(3)	site	8s(2g)[h]	4n(1c)[i]
	x	0.12(1)	0.33(2)
	y	0.00	0.50
	z	0.384(3)	0.00
	B	1.6(12)	0.2(9)

(continued)

Table 7. (Continued)

a(Å)	3.813(2)	3.818(1)
c(Å)	15.76(1)	11.829(2)

[a] Space group P4/mmm
[b] Reference 33.
[c] Site fully occupied by Pb.
[d] Site composition fixed 0.63 Pb + 0.37 Cu.
[e] Site composition fixed at 0.75 Y + 0.25 Ca.
[f] Site composition fixed at 0.85 Y + 0.15 Ca.
[g] Site composition fixed at 0.97 Cu + 0.03 Pb.
[h] Atom displaced from 2g sites to 8s sites with quarter occupancy.
[i] Atom displaced from partially occupied 1c sites to 4n sites with a quarter of the refined occupancy.

Table 8. Calculated x-ray powder diffraction patterns for the compounds in Tables 1 through 7. Patterns were calculated using LAZY PULVERIX (47) with copper radiation ($\lambda = 1.5418$ Å). Not all lines are given. Calculated intensities less than 1 are not included.

$YBa_2Cu_3O_6$

2θ	d	h	k	l	I_{calc}
7.49	11.804	0	0	1	11
15.01	5.902	0	0	2	5
22.60	3.935	0	0	3	5
23.09	3.852	1	0	0	9
27.65	3.226	1	0	2	14
32.53	2.753	1	0	3	100
32.88	2.724	1	1	0	60
33.77	2.654	1	1	1	1
36.33	2.473	1	1	2	3
38.12	2.361	0	0	5	8
38.43	2.343	1	0	4	15
40.27	2.240	1	1	3	12
45.04	2.013	1	0	5	1
45.31	2.002	1	1	4	3
46.14	1.967	0	0	6	6
47.19	1.926	2	0	0	32
49.80	1.831	2	0	2	1
51.21	1.784	1	1	5	6
52.21	1.752	1	0	6	1
52.93	1.730	2	0	3	2
53.17	1.723	2	1	0	2
54.41	1.686	0	0	7	3
55.57	1.654	2	1	2	4
57.81	1.595	1	1	6	12
58.49	1.578	2	1	3	34
59.88	1.545	1	0	7	4

(continued)

YBa$_2$Cu$_3$O$_7$

2θ	d	h	k	l	I$_{calc}$
7.57	11.681	0	0	1	3
15.17	5.840	0	0	2	3
22.84	3.894	0	0	3	8
22.88	3.886	0	1	0	10
23.27	3.823	1	0	0	6
24.13	3.688	0	1	1	1
27.57	3.236	0	1	2	5
27.89	3.199	1	0	2	10
30.61	2.920	0	0	4	1
32.55	2.751	0	1	3	90
32.83	2.728	1	0	3	100
32.86	2.725	1	1	0	97
33.77	2.654	1	1	1	2
36.38	2.470	1	1	2	5
38.54	2.336	0	0	5	12
38.56	2.335	0	1	4	6
38.80	2.321	1	0	4	4
40.40	2.233	1	1	3	27
45.29	2.002	0	1	5	1
45.52	1.993	1	1	4	1
46.66	1.947	0	0	6	17
46.75	1.943	0	2	0	30
47.57	1.912	2	0	0	29
51.52	1.774	1	1	5	7
52.58	1.741	0	1	6	2
52.64	1.739	0	2	3	2
52.77	1.735	1	0	6	1
52.85	1.732	1	2	0	2
53.39	1.716	2	0	3	2
53.41	1.715	2	1	0	2

(continued)

2θ	d	h	k	l	
55.03	1.669	0	0	7	2
55.31	1.661	1	2	2	3
55.86	1.646	2	1	2	2
58.24	1.584	1	1	6	31
58.30	1.583	1	2	3	36
58.83	1.570	2	1	3	32

$DyBa_2Cu_4O_8$

2θ	d	h	k	l	I_{calc}
6.49	13.619	0	0	2	20
13.00	6.809	0	0	4	10
19.55	4.540	0	0	6	2
23.13	3.846	1	0	0	1
23.20	3.834	0	1	1	9
24.04	3.701	1	0	2	13
25.00	3.562	0	1	3	5
26.17	3.405	0	0	8	4
26.62	3.349	1	0	4	36
28.27	3.157	0	1	5	3
30.46	2.934	1	0	6	37
32.62	2.745	0	1	7	100
32.88	2.724	0	0	10	5
32.99	2.715	1	1	1	96
34.31	2.614	1	1	3	28
35.20	2.549	1	0	8	67
37.72	2.385	0	1	9	10
39.71	2.270	0	0	12	9
40.37	2.234	1	1	7	17
43.37	2.086	0	1	11	1
44.71	2.027	1	1	9	8
46.69	1.946	0	0	14	20
46.92	1.937	0	2	0	35

(continued)

2θ	d	h	k	l	I_calc
47.27	1.923	2	0	0	34
48.90	1.863	0	2	4	1
49.24	1.851	2	0	4	1
49.72	1.834	1	1	11	13
53.29	1.719	2	1	1	2
53.40	1.716	1	2	2	2
53.85	1.702	0	0	16	3
54.20	1.692	2	1	3	1
54.51	1.683	0	2	8	1
54.76	1.676	1	2	4	9
54.83	1.674	2	0	8	1
55.27	1.662	1	1	13	5
55.93	1.644	0	1	15	3
55.99	1.642	2	1	5	1
56.97	1.616	1	2	6	12
58.48	1.578	0	2	10	1
58.61	1.575	2	1	7	36
58.78	1.571	2	0	10	1
59.37	1.557	1	0	16	4
59.99	1.542	1	2	8	27

$Y_2Ba_4Cu_7O_{14}$

2θ	d	h	k	l	I_calc
3.51	25.145	0	0	2	2
7.03	12.573	0	0	4	16
14.09	6.286	0	0	8	7
21.20	4.191	0	0	12	4
23.06	3.858	0	1	1	8
23.10	3.851	1	0	0	3
23.37	3.807	1	0	2	4
23.60	3.770	0	1	3	1
24.17	3.682	1	0	4	3

(continued)

24.65	3.611	0	1	5	2
24.78	3.592	0	0	14	5
25.45	3.499	1	0	6	6
26.16	3.406	0	1	7	2
27.15	3.284	1	0	8	20
28.05	3.181	0	1	9	5
29.21	3.058	1	0	10	2
31.55	2.836	1	0	12	40
32.03	2.794	0	0	18	2
32.74	2.736	0	1	13	100
32.86	2.725	1	1	1	95
33.26	2.694	1	1	3	8
34.03	2.634	1	1	5	13
34.13	2.627	1	0	14	51
35.17	2.552	1	1	7	4
35.71	2.515	0	0	20	1
36.64	2.453	1	1	9	1
36.91	2.435	1	0	16	13
38.30	2.350	0	1	17	9
39.42	2.286	0	0	22	10
39.86	2.261	1	0	18	1
40.44	2.230	1	1	13	21
42.72	2.117	1	1	15	1
42.96	2.105	1	0	20	1
45.20	2.006	1	1	17	4
46.97	1.935	0	2	0	33
46.98	1.934	0	0	26	17
47.20	1.926	2	0	0	32
47.87	1.900	1	1	19	1
49.28	1.849	0	2	8	1
49.51	1.841	2	0	8	1
50.71	1.800	1	1	21	10
51.18	1.785	0	1	25	1
52.07	1.756	0	2	12	1

(continued)

52.28	1.750	2	0	12	1
52.97	1.729	1	2	0	1
52.98	1.729	1	0	26	1
53.10	1.725	1	2	2	1
53.16	1.723	2	1	1	2
53.51	1.713	1	2	4	1
53.82	1.703	0	2	14	1
53.97	1.699	2	1	5	1
54.03	1.697	2	0	14	1
54.17	1.693	1	2	6	1
54.69	1.678	0	1	27	2
54.76	1.676	0	0	30	3
55.10	1.667	1	2	8	5
55.81	1.647	2	1	9	1
56.27	1.635	1	2	10	1
56.86	1.619	1	1	25	8
57.68	1.598	1	2	12	13
57.99	1.591	0	2	18	1
58.19	1.585	2	0	18	1
58.31	1.583	0	1	29	1
58.63	1.575	2	1	13	35
59.33	1.558	1	2	14	19

La_2CuO_4

2θ	d	h	k	l	I_{calc}
13.47	6.575	0	0	2	3
24.32	3.659	1	1	1	25
27.12	3.288	0	0	4	12
31.10	2.876	1	1	3	100
33.14	2.703	2	0	0	33
33.37	2.685	0	2	0	33
35.92	2.500	2	0	2	1

(continued)

2θ	d	h	k	l	I_{calc}
36.13	2.486	0	2	2	1
41.19	2.192	0	0	6	10
41.73	2.164	1	1	5	23
43.34	2.088	2	0	4	14
43.52	2.080	0	2	4	14
47.74	1.905	2	2	0	32
53.85	1.702	2	0	6	5
54.01	1.698	0	2	6	5
54.13	1.694	3	1	1	3
54.44	1.685	1	3	1	3
54.46	1.685	1	1	7	13
55.77	1.648	2	2	4	7
55.94	1.644	0	0	8	5
57.93	1.592	3	1	3	17
58.23	1.584	1	3	3	17

$La_{1.92}Sr_{0.08}CuO_4$

2θ	d	h	k	l	I_{calc}
13.43	6.593	0	0	2	3
24.39	3.649	1	0	1	22
27.05	3.297	0	0	4	12
31.13	2.873	1	0	3	100
33.37	2.685	1	1	0	65
36.12	2.487	1	1	2	1
41.07	2.198	0	0	6	11
41.70	2.166	1	0	5	21
43.47	2.082	1	1	4	27
47.91	1.899	2	0	0	31
53.91	1.701	1	1	6	11
54.37	1.688	1	0	7	12
54.48	1.684	2	1	1	5

(continued)

2θ	d	h	k	l	I_{calc}
55.77	1.648	0	0	8	5
55.88	1.645	2	0	4	7
58.25	1.584	2	1	3	33

$Bi_2Sr_2CuO_6$

2θ	d	h	k	l	I_{calc}
7.18	12.311	0	0	2	13
16.89	5.250	0	1	1	5
19.75	4.496	0	1	3	1
21.66	4.104	0	0	6	17
23.72	3.751	1	1	1	4
24.52	3.631	0	1	5	1
25.86	3.445	1	1	3	74
29.01	3.078	0	0	8	27
29.72	3.006	1	1	5	100
30.37	2.943	0	1	7	2
33.34	2.687	0	2	0	33
33.42	2.681	2	0	0	34
34.15	2.625	0	2	2	2
34.23	2.620	2	0	2	2
34.77	2.580	1	1	7	8
36.49	2.463	0	2	4	1
36.49	2.462	0	0	10	6
37.67	2.388	2	1	1	1
38.17	2.358	1	2	2	1
40.11	2.248	0	2	6	1
40.18	2.245	2	0	6	2
40.30	2.238	1	2	4	2
43.66	2.073	1	2	6	1
44.14	2.052	0	0	12	3
44.77	2.024	0	2	8	15
44.83	2.022	2	0	8	17

(continued)

2θ	d	h	k	l	I_{calc}
47.13	1.928	1	1	11	15
47.93	1.898	2	2	0	21
50.26	1.815	0	2	10	2
50.31	1.814	2	0	10	1
51.12	1.787	0	3	1	1
53.17	1.723	2	2	6	2
54.11	1.695	1	1	13	8
55.21	1.664	1	3	3	4
55.31	1.661	3	1	3	5
56.42	1.631	0	2	12	1
56.47	1.629	2	0	12	1
57.01	1.615	2	2	8	8
57.37	1.606	1	3	5	8
57.47	1.604	3	1	5	10
57.76	1.596	0	3	7	1

$Bi_2Sr_2CaCu_2O_8$

2θ	d	h	k	l	I_{calc}
5.72	15.452	0	0	2	32
16.62	5.333	0	1	1	3
17.22	5.151	0	0	6	1
18.51	4.792	0	1	3	1
23.02	3.863	0	0	8	24
23.45	3.794	1	1	1	4
24.84	3.584	1	1	3	28
26.04	3.422	0	1	7	1
27.43	3.251	1	1	5	100
28.89	3.090	0	0	10	16
30.84	2.900	0	1	9	1
30.94	2.890	1	1	7	56
33.09	2.707	0	2	0	37
33.19	2.700	2	0	0	28

(continued)

33.61	2.666	0	2	2	5
33.70	2.659	2	0	2	2
34.84	2.575	0	0	12	13
35.13	2.555	1	1	9	16
37.53	2.396	0	2	6	1
37.62	2.391	1	2	2	1
39.00	2.309	1	2	4	1
39.82	2.264	1	1	11	1
40.70	2.217	0	2	8	4
40.78	2.213	2	0	8	2
41.22	2.190	1	2	6	1
44.49	2.036	0	2	10	14
44.57	2.033	2	0	10	11
44.90	2.019	1	1	13	9
47.05	1.932	0	0	16	3
47.57	1.912	2	2	0	24
48.81	1.866	0	2	12	5
48.88	1.863	2	0	12	4
50.31	1.814	1	1	15	18
50.67	1.802	0	3	1	1
53.48	1.713	2	2	8	4
54.33	1.688	1	3	3	2
54.46	1.685	3	1	3	1
55.73	1.650	1	3	5	10
55.85	1.646	3	1	5	6
56.01	1.642	1	1	17	4
56.62	1.626	2	2	10	4
57.71	1.598	0	3	9	1
57.77	1.596	1	3	7	6
57.89	1.593	3	1	7	4
58.72	1.572	0	2	16	2
58.78	1.571	2	0	16	1
59.85	1.545	0	0	20	4

(continued)

$Bi_2Sr_2Ca_2Cu_3O_{10}$

2θ	d	h	k	l	I_{calc}
4.76	18.550	0	0	2	100
9.54	9.275	0	0	4	1
14.32	6.183	0	0	6	2
16.55	5.356	0	1	1	3
17.89	4.959	0	1	3	2
19.14	4.638	0	0	8	7
23.36	3.807	1	1	1	6
23.99	3.710	0	0	10	18
24.34	3.656	1	1	3	11
26.20	3.402	1	1	5	47
27.19	3.280	0	1	9	1
28.77	3.103	1	1	7	80
28.88	3.092	0	0	12	4
31.25	2.863	0	1	11	1
31.91	2.805	1	1	9	50
33.10	2.707	0	2	0	37
33.10	2.707	2	0	0	38
33.46	2.678	2	0	2	9
33.46	2.678	0	2	2	8
33.82	2.650	0	0	14	20
35.47	2.531	1	1	11	27
36.23	2.479	0	2	6	2
36.23	2.479	2	0	6	2
37.22	2.416	2	1	1	1
37.47	2.400	1	2	2	1
38.43	2.342	1	2	4	1
38.51	2.338	2	0	8	3
38.51	2.338	0	2	8	2
38.84	2.319	0	0	16	4
40.00	2.254	1	2	6	1
41.29	2.187	2	0	10	2

(continued)

41.29	2.187	0	2	10	2
42.11	2.146	1	2	8	1
43.57	2.077	1	1	15	6
43.93	2.061	0	0	18	1
44.49	2.036	0	2	12	7
44.49	2.036	2	0	12	7
47.51	1.914	2	2	0	35
47.78	1.904	2	2	2	1
47.99	1.896	1	1	17	7
48.05	1.894	2	0	14	9
48.05	1.894	0	2	14	9
49.11	1.855	0	0	20	7
50.65	1.802	0	3	1	1
51.16	1.785	0	3	3	1
51.67	1.769	2	2	8	1
51.93	1.761	0	2	16	2
51.93	1.761	2	0	16	2
52.62	1.739	1	1	19	5
53.60	1.710	3	1	1	1
53.90	1.701	2	2	10	5
54.09	1.696	1	3	3	1
54.09	1.696	3	1	3	1
54.82	1.675	1	2	16	1
55.06	1.668	1	3	5	4
55.06	1.668	3	1	5	6
55.60	1.653	0	3	9	1
56.09	1.640	0	2	18	1
56.09	1.640	2	0	18	1
56.49	1.629	1	3	7	9
56.49	1.629	3	1	7	12
56.56	1.627	2	2	12	2
57.45	1.604	1	1	21	4
57.97	1.591	0	3	11	1
58.37	1.581	3	1	9	8
58.37	1.581	1	3	9	6

(continued)

59.59	1.552	2	2	14	10
59.83	1.546	0	0	24	2

Tl$_2$Ba$_2$CuO$_6$

2θ	d	h	k	l	I$_{calc}$
7.61	11.620	0	0	2	16
15.25	5.810	0	0	4	3
22.96	3.873	0	0	6	1
25.75	3.459	1	0	3	21
30.07	2.972	1	0	5	100
30.78	2.905	0	0	8	5
32.76	2.734	1	1	0	58
33.68	2.661	1	1	2	6
38.75	2.324	0	0	10	14
40.38	2.233	1	1	6	3
42.08	2.147	1	0	9	6
45.57	1.991	1	1	8	11
47.01	1.933	2	0	0	23
47.69	1.907	2	0	2	1
49.14	1.854	1	0	11	1
51.62	1.771	1	1	10	18
52.94	1.730	2	0	6	1
54.37	1.687	2	1	3	5
55.35	1.660	0	0	14	2
56.73	1.623	1	0	13	8
56.81	1.620	2	1	5	32
57.25	1.609	2	0	8	4

(continued)

Tl$_2$Ba$_2$CaCu$_2$O$_8$

2θ	d	h	k	l	I$_{calc}$
6.03	14.659	0	0	2	72
12.07	7.330	0	0	4	2
18.15	4.886	0	0	6	5
24.83	3.586	1	0	3	2
27.69	3.221	1	0	5	79
30.49	2.932	0	0	10	1
31.54	2.836	1	0	7	100
32.86	2.726	1	1	0	78
33.44	2.680	1	1	2	16
36.10	2.488	1	0	9	2
36.79	2.443	0	0	12	21
37.79	2.381	1	1	6	1
41.17	2.192	1	0	11	7
41.28	2.187	1	1	8	3
43.20	2.094	0	0	14	4
45.43	1.996	1	1	10	6
46.66	1.947	1	0	13	4
47.15	1.928	2	0	0	35
47.58	1.911	2	0	2	2
50.14	1.819	1	1	12	25
50.93	1.793	2	0	6	1
55.32	1.661	1	1	14	7
55.56	1.654	2	1	5	21
56.50	1.629	0	0	18	3
57.19	1.611	2	0	10	1
57.84	1.594	2	1	7	33
58.64	1.574	1	0	17	8

(continued)

$Tl_2Ba_2Ca_2Cu_3O_{10}$

2θ	d	h	k	l	I_{calc}
4.93	17.940	0	0	2	100
9.86	8.970	0	0	4	3
14.81	5.980	0	0	6	2
19.79	4.485	0	0	8	4
26.27	3.393	1	0	5	20
29.01	3.078	1	0	7	67
32.32	2.769	1	0	9	55
32.90	2.722	1	1	0	60
33.29	2.692	1	1	2	16
34.43	2.605	1	1	4	1
35.01	2.563	0	0	14	14
36.09	2.489	1	0	11	5
38.69	2.327	1	1	8	2
40.20	2.243	1	0	13	5
40.21	2.243	0	0	16	8
41.64	2.169	1	1	10	2
44.60	2.032	1	0	15	3
45.03	2.013	1	1	12	3
47.21	1.925	2	0	0	30
47.50	1.914	2	0	2	2
48.80	1.866	1	1	14	15
49.23	1.851	1	0	17	1
51.67	1.769	2	0	8	1
52.90	1.731	1	1	16	11
54.09	1.695	1	0	19	1
54.83	1.674	2	1	5	4
56.37	1.632	2	1	7	19
56.42	1.631	0	0	22	3
57.28	1.608	1	1	18	1
58.38	1.581	2	1	9	18
59.16	1.562	1	0	21	5

(continued)

$Tl_2Ba_2Ca_3Cu_4O_{12}$

2θ	d	h	k	l	I_{calc}
4.20	21.026	0	0	2	100
8.41	10.513	0	0	4	5
12.63	7.009	0	0	6	1
16.87	5.256	0	0	8	2
21.13	4.205	0	0	10	2
25.41	3.505	1	0	5	5
27.49	3.245	1	0	7	31
30.05	2.974	1	0	9	48
32.85	2.727	1	1	0	47
32.99	2.715	1	0	11	23
33.13	2.704	1	1	2	12
33.97	2.639	1	1	4	2
34.11	2.628	0	0	16	9
36.25	2.478	1	0	13	9
38.54	2.336	0	0	18	7
39.38	2.288	1	1	10	2
39.75	2.268	1	0	15	6
41.98	2.152	1	1	12	1
43.02	2.103	0	0	20	2
43.46	2.082	1	0	17	2
44.90	2.019	1	1	14	2
47.14	1.928	2	0	0	25
47.35	1.920	2	0	2	2
47.36	1.920	1	0	19	2
48.08	1.892	1	1	16	10
51.51	1.774	1	1	18	9
52.19	1.753	2	0	10	1
54.30	1.689	2	1	5	1
55.16	1.665	1	1	20	3
55.43	1.658	2	1	7	8
55.63	1.652	1	0	23	2

(continued)

56.92	1.618	2	1	9	14
56.93	1.617	0	0	26	3
58.74	1.572	2	1	11	8
59.46	1.555	2	0	16	6

TlBa$_2$CuO$_5$

2θ	d	h	k	l	I$_{calc}$
9.22	9.590	0	0	1	5
18.50	4.795	0	0	2	1
24.92	3.573	1	0	1	24
27.91	3.197	0	0	3	10
29.76	3.002	1	0	2	100
32.90	2.722	1	1	0	56
34.24	2.619	1	1	1	5
36.53	2.459	1	0	3	2
37.51	2.398	0	0	4	16
43.67	2.073	1	1	3	17
44.52	2.035	1	0	4	2
47.21	1.925	2	0	0	24
50.74	1.799	1	1	4	18
53.36	1.717	1	0	5	13
54.12	1.695	2	1	1	5
55.74	1.649	2	0	3	5
56.81	1.621	2	1	2	30
57.67	1.598	0	0	6	1

(continued)

TlBa₂CaCu₂O₇

2θ	d	h	k	l	I_calc
6.93	12.754	0	0	1	32
20.89	4.251	0	0	3	5
23.06	3.857	1	0	0	1
24.11	3.692	1	0	1	8
27.02	3.300	1	0	2	97
27.98	3.189	0	0	4	7
31.31	2.857	1	0	3	100
32.84	2.727	1	1	0	86
33.60	2.667	1	1	1	17
35.18	2.551	0	0	5	29
36.56	2.458	1	0	4	13
42.49	2.128	1	0	5	2
42.53	2.126	0	0	6	2
43.67	2.073	1	1	4	23
47.12	1.929	2	0	0	43
47.69	1.907	2	0	1	2
48.89	1.863	1	1	5	28
48.92	1.862	1	0	6	8
50.06	1.822	0	0	7	1
52.07	1.756	2	0	3	1
53.61	1.709	2	1	1	2
54.75	1.677	1	1	6	4
55.16	1.665	2	1	2	26
55.70	1.650	2	0	4	5
55.80	1.647	1	0	7	16
57.67	1.598	2	1	3	33
57.84	1.594	0	0	8	3

(continued)

TlBa$_2$Ca$_2$Cu$_3$O$_9$

2θ	d	h	k	l	I$_{calc}$
5.55	15.913	0	0	1	78
22.35	3.978	0	0	4	8
23.08	3.853	1	0	0	1
23.76	3.745	1	0	1	7
25.69	3.468	1	0	2	49
28.04	3.183	0	0	5	6
28.63	3.117	1	0	3	100
32.35	2.768	1	0	4	74
32.87	2.725	1	1	0	95
33.37	2.685	1	1	1	18
33.80	2.652	0	0	6	32
36.62	2.454	1	0	5	34
37.10	2.424	1	1	3	1
39.65	2.273	0	0	7	8
41.33	2.185	1	0	6	1
43.74	2.070	1	1	5	19
46.37	1.958	1	0	7	4
47.18	1.927	2	0	0	51
47.54	1.913	2	0	1	1
47.86	1.900	1	1	6	26
51.70	1.768	0	0	9	2
51.72	1.768	1	0	8	11
52.42	1.746	1	1	7	1
52.80	1.734	2	0	4	1
54.49	1.684	2	1	2	8
55.78	1.648	2	0	5	1
56.12	1.639	2	1	3	23
57.33	1.607	1	0	9	13
57.95	1.591	0	0	10	5
58.36	1.581	2	1	4	20
59.28	1.559	2	0	6	18

(continued)

TlBa₂Ca₃Cu₄O₁₁

2θ	d	h	k	l	I_calc
4.65	19.010	0	0	1	100
9.30	9.505	0	0	2	4
18.67	4.753	0	0	4	1
23.10	3.850	1	0	0	3
23.40	3.802	0	0	5	8
24.95	3.568	1	0	2	23
27.10	3.290	1	0	3	69
28.16	3.168	0	0	6	1
29.87	2.992	1	0	4	87
32.90	2.722	1	1	0	98
32.98	2.716	0	0	7	27
33.11	2.705	1	0	5	46
33.24	2.695	1	1	1	21
34.26	2.617	1	1	2	4
36.74	2.446	1	0	6	45
37.86	2.376	0	0	8	12
40.65	2.219	1	0	7	3
40.76	2.213	1	1	5	2
42.81	2.112	0	0	9	2
43.84	2.065	1	1	6	19
44.82	2.022	1	0	8	1
47.21	1.925	2	0	0	59
47.28	1.923	1	1	7	22
47.47	1.915	2	0	1	2
47.85	1.901	0	0	10	1
49.20	1.852	1	0	9	7
51.01	1.790	1	1	8	14
52.98	1.728	0	0	11	1
53.34	1.717	2	0	5	3
53.78	1.705	1	0	10	11
54.13	1.694	2	1	2	5

(continued)

2θ	d	h	k	l	I_{calc}
55.03	1.669	1	1	9	3
55.29	1.662	2	1	3	18
55.89	1.645	2	0	6	1
56.88	1.619	2	1	4	27
58.24	1.584	0	0	12	6
58.54	1.577	1	0	11	9
58.80	1.571	2	0	7	17
58.88	1.568	2	1	5	16

$(Tl_{0.5}Pb_{0.5})Sr_2CuO_5$

2θ	d	h	k	l	I_{calc}
9.85	8.980	0	0	1	2
19.77	4.490	0	0	2	1
25.86	3.446	1	0	1	31
29.85	2.993	0	0	3	13
31.17	2.870	1	0	2	100
33.98	2.638	1	1	0	62
35.46	2.531	1	1	1	4
38.56	2.335	1	0	3	4
40.17	2.245	0	0	4	13
45.85	1.979	1	1	3	22
47.25	1.924	1	0	4	4
48.82	1.866	2	0	0	26
50.84	1.796	0	0	5	1
53.60	1.710	1	1	4	15
56.06	1.641	2	1	1	7
56.90	1.618	1	0	5	14
58.28	1.583	2	0	3	7
59.06	1.564	2	1	2	31

(continued)

$(Tl_{0.5}Pb_{0.5})Sr_2CaCu_2O_7$

2θ	d	h	k	l	I_{calc}
7.33	12.060	0	0	1	22
14.69	6.030	0	0	2	4
22.11	4.020	0	0	3	13
23.47	3.791	1	0	0	7
24.62	3.617	1	0	1	18
27.80	3.209	1	0	2	52
29.63	3.015	0	0	4	6
32.46	2.758	1	0	3	100
33.43	2.681	1	1	0	76
34.27	2.617	1	1	1	12
36.69	2.450	1	1	2	1
37.28	2.412	0	0	5	18
38.14	2.360	1	0	4	8
40.44	2.230	1	1	3	1
44.52	2.035	1	0	5	5
45.11	2.010	0	0	6	6
45.26	2.003	1	1	4	19
48.00	1.896	2	0	0	38
50.93	1.793	1	1	5	17
51.46	1.776	1	0	6	10
53.16	1.723	0	0	7	1
53.44	1.715	2	0	3	2
54.09	1.695	2	1	0	1
54.67	1.679	2	1	1	2
56.37	1.632	2	1	2	11
57.29	1.608	1	1	6	8
57.42	1.605	2	0	4	3
58.88	1.569	1	0	7	10
59.14	1.562	2	1	3	29

(continued)

$(Tl_{0.5}Pb_{0.5})Sr_2Ca_2Cu_3O_9$

2θ	d	h	k	l	I_{calc}
5.80	15.232	0	0	1	44
11.62	7.616	0	0	2	5
17.47	5.077	0	0	3	5
23.36	3.808	1	0	0	9
23.36	3.808	0	0	4	16
24.09	3.694	1	0	1	11
26.16	3.406	1	0	2	34
29.32	3.046	1	0	3	60
29.32	3.046	0	0	5	2
33.27	2.693	1	0	4	100
33.27	2.693	1	1	0	90
33.80	2.652	1	1	1	14
35.36	2.539	1	1	2	3
35.36	2.539	0	0	6	17
37.82	2.379	1	0	5	21
41.05	2.199	1	1	4	1
41.50	2.176	0	0	7	10
42.81	2.112	1	0	6	3
44.93	2.018	1	1	5	17
47.77	1.904	2	0	0	49
47.77	1.904	0	0	8	2
48.16	1.889	1	0	7	7
49.33	1.847	1	1	6	15
53.83	1.703	2	1	0	1
53.83	1.703	2	0	4	4
53.83	1.703	1	0	8	9
54.19	1.692	1	1	7	10
54.19	1.692	0	0	9	2
54.19	1.692	2	1	1	1
55.27	1.662	2	1	2	6

(continued)

2θ	d	h	k	l	I_calc
57.04	1.615	2	1	3	14
57.04	1.615	2	0	5	1
59.46	1.555	1	1	8	3
59.46	1.555	2	1	4	28
59.80	1.547	1	0	9	6

$Pb_2Sr_2(Y_{0.75}Ca_{0.25})Cu_3O_8$

2θ	d	h	k	l	I_{calc}
5.61	15.760	0	0	1	28
11.23	7.880	0	0	2	2
16.88	5.253	0	0	3	1
22.57	3.940	0	0	4	9
23.33	3.813	1	0	0	10
24.01	3.706	1	0	1	21
25.96	3.432	1	0	2	2
28.31	3.152	0	0	5	9
28.93	3.086	1	0	3	12
32.68	2.740	1	0	4	100
33.23	2.696	1	1	0	67
33.72	2.658	1	1	1	9
34.13	2.627	0	0	6	4
37.00	2.429	1	0	5	1
37.49	2.399	1	1	3	1
40.05	2.251	0	0	7	2
40.54	2.225	1	1	4	3
41.76	2.163	1	0	6	12
44.20	2.049	1	1	5	19
46.07	1.970	0	0	8	7
47.70	1.907	2	0	0	30
48.07	1.893	2	0	1	1
48.38	1.881	1	1	6	2

(continued)

52.27	1.750	1	0	8	3
52.99	1.728	1	1	7	3
53.39	1.716	2	0	4	3
53.75	1.705	2	1	0	2
54.09	1.695	2	1	1	6
55.10	1.667	2	1	2	1
56.40	1.631	2	0	5	5
56.76	1.622	2	1	3	4
57.95	1.591	1	0	9	7
57.98	1.591	1	1	8	13
58.57	1.576	0	0	10	3
59.02	1.565	2	1	4	35
59.95	1.543	2	0	6	2

8.0 REFERENCES

1. Tokura, Y., Takagi, H., and Uchida, S., *Nature* 337:345 (1989).
2. James, A. C. W. P., Zahurak, S. M., and Murphy, D. W., *Nature* 338:240 (1989).
3. For reviews, see:
   (a) Sleight, A. W., *Science* 242:1519 (1988).
   (b) Sleight, A. W., Subramanian, M. A., and Torardi, C. C., *Mater. Res. Soc. Bull.* XIV:45 (1989).
4. Hibble, S. J., Cheetham, A. K., Chippindale, A. M., Day, P., and Hriljac, J. A., *Physica C* 156:604 (1988).
5. Torardi, C. C., Subramanian, M. A., Calabrese, J. C., Gopalakrishnan, J., Morrissey, K. J., Askew, T. R., Flippen, R. B., Chowdhry, U., and Sleight, A. W., *Science* 240:631 (1988).
6. Jung, D., Whangbo, M.-H., Herron, N., and Torardi, C. C., *Physica C*, 160:381 (1989).
7. Beno, M. A., Soderholm, L., Capone, D. W., Hinks, D. G., Jorgensen, J. D., Grace, J. D., Schuller, I. K., Segre, C. U., and Zhang, K., *Appl. Phys. Lett.* 51:57 (1987).
8. McCarron, E. M., Crawford, M. K., and Parise, J. B., *J. Solid State Chem.* 78:192 (1989).
9. Marsh, P., Fleming, R. M., Mandich, M. L., DeSantolo, A. M., Kwo, J., Hong, M., and Martinez-Miranda, L. J., *Nature* 334:141 (1988).
10. Hazen, R. M., Finger, L. W., and Morris, D. E., *Appl. Phys. Lett.* 54:1057 (1989).
11. Bordet, P., Chaillout, C., Chenavas, J., Hodeau, J. L., Marezio, M., Karpinski, J., and Kaldis, E., *Nature* 334:596 (1988).
12. Pooke, D. M., Buckley, R. G., Presland, M. R., and Tallon, J. L., *Phys. Rev. B* 41:6616 (1990).
13. Grande, V. B., Muller-Buschbaum, H., and Schweizer, M., *Z. Anorg. Allg. Chem.* 428:120 (1977).
14. Jorgensen, J. D., Dabrowski, B., Pei, S., Hinks, D. G., Soderholm, L., Morosin, B., Schirber, J. E., Venturini, E. L., and Ginley, D. S., *Phys. Rev. B* 38:11337 (1988).
15. Torardi, C. C., Subramanian, M. A., Calabrese, J. C., Gopalakrishnan, J., McCarron, E. M., Morrissey, K. J., Askew, T. R., Flippen, R. B., Chowdhry, U., and Sleight, A. W., *Phys. Rev. B* 38:225 (1988).

16. Subramanian, M. A., Torardi, C. C., Calabrese, J. C., Gopalakrishnan, J., Morrissey, K. J., Askew, T. R., Flippen, R. B., Chowdhry, U., and Sleight, A. W., *Science* 239:1015 (1988).
17. Torardi, C. C., Parise, J. B., Subramanian, M. A., Gopalakrishnan, J., and Sleight, A. W., *Physica C* 157:115 (1989).
18. Gao, Y., Lee, P., Coppens, P., Subramanian, M. A., and Sleight, A. W., *Science* 241:954 (1988).
19. Le Page, Y., McKinnon, W.R., Tarascon, J.-M., and Barboux, P., *Phys. Rev. B* 40:6810 (1989).
20. Lee, P., Gao, Y., Sheu, H. S., Petricek, V., Restori, R., Coppens, P., Darovskikh, A., Phillips, J. C., Sleight, A. W., and Subramanian, M. A., *Science* 244:62 (1989).
21. Gai., P. L., Subramanian, M. A., Gopalakrishnan, J., and Boyes, E. D., *Physica C*, 159:801 (1989).
22. Subramanian, M. A., Calabrese, J. C., Torardi, C. C., Gopalakrishnan, J., Askew, T. R., Flippen, R. B., Morrissey, K. J., Chowdhry, U., and Sleight, A. W., *Nature* 332:420 (1988).
23. Parise, J. B., Torardi, C. C., Subramanian, M. A., Gopalakrishnan, J., Sleight, A. W., and Prince, E., *Physica C* 159:239 (1989).
24. Parkin, S. S. P., Lee, V. Y., Nazzal, A. I., Savoy, R., Beyers, R., and LaPlaca, S. J., *Phys. Rev. Lett.* 61:750 (1988).
25. Haldar, P., Chen, K., Maheswaran, B., Roig-Janicki, A., Jaggi, N. K., Markiewicz, R. S., and Giessen, B. C., *Science* 241:1198 (1988).
26. Morosin, B., Ginley, D. S., Hlava, P. F., Carr, M. J., Baughman, R. J., Schirber, J. E., Venturini, E. L., and Kwak, J. F., *Physica C* 152:413 (1988).
27. Subramanian, M. A., Parise, J. B., Calabrese, J. C., Torardi, C. C., Gopalakrishnan, J., and Sleight, A. W., *J. Solid State Chem.* 77:192 (1988).
28. Torardi, C. C., and Subramanian, M. A., unpublished results.
29. Subramanian, M. A., unpublished results.
30. Subramanian, M. A., Torardi, C. C., Gopalakrishnan, J., Gai, P. L., Calabrese, J. C., Askew, T. R., Flippen, R. B., and Sleight, A. W., *Science* 242:249 (1988).
31. Parise, J. B., Gai, P. L., Subramanian, M. A., Gopalakrishnan, J., and Sleight, A. W., *Physica C* 159:245 (1989).

32. Cava, R. J., Batlogg, B., Krajewski, J. J., Rupp, L. W., Schneemeyer, L. F., Siegrist, T., van Dover, R. B., Marsh, P., Peck, W. F., Gallagher, P. K., Glarum, S. H., Marshall, J. H., Farrow, R. C., Waszczak, J. V., Hull, R., and Trevor, P., *Nature* 336:211 (1988).

33. Subramanian, M. A., Gopalakrishnan, J., Torardi, C. C., Gai, P. L., Boyes, E. D, Askew, T. R., Flippen, R. B., Farneth, W. E., and Sleight, A. W., *Physica C* 157:124 (1989).

34. Torardi, C. C., McCarron, E. M., Bierstedt, P. E., and Sleight, A. W., *Solid State Comm.* 64:497 (1987).

35. Cava, R. J., Marezio, M., Krajewski, J. J., Peck, W. F., Santoro, A., and Beech, F., *Physica C* 157:272 (1989).

36. Dmowski, W., Toby, B. H., Egami, T., Subramanian, M. A., Gopalakrishnan, J., and Sleight, A. W., *Phys. Rev. Lett.* 61:2608 (1988).

37. Parise, J. B., Gopalakrishnan, J., Subramanian, M. A., and Sleight, A. W., *J. Solid State Chem.* 76:432 (1988).

38. Ren, J., Jung, D., Whangbo, M.-H., Tarascon, J.-M., LePage, Y., McKinnon, W. R., and Torardi, C. C., *Physica C* 159:151 (1989).

39. See for example, Massidda, S., Yu, J., and Freeman, A. J., *Physica C* 152:251 (1988).

40. Whangbo, M.-H., Kang, D. B., and Torardi, C. C., *Physica C* 158:371 (1989).

41. Brown, I.D., and Altermatt, D., *Acta Crystallogr. B* 41:244 (1985).

42. Whangbo, M.-H., and Torardi, C.C., *Science* 249:1143 (1990).

43. Wang, H. H., Geiser, U., Thorn, R. J., Carlson, K. D., Beno, M. A., Monaghan, M. R., Allen, T. J., Proksch, R. B., Stupka, D. L., Kwok, W. K., Crabtree, G. W., and Williams, J. M., *Inorg. Chem.* 26:1190 (1987).

44. Torardi, C. C., Subramanian, M. A., Gopalakrishnan, J., and Sleight, A. W., *Physica C* 158:465 (1989).

45. Maeda, A., Noda, K., Uchinokura, K., and Tanaka, S., *Jpn. J. Appl. Phys.* 28:L576 (1989).

46. Kim, J. S., Swinnea, J. S., Tao, Y. K., and Steinfink, H., *J. Less-Comm. Met.* 156:347 (1989).

47. Yvon, K., Jeitschko, W., and Parthe, E., *J. Appl. Crystallogr.* 10:73 (1977).

48. Yvon, K., and Francois, M., Z. *Phys. B - Condensed Matter* 76:413 (1989).

14

Chemical Characterization of Oxide Superconductors by Analytical Electron Microscopy

Anthony K. Cheetham and Ann M. Chippindale

1.0 INTRODUCTION

The discovery of high-temperature superconductivity in a number of mixed-metal oxides of copper has led to an unprecedented level of activity in the study of their syntheses, properties and applications. However, this exciting development has also highlighted some fundamental inadequacies in the techniques that are available for the chemical characterisation of such oxides. It is perhaps not surprising that, in the frenetic atmosphere that surrounded the first discoveries in the high-T_c area, incorrect stoichiometries were initially ascribed to several of the new superconducting phases (1)(2), but it is remarkable that controversy should continue to surround the compositions of phases in the Bi-Sr-Ca-Cu-O system and related materials, over 18 months after they were first reported (3). The "2212" phase, for example, is most frequently ascribed the stoichiometry $Bi_2Sr_2CaCu_2O_{8+\delta}$ but others, including ourselves (4), have argued forcibly that the correct description should be $Bi_2Sr_{2-x}CaCu_2O_8$. This is not merely a question of semantics; rather, it represents a serious inadequacy in our understanding of the mechanism by which the hole concentration is controlled in these materials.

How can it arise that we seem unable to differentiate between the compositions, say, $Bi_2Sr_2CaCu_2O_{8.33}$ and $Bi_2Sr_{1.67}CaCu_2O_8$ (i.e. for a formal copper oxidation state of +2.33)? Under normal circumstances, the crystal structure of such a material would give a definitive answer to this question. In these particular systems however, there is a complex, incommensurate superlattice that has hampered progress in solving the structure with the necessary degree of precision. The first powder-diffraction refinements of the average structure (i.e. based upon the sub-cell intensities) were unable to find evidence for either oxygen interstitials or cation vacancies (5)(6), and subsequent studies of the superstructure by single-crystal X-ray methods (7) and combined X-ray and neutron powder techniques (8) yielded conflicting results. The former, which involved a study of the cation distribution by anomalous scattering with synchrotron X-rays, provides strong support for the cation vacancy mechanism, but the latter claims to find clear evidence for oxygen interstitials.

A more direct approach is to determine the stoichiometry by chemical analysis, but here one is hindered because samples of these materials are almost always polyphasic. Bulk analytical methods are clearly inappropriate, but an unambiguous answer should be forthcoming from, say, microprobe analysis. Several authors have reported such data, and they tend to support the cation deficiency model (see Table 1), but the results have been regarded with some scepticism and have tended to be overlooked in favour of the stoichiometric (e.g. 2212) composition. Certainly there are grounds for suspicion, because the spatial resolution of conventional microprobe analysis is often no better than about 1 micron, and the particle sizes of the different phases in high-T_c mixtures are frequently substantially smaller. Under these circumstances, analytical methods based upon transmission electron microscopy offer considerably better spatial resolution. This chapter will focus on the use of analytical electron microscopy, and, in particular, high-resolution X-ray emission spectroscopy, as a means of obtaining quantitative chemical composition data from *individual* crystallites within polyphasic mixtures of superconducting oxides and related materials.

Table 1:   A Survey of Microprobe Analysis Results for the 2212 Bi-Sr-Ca-Cu-O Superconductor

Stoichiometry	Ref.
$Bi_2Sr_{1.4}CaCu_2O_y$	(9)
$Bi_2(Sr,Ca)_{2.50}Cu_2O_y$	(10)
$Bi_2Sr_{1.7}CaCu_2O_y$	(11)
$Bi_2Sr_{1.7}Ca_{0.8}Cu_2O_y$	(12)
$Bi_{2.11}Sr_{1.34}Ca_{1.53}Cu_{2.03}O_y$	(13)
$Bi_{2.15}Sr_{1.68}Ca_{1.17}Cu_2O_y$	(14)

2.0 ANALYTICAL ELECTRON MICROSCOPY: A BRIEF SURVEY

During the 1970's, advances in the design of transmission electron microscopes led to the widespread availability of instruments with X-ray detectors suitable for monitoring X-ray emission in the energy range 1-20 keV. With such an instrument, X-ray spectra of individual crystallites as small as 500Å can be obtained routinely with a spatial resolution, which is determined by the lateral diffusion of the beam, of down to 100Å in the case of very thin crystallites. This capability offers exciting possibilities for the study of materials and we shall review some of the achievements and prospects in this area and consider the precision and accuracy with which quantitative analysis can be obtained from small crystals.

The quantitative analysis of thick specimens, as carried out in the electron microprobe, is well documented (15). The concentration, c_x, of the element x present in a specimen is related to the intensity, I_x, of a characteristic emission line of x by the equation

$$c_x = k\ I_x/C_Z\ C_A C_F \qquad (1)$$

where k is a constant and C_Z, C_A and C_F are corrections for atomic-number effects, absorption, and fluorescence, respectively. These corrections are a consequence of the long path lengths that are traversed both by the electron beam and the emitted X-rays.

In the thin-crystal limit, on the other hand, such corrections should be negligible and for two elements x and y present in the sample we can write:

$$c_x / c_y = k_{xy} I_x / I_y \qquad (2)$$

This approximation, known as the "ratio method" (16), is particularly attractive for applications in solid-state chemistry because it should apply under the normal working conditions of a transmission electron microscope. If the approximation holds, then a determination of k_{xy} using any well-characterised compound containing x and y will then afford a simple method for measuring the x:y ratio in any other compound. This approach will be illustrated below with the results obtained for some standards related to the 2212 superconductor. Alternatively, k_{xy} can be calculated from the appropriate ionisation cross sections, fluorescence yields, and parameters relating to the efficiency of the X-ray detector (17), although it should be noted that the value of k_{xy} will vary somewhat from one instrument to another.

The ratio method was first applied successfully to the K lines of light elements (Z>10), especially the first transition series, by mineralogists (16) and metallurgists (18), and to a range of L and M lines of heavier elements by solid-state chemists (19). During the 1980's, it has been applied with considerable success to a wide range of materials, including alloys (20-22), semiconductors (23), metal sulphides (24), phosphates (25), silicate minerals (26), and, of particular relevance to high-temperature superconductors, mixed-metal oxides (27-29). Quantitative analysis has even been obtained for light elements such as oxygen (30), in spite of the severe X-ray absorption that is encountered with the low energy K lines of these elements. The key to obtaining reliable analyses lies in ensuring that the crystallites being analysed conform to the thin-crystal limit required by the ratio method. In the following section, therefore, we describe the current practice in our Laboratory at Oxford and present some results obtained on materials related to the high-temperature superconductors.

3.0 EXPERIMENTAL METHOD

All the X-ray emission spectra reported in this work were measured at 200 keV in a JEOL 2000FX TEMSCAN analytical electron microscope. In such an instrument, it is possible to switch rapidly between the various modes of the instrument: TEM, SEM, STEM and EDX. In theory therefore, diffraction patterns and high-resolution lattice images can be obtained for analysed crystallites, although this is often impracticable. The Tracor Northern high-angle X-ray detector (take-off angle ~70°) has a cross section of 30 mm^2 and an energy resolution of approximately 150 eV. For materials that do not contain copper, samples in the form of finely divided powders dispersed in chloroform are deposited on 3-mm copper grids coated with holey carbon films (31). The resulting X-ray spectra will, of course, contain the Cu K lines, but the intensities of the latter can be minimised by analysing crystals that are near to the centre of the grid squares. In the case of the copper-containing superconductors and related phases, the Cu emissions are eliminated by using a titanium or nylon grid with a graphite holder. In our particular set up, the specimen is held at 0° tilt, although the optimum degree of tilt towards the X-ray detector and the detector take-off angle do vary from instrument to instrument. A typical X-ray emission spectrum is shown in Figure 1.

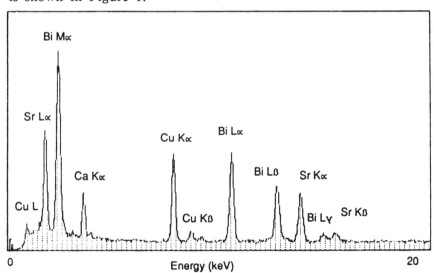

Figure 1: X-ray emission spectrum for "2212" Bi-Sr-Ca-Cu-O phase

Using the magnifying power of the microscope, an individual crystallite can be selected in the SEM or TEM mode and its thickness gauged from its size, the X-ray count rate and, where possible, the intensity ratio between the two lines with high and low absorption cross sections, respectively (see below). Count rates in excess of 1500 counts/s were avoided in order to minimise saturation effects and coincidence corrections in the processor; the counting time was 100s at a magnification of, typically, 50,000X in the TEM mode, with a spot size of approximately 400Å. For data analysis on more beam-sensitive crystals, area scans (approximately 0.003 mm^2) were preferred to spot analyses because this reduces specimen damage and contamination. It is also more efficient to let the instrument average over an area rather than measure several discrete points. If the crystals are believed to be inhomogeneous and the elemental distribution is of interest, then area scans are obviously unsuitable. Typically, 10 thin crystals of each monophasic sample were examined and, of course, more in the case of polyphasic mixtures.

X-ray emission intensities were obtained from the spectra by using a digital filtering method which is available in the standard TN5500 software. For light elements up to, say, bromine, the only lines suitable for analysis are the Kα and Kβ emissions, and the Kα's were used as in the work of Cliff and Lorimer (16). Where overlap occurred between the Kα of one element and the Kβ of another, an experimentally determined correction was used; our corrections were in good agreement with those measured by Heinrich et al. (32).

For elements of the second transition series and beyond, a choice of lines is often available, with, for example, K and L lines in the second transition series and M and L lines in the third. The high-energy lines were normally used for the analysis because of their lower absorption cross sections (33), and Lα:Kα or M:Lα ratios were used to monitor crystal thickness (see below). The most straightforward analyses are those involving elements with approximately the same atomic numbers, when the K:K or L:L ratios may be used and the absorption cross sections are normally of similar magnitude (although difficulties can arise if an emission from one element leads to fluorescence from another). For elements that are far removed from each other in the periodic table, mixed ratios are more appropriate. The optimum precision is found in cases where it is practicable to use only high-energy lines, say E>10 keV.

4.0 DATA COLLECTION AND ANALYSIS OF STANDARDS

For our AEM studies of the "2212" phase in the Bi-Sr-Ca-Cu-O system, it was necessary to obtain k-values for the appropriate elements. The most convenient choice was to determine values relative to copper, and for this purpose we prepared standard samples of Bi_2CuO_4 (34), $SrCuO_2$ (35) and Ca_2CuO_3 (35). The purity of each standard was confirmed by powder X-ray diffraction. The raw X-ray emission intensity data are shown in Table 2, which also contains calibration data for yttrium- and barium-containing systems. Specimen thickness was assessed by monitoring the Bi Mα/Bi Lα ratio for Bi_2CuO_4 and the Sr Lα/Sr Kα ratio for $SrCuO_2$. The justification for this is apparent from the absorption coefficients given in Table 3. For example, the coefficients for the Bi Mα line are 5-10 times greater than those for Bi Lα, and consequently the Mα/Lα ratio is very sensitive to thickness. Data were used in the analysis only if this ratio was greater than 1.8, a value that was determined empirically. We estimate that this corresponds to a thickness of 2500Å. It was not possible to monitor the thickness of our Ca_2CuO_3 in this way because there is no suitable pair of emissions for calcium or copper. It was therefore necessary to gauge the thickness from the appearance of the sample in cross section and the X-ray count rate; thicker samples will give count rates in excess of, say, 2000 counts/s.

The reproducibility of the results from the standards is clear from the narrow spread of intensity ratios in Table 2. Readers are reminded, however, that the k-values will vary somewhat from one instrument to another because they are dependent not only upon the appropriate ionisation cross sections and fluorescence yields, but also upon the efficiency of the detection system. Values that are tabulated in commercial software packages should be used with caution.

Table 2: X-ray emission intensity ratios for standard samples of (a) Bi_2CuO_4, (b) $SrCuO_2$, (c) Ca_2CuO_3, (d) $BaCuO_2$ and (e) $Y_2Cu_2O_5$

(a) Bi_2CuO_4

	BiLα/CuKα	BiMα/BiLα
1.	1.696	1.846
2.	1.817	1.875
3.	1.869	2.031
4.	1.774	1.902
5.	1.737	2.044
6.	1.638	1.994
7.	1.740	2.044
8.	1.833	1.969
9.	1.768	2.013

Mean = 1.764(24)
k = 1.13(2)

(b) $SrCuO_2$

	SrKα/CuKα	SrLα/SrKα
1.	0.617	1.822
2.	0.623	2.017
3.	0.645	1.894
4.	0.628	1.989
5.	0.624	1.873
6.	0.622	2.014
7.	0.624	1.965
8.	0.598	1.933
9.	0.606	1.775

Mean = 0.621(4)
k = 1.61(1)

(continued)

(c) **Ca$_2$CuO$_3$**

CaKα/CuKα

1. 2.289
2. 2.350
3. 2.241
4. 2.046
5. 2.186
6. 2.132
7. 2.286
8. 2.199
9. 2.115
10. 2.138

Mean = 2.198(30)
k   = 0.91(1)

(d) **BaCuO$_2$**

BaLα/CuKα

1. 1.006
2. 1.018
3. 0.943
4. 1.029
5. 0.929
6. 0.943
7. 0.943
8. 0.962
9. 0.994
10. 1.005
11. 0.947

Mean = 0.974(11)
k   = 1.03(1)

(e) **Y$_2$Cu$_2$O$_5$**

YKα/CuKα

1. 0.567
2. 0.573
3. 0.585
4. 0.547
5. 0.563
6. 0.589
7. 0.594
8. 0.543
9. 0.581

Mean = 0.571(6)
k   = 1.75(2)

YLα/YKα

1. 1.871
2. 1.805
3. 1.897
4. 1.898
5. 1.932
6. 1.968
7. 1.819
8. 1.818
9. 1.890

Table 3: Mass absorption coefficients ($cm^2 g^{-1}$) in the system Bi-Sr-Ca-Cu-O

Line	Abs. by Bi	Abs. by Sr	Abs. by Ca	Abs. by Cu
BiMα	941.6	3148.7	440.0	1419.1
BiLα	113.4	52.7	70.0	182.6
SrLα	1806.1	990.4	980.1	3161.2
SrKα	134.0	25.4	33.7	88.6
CaKα	1332.9	997.3	139.4	449.5
CuKα	246.5	119.1	158.6	53.7

5.0 ANALYSIS OF THE "2212" COMPOUND (36)

A sample of the "2212" compound was prepared from a mixture of Bi_2O_3, $SrCO_3$, $CaCO_3$ and CuO, using the method of Maeda et al. (3). Intimate mixtures of the starting materials were calcined at 1073 K for 2 hours, reground and pelletised, and heated at 1133 K for a further 12 hours. The product was cooled at a rate of ~1° min. Powder X-ray diffraction patterns indicated that the sample was polyphasic, with the so-called "2212" phase predominating. Four-probe ac resistivity measurements showed metallic behaviour with a nearly linear temperature coefficient from 300 K down to the superconducting onset at 98 K; zero resistance was reached at 88 K. The composition of the "2212" phase was determined by AEM in the manner described above. Fifteen crystallites of the predominant phase were analysed, and Bi/Cu, Sr/Cu and Ca/Cu ratios were obtained by using the k-values given in Table 2; results are shown in Table 4. The mean composition is estimated to be $Bi_{2.06}Sr_{1.68}Ca_{0.98}Cu_2O_y$. This finding is in rather good agreement with most of the microprobe results given in Table 1, and it is on this basis that we have argued (4) that the "2212" phase is more correctly described by the composition $Bi_2Sr_{2-x}CaCu_2O_8$. The Ca/Cu ratios given in Table 4 show a rather large range, and it would seem likely, as well as chemically reasonable, that calcium as well as strontium vacancies may be present under some circumstances. There is also evidence in this and other work that the "2212" compound tends to be slightly bismuth rich.

Table 4: Analytical electron microscopy results for a sample of "$Bi_2Sr_2CaCu_2O_8$", obtained using calibration constants (Table 2).

Bi/Cu	Sr/Cu	Ca/Cu
1.00	0.92	0.49
1.06	0.80	0.46
1.02	0.78	0.55
1.06	0.80	0.53
1.02	0.85	0.49
1.11	0.82	0.52
1.01	0.81	0.46
1.06	0.84	0.48
1.08	0.88	0.45
1.03	0.84	0.45
0.95	0.77	0.53
1.01	0.82	0.57
0.99	0.80	0.48
1.07	0.91	0.42
1.04	0.88	0.47

Mean values: 1.03(1)   0.84(1)   0.49(1)

6.0 DISCUSSION

We have described in some detail the means by which the chemical composition of a superconducting phase may be determined in the presence of other phases. Our approach is quite general, and we have applied it to a number of systems in the high-T_c area (see Table 5). The important question is: what credence can be placed on such results?

There are a number of potential difficulties with our method. In the first instance, we must ask whether the ratio method (equation (2)) is reliable under all circumstances if the crystals are sufficiently thin. We believe that in the vast majority of instances this will be the case, but we have encountered situations in which the emission intensity from one element appears to be increased because of fluorescence from a line of another element at a slightly higher energy. For example, the oxygen K line appears to be enhanced by fluorescence due to the barium M line in Ba_2ReO_5 (30). This situation is predictable in the sense that it will be reflected in a very high absorption cross section for the higher energy line, and it is not expected to be a problem in the analysis of the "2212" compound. Second, there is the possibility that the data will be affected by the orientation of the beam with respect to the crystallite under study (15). This is important in microprobe analysis, but we have never encountered difficulties of this nature with thin crystals. A more serious problem, however, is that the data may be reliable for the crystallites that have been examined, but that such thin crystals are not representative of the sample as a whole. For example, in well-characterised samples of the 1:2:3 compound, $YBa_2Cu_3O_{7-x}$, we have found that the thinner crystallites, which are present in rather small numbers in this case, tend to be rich in barium. This may be a consequence of surface hydrolysis reactions, which are likely to be more severe for thin crystals. A further problem that may lead to errors arises when the emission lines of interest are too close to be properly resolved by the solid-state X-ray detector. This situation is found in the

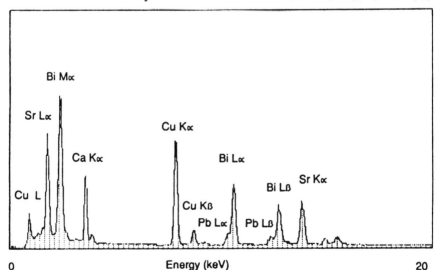

Figure 2: X-ray emission spectrum for "$(Bi,Pb)_2Sr_2Ca_2Cu_3O_{10}$" showing overlap of Pb and B:L lines

lead-doped Bi-Sr-Ca-Cu-O system and may lead to uncertainties in the Bi/Pb ratio (Table 5; Figure 2). Finally, the analysis may be rendered less reliable by the absence of a suitable standard. The high-T_c phases in the Tl-Ba-Ca-Cu-O system illustrate this point, since there appears to be no phase in the Tl-Cu-O system that can be used as a standard for the determination of Tl/Cu ratios. A k-value has been estimated indirectly by interpolation between values for Pt/Cu and Bi/Cu (37), but clearly this introduces a new element of uncertainty into the results.

Table 5: Idealised and actual compositions determined by AEM

Idealised Composition	Actual Composition	Ref.
$Bi_2Sr_2CaCu_2O_8$	$Bi_2Sr_{1.67}CaCu_2O_y$	(36)
$Tl_2Ba_2CaCu_2O_8$	$Tl_{1.76}Ba_2Ca_{0.7}Cu_2O_y$	(37)
$Tl_2Ba_2Ca_2Cu_3O_{10}$	$Tl_{1.83}Ba_2Ca_{1.44}Cu_3O_y$	(37)
$TlBa_2Ca_2Cu_3O_9$	$Tl_{1.08}Ba_2Ca_{1.95}Cu_3O_y$	(37)
$Tl_2Ba_2CuO_6$	$Tl_{1.71}Ba_{1.82}CuO_y$	(38)
$(Bi,Pb)_2Sr_2Ca_2Cu_3O_{10}$	$Bi_{1.83}Pb_{0.30}Sr_{2.04}Ca_{1.68}Cu_3O_y$	(38)

The previous paragraph may give the impression that analytical electron microscopy is fraught with pitfalls, but in practice our experience has been that the technique remains the best method for examining polyphasic mixtures and is reliable in the vast majority of cases. For example, on many occasions the correctness of the analytical results has been confirmed by subsequent crystallographic studies. A recent study of a novel mixed-metal oxide in the system La-Mo-O provides a nice illustration (39). In the case of the "2212" compound, the crucial question surrounds the apparent deficiency of strontium. There is no evidence from the Bi/Cu or Ca/Cu ratios to suggest that our data are suffering from any systematic errors, and the absorption cross sections for the Sr and Cu $K\alpha$ lines are sufficiently small that the Sr/Cu ratio can be estimated with a high degree of confidence. In the meantime, the controversy surrounding the defect structure of the "2212" compound continues to attract attention, but we are confident that further work by other methods will finally confirm the importance of cation vacancies in this system and vindicate the analytical electron microscopy technique.

ACKNOWLEDGMENT

We wish to thank the SERC for financial support and one of us (AMC) would like to thank New College, Oxford, for a Research Fellowship.

7.0 REFERENCES

1. Bednorz, J.G. and Muller, K.A., *Z. Phys. B: Cond. Matt.* 64:189 (1986).
2. Wu, M.K., Ashburn, J.R., Torng, C.J., Hor, P.H., Meng, R.L., Gao, L., Huang, Z.J., Wang, Y.Q. and Chu, C.W., *Phys. Rev. Lett.* 58:908 (1987).
3. Maeda, A.H., Tanaka, Y., Fukutomi, N. and Asano, T., *Jap. J. Appl. Phys.* 27:2 (1988).
4. Cheetham, A.K., Chippindale, A.M. and Hibble, S.J., *Nature* 333:21 (1988).
5. Bordet, P., Capponi, J.J., Chaillout, C., Chenevas, J., Hewat, A.W., Hewat, E.A., Hodeau, J.L., Marezio, M., Tholence, J.L. and Tranqui, D., *Physica C* 156:189 (1988).
6. Subramanian, M.A., Torardi, C.C., Calabrese, J.C., Gopalakrishnan, J., Morrissey, K.J., Askew, T.R., Flippen, R.B., Chowdhry, U. and Sleight, A.W., *Science* 239:1015 (1988).
7. Lee, P., Gao, Y., Sheu, H.S., Petricek, V., Restori, R., Coppens, P., Darovskikh, A., Phillips, J.C., Sleight, A.W. and Subramanian, M.A., *Science* 244:62 (1989).
8. Yamamoto, A., Onoda, M., Takayama-Muromachi, E., Izumi, F., Ishigaki, T. and Asano, H., to be published.
9. Takagi, H., Eisaki, H., Uchida, S., Maeda, A., Tajima, S., Uchinokura, K. and Tanaka, S., *Nature* 332:236 (1988).
10. Takayama-Muromachi, E., Uchida, Y., Ono, A., Izumi, F., Onoda, M., Matsui, Y., Kosuda, K., Takekawa, S. and Kato, K., *Jpn. J. Appl. Phys.* 27(3):L365 (1988).
11. Torrance, J.B., Tokura, Y., LaPlaca, S.J., Huang, T.C., Savoy, R.J. and Nazzal, A.I., *Solid State Commun.* 66(7): 703 (1988).
12. Zandbergen, H.W., Huang Y.K., Menken, M.J.V., Li, J.N., Kadowaki, K., Menovsky, A.A., van Tendeloo, G. and Amelinckx, S., *Nature* 332:620 (1988).

13. Morgan, P.E.D., Ratto, J.J., Housley, R.M. and Porter, J.R., in: *MRS Spring 1988 Meeting Proceedings: Better Ceramics Through Chemistry III*, Materials Research Society, Reno, NV (1988).
14. Hazen, R.M., Prewitt, C.T., Angel, R.J., Ross, N.L., Finger, L.W., Hadidiacos, C.G., Veblen, D.R., Heaney, P.J., Hor, P.H., Meng, R.L., Sun, Y.Y., Wang, Y.Q., Xue, Y.Y., Huang, Z.J., Gao, L., Bechtold, J. and Chu, C.W., *Phys. Rev. Lett.* 60:1174 (1988).
15. Reed, S.J.B., *Electron Microprobe Analysis*, Cambridge, CUP (1975).
16. Cliff, G. and Lorimer, G.W., *J. Microscopy* 105:205 (1975).
17. Goldstein, J.I., Costley, J.L., Lorimer, G.W. and Reed, S.J.B., in: *Proc. of the Workshop on Analytical Electron Microscopy*, Scanning Electron Microscopy (SEM/77) (O. Joharis, ed.), Vol 1, p 315, IIT Research Institute: Chicago Il (1977).
18. Chowdhury, A.J.S., Cheetham, A.K. and Cairns, J.A., *J. Catal.* 95:353 (1985).
19. Cheetham, A.K. and Skarnulis A.J., *Anal. Chem.* 53:1060 (1981).
20. Cheetham, A.K., *Nature* 288:469 (1980).
21. Hero, H., Sorbroden, E. and Gjonnes, J., *J. Mat. Sci.* 22: 2542 (1987).
22. Liu, J.C., Mayer J.W. and Barbour J.C., *J. Appl. Phys.* 64: 651 (1988).
23. Chen, S.H., Carter C.B. and Palmstron C.J., *J. Mat. Res.* 3:1385 (1988).
24. Cheetham, A.K., Cole, A.J. and Long, G.J., *Inorg. Chem.* 20:2747 (1981).
25. Long, G.J., Cheetham, A.K. and Battle, P.D., *Inorg. Chem.* 22:3012 (1983).
26. Livi, K.J.T. and Veblen, D.R., *Amer. Min.* 72:113 (1987).
27. Cheetham, A.K. and Thomas, D.M., *J. Solid State Chem.* 71:61 (1987).
28. Forghany, S.K.E., Cheetham, A.K. and Olsen, A., *J. Solid State Chem.* 71:305 (1987).

29. Hibble, S.J., Cheetham, A.K. and Cox, D.E., *Inorg. Chem.* 26:2389 (1987).

30. Cheetham, A.K., Skarnulis, A.J., Thomas, D.M. and Ibe, K., *J. Chem. Soc. Chem. Commun.* 1603 (1984).

31. Baumeister, W. and Seredynski, J., *Micron* 7:49 (1976).

32. Heinrich, K.F.J., Fiori, C.E. and Myklebust, R.L., *J. Appl. Phys.* 50:5589 (1979).

33. Heinrich, K.F.J. in: *The Electron Microprobe* (McKinley, T.D., Heinrich, K.F.J. and Wittry, D.B., eds), p 350, Wiley, N.Y. (1966).

34. Boivin, J.C., Trehoux, J. and Thomas, D., *Bull. Soc. Chim. Fr. Miner. Cristallogr.* 99:193 (1976).

35. Teske, C.L. and Muller-Bushbaum, H., *Z. Anorg. Allg. Chem.* 379:234 (1970).

36. Chippindale, A.M., Hibble, S.J., Hriljac, J.A., Cowey, L., Bagguley, D.M.S., Day, P. and Cheetham, A.K., *Physica C* 152:154 (1988).

37. Hibble, S.J., Cheetham, A.K., Chippindale, A.M., Day, P. and Hriljac, J.A., *Physica C* 156:604 (1988).

38. Cheetham, A.K., Chippindale, A.M., Hibble, S.J. and Woodley, C.J., *Phase Transitions* 19:223 (1989).

39. Hibble, S.J., Cheetham, A.K., Bogle, A.R.L, Wakerley, H.R. and Cox, D.E., *J. Am. Chem. Soc.* 110:3295 (1988).

15

Electron Microscopy of High Temperature Superconducting Oxides

Pratibha L. Gai

1.0 INTRODUCTION

All of the high temperature copper-oxide based superconductors discovered so far, have copper-oxygen sheets as a common structural feature, and these are believed to play a key role in high temperature superconductivity. Following the discovery of high temperature superconductivity in the lanthanum based ceramic oxides (1) there have been numerous reports on a variety of superconducting compositions containing lanthanum, yttrium, bismuth and thallium (2-6). The structures of the materials are related to perovskite structures with the general formula ABO_3. In general, high temperature superconductors (HTSC) consist of intergrowths of rock salt (AO) and perovskite (ABO_3) structures forming layers of $(AO)_n (ABO_3)_m$, where B = Cu. Superconductivity is believed to depend on the average formal oxidation state (or, the number of electron holes) in the CuO_2 sheets. For the materials to superconduct, the formal valence of copper should be greater than 2, i.e., the copper ions are in both 2^+ and 3^+ formal oxidation states. Changes can be introduced in the average formal oxidation state of the CuO_2 sheets by substitutional chemistry at the cation sites, or by the incorporation of excess oxygen, modifying the microstructure and superconducting properties of these fascinating materials.

Extensive microstructural and microchemical variations occur in the bulk superconductors which are often nonstoichiometric. The

562 Chemistry of Superconductor Materials

flexibility of the perovskite structure, and the variable coordination of copper, result in fine stoichiometric variations which affect the critical transition temperature (T_c) dramatically. In addition, the configuration of local coordination of copper by the surrounding oxygen ions also plays a major role in determining the electrical behavior. Copper can exist in six (octahedral), five (pyramidal) and four (square) coordinations. The chemical and structural heterogeneity, the origin of defect structures and their role in solid state reactions, and the multiphasic nature of many of the HTSC materials are only fully revealed by electron microscopy (EM) and related high spatial resolution microanalytical techniques. EM is essential to characterize fully the microstructure and chemical compositions of the constituent phases of new materials, complementing the molecular science information obtained with x-ray and neutron diffraction from relatively large amounts of material, and therefore inevitably with some averaging. EM is the only technique which allows single-crystal characterization from multi-phasic materials, which is impossible or difficult to obtain from other techniques. Mixed phases and interfaces can be studied directly on the atomic scale, to elucidate critical microstructure-chemistry relationships on which the copper oxidation state and superconducting properties depend. The nature of grain-boundary interfaces, which are a key element in many areas of the materials technology, is also important, since the superconducting coherence lengths - the distance over which two holes can be bound in Cooper pairs - are small. Microchemical analysis of the same regions is valuable to put the structural information into a context which can be related directly to the preparative procedures and any subsequent treatment of the sample. This chapter attempts to elucidate briefly the role of EM in the microstructural characterization of the bulk HTSC compounds.

2.0 TECHNIQUES OF HIGH RESOLUTION TRANSMISSION ELECTRON MICROSCOPY (HREM), TRANSMISSION EM (TEM) DIFFRACTION CONTRAST, AND ANALYTICAL EM (AEM)

Electron microscopy is basically a diffraction technique in which periodic crystals diffract electrons, according to Bragg's Law, to form a diffraction pattern. This may be regarded as the Fourier transform of the crystal. An image is formed by an inverse Fourier transform in the objective lens. The interaction of electrons with matter results primarily in elastic scattering, giving rise to diffraction effects; inelastic scattering and absorption effects are also observed.

2.1 HREM

HREM is increasingly being used for the direct imaging of crystal lattice structures at the atomic level (7)(13). In this technique, two or more Bragg reflections are used for imaging. The image contrast is controlled by the contrast transfer function $T(\Theta)$, containing phase contrast sinusoidal terms modified by an attenuating envelope function $f(\Theta)$ which depends on the coherence of the incident electrons:

$$T(\Theta) = f(\Theta) \sin \{ \pi D\Theta^2/\lambda + \pi C_s \Theta^4/2\lambda \},$$

where Θ is scattering angle, D is objective lens defocus, λ = wave length, and C_s is objective lens spherical aberration coefficient. It is a real space technique and the resolving power of modern electron microscopes can provide the solid state scientist with direct visualization of a crystal's local structure at the atomic level. In addition, very small amounts of samples (typically, only a few thousand contiguous unit cells) are sufficient for structural analysis, which is a major advantage. This technique has been imaginatively pioneered in solid state chemistry (8-11).

In the case of complex structures, however, complementary image simulations, using the dynamical theory of electron diffraction (7), are often essential for the ultrastructural analysis, so that experimental lattice images can be matched with theory. This is because the HREM image contrast depends very sensitively on EM operating parameters such as defocus, astigmatism and lens aberrations; as well as the sample thickness within which the electrons may undergo multiple scattering. Nevertheless, it is possible to use HREM to study reactions in solids and defect structures (12), including extended defects such as topologically compatible coherent intergrowths between two structures with different compositions (9)(11). The state-of-the-art EM has achieved resolutions of ~1.7-2.3 Å at an operating voltage of 200keV, and ~1.3-1.7 Å at 400keV, providing access to typical interatomic distances (13). The point to point resolution (d_o) of an EM depends on the spherical aberration coefficient (C_s) of the objective lens and the wavelength (λ) of the electrons, with a relationship $d_o = 0.64\, C_s^{1/4}\, \lambda^{3/4}$. Under optimum defocus conditions ($\Delta f = -1.2\,(C_s \lambda)^{1/2}$), and for a very thin crystal, the image contrast can be directly related to the projected structure of the atomic columns, with the dark regions corresponding to the contrast from the heavier atoms. A change in focus of $\Delta f/2$ reverses the contrast in lattice images (7)(14) turning dark into light regions to coincide with the positions of atom columns. Δf values for 200 keV ($C_s \approx 1.2$ mm), 300 keV ($C_s \approx 1$ mm), and 400 keV ($C_s \approx 1$ mm) are $\approx$ -66 nm, -53 nm and -48 nm respectively. The images are in general recorded near the Δf, unless stated otherwise. For electron microscopy, powders dispersed in chloroform are supported on carbon-filmed beryllium or aluminium grids. Alternately, the bulk compacts are thinned by argon ion milling to make them electron beam transparent. Normally, a double tilting specimen stage is used in the EM to set the crystals at the desired zone-axis orientations which yield informative, high symmetry projections.

2.2 TEM Diffraction Contrast

This is the conventional, classical TEM imaging technique where the Bragg condition is satisfied for a single diffracted beam and the image is formed using either the transmitted beam (bright field), or the diffracted beam (dark field) (15). Large areas (up to several microns) can be accessed for imaging. This diffraction contrast technique is used to provide information about the nature and the role of crystal imperfections or defects in solid state reactions in the materials, e.g., to characterize displacements at dislocations. These extended defects often create regions within crystal structures which are imperfect in register and thus involve characteristic displacements. This may be visualized by considering the plastic deformation of a crystal.

2.3 High Spatial Resolution Analytical EM

Analytical electron microscopy (AEM) complements the

microstructural and crystal structure information obtained from TEM and HREM with small probe chemical microanalysis using electron stimulated x-ray emission and energy dispersive x-ray spectroscopy (EDX), together with convergent beam electron diffraction (CBDP) for crystallography. Typical probe sizes for analysis can be <10 nanometer (nm) in diameter. Recent results using a dedicated field emission gun scanning transmission electron microscope (FEG STEM) have demonstrated analysis at the unit cell level with a resolution of only 1-2 nm (17c).

For any heterotype solid solution, or a nonstoichiometric compound, EDX analysis in the AEM on a large number of crystals is required. In a typical laboratory situation 30 to 40 crystals are routinely analyzed for each preparation. This sampling is adequate to establish trends in stoichiometric variations in a heterogeneous material. Fine gradations in compositions of a seemingly 'phase-pure' material by the criterion of bulk diffraction techniques, can also be revealed. For quantitative microanalysis, a ratio method for thin crystals (16) is used, given by the equation:

$$C_A/C_B = K_{AB}\, I_A/I_B$$

where C_A and C_B are concentrations of the elements A and B, and I_A and I_B are the background subtracted peak intensities for A and B measured experimentally from the x-ray spectra. The sensitivity factor K_{AB} is determined using standards. Well-characterized single phase binary standards are used for calibrations and quantification of analyses of complex oxides (16)(17). With the superconducting oxide compound, the data are normalized to the CuK_α peak. For bulk materials more complex (ZAF) correction procedures are required for atomic number (Z), absorption (A) and x-ray fluorescence (F). AEM also allows great flexibility in phase identification since individual crystallites can be analyzed using complementary EDX and electron diffraction from selected small regions of the samples (selected area diffraction patterns or SADP), or CBDP (18). Microdiffraction using convergent (or focused) probe (CBDP) is particularly attractive for local structure determination. In CBDP, the radius of the first higher order Laue zone (HOLZ) ring, G, is related to the periodicity along the zone axis (c), by $G^2 = 2/\lambda c$, where λ is the wave length of the incident electrons. CBDP can provide reciprocal space data in all three dimensions (x, y and z), with a typical lateral resolution of <50 nm and sometimes <10 nm. Thus, AEM can provide real space imaging, crystallographic and microcompositonal information on a very fine scale. Furthermore, with careful attention to experimental errors, AEM can be used to study partial occupancies within cation, and under special conditions, anion sites. In addition, scanning EM (SEM) can be used to characterize the topography of bulk compacts.

In the following sections examples of applications of HREM structure imaging and high spatial resolution EDX microanalysis often from the same areas, to the 'hole' superconductors are given.

3.0 MICROSTRUCTURAL AND STOICHIOMETRIC VARIATIONS

3.1 Substitutional Effects in La-Based Superconductors

The breakthrough in HTSC was the discovery of superconductivity at ~40K in the La-Cu-O system partially substituted by Ba (1). Subsequently, reports appeared on the structural, physical and electronic properties of La_2CuO_4 substituted by Sr (5)(19)(20). The parent La_2CuO_4 has a distorted K_2NiF_4 structure with an orthorhombic cell (a = 5.36 Å, b = 5.41 Å and c = 13.17 Å). It consists of corner-shared CuO_6 octahedra intersected by La-O layers forming single $(CuO_2)^{2-}$ sheets alternating with $(La_2O_2)^{2+}$ double layers with the rock-salt structure (21) (Figure 1a, which shows the projection, tilted along [110] from [010]). The Cu-O sheets are buckled and the octahedra are alternately rotated. Partial substitutions, for example of Sr^{2+} for La^{3+} in $La_{2-x}Sr_xCuO_4$, oxidizes the CuO_2 sheets and raises the average formal copper oxidation state. A tetragonal, oxygen deficient K_2NiF_4 structure is obtained for x >0.05 (Figure 1b, which is modeled on Figure 1a). Superconductivity is realized for 0.05 < x < 0.25 (19). The samples prepared from the constituent oxides in an oxygen environment at 1100°C show that the T_c of $La_{2-x}Sr_xCuO_4$ varies with Sr content, approaching a maximum value of T_c ~37K at x = 0.15 and falling off sharply on either side of this value of x (19)(22)(23). This behavior is shown in Figure 2. The T_c values are from a superconducting quantum interference device (SQUID). The rise in T_c towards x = 0.15 is attributed to an increase in the formal oxidation state of copper, or the hole content.

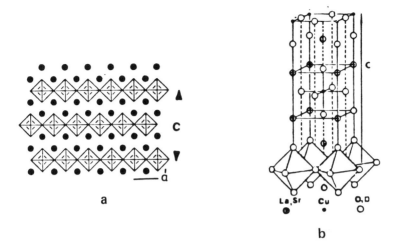

Figure 1: Idealized crystal structures of: (a) La_2CuO_4 tilted along [110] from [010] directions. (b) Oxygen defect K_2NiF_4 structure of $La_{2-x}Sr_xCuO_4$ (x >0.05), in [010], with c-axis vertical, and is related to Figure 1a: $a' = a_{orth}/\sqrt{2}$.

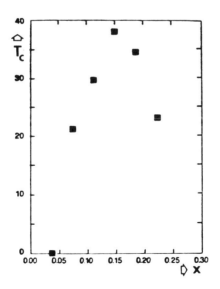

Figure 2: Variations of T_c with x in $La_{2-x}Sr_xCuO_4$.

568 Chemistry of Superconductor Materials

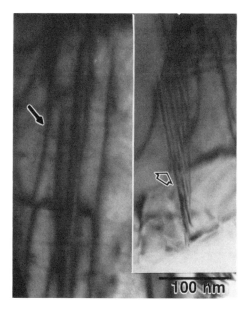

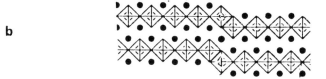

Figure 3: (a) Extended dislocation structures (dark lines) in a $La_{2-x}Sr_xCuO_4$ grain (x = 0.15), which can be interpreted as shear defects, involving CuO_2 layers. (TEM diffraction contrast image). Stacking fault contrast is inset. (b) A possible model for shear defects (c) Partial disintegration of grain giving rise to chemical inhomogeneity is observed for x = 0.2.

c

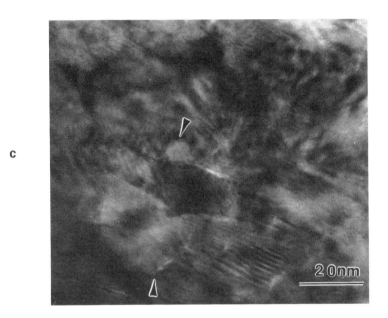

Figure 3: (continued)

The microstructural and microchemical variations associated with Figure 2 show extended defect structures in the matrix which increase in concentration up to x = 0.15, and deficiencies in Sr (up to ≈ 10%) and La cations, respectively (23). EDX also showed a few grains of the unreacted component oxides. An example of the dislocations using TEM diffraction contrast is shown in Figure 3a. They could be formed by a shear mechanism along a CuO_6 octahedron edge, with a displacement vector of the type $<a/2, o, c/6>$ which converts the CuO_6 from being corner-shared to edge-shared octahedra (Fig. 3b). The presence of the CuO_2 shear defects implies vacancies (point defect disorder) namely, oxygen deficiency in the CuO_2 lattice, and A-type cation deficiency.. Thus, along the defect lines there are changes in the coordination of oxygen to copper, as well as deviations from the ideal stoichiometry of both the oxygen and metal atoms. The mechanism accommodates the supersaturation of the point defects in the material. There will, however, be an equilibrium background concentration of the point defects in the material. The extended dislocations bound the resulting planar stacking faults which lie in {010} planes (inset in Fig. 3a). The electron diffraction patterns show streaking in directions normal to the defects, suggesting that the defects are not ordered. If the defects were ordered, a crystallographically defined compound might be obtained. The defects are observed only in the Sr-doped material and not in the stoichiometric parent compound. Samples with x = 0.2 exhibit chemical inhomogeneity as shown in Fig. 3c (without any discernible pattern of extended defects). Because the defects disrupt the continuity of the CuO_2 lattice, they can have a direct influence on the bulk superconducting properties. Indeed, initial results indicate enhanced flux pinning in the presence of the defects providing a chemical alternative to irradiating the samples (23b). It is possible that the coherent intergrowth structures reported in the La-Cu-O series (24) are formed by shear and ordering. This aspect is discussed further in Section 4.0.

3.2 Y-Based Superconductors

Yttrium-barium-copper oxide compounds $YBa_2Cu_3O_{6+x}$ (x = 0.6-1), generally abbreviated to 123, with a T_c of ~90K which is above the temperature of liquid nitrogen, are of great importance (2). As noted earlier, T_c is governed principally by the crystal structure whilst the critical current (J_c) also depends on the microstructure, grain-boundary character in terms of appropriate flux-pinning centers, porosity and interconnectivity of the material. The materials possess a high degree of anisotropy by virtue of their crystal structure and chemical bonding. They can be prepared from carbon-free precursors (to avoid contamination from carbon): yttrium oxide, barium oxide and copper oxide in the appropriate stoichiometric proportions (30). During sintering and densification of the bulk oxides, the free surfaces are replaced by grain boundaries and the interfacial area is reduced by grain growth. The driving force for sintering is the elimination of surface, resulting in a reduction of total energy. Two possible mass transport mechanisms occur during sintering: plastic flow, and diffusion. The hardness, sintering and anisotropic properties of the materials indicate diffusion as the dominant transport mechanism. Further, grain boundary diffusion, and lattice diffusion may dominate the different stages (e.g. the initial shrinkage and the final stage) of the sintering process. Interfaces forming any weak links between superconducting grains, or affecting grain connectivity, need to be characterized. For progress towards practical applications, it is necessary to understand and control the microstructure.

The crystal structure of 123 is orthorhombic, with a=3.82 Å, b=3.88 Å and c=11.67 Å (25). The schematic representation of the perfect structure is shown in Figure 4a. Above 600°C, 123 has a tetragonal structure ($YBa_2Cu_3O_6$) with a=3.872 Å and c=11.738 Å (26b). Below 600°C the transformation of the crystal structure results from the ordering of oxygen vacancies on the copper planes located between the barium planes (26). The 123 phase is often twinned and the twinning defects accompanying the tetragonal-orthorhombic transformation have been studied by EM (27)(28)(30). The lattice of the twin is a mirror image of that of the parent crystal structure as shown in the schematic in Figure 4b. Twinning is generally on {110} planes (Figure 4c) and results from the lowering of the free energy of the system associated with the strain from the transformation. Twin boundaries whose thickness is in the range of coherence lengths (typically ~3nm in the a-b plane and ~0.4nm along c for this material) are believed to be flux pinning centers, although this needs to be further investigated. The high ionic mobility of oxygen in the 123 compound allows the existence of a wide range of compositions with

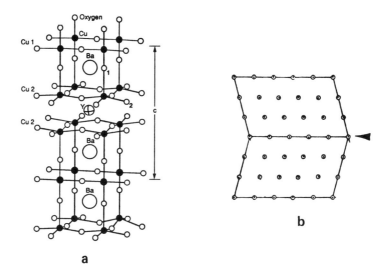

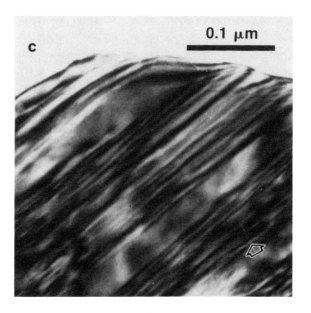

Figure 4: (a) Schematic of the perfect crystal structure of 123. (b) Schematic of a twin. (c) Low magnification TEM (diffraction contrast) image of twins observed on (110) planes. (d) HREM structure image of 123 in [010] at a defocus of -70 nm. Atoms are white and individual atom columns are shown. The corresponding selected area

(continued)

Electron Microscopy of High Temperature Superconducting Oxides 573

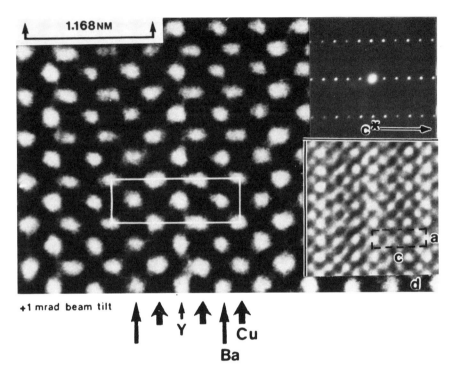

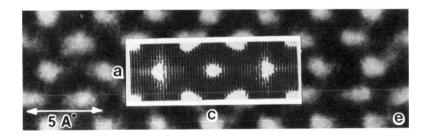

electron diffraction pattern (SADP) and the structure image at optimum defocus are inset: atoms columns are black. (e) The correspondence between the observed and calculated (inset) images is shown. The computer-simulated image uses dynamical theory on electron diffraction (Ref. 7a), and EM parameters (Ref. 14) and is for a sample of thickness ~5nm.

oxygen contents from O_6 to O_7. Insights into the microstructure of 123 were obtained by HREM (28-30). Figure 4d shows an HREM structure image on the atomic scale which is consistent with the x-ray and neutron data. The image is at a defocus value of -70nm at 400 keV and the columns of atoms are white confirmed by image simulations (Figure 4e). The corresponding electron diffraction pattern and the structure image at optimum defocus are inset in Figure 4d: the unit cell is outlined and the black dots are columns of atoms. HREM coupled with image calculations to explore oxygen arrangement in 123 has been reported (29) but the difficulties in interpretation of such images have also been outlined (31). This is because the scattering cross section of oxygen for electrons is low, and the contrast might be affected by EM parameters and sample thickness. Careful sample preparation for EM is also essential to avoid errors and artifacts.

Controlled processing of the air sensitive materials, combined with parallel characterization, provide insights into the microstructure and grain boundary character (30)(32). Figure 5a and 5b show grain boundary (gb) contact recorded using the SEM of a bulk compact and HREM respectively in a sample annealed in oxygen (30). Bridges between different grains (at A), surface particles (Cu-oxide composition) and grain boundary contact are revealed in Figure 5a. For HREM interface imaging, it is generally difficult to set up optimum zone axis conditions simultaneously for all the grains involved. It is often necessary to work with a compromise orientation which produces lattice resolution at an acceptable level, but which unfortunately may introduce some interpretational difficulties due to image overlap. The HREM image at the atomic level from a microtomed bulk sample (Figure 5b) reveals crystal to crystal connectivity at a high angle grain boundary between grains A and B, with no intervening amorphous layers. Grain boundaries with large misorientations have been shown to be detrimental to J_c (33). EDX reveals a small proportion ($\approx 3\%$) of the non-superconducting Y_2BaCuO_5 (211) second phase in the 123 preparations. Submicrometer superconducting 123 particles have also been prepared by a low temperature synthetic route using solution-derived carbon-free precursors (34). Recently, reports (35a) describe the role of clean boundaries in weak links and the controlled fabrication of bulk 123 to obtain textured HTSC 123 phases, incorporating particles of 211 and copper oxide as flux pinning

Electron Microscopy of High Temperature Superconducting Oxides 575

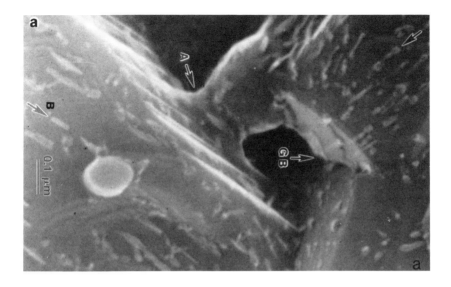

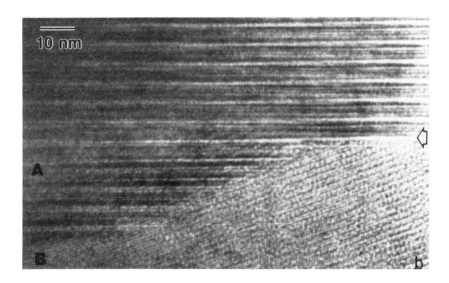

Figure 5: (a) High magnification SEM image showing extensive grain boundary (GB) contact. Growth bridges at A and surface particles (B) are shown. (b) HREM image of a high angle grain boundary in 123 with grain to grain connectivity. Grain A is in [120] and B in [001] orientation.

centers to achieve high J_c (up to 10^4 A/cm^2 at 77K and 1T magnetic field), i.e., an order of magnitude higher than the better current densities achieved for the materials so far. This represents an essential first step towards controlling microstructure for certain applications. Electron diffraction measurements on YBa$_2$Cu$_3$O$_x$ with $6.7 < x < 6.9$ show that T_c increases gradually from 60K to 90K and the nature of oxygen ordering evolves smoothly with x (35b). Silver-coating of 123 is found to enhance stability. Structures and compositions of YBa$_2$Cu$_4$O$_8$ (124), where CuO$_2$ sheets are interconnected by Cu-O double chains, have also been examined (36).

3.3 Substitution of Ca in Tetragonal YBa$_2$Cu$_3$O$_6$

Defects and Ordering: Substitution of Ca^{2+} on Y^{3+} sites changes the tetragonal 123-O$_6$ which is an antiferromagnetic insulator into a superconductor with a T_c onset at ~50K (37). However, substitutional chemistry is not always straightforward and the excess production of holes eventually leads to metallic properties. The role of EM in understanding the complexity of substitution and the limits of solid-solution formation is therefore critical. Examples of microstructural changes associated with various Ca levels in Y$_{1-x}$Ca$_x$Ba$_2$Cu$_3$O$_6$ materials elucidate these (38), and are given in the following. Further, they provide insights into nucleation, growth and ordering of defects leading to intergrowth structures.

Samples of Y$_{1-x}$Ca$_x$Ba$_2$Cu$_3$O$_6$ are prepared from the binary oxides, by first oxidizing the materials in air at ~950°C and reducing them in flowing argon at ~725°C (38). Samples with a nominal Ca content of $0.05 < x < 0.25$ exhibit a 'mottled' contrast from precipitates indicative of extensive disorder, evidence for which is also apparent in the corresponding SADP which is inset (Figure 6a). Using EDX to probe the samples indicates a variable Ca content. The Ba/Cu ratio is maintained at ~0.67: establishing that substitution of Ca occurs only at the Y-sites. The main impurities can also be revealed by EDX. The material with Ca ~0.3 indicates partial long range order with planar interface features (Figure 6b). These are shown at a higher magnification, in the [010] orientation, in Figure 6c. These planar faults are formed by an offset of 1/2 [100] between crystal blocks on either side of the fault in the [010] image. This leads to a structure of the Ca-doped 124 (YBa$_2$Cu$_4$O$_8$) type (38). The associated dislocations

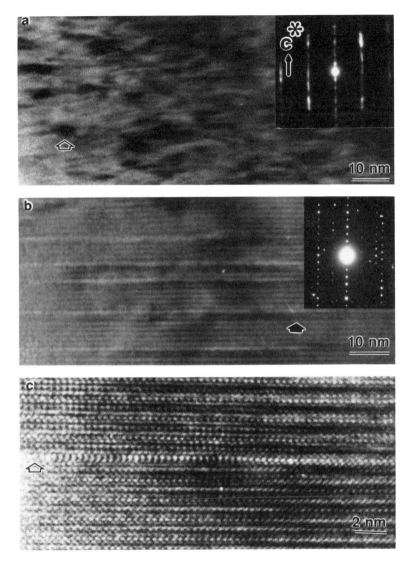

Figure 6: (a) "Mottled" contrast from the precipitate-like features indicating extensive disorder in an HREM image of $Y_{1-x}Ca_xBa_2Cu_3O_6$ for $0.05 < x < 0.25$. Inset shows SADP in [010] orientation showing streaking along c^* direction. (b) Planar interfaces observed in samples with Ca ~0.3. The SADP (inset) indicates ordering. (c) A planar interface defect at higher magnification in [010]. The fault is formed by a fault displacement vector of 1/2 [100] leading to Ca-doped 124 intergrowths.

(defects) have a characteristic displacement vector of the type <a/2, o, c/6>. EDX microanalysis from the interface regions show increased Cu content, presumably associated with the faults. Thus, it may be possible to prepare the Ca-doped 124 at ambient pressures. Further, EDX indicates an average compositional level of x ~0.24, with minority phases of unreacted 123 and CaO. Samples with x = 0.5 exhibit chemical inhomogeneity and disintegration of the structure. This would indicate that a solid solution limit and a homogeneous distribution of Ca is reached at Ca ~0.3 per formula unit. The material therefore possesses a very limited superconducting stoichiometry range.

Variations in unit cell parameters and interatomic distances are also consistent with the substitution of Ca^{2+} for Y^{3+}, as observed in the EM, causing hole-doping only in the copper-oxygen sheets, and that the Cu-O chains, once believed to be necessary for HTSC, only facilitate this oxidation (37). The flux exclusion signal increases up to ~0.3 and then decreases. The a-unit cell parameter decreases as x increases, shown plotted as a function of x in Figure 7a (38). This is consistent with the removal of electrons from the antibonding $d_{x^2-y^2}$ orbitals in the Cu-O sheets, and the resultant shortening of the Cu_2-O_2 bond length (Figure 7b). The interatomic distances for Cu_2-O_2 for the doped (Y/Ca) and the undoped (Y) are 1.934 Å and 1.943 Å, respectively as obtained from neutron diffraction (38). Thus, the CuO_2 sheets are more flattened in Y/Ca. The distances for Cu_1-O_1 remain the same (i.e., 1.798 Å for Y/Ca and 1.8 Å for Y). Removal of electrons from the orbitals in the plane of CuO_2 sheets stabilizes the Cu-O bond which leads to flattening. The O_2-O_2 distance between the sheets increases for Y/Ca to accommodate the substitution of the larger Ca^{2+} for Y^{3+}. The c-unit cell parameter is found to be ~11.81 Å for x ~0.3. The Cu_2-Cu_2 distance decreases (3.27 Å for Y/Ca and 3.293 Å for Y). This distance and the Y/Ca-Cu_2 distance may dictate the amount of Ca that can be substituted. The Cu_1-O_1 distance does not vary, indicating a constant amount of Cu_1. The substitution of Ca^{2+} for Y^{3+} causing oxidation only in the sheets is also apparent in the structurally related $Pb_2Sr_2Y_{1-x}Ca_xCu_3O_8$ (39).

3.4 Bi-Based Superconductors

Structural Modulations: The discovery of superconductivity

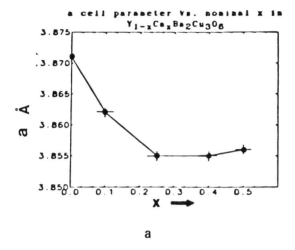

a

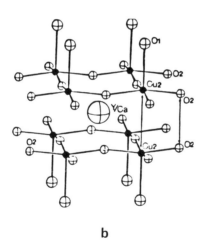

b

Figure 7: (a) Variation of the a-axis unit cell parameter, (related to the in-plane Cu(2)-O(2) distance) decreases as x increases in $Y_{1-x}Ca_x Ba_2Cu_3O_6$. (b) Portion of the $YBa_2Cu_3O_6$ structure showing the site of Ca-substitution; Cu_2 and O_2 atoms are indicated.

in the Bi-Sr-Ca-Cu-O system (3), with a general formula of the homologous series with the idealized formulations $Bi_2Sr_2Ca_{n-1}Cu_n$-O_{2n+4} (n = 1,2,3) is significant, since all the copper atoms lie in the CuO_2 planes (40). The n = 1, 2 and 3 phases have T_c values of ~10K, 90K and 110K respectively, determined both from magnetic flux exclusion and electrical resistivity measurements. All of the Bi-based superconductors exhibit superstructure modulations which are often incommensurate with the basic structure. The relationship of structural modulations with the oxygen content, and their role on the superconducting properties, are of considerable interest. A number of these modulations were revealed first by high resolution electron microscopy (41-43). HREM in appropriate zone-axis orientations, usually with the c-axis in the (x,y) image plane, was used to directly map the modulations.

The average structure, ignoring the modulations, has been reported from x-ray diffraction (44)(40). The structure is orthorhombic with a = 5.39 Å, b = 5.41 Å and c= 30.8 Å. The space group Amaa is confirmed by CBDP (45). The structure consists of perovskite layers separated by $(BiO)_2$ lamellae related to, but different from, those found in Aurivillius phases (Figure 8; based on Reference 44b). In $Bi_2Sr_2CaCu_2O_8$ (or 2212), each Bi_2O_2 layer consists of two BiO planes with rock salt arrangement of Bi and O atoms. Each perovskite layer consists of two CuO_2, two SrO and one Ca layers. The stacking sequence along the c-axis is as follows:

$$BiO-SrO-CuO_2-Ca-CuO_2-SrO-BiO$$

The n = 3 phase has additional Ca and CuO_2 layers. In general, the samples are prepared by solid state reactions of the component oxides in appropriate amounts at ~900°C in air. The compounds are micaceous (plate-like) in nature. The positioning of the lone pair of electrons for Bi^{3+} in the Bi_2O_2 layers results in large Bi-O interlayer spacing (of the order of ~3.2 Å). This means that the interlayer bonding is weaker and as a consequence, there is a preference for cleavage on (001) planes. It is often difficult to prepare single phase materials and extensive intergrowths are found to occur. AEM compositional analysis of 2212 shows that the phase contains strontium deficiency (Table 1). For quantitative EDX, calibrations are performed using standards of Bi_2CuO_4, $SrCuO_2$ and Ca_2CuO_3. It is

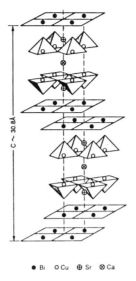

● Bi ○ Cu ⊕ Sr ⊗ Ca

Figure 8: Idealized structure of $Bi_2Sr_2CaCu_2O_8$ (2212) showing perovskite layers separated by Bi_2O_2 lamellae.

Table 1: '2212' Bi-Sr-Ca-Cu-O
Quantification of Chemical Composition analyses:
Spectra normalized to CuK_α Peak:

Bi/Cu	Sr/Cu	Ca/Cu
1.00	0.845	0.45
0.96	0.785	0.445
0.93	0.80	0.465
0.977	0.78	0.44
0.85	0.74	0.386
0.90	0.71	0.39
0.91	0.82	0.45
1.01	0.78	0.45
1.10	1.09	0.658
1.08	0.806	0.497
0.978	0.87	0.42

Average composition: $Bi_{1.95}Sr_{1.64}Ca_{0.92}Cu_2O_x$

suggested that the strontium defects might create mixed formal copper oxidation states (46) assuming, however, that the anion sublattice is fully occupied. The concentration of holes therefore, is not uniquely determined by the presence of cation deficiency. The hole concentration can only be obtained unambiguously from measurements of the oxidizing power per unit copper, considering both the cation and anion stoichiometries. This will be discussed further.

A number of models have been proposed to explain the crystal chemistry associated with the structural modulations. The following paragraphs survey the current understanding of the origin of the modulations:

The structural modulations in the Bi-compounds are very strong, and the local structure deviates significantly from the average structure. The a-axis exhibits an incommensurate modulation with a periodicity ~4.7 in 2212. The [001] electron diffraction pattern and the corresponding atomic image revealing the modulated structure are shown in Figure 9a. The spacing between the columns of atoms along the a and b axes in the image is ~2.7 Å. The modulations, leading to wavy patterns, are clearly visible in the structure image in the [010] orientation shown in Figure 9b. The image also shows modulations within the perovskite blocks, with the atom positions displaced from their ideal positions within the blocks (42). The heaviest atoms are in the Bi_2O_2 layers (arrowed). The corresponding SADP is inset in Figure 9b. The modulations are schematically represented in Figure 10a and show repeated shortening and widening of Bi-Bi distances. The modulations appear to be sinusoidal to a first approximation (42). In the case of a pure sinusoidal modulation, only the first order satellites would be visible in the electron diffraction pattern. Figure 10b shows the modulated wave, where Bi-Bi distances (from Figure 9b) are plotted against the number of Bi-planes.

In general, twinning of the modulated structures is present. Figure 11 shows a [001] 90° rotation twin in the 2212 sample. On the left part (A), the modulation is in the plane of view and on the right (B), it is along the viewing direction. No misfit dislocations are observed at the twin plane.

The grain boundary interfaces are clean and essentially flat on the atomic scale, but contain large misorientations, as shown in Figure 12 (48). Grain B is [110] 2212 phase, and grain C is strontium-copper oxide.

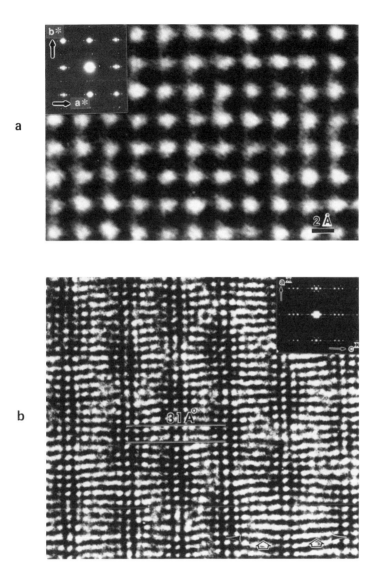

Figure 9: (a) Electron diffraction pattern (inset) with incommensurate superstructure modulation and structure image in [001] orientation. (b) Structure image of 2212 in [010] projection near the optimum defocus. Corresponding SADP is inset. Structural modulations are observed along the Bi_2O_2 (arrowed) as well as within the perovskite blocks. The image also shows intergrowths with some of the Bi-layers separated by $\approx$ 19 Å.

584 Chemistry of Superconductor Materials

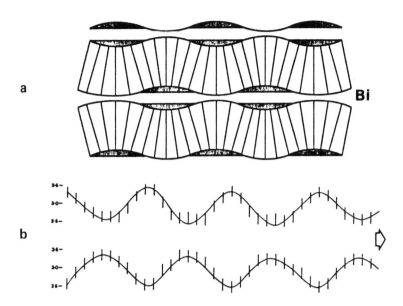

Figure 10: (a) Schematic representation of structural modulations showing repeated shortening and widening of Bi-Bi distances. (b) Modulations in Bi-Bi separation in adjacent layers, indicating a sinusoidal variation.

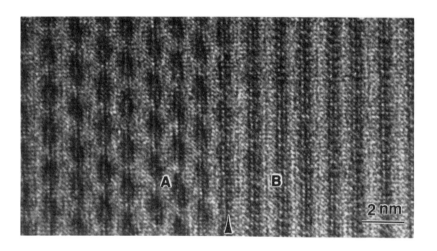

Figure 11: [001] 90° rotation twin in 2212. Modulation variant is in the plane of view (A) and along the viewing direction (B). A gradual decrease in the modulation towards the twin plane is seen.

Electron Microscopy of High Temperature Superconducting Oxides 585

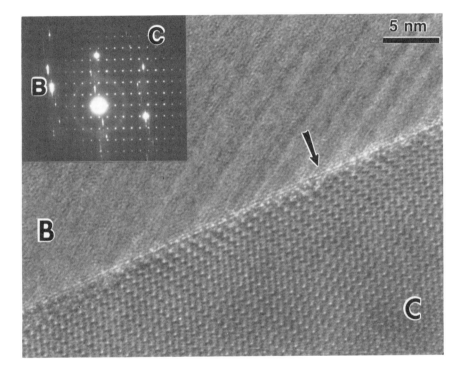

Figure 12: An interface in 2212 showing atomically sharp high angle grain boundary with grain to grain connectivity. The diffraction pattern from the grains is inset. Grain B is 2212 phase and grain C is strontium-copper oxide.

Interpretations of Structural Modulations: The following paragraphs summarize the reported models concerning the origin of the modulations and report some of the recent experiments. The origin of the modulations, the role of anion stoichiometry and of the cation substitution affecting the modulations and T_c are not yet fully resolved and need further experimental clarification.

Proposed models for the modulations include: (a) one of the earlier interpretations suggesting the rigid rotation of CuO_5 polyhedra resulting from, for example, small differences in the equilibrium bond lengths between the Bi-O and the Cu-O layers (i.e., due to a size mismatch) (42); (b) ordering of Bi-depleted and Bi-concentrated zones and site exchange of cations (43); (c) extra oxygen in the BiO layers and orientations of the lone pairs of electrons of Bi^{3+} consistent with HREM and image calculations (47); and (d) the presence of strontium vacancies (46). These have been summarized in the literature (47).

Oxygen Interstitials: Recently, based on (c), it has been proposed that the addition of one in ten oxygen atoms in the Bi-0 planes (periodic insertion of one oxygen atom pair in Bi_2O_2 layers, after every five unit cells along the a-axis) and the displacement of the surrounding atoms cause the superstructure modulations in the Bi-compounds with structural modulations of ~5 (49). Electron diffraction studies on the substituted Bi-compounds are also reported showing the dependence of the modulations on the compositions, i.e., partial substitution of Sr or Ca by a trivalent rare earth element (50)(51), Cu by another 3d metal (51), as well as Bi by Pb (52). In the $Bi_{2-x}Pb_xSr_2Ca_2Cu_2O_y$ systems, the superstructure modulations gradually disappear with increased Pb content, resulting in higher T_c (of up to 110K). This apparent increase in T_c might be due to the altered electronic states caused by the change in the modulation, or to lower oxygen content in the Bi-Pb-O layers, or to a change in the hole concentration caused by the Pb substitution (52). In the $Bi_2Sr_2M_xCu_{1-x}O_y$ (M = Co, Fe) structure, the substitution of Cu by Co reduces the periodicity of the commensurate modulation from 5 to 4 (with one extra row of oxygen inserted for every 8 rows of Bi atoms); for M = Fe the periodicity of the incommensurate modulation varies as a function of Fe, with drastic reduction in T_c (51). The relationship between the structural modulations, compositions and magnetic properties of these materials is also of considerable interest, and more work on the Bi-compounds is necessary to better understand such relationships. Recent muon spin resonance results (53a) have shown magnetic correlations in $Bi_2Sr_{3-x}Y_xCu_2O_8$ which are analogous to those observed previously in La-Sr-Cu-O and 123 systems (63b). The metal-insulator boundary has been found in these systems where the insulating state is associated with antiferromagnetism and long range magnetic order is apparently destroyed through Cu^{III} doping before superconductivity is established. Compositions in the Bi-Y-Sr-Ca-Cu-O system traversing the insulator-superconducting boundary have been investigated (53a).

To examine some of the hypotheses described above concerning superstructure modulations in the Bi-Sr-Ca-Cu-O materials, 2212 samples reacted under different environments have been investigated by EM (54): (a) sample annealed (reduced) in nitrogen at 400°C for 12 hours and slow cooled, with a T_c value of 90K and (b) sample annealed in oxygen at 400°C for 12 hours and slow cooled, with a T_c value of 70K. The SADP from [001] orientations are shown in Figure 13a,b. Both the samples exhibit modulations. A preliminary analysis of the two figures indicates that the nitrogen reduced sample has a commensurate superstructural modulation (with a periodicity of 5) and the oxygenated sample has an incommensurate modulation (with ~4.5). The lattice spacings are also different. The Cu-O distance of the oxygenated sample also show a slight decrease in the x-ray data relative to the reduced sample. This indicates a complex dependence of the periodicity of the modulations on the oxygen content and T_c. A CBDP from the oxygenated sample in the [001] orientation is also shown in Figure 13c. The first ring relating to the higher order Laue Zone (HOLZ), used to provide information in the third dimension, is shown by the arrow. The measurements from the pattern indicate a slight decrease in the c-lattice parameter and the images show an increase in the modulation amplitude. The oxygenated sample is relatively free of intergrowths. The EDX data shows cation deficiencies of Sr and Ca, which do not appear to affect the periodicity of the modulation. Further, the samples prepared entirely in N_2 exhibit commensurate modulations (54), which may mean, either (a) the modulations are intrinsic to the materials, or (b) it is difficult to extract the extra oxygen even after prolonged reduction.

Although the full implications of these data are yet to be analyzed, it seems likely that the lattice or size mismatch between the Bi-O and the perovskite layers plays a key role in constituting the

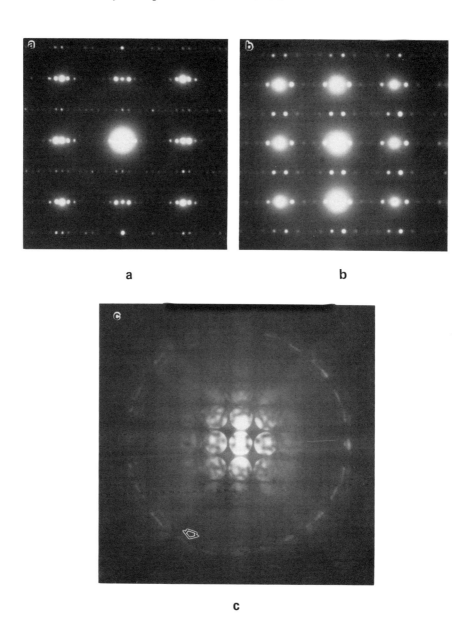

Figure 13: SADP in [001] orientation for (a) the sample annealed in (a) N2; (b) in O2. The patterns are at the same magnification. (c) [001] convergent beam diffraction pattern of the sample annealed in oxygen. HOLZ is arrowed.

driving force for the modulations. Further, a comparison of the T_c values of the two samples implies that the interstitial oxygen causes a decrease in T_c, presumably because of the over-doping of holes, or the over-oxidation of the CuO_2 sheets. (This is discussed further in Section 4.0). Partial oxidation of Bi^{3+} to B^{5+} is possible; however x-ray photoelectron (XPS) measurements (51) have shown that Bi is only in 3+ state in these cuprates.

4.0 Tl-BASED SUPERCONDUCTORS

The discovery of thallium containing superconductors (4) was another important development. Several superconducting phases exist and consist of intergrowths of rock salt (Tl-O) and perovskite layers. They have been reported with zero resistance and Meissner effect up to 125K, i.e., with the highest critical temperatures discovered so far.

Two homologous series exist, consisting of single and double thallium layers:

$TlBa_2Ca_{n-1}Cu_nO_{2n+3}$, (n = 1-6); and $Tl_2Ba_2Ca_{n-1}Cu_nO_{2n+4}$, (n = 1-4) (e.g., 55). The phases $Tl_2Ba_2CuO_6$, (221); $Tl_2Ba_2CaCu_2O_8$ (2212); $Tl_2Ba_2Ca_2Cu_3O_{10}$, (2223); $Tl_2Ba_2Ca_3Cu_4O_{12}$ (2234) and $TlBa_2Ca_4Cu_5O_{13}$ (1245) have T_c values of ~85K, 98-110K, 105-120K, 119K and 114K, respectively.

In addition, bulk superconductivity has recently been discovered at Du Pont in the (Tl,Pb)-based oxides (56) with the general formula, $(Tl,Pb)Sr_2Ca_{n-1}Cu_nO_{2n+3}$; (n = 2,3). The T_c values of 90K and 122K are obtained for the n=2 and n=3 phases, respectively.

The basic structures of Tl-compounds are tetragonal body centered layer structures with layers normal to the c-axis. The lattice constants are a = 3.8Å, and c = 29.4Å (for 2212), 35.4Å (for 2223), 22Å (for 1245) and so on. The common structural features are Tl-O layers sandwiched by Ba-O sheets and blocks of Cu-O layers separated by Ca intermediate layers. The structures differ from each other by the number of Cu-O layers. For example, the number of Cu-O layers in the 2212 phase is two; three in 2223 and five in the 1245 structure. A schematic representation of the 1245 structure is shown in Figure 14. The samples are normally prepared from the component oxides in sealed gold tubes to avoid problems with the toxicity and loss of Tl

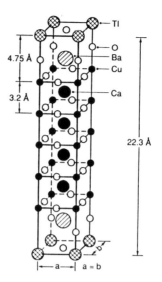

Figure 14: A schematic representation of the idealized $TlBa_2Ca_4Cu_5O_{13}$ structure.

Table 2: Quantification of EDX
Nominal Composition: $Tl_2Ba_2Ca_3Cu_4O_{12}$
Spectra Normalized to the Copper K_α Peak

Tl/Cu	Ba/Cu	Ca/Cu
0.4182	0.4801	0.6346
0.4292	0.5063	0.6359
0.4212	0.3888	0.6574
0.4248	0.4428	0.6710
0.4898	0.5198	0.6201
0.4151	0.4275	0.7663
0.4363	0.4445	0.6898
0.4423	0.5072	0.6523
0.4059	0.6691	0.7506
0.4426	0.4743	0.7790

Giving the Average Formula: $Tl_{1.78}Ba_{1.96}Ca_{2.72}Cu_4O_y$ with Tl and Ca deficiencies.

(56). Alternative chemical techniques also exist (57).

Major challenges in the complex Tl-series are the structural and compositional control and the contributions of structural irregularities (such as cation disorder, oxygen vacancies, syntactic intergrowths and strain effects due to lattice constant changes) to the superconducting properties. HREM and related techniques can provide some of the answers. In this section, it would be exceedingly difficult to give a comprehensive account of all the EM contributions. However, some of the major structural features revealed by EM are highlighted.

4.1 Tl-Ba-Ca-Cu-O Superconductors

By means of HREM and electron diffraction a variety of intergrowth defects have been characterized in both the single and double layer Tl-systems. In some (Tl,Ba) systems weak superstructure modulations have been reported (47)(48)(58).

2234 and 1245 Superconductors; Site Disorder and Intergrowth Defects: *2234:* The bulk topography of a typical sintered compacted pellet by SEM is shown in Figure 15a, illustrating the grain connectivity. The morphology shows angular crystals several micrometers in extent and evidence of less reactive sintering. HREM and SADP of the compound along [010] are shown in Figure 15b and c, respectively (48). The SADP confirms the lattice constants of a = 3.8Å and c = 42Å, obtained from x-ray diffraction (59). An EDX spectrum of the phase is shown in Figure 15d.

In addition, a high density of coherent intergrowths is observed, as shown in Figure 16a with the corresponding electron diffraction pattern. The weak superlattice reflections are present in some electron diffraction patterns. The intergrowth defects contain a different number of CuO_2 sheets and thus alter the local chemical composition. High precision microanalysis on the nanometer scale from two layers (Figure 16b) illustrates the difference in Cu-content as shown in Figure 16c and d respectively. EDX from several grains reveals cation site disorder as shown in Table 2. Small deficiencies in Tl and Ca are observed, giving an average composition $Tl_{1.78}Ba_{1.96}Ca_{2.72}Cu_4O_y$. This would increase the formal copper oxidation state to ~2.33, if the anion sites are fully occupied (48). Ba stoichiometry

592　Chemistry of Superconductor Materials

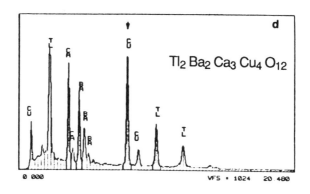

Figure 15: (a) SEM of the bulk $Tl_2Ba_2Ca_3Cu_4O_{12}$ (2234). (b) Lattice image of 2234 in [010] with SADP (c) inset. (d) EDX spectrum.

Electron Microscopy of High Temperature Superconducting Oxides 593

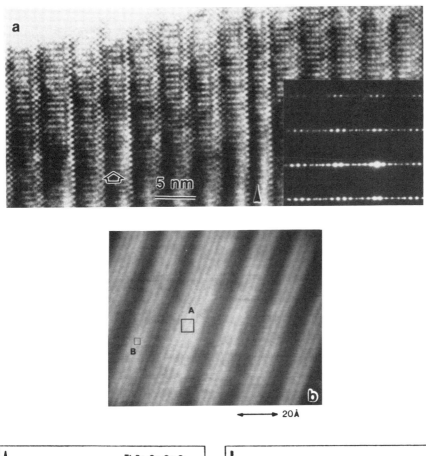
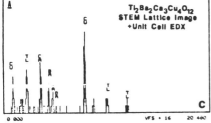
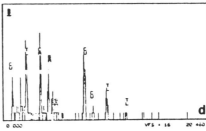

Figure 16: (a) High density of intergrowths with various perovskite slab widths (arrowed), with the SADP (inset). (b) High resolution lattice image from FEG STEM showing layers with different perovskite slabs A and B. (c) and (d) EDX nanometer analysis from A and B, showing increased Cu content in A with more Cu-O layers.

is essentially ideal, and partial exchange between Tl and Ca is possible.

EM also elucidates the nature and composition of the grain boundary interfaces in these materials for the first time (48). Figure 17a shows an HREM image of a grain boundary at a triple point. The three grains are identified by EDX and diffraction. The triangular amorphous region at the interface triple point consists of mostly copper (presumably copper oxide) as shown in Figure 17b and the amorphous region is ~10nm in extent. This is in contrast to the observations in Y- and Bi-superconductors where the grain boundaries are clean. The presence of chemically different grains indicate phase separation, probably during sintering. Furthermore, the absence of dislocations at the boundary is consistent with the rigid close packed structures of these ceramics and the limited extent of strain fields.

1245 Superconductor: The layer stacking of the compound is elucidated in the HREM image along [010] in Figure 18a (48). The unit cell is outlined and the number of Cu-layers is indicated by the arrows. The corresponding selected area diffraction pattern is inset, with lattice constants a = 3.85Å and c = 22.3Å, in agreement with x-ray diffraction (59). The presence of coherent, yet nonstoichiometric, intergrowths, with four and six Cu-O layers, is frequently observed. The 4-layer intergrowths of the 1234 structure in the 1245 are shown in Figure 18b with the associated structural schematic. Intergrowths of up to nine Cu-O layers have been reported, indicating the possibility of even higher order Cu-O layers in the system.

The concept of topologically compatible coherent intergrowth defects in nonstoichiometric solids is of considerable significance, both from the structural and thermodynamic viewpoints. Structural compatibility enables the elements of two distinct structures to coexist as coherently intergrown domains as shown in the HREM images. Coherence can be achieved by local ordering of defects, e.g., point defect ordering (vacancy or interstitial ordering), or by substitutional ions within coherent crystal structures, or by a polyhedral, or compound shear mechanism. This would also in principle lead to models for defects. The intergrowths are revealed reliably only by HREM. If the intergrowths are random, the material is a nonstoichiometric and a non-equilibrium phase. If they are ordered in a regular fashion, a new compound is produced. In the absence of other defects it should be possible to account for the stoichiometric variations from the ideal stoichiometry in terms of coherent intergrowths. Dislocation

Electron Microscopy of High Temperature Superconducting Oxides 595

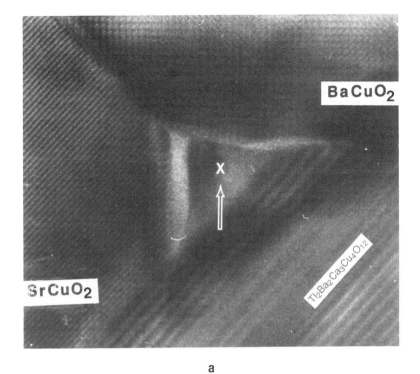

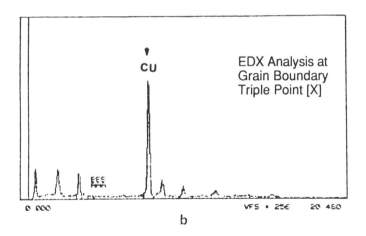

Figure 17: (a) Grain boundary interface with amorphous region at the triple point. (b) EDX from the area x, showing mostly copper (oxide).

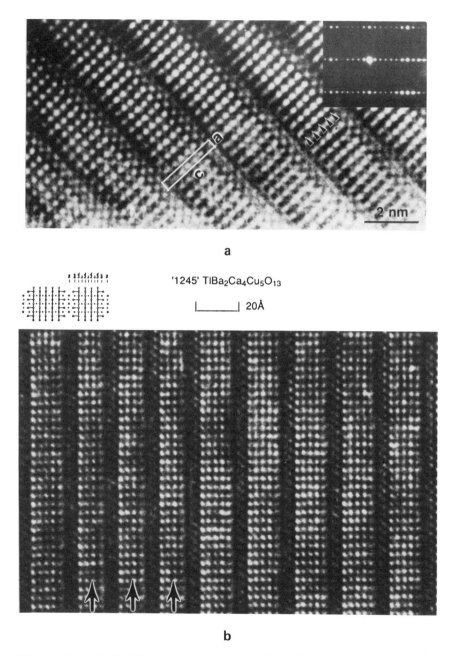

Figure 18: (a) HREM lattice image of $TlBa_2Ca_4Cu_5O_{13}$ (1245) with SADP inset. The unit cell is outlined. (b) Coherently intergrown 4 layer intergrowths (arrowed) of 1234 structure in 1245.

structures facilitating insertion or removal of layers are also present (48).

As far as the relationship between T_c and the number (n) of Cu-O layers in the perovskite structure is concerned, experiments on both the Tl_1 and Tl_2-series show that T_c increases with n up to n = 4 and n = 3, respectively. Further increases in n do not raise T_c. This is related to the optimum number of holes permissible in the CuO_2 sheets for maximum T_c, and to the in-plane copper-oxygen bond lengths (60). Further, a comparison of the electronic structures of the Tl_2 and Tl_1-series obtained from tight-binding calculations (60b) show the Tl-6s block bands for the Tl_2O_4 double layers, to be below the Fermi level, and above for the TlO_3 single layers. This means that TlO_3 layers do not remove electrons from the $d_{x^2-y^2}$ bands of the CuO_2 sheets and the stoichiometric Tl_1-series has a high enough formal Cu-oxidation state and can superconduct without the introduction of additional holes. From EM the density of intergrowths in the 1245 is low relative to the 2234 or 1234 compounds. Empirically, this might imply that a lower Ca content leads to more intergrowths. However, the existence of such defects must depend critically on the synthesis conditions, the energies of formation of the defects and configurational entropy associated with them due to the number of ways in which they can be distributed in the material. Microanalysis reveals stoichiometric variations in cation sites in the unit cell, providing a mechanism for creating formal mixed valence copper. However, a more systematic study of oxygen stoichiometry is required. In principle, titration methods can be used to provide this information. However, difficulties may arise for the thallium phases since the redox chemistry of Tl conflicts with that of Cu.

4.2 (Tl,Pb)-Sr-Ca-Cu-O Superconductors: $(Tl,Pb)Sr_2CaCu_2O_7$ (n=2) and $(Tl,Pb)Sr_2Ca_2Cu_3O_9$ (n=3)

These compounds are isostructural with the corresponding (Tl,Ba) compositions, possessing two or more CuO_2 sheets separated by (Tl,Pb)-O and Sr_2O_2 sheets. The (Tl,Pb) system possesses a narrow composition range. The phases n=2 and 3 are line phases, and thus correspond to the equilibrium states in the (Tl,Pb) system. HREM image corresponding to the $(Tl,Pb)Sr_2Ca_2Cu_3O_9$ (n = 3) phase is shown in Figure 19a. An occasional intergrowth with four Cu layers can be

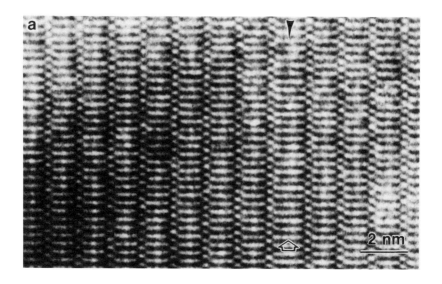
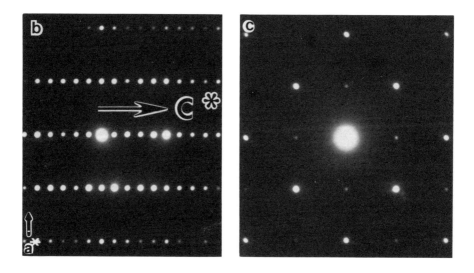

Figure 19: (a) HREM image of $(Tl,Pb)Sr_2Ca_2Cu_3O_9$ showing a 4-layer intergrowth (arrowed). (b) SADP in [010]. (c) [001] SADP with no superstructure spots, indicating that structural modulations are not present. (d) EDX spectrum showing the presence of Tl and Pb. Quantification is shown in Table 3. (e) Flux exclusion behavior for n=2 and n=3.

Electron Microscopy of High Temperature Superconducting Oxides 599

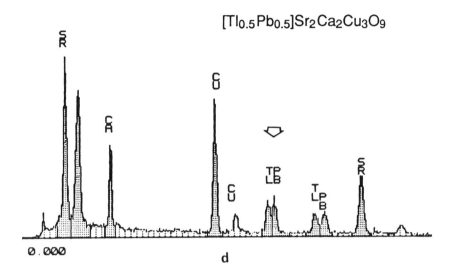

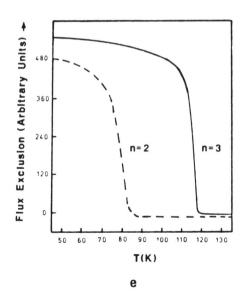

Figure 19: (continued)

seen in the image (arrowed) suggesting the possibility of higher order Cu-O perovskite blocks in the (Tl,Pb) system. Figure 19b and c show an [010] and [001] SADP, respectively for the n = 3 phase (61). The significant information obtained from EM is that the materials are relatively free of intergrowths compared to the (Tl,Ba) systems, and have a smoother microstructure, (with larger grain sizes up to ~200 μm). This indicates that Pb stabilizes the compositions. In addition electron diffraction experiments with tilting to various orientations, including the [001] shown in Figure 19c and [110], do not reveal superstructure modulations of the type observed in some (Tl,Ba) and Bi-based superconductors. Furthermore, EDX microcomposition analysis based on the characteristic x-ray lines (Figure 19d) provides a powerful tool enabling Tl and Pb, which are adjacent elements in the Periodic table, to be clearly distinguished. This distinction is not attainable with either x-ray or neutron diffraction techniques which depend on relative structure factor intensities. Quantification of EDX analyses reveals site disorder in the (Tl,Pb) and Ca-sites giving average compositions of $(Tl_{0.4}Pb_{0.4})Sr_{1.94}Ca_{0.8}Cu_2O_x$ for the n = 2 phase and $(Tl_{0.4}Pb_{0.4})Sr_{1.95}Ca_{1.67}Cu_3O_x$ for the n = 3 phase (Table 3), as confirmed by neutron diffraction. The formal Cu-oxidation states are deduced to be 2.78 and 2.65 for the phases, with essentially full occupancy for Sr and assuming full anion site occupancy. Flux exclusion behavior of the phases is shown in Figure 19(e). The grain interfaces in the (Tl,Pb) materials contain narrow (~2 nm) amorphous Cu-oxide layers. These are of the order of coherence lengths in the superconducting oxides.

The weak modulations in the (Tl,Ba) system have been discussed based on various hypotheses, including partial reduction of Tl^{3+} to Tl^{1+}, insertion of extra oxygen in the Tl-O layers, orientation of a lone pair, or ordering in the rock salt layers where the atoms are displaced from their ideal position to create a more favorable bonding environment (47)(58)(6). However, in the (Tl,Pb) system, X-ray Absorption Near Edge Spectroscopic (XANES) studies are consistent with the formal oxidation states of Tl^{3+} and Pb^{4+} (61). The outer electronic configurations of Tl^{3+} and Pb^{4+} are similar, thus facilitating smoother incorporation of Pb atoms into the structure. The existence of correlated atomic displacements in the (Tl,Pb)-O layers is suggested. Based on the average tetragonal structure, these result from chains or pairs of (Tl,Pb) and O atoms; local ordering of the pairs is only

Table 3. Chemical composition analysis n = 3

Tl/Cu	(Tl+Pb)/Cu	Ca/Cu	Sr/Cu
0.13	0.27	0.54	0.63
0.12	0.26	0.567	0.63
0.12	0.246	0.60	0.60
0.136	0.27	0.58	0.62
0.125	0.255	0.51	0.64
0.145	0.27	0.54	0.66
0.15	0.29	0.56	0.77

Table 4. Microchemical analyses using the nanoprobe in the STEM for a sample of nominal composition $Tl_{1.6}Cd_{0.4}Ba_2CuO_y$

Tl/Bu	Cd/Cu	Tl+Cd/Cu	Ba/Cu
1.50	0.38	1.88	1.73
1.69	0.31	2.00	1.86
1.47	0.41	1.88	1.60
1.51	0.37	1.88	1.87
1.72	0.32	2.04	1.85
1.55	0.25	1.80	1.85
1.41	0.42	1.83	2.10
1.55	0.33	1.88	2.05
1.58	0.34	1.92	1.95
1.65	0.33	1.98	1.92

Mean values[a]

| 1.56(10) | 0.35(5) | 1.91(8) | 1.88(14) |

[a]The standard deviations (in brackets) of the mean ratios (with respect to an assumed Cu stoichiometry of 1.0 per formula unit) are calculated from the variance, $\sigma^2 = [\Sigma_i(<r>-r_i)^2/n-1]$ where $<r>$ is the mean, and r_i is an individual analysis and the sum (Σ) is over the n values.

short range (6). It seems likely that the cooperative displacements and the extent of local order within this unit with lower oxygen content might not be sufficient to induce structural modulations of the type observed in some (Tl,Ba) systems. Since the (Tl,Pb) and the corresponding (Tl,Ba) systems have similar T_c values, structural modulations appear to be secondary to superconductivity in these systems. Substitution of Pb might result in intrinsic flux pinning sites due to local variations in free energy terms associated with the substitution. Further, the increased stability of the (Tl,Pb) systems may be related to the enhanced bond strengths in the (Tl,Pb)-O layers.

Substitutional chemistry in Tl-compounds over a wide composition range is also of interest and superconducting phases isostructural with the Tl-series can be prepared by a suitable choice of cations and characterized by electron microscopy (62)(63). Substitutions of Cd in $Tl_{2-x}Cd_xBa_2CuO_{6+\delta}$ (62) show the T_c dependence on hole concentration, which can be adjusted either by cation doping (x) or by oxygen excess (δ). Metallic behavior is observed for x > 0.6. EDX confirms that Cd^{2+} substitutes exclusively for Tl (Table 4). Figure 20 shows SEM (a) and an HREM image (b) for the system. The HREM image shows a well ordered structure. The corresponding electron diffraction pattern is inset. The HREM and EDX on the nanometer scale show smooth incorporation of Cd in the Tl-layers. A comparison with the undoped structure shows a decrease in the Cu-O bond length correlating with the oxidation of the CuO_2 sheets. Based on the variations a-axial dimension as a function of x, in combination with single crystal x-ray analysis, the optimum 'hole' concentration on the CuO_2 sheets is estimated to be 0.3 ± 0.1 per Cu (62), which gives the highest T_c of ~92K. An increase in the c-axis is also observed, consistent with the substitution of the larger Cd^{2+} for Tl^{3+}. This material also exhibits a wide range of superconducting stoichiometry compared to $Y_{1-x}Ca_xBa_2Cu_3O_6$ and $La_{2-x}Sr_xCuO_4$. It seems likely, therefore, that the range of doping levels over which bulk high temperature superconductivity is observed is material specific.

In conclusion, electron microscopy techniques reveal considerable evidence for variability in composition and defect structures, such as cation and anion vacancies, extended defects, substitutional ions and oxygen interstitials; these are common to all of the high temperature superconducting oxides. These defects play an important role in controlling the carrier concentrations and therefore the

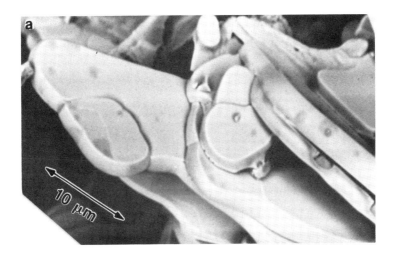

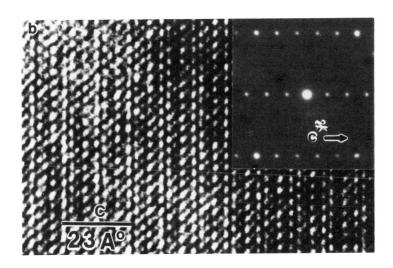

Figure 20: (a) SEM of $Tl_{2-x}Cd_xCuO_{6+\delta}$ plates. (b) HREM with the corresponding electron diffraction pattern inset.

superconducting properties. Understanding of the microstructural heterogeneity should yield a better fundamental understanding of the structure-property relationships in the HTSC oxides and facilitate a more plausible theoretical explanation of the phenomenon of high temperature superconductivity.

5.0 REFERENCES

1. Bednorz, J.G. and Muller, K.A., *Z. Phys.* B64:189 (1986).

2. Wu, M.K., Ashburn, J.R., Trong, C.J., Hor, P.H., Meng, R.L., Gao, L., Huang, Z.J., Wang, O., and Chu, C.W., *Phys. Rev. Lett.* 58:908 (1987).

3. Maeda, H., Tanaka, Y., Fukutomi, M., and Asano, T., *Jpn. J. Appl. Phys.* 27:L209 (1988).

4. Sheng, Z.Z. and Hermann, A.M., *Nature* 332:138 (1988).

5. Cava, R.J., Van Dover, R. B., Batlogg, B., and Reitman, E. A., *Phys. Rev. Lett.* 58:408 (1987).

6. Sleight, A.W., *Science* 242:1519 (1988).

7. Cowley, J.M., *Diffraction Physics*, North Holland, Amsterdam (1981).

8. Anderson, J.S., in *Nonstoichiometry*, ed. A. Rabenau, North Holland, Amsterdam (1970); *J. Chem. Soc. Dalton Trans.* 1107 (1973).

9. Wadsley, A.D., *Nonstoichiometric Compounds*, ed: L. Mandelcorn, Academic Press (1964).

10. Thomas, J.M., *J.C.S. Faraday* 70:1691 (1974).

11. Allpress, J.G., *J. Solid St. Chem.* 1:66 (1970).

12. Gai, P.L. and Anderson, J.S., *Acta Cryst.* 32:157 (1978).

13. (a) Boyes, E.D., Watanabe, E., Skarnulis, A.J., Hutchison, J., Gai, P.L., Jenkins, M. and Naruse, M., *Inst. Phys. Conf. Proc. Ser.* 52:445 (1980) (London, U.K.); *Jeol News* 24E:9 (1986). (b) Boyes, E.D., *J. Microscopy* 127:321 (1982); *Nature* 304:289 (1983).

14. Gai, P.L., Goringe, M.J., and Barry, J.C., *J. Microscopy* 142:9 (1986).

15. Hirsch, P.B., Howie, A., Nicholson, R.B., Pashley, D.W. and Whelan, M.J., *Electron Microscopy of Thin Crystals*. R. Kreiger, New York, (1977).

16. Cliff, G., and Lorimer, G.W., *J. Microscopy* 103:203 (1975).

17. (a) Hren, J., Goldstein, J.I., Joy, D.C. *Introduction to Analytical Electron Microscopy* (1979), Plenum Press. (b) Gai, P.L. and Boyes, E.D., *Analytical Electron Microscopy*, ed: R. Geiss, San Francisco Press (1981). (c) Boyes, E.D. *Institute of Physics*, London, Series 98, Chapter 12 (1989).

18. Steeds, J.W., *Introduction to Analytical EM*, ed: Hren, J., Goldstein, J.I., and Joy, D.C., 387 Plenum (1979).

19. Torrance, J.B., Tokuna, Y., Nazzal, A.I., Gezinge, A., Huang, T.C., and Parkin, S.S.P., *Phys. Rev. Lett.* 61:1127 (1988).

20. (a) Day, P., Rosseinsky, M., Prassides, K., David, W., Moze, O., Soper, *A. J. Phys. C* 20:429 (1987). (b) Tarascon, J.M., et al., *Science* 235:1373 (1987).

21. (a) Longo, J.M. and Raccah, P.M., *J. Solid St. Chem.* 6:526 (1973). (b) Michel, C. and Raveau, B., *Rev. de Chimie Minerale.* 21:407 (1984).

22. Crawford, M.K., Farneth, W.E. and McCarron, E.M., *Phys. Rev. B.* (in press).

23. (a) Gai, P.L. and McCarron, E.M., *Science* 247:553 (1990). (b) Gai, P.L. and Kunchur, M., 1990, unpublished results.

24. Davies, A.H. and Tilley, R.J.D., *Nature* 326:859 (1987).

25. (a) Capponi, J.J., Chaillout, C., Hewat, A.W., Lejay, P., Marezio, M., Nguyen, N., Raveau, B., Souberoux, G.L., Tholence, J.L., and Thournier, R., *Europhys. Lett.* 3:1201 (1987). (b) Hazen, R.M., Finger L.W., Angel, R., Prewitt, C.T., et al., *Phys. Rev.* B35:7238 (1987).

26. (a) Strobel, P., Capponi, J.J., Marezio, M., and Tholence, J.L., *Nature* 327:306 (1987); (b) Bordet, P., et al., *Nature* 327:687 (1987).

27. (a) Van Tendeloo, G. and Amelinckx, S., *Phys. Stat. Sol.*, 103:1 (1987). (b) Eaglesham, D.J., *Appl. Phys. Lett.* 51:549 (1987).

28. Beyers, R., Lim, G., Engler, E.M., Savoy, R.J., Shaw, T.M., Dinger, T.R., Gallagher, W.J., Sandstrom, R.L., *J. Appl. Phys. Lett.* 50:1918 (1987).

29. Ourmazd, A., Rentscher, J.A., Spence, J.C.H., O'Keeffe, M., Graham, R.J., Johnson, D.W., and Rhodes, W.W., *Nature* 327:308 (1987); ibid. 329:425 (1987).

30. (a) Gai, P.L., Male, S.E., and Boyes, E.D., *Proc. 40, Superconducting Ceramics*, (1987) p. 109; British Ceram. Proc., ed: R. Freer, Inst. Ceram. U. K. 109, (1987-88). (b) *EMAG 1987, Inst. Phys., Proc. Ser.* 90 London, U.K. (1987).

31. Gibson, J.M., *Nature* 329:763 (1987).

32. Nahahara, S., Fisanick, G.J., Yan, M., Van Dover, R.B., Boone, T., *Appl. Phys. Lett.* 53:21 (1988).

33. Mannhart, J., Dimos, D., and Chaudhuri, P., *Phys. Rev. Lett.* 61:2476 (1988).

34. Horowitz, H.S., McLain, S.J., Sleight, A.W., Druliner, J.D., Gai, P.L., Vankavelaar, M.J., Wagner, J.L., Biggs, B.D., and Poon, S.J., *Science* 243:66 (1989).

35. (a) Murukami, M., Morita, M., Doi, K., Miyamoto, K. and Hamada, H., *Jpn. J Appl. Phys.* 28:332 (1989). (b) Beyers, R., Ahn, B., Gorman, G., Lee, V., Parkin, S.S.P., Ramirez, M.L., Roche, K.P., Vazquez, S.E., Huggins, R.A., *Nature* 340:619 (1989).

36. (a) Zandbergen, H.W., Gronsky, P., Wang, K., and Thomas, G., *Nature* 331:569 (1988); (b) Marsh, P., Fleming, R.M., Mandich, M.L., Desantolo, A., Kwo, J., Hans, M., and Martinez-Miranda, L.J., *Nature* 334:141 (1988).

37. McCarron, E.M., Crawford, M.K., and Parise, J.B., *J. Solid St. Chem.* 78:192 (1989).

38. Parise, J.B., Gai, P.L., Crawford, M.K., McCarron, E.M., *Mat. Res. Soc. Proc. High Tc Supercond.* San Diego, Proc. 156, p. 105 (ed: J. Jorgensen, et al.) (1989).

39. (a) Cava, R.J. et al., *Nature* 336:211 (1988); (b) Subramanian, M.A., Gopalakrishnan, J., Torardi, C.C., Gai, P.L., Boyes, E.D., Askew, T., Flippen, R.B., Farneth, W.E., and Sleight, A.W., *Physica C* 157:124 (1989).

40. Subramanian, M.A., Torardi, C.C., Calabrese, J.C., Gopalakrishnan, J., Morrissey, K.J., Askew, T.R., Flippen, R.B., Chowdhry, U., and Sleight, A.W., *Science* 239:1015 (1988).

41. Shaw, T.M., Shivshankar, S.A., LaPlaca, S.J., Cuomo, J.J., McGuire, T.R., Ray, R.A., Kellher, K.H., and Yee, D.S., *Phys. Rev.* B37:9856 (1988).

42. Gai, P.L. and Day, P., *Physica* C152:335 (1988).

43. Matsui, Y., Maeda, H., Tanaka, and Horiuchi, S., *Jpn. J. Appl. Phys.* 27:361 (1988).

44. (a) Torrance, J.B., Tokura, Y., LaPlaca, S.J., Huang, T., Savoy, R. and Nazzal, J.I., *Solid St. Commun.* 2:47 (1988). (b) Bordet, P., Capponi, J.J., Chailout, C., Chenavas, J., Hewat, A.W., Hewat, A.E., Hodbau, S.L., Marezio, M., Tholence, J.L., and Tranqui, D., *Physica C* 153-632 (1988).

45. Withers, R.L., Anderson, J.S., Hyde, B.G., Thompson, J.G., Wallenberg, L.R., Fitzgerald, J.D., and Stewart, S.M., *J. Phys. C* 32:417 (1988).

46. Cheetham, A.K., Chippindale, A M., and Hibble, S.J., *Nature* 333:221 (1988).

47. Zandbergen, H.W., Groen, W.A., Mijlhoff, F.C., Van Tendeloo, G. and Amelinckx,, S., *Physica C* 156:325 (1988).

48. Gai, P.L., Subramanian, M.A., Gopalakrishnan, J., and Boyes, E.D., *Physica C* 159:801 (1989).

49. (a) Hewat, E.A., Capponi, J.J., and Marezio, M., *Physica C* 157:502 (1989). (b) Le Page, Y., MicKinnon, W. R., Tarascon, J.M., Barboux, P., *Phys. Rev.* (in press).

50. (a) Torardi, C.C., Parise, J.B., Subramanian, M.A., Gopalakrishnan, J., and Sleight, A.W., *Physica C* 157:115 (1989). (b) Subramanian, M.A., Torardi, C.C., Gopalakrishnan, J., Gai, P.L., Calabrese, J.C., Askew, T., Flippen, R., and Sleight, A.W., *Advances in Superconductivity*, eds.: Kitazawa, K. and Ishizuro, I., Springer-Verlag, Tokyo (1989).

51. Tarascon, J.M., Barboux, P., Hull, G., Ramesh, R., Greene, L., Hegde, M., McKinnon, W.R., *Phys. Rev. B*. 39:4316 (1989); 39:11857 (1989).

52. (a) Fukushima, N., Niu, H., Nakamura, S., Takeno, S., and Ando, K., *Physica C* 159:777 (1989). (b) Green, S.M., et al., *Phys. Rev.* (1989).

53. (a) Uemura, Y.J., Yang, B.X., Subramanian, M.A., Strezelecki, A.R., Gopalakrishnan, J., and Sleight, A.W., et al.; *J. de Physique*, Collogue July (1988). (b) Uiemura, Y.J., et al., *Phys. Rev. Lett.* 59:1045 (1987).

54. Gai, P.L., Subramanian, M.A., and Sleight, A.W., *Proc. XII Int'l. Cong. on EM*, 4:36 (1990).

55. Torardi, C.C., Subramanian, M.A., Gopalakrishnan, J., Calabrese, J.C., Morissey, K.M., Askew, T., Flippen, R.B., Chowdhry, U., and Sleight, A.W., *Science* 240:631 (1988).

56. Subramanian, M.A., Torardi, C.C., Gopalakrishnan, J., Gai, P.L., Calabrese, J.C., Askew, T.R., Flippen, R.B., and Sleight, A.W., *Science* 242:249 (1988).

57. Ginley, D.S., Kwak, J.F., Hellmer, R.P., Baughman, R.J., Venturini, E.L., Mitchell, M.A., Morosin, B., *Physica C* 156:592 (1988).

58. Hewat, A.W., Bordet, P., Capponi, J.J., Chailout, C., Chenvas, J., Godinho, M., Hewat, E.A., Hodeau, J., and Marezio, M., *Physica C* 156:369 (1988).

59. (a) Ihara, H., Sugise, R., Hayashi, K., Terada, N., Hirabayashi, M., Negishi, A., Atoda, N., Oyanagi, H., Shimomura, T., and Ohash, S., *Phys. Rev. B* 38:11952 (1988). (b) *Adv. in Superconductivity*, eds: Kitazawa, K., Ishizuro, I., Springer-Verlag, Tokyo (1989).

60. (a) Whangbo, M.H., Kang, D.B., and Torardi, C.C., *Physica C* 158:371 (1989). (b) Jung, et al., *Physica C* 160:381 (1989).

61. Parise, J.B., Gai, P.L., Subramanian, M.A., Gopalakrishnan, J., and Sleight, A. W., *Physica C* 159:245 (1989).

62. Parise, J.B., Herron, N., Crawford, M.K., and Gai, P.L., *Physica C* 159:255 (1989).

63. Subramanian, M.A., Gai, P.L., and Sleight, A.W., *Mat. Res. Bull.* 25:101 (1990).

16

Oxidation State Chemical Analysis

Daniel C. Harris

1.0 SUPERCONDUCTORS EXIST IN VARIABLE OXIDATION STATES...

A hallmark of high temperature superconductors is that they contain elements whose oxidation states vary over a range of values. The archetypical family of compounds whose superconductivity was discovered by Müller and Bednorz (1) has the formula $La_{2-x}(Sr,Ba)_x CuO_{4+\delta}$ (x = 0 to 0.3, δ is a small positive number), (2) in which the copper oxidation state depends on the alkaline earth content and exact oxygen stoichiometry. For example, the composition $La_{1.85}Sr_{0.15}CuO_{4.03}$ can be described as $(La^{3+})_{1.85}(Sr^{2+})_{0.15}(Cu^{2+})_{0.85}(Cu^{3+})_{0.17}(O^{2-})_{4.03}$ in which the sum of cation charge is +8.06 and the sum of anion charge is -8.06. Since other oxidation states of La and Sr are extremely unlikely, the variable oxidation state is assigned to Cu, whose +3 state is rare, but not unknown (3). By altering the Sr content in this formula, the relative amounts of Cu^{2+} and Cu^{3+} change. The charge carriers in these oxidized compounds are mobile holes. Another class of high temperature superconductors containing mobile electrons is exemplified by $(Nd^{3+})_{1.85}(Ce^{4+})_{0.15}(Cu^+)_{0.29}(Cu^{2+})_{0.71}(O^{2-})_{3.93}$ (4), whose postulated (5) quadrivalent Ce and variable oxygen content require that some of the Cu be in the +1 oxidation state.

The most heavily studied high temperature superconductor is $YBa_2Cu_3O_{7-x}$ (x = 0 to 1), whose Cu oxidation state is determined by the oxygen content. The parent structure, $YBa_2Cu_3O_7$, contains layers of Cu-O sheets and layers with Cu-O chains (6). Heating removes oxygen atoms from the chains (7). The composition $YBa_2Cu_3O_7$ can be thought of as $(Y^{3+})(Ba^{2+})_2(Cu^{2+})_2(Cu^{3+})(O^{2-})_7$, with one-third of its Cu in the +3 state. When the oxygen content is reduced to 6.5, all Cu is formally in the +2 state and the material is a semiconductor. Upon further reduction to $YBa_2Cu_3O_6$, one third of the Cu is in the +1 state, but the material remains semiconducting. In general, the electronic (8) and crystallographic (9) properties of high temperature superconductors are strong functions of their oxygen contents and valence states.

...AND WE DON'T KNOW WHAT IS OXIDIZED

A striking chemical behavior of $YBa_2Cu_3O_7$ is the vigorous evolution of gas when the solid is added to acid or water (10)(11). It was first thought that this represented oxidation of water to O_2 by Cu^{3+}, but it was conclusively shown with ^{18}O-enriched superconductor that the evolved O_2 is derived from the solid, not the solvent (12)-(14):

$$YBa_2Cu_3{}^{18}O_7 + 12H^+ \rightarrow Y^{3+} + 2Ba^{2+} + 3Cu^{2+} + 6H_2{}^{18}O + \tfrac{1}{2}{}^{18}O_2 \quad (1)$$

This result casts doubt on the existence of Cu^{3+} in $YBa_2Cu_3O_7$. An alternative formulation is $(Y^{3+})(Ba^{2+})_2(Cu^{2+})_3(O^{2-})_6(O_2{}^{2-})_{0.5}$, with a cation charge of +13 and an anion charge of -13. However, other experiments cast similar doubt on the existence of peroxide ($O_2{}^{2-}$) in $YBa_2Cu_3O_7$ (13)(15)(16). It seems likely that any description of localized holes in $YBa_2Cu_3O_7$ is simplistic, and the solid is better described by holes in electronic bands. The nature of the ephemeral redox-active species liberated when the superconductor dissolves is unknown, and not critical to the discussion of analytical methods in this chapter. It is possible that redox chemistry occurs on the surface of the dissolving superconductor, and no unusual species are ever liberated. In any case, for the sake of balancing chemical equations, it is permissible to write the redox chemistry of $YBa_2Cu_3O_7$ in terms of Cu^{3+} and recognize that this is not a true representation of the mechanisms of the redox reactions.

2.0 ANALYSIS OF SUPERCONDUCTOR OXIDATION STATE BY REDOX TITRATION

Immediately after the discovery of $YBa_2Cu_3O_{7-x}$, numerous groups employed **iodometric titration** procedures to measure the effective oxidation state of the material, and therefore the value of x. The procedure described below involves two different titrations (10)(17)(18) and is more accurate than a procedure in which the first titration is omitted (19). Experiment A measures the total copper content of the superconductor and Experiment B measures the total charge of the copper. The two experiments, together, give the average oxidation state of copper.

In Experiment A, $YBa_2Cu_3O_{7-x}$ is dissolved in dilute $HClO_4$, in which all copper is reduced to Cu^{2+} (eq. 1). The total copper content is determined by addition of iodide

$$Cu^{2+} + \frac{5}{2}I^- \rightarrow CuI(s) + \frac{1}{2}I_3^- \quad (2)$$

and titration of the liberated iodine (actually I_3^-) with standard thiosulfate:

$$\frac{1}{2}I_3^- + S_2O_3^{2-} \rightarrow \frac{3}{2}I^- + \frac{1}{2}S_4O_6^{2-} \quad (3)$$

Each mole of Cu in $YBa_2Cu_3O_{7-x}$ requires one mole of $S_2O_3^{2-}$ in Experiment A.

In Experiment B, $YBa_2Cu_3O_{7-x}$ is dissolved in dilute $HClO_4$ containing I^-. Cu^{3+} (or whatever is really the oxidized species) oxidizes one equivalent of I^-, and then all of the Cu^{2+} oxidizes more I^- as the copper is reduced to $CuI(s)$:

$$Cu^{3+} + \frac{3}{2}I^- \rightarrow Cu^{2+} + \frac{1}{2}I_3^- \quad (4)$$

$$Cu^{2+} + \frac{5}{2}I^- \rightarrow CuI(s) + \frac{1}{2}I_3^- \quad (5)$$

The difference between the moles of thiosulfate per gram of superconductor in the two experiments equals the moles of Cu^{3+} per gram of superconductor.

Suppose that mass m_A is analyzed in Experiment A and volume V_A of standard thiosulfate is required for titration. Let the corresponding quantities in Experiment B be m_B and V_B. Let the average oxidation state of Cu in the superconductor be 2+p. Then p is given by (18)

$$p = \frac{(V_B/m_B) - (V_A/m_A)}{V_A/m_A} \qquad (6)$$

and x in the formula $YBa_2Cu_3O_{7-x}$ is related to p as follows:

$$x = \frac{7}{2} - \frac{3}{2}(2+p) \qquad (7)$$

For example, if the superconductor contains 1 Cu^{3+} and 2 Cu^{2+}, the average oxidation state of copper is 7/3 and p is 1/3. Setting p = 1/3 in Eq. 7 gives x = 0 and the formula is $YBa_2Cu_3O_7$. Eq. 6 does not depend on the metal stoichiometry being exactly Y:Ba:Cu = 1:2:3, but Eq. 7 does require this exact stoichiometry.

A significant limitation on the accuracy of the iodometric procedure is that Eq. 6 requires the subtraction of one large number from another. Relatively small uncertainties in each quantity lead to a relatively large uncertainty in the difference, and hence in the value of p. Both titrations are heterogeneous, with suspended solid CuI making it somewhat difficult to measure the end point. An advantage of eq. 6 is that errors in the standardization of the thiosulfate cancel each other. In fact, it is not necessary to standardize the thiosulfate at all to apply eq. 6. However, the determination of absolute copper content in Experiment A requires accurately standardized thiosulfate. The results of Experiment A should agree closely with the expected copper content of $YBa_2Cu_3O_{7-x}$.

A sensitive, elegant **citrate-complexed copper titration** (20) that directly measures the Cu^{3+} content eliminates some of the experimental errors associated with the previous analysis. The sample is first dissolved in a closed container with 4.4 M HBr, in which Cu^{3+} oxidizes Br^- to Br_3^-.

$$Cu^{3+} + \frac{11}{2} Br^- \rightarrow CuBr_4^{2-} + \frac{1}{2} Br_3^- \qquad (8)$$

Further reduction of Cu(II) does not occur. The solution is transferred to a vessel containing excess iodide, excess citrate and enough ammonia to neutralize most of the acid. Cu^{2+} is complexed by citrate and is not reduced to CuI. The Br_3^- from eq. 8 oxidizes I^- to an equivalent amount of I_3^-,

$$Br_3^- + 3I^- \rightarrow 3Br^- + I_3^- \qquad (9)$$

which is then titrated with standard thiosulfate. The solution remains homogeneous, and the moles of thiosulfate required are equal to the moles of Cu^{3+} in the sample. If R is defined as the moles of Cu^{3+} per gram of superconductor, then x in the formula $YBa_2Cu_3O_{7-x}$ is given by

$$x = \frac{1 - 666.20 R}{2 - 15.9994 R} \qquad (10)$$

where 666.20 is the formula mass of $YBa_2Cu_3O_7$ and 15.9994 is the atomic mass of oxygen. A general formula analogous to eq. 10, but applicable to other superconductors, is found in reference 20.

The citrate-complexed copper procedure can be modified to analyze samples with oxygen content in the range 6.0 - 6.5, which formally contain Cu^+ and Cu^{2+}. A small excess of standard Br_2 is added to the 4.4 M HBr in which the solid is to be dissolved. Cu^+ reduces Br_2 to Br^-, decreasing the quantity of I_2 produced in eq. 9.

Our experience with the two-titration iodometric procedure shows that the uncertainty in oxygen content of $YBa_2Cu_3O_{7-x}$ is approximately ± 0.04 (e.g. $YBa_2Cu_3O_{6.85 \pm 0.04}$). Nazzal et al. (18) find an uncertainty of ± 0.03. These uncertainties are reduced to near ± 0.01 in the citrate-complexed copper procedure. The relative uncertainty of the Cu^{3+} content when the oxygen content is close to 6.5 is much smaller for the citrate-complexed copper procedure than the two-titration iodometric procedure. Application of the citrate-complexed copper procedure to compounds such as $La_{1.8}Sr_{0.2}CuO_{4+x}$, La_2CuO_{4+x} and La_2NiO_{4+x} gives three decimal place precision for the oxygen content (e.g. $La_2CuO_{4.004}$) (15).

Another popular titrimetric analysis of superconductor oxidation state is based on the oxidation of ferrous ion (21)-(24):

$$Cu^{3+} + Fe^{2+} \rightarrow Cu^{2+} + Fe^{3+} \quad (11)$$

The superconductor is dissolved in 2.6 M H_3PO_4 solution containing standard 0.1 M Fe^{2+} under an inert atmosphere. The Fe^{2+} remaining after reaction 11 is titrated with standard dichromate or permanganate. For the most complete details, see reference 21. Samples containing excess Cu^+ (i.e., oxygen content < 6.5) are dissolved in 2.6 M H_3PO_4 solution containing standard 0.1 M Fe^{3+}, which is reduced to Fe^{2+} that can be titrated with dichromate (22).

A method that can distinguish between the oxidation states Cu^{3+} and Bi^{5+} in bismuth-based superconductors has been described (33). Bi^{5+} (BiO_3^-) can oxidize Mn^{2+} to MnO_4^-, while Cu^{3+} is not a strong enough oxidizing agent to accomplish this task.

$$5BiO_3^- + 2Mn^{2+} + 14H^+ \rightarrow 5Bi^{3+} + 2MnO_4^- + 7H_2O \quad (12)$$

The superconductor (100 mg) is dissolved in 10 mL of 0.025 M $Mn(NO_3)_2$ in 3 M HNO_3, cooled to room temperature, diluted with 70 mL of cold water, and treated with 10 mL of standard 0.1 M Fe^{2+} in 1 M H_2SO_4. Permanganate produced in reaction 12 oxidizes Fe^{2+} to Fe^{3+}. A mixture of concentrated acids (1.5 mL H_2SO_4 + 1.5 mL H_3PO_4 + 7 mL H_2O) is added and the remaining unreacted Fe^{2+} is titrated with standard 0.017 M potassium dichromate using sodium 4-diphenylamine sulfonate indicator. In a second experiment, the superconductor is dissolved in acid containing standard Fe^{2+} and the unreacted Fe^{2+} is determined by dichromate titration. The Fe^{2+} consumed by superconductor is equivalent to the total Cu^{3+} amd Bi^{5+}. By this method, it was found that the superconductors $BiCaSrCu_2O_x$ and $Bi_4Ca_3Sr_3Cu_5O_x$ contain Bi^{5+}, but little Cu^{3+}, since the moles of Fe^{2+} consumed in both parts of the procedure were nearly equal.

2.1 Iodometric Titration Procedure

0.03 M $Na_2S_2O_3$ Solution. Dissolve 3.7 g of $Na_2S_2O_3 \cdot 5H_2O$ plus 0.05 g of Na_2CO_3 in 500 mL of freshly boiled water. Add 3 drops of chloroform (a preservative) and store in a tightly capped amber bottle. The solution should be restandardized after several weeks.

Starch Indicator. Add to 90 mL of boiling water a slurry containing 1 g of soluble starch and 1 mg of HgI_2 (a preservative). The resulting clear solution is stable for weeks.

Standard Cu. In a fume hood, add 15 mL of water and 3 mL of 70% nitric acid to 0.5 - 0.6 g of accurately weighed reagent Cu wire in a 100 mL volumetric flask and boil gently to dissolve the wire. Add 1.0 g of urea or 0.5 g of sulfamic acid and boil 1 min to destroy HNO_2 and oxides of nitrogen that would interfere with the iodometric titration. Cool to room temperature and dilute to 100 mL with 1.0 M HCl.

Standardization of $Na_2S_2O_3$ with Cu. To prevent air oxidation of iodide in the acidic solution, use a 180 mL tall-form beaker (or a 150 mL standard beaker) loosely fitted with a 2-hole stopper. One hole serves as inlet for a brisk flow of N_2 or Ar that leaks out the side of the stopper. The other hole is used for the buret. Pipet 10.00 mL of standard Cu solution into the beaker and flush with inert gas. Remove the cork briefly to add 10 mL of water containing 1.0-1.5 g of freshly dissolved KI and begin magnetic stirring. Titrate with $Na_2S_2O_3$ from a 50 mL buret, adding 2 drops of starch just before the last trace of I_2 disappears. Premature addition of starch leads to irreversible binding of I_2 to the starch, and makes the end point harder to detect.

Experiment A. Dissolve an accurately weighed 150-200 mg sample of powdered $YBa_2Cu_3O_{7-x}$ in 10 mL of 1.0 M $HClO_4$ in a titration beaker in a fume hood. [CAUTION: Perchloric acid is recommended because it is inert to reaction with superconductor. Solutions of $HClO_4$ should not be boiled to dryness because of their explosion hazard. We have used HCl instead of $HClO_4$ with no significant interference in the analysis.] Boil gently for 10 min, cool to room temperature, cap with the two hole stopper - buret assembly, and begin inert gas flow. Add 10 mL of water containing 1.0 - 1.5 g of KI and titrate rapidly with magnetic stirring as described above.

Experiment B. Place an accurately weighed 150-200 mg sample of powdered $YBa_2Cu_3O_{7-x}$ in the titration beaker and begin inert gas flow. Add 10 mL of 1.0 M $HClO_4$ and 0.7 M KI and stir magnetically for 1 min. Add 10 mL of water and complete the titration.

2.2 Citrate-Complexed Copper Titration Procedure

Place an accurately weighed 20-50 mg sample of superconductor in a 4-mL screw cap vial with a Teflon cap liner and add 2.00 mL of ice cold 4.4 M HBr by pipet. (The HBr is prepared by diluting 50 mL of 48% HBr to 100 mL.) Cap tightly and stir or gently agitate the sample for 15 min as it warms to room temperature. Cool the solution back to 0°C and carefully transfer it to a titration beaker (as in the $Na_2S_2O_3$ standardization above) containing an ice cold freshly prepared mixture of 0.7 g of KI, 20 mL of water, 5 mL of 1.0 M trisodium citrate and approximately 0.5 mL of 28% ammonia. The exact amount of ammonia should be enough to neutralize all but 1 mmol of acid present in the reaction. When calculating the acid content, remember that each mole of $YBa_2Cu_3O_{7-x}$ consumes 2(7-x) moles of HBr. Wash the vial with 3 1-mL aliquots of 2 M KBr to complete the quantitative transfer to the beaker. Add 0.1 mL of 1% starch solution and titrate with 0.1 M standard $Na_2S_2O_3$ under a brisk flow of inert gas using a 250 µL Hamilton syringe to deliver titrant. The end point is marked by a change from the dark blue-black starch-iodine color to the lighter blue-green hue of the cupric ion. A blank reaction must be run with $CuSO_4$ in place of superconductor. The moles of Cu in the blank should be the same as the moles of Cu in the superconductor. In a typical experiment, 30 mg of $YBa_2Cu_3O_{6.88}$ required approximately 350 µL of $Na_2S_2O_3$ and the blank required 10 µL of $Na_2S_2O_3$.

3.0 REDUCTIVE THERMOGRAVIMETRIC ANALYSIS

When $YBa_2Cu_3O_{7-x}$ is heated to 1000°C in a flowing atmosphere containing H_2, [CAUTION: The H_2 should be diluted with 10 volumes of Ar to reduce the hazard of an explosion. A 10-turn 5-cm-diameter coil of glass tubing should be inserted next to the H_2 tank to prevent flashback of the tank. The thermal analyzer must be purged completely so there are no dead volumes of air and the exit gas must be exhausted carefully.] reduction to BaO, Y_2O_3 and Cu occurs: (10)(25)

$$YBa_2Cu_3O_{7-x} + (\frac{7}{2} - x)H_2 \rightarrow 2BaO +$$
$$\frac{1}{2}Y_2O_3 + 3Cu + (\frac{7}{2} - x)H_2O \quad (13)$$

By measuring the decrease in sample mass, the oxygen stoichiometry can be deduced from the formula

$$x = \frac{\frac{7}{2}(15.9994) - 666.20\,F}{15.9996\,(1-F)} \quad (14)$$

where F is the fraction of mass lost ([initial mass - final mass]/initial mass), 666.20 is the formula mass of $YBa_2Cu_3O_7$ and 15.9994 is the atomic mass of oxygen. A similar analytical procedure can be applied to other superconductors. For example, $Pb_2Sr_2YCu_3O_8$ gives Pb, SrO, Y_2O_3 and Cu upon heating to 600°C in H_2/N_2 (26). Evaporation of Pb occurs above 600°C and slow mass loss continues.

The reductive thermogravimetric trace for $YBa_2Cu_3O_{7-x}$ in Figure 1 shows that the reaction begins near 400°C and proceeds in several steps until it ends near 900°C. The final mass may be taken at 1000°C, but it is better to eliminate convection effects on the apparent mass by cooling the sample back to room temperature before recording the final mass. In the experiment in Figure 1 the flowing gas was not dried and the mass increase below 500°C probably corresponds to $Ba(OH)_2$ formation. CO_2 and O_2 impurities in the flow will also cause a mass increase. For the experiment in Figure 1, the temperature for the final mass reading was 550°C.

The accuracy of carefully performed reductive thermogravimetric analysis can be comparable to that of the double iodometric titration, if a large sample is used or if a micro-thermogravimetric analyzer is available. Neither procedure is very useful when the amount of Cu^{3+} is small (oxygen stoichiometry near 6.5). In this case the only accurate procedure is the citrate-complexed copper titration. To estimate the accuracy of reductive thermogravimetric analysis, consider the following example. For $YBa_2Cu_3O_7$ the loss of mass in reductive thermogravimetric analysis is 8.41%. For $YBa_2Cu_3O_{6.5}$ the loss is 7.29%. Oxygen content in the range 6.5 to 7 will give a mass loss in the range 7.29% to 8.41%. Suppose that a 20 mg sample gives

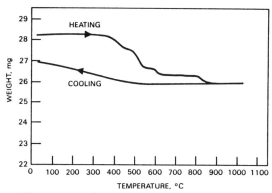

Figure 1: Thermogravimetric analysis of $YBa_2Cu_3O_{7-x}$ in flowing H_2 (8% by volume in Ar) heated at a rate of 10°C/min, The sample was held at 1000°C for 30 min. (From Reference 10. With Permission.)

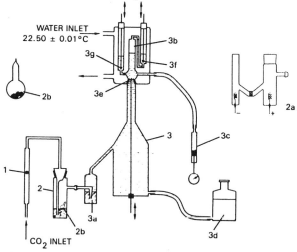

Figure 2: Apparatus used to measure volume of O_2 released when superconductor is dissolved in acid (27) Key: (1) flowmeter for CO_2 stream, (2) reaction ampoule with HNO_3, (2a) electrolysis chamber for calibration (substituted in place of reaction ampoule), (2b) sealed, evacuated capsule containing superconductor to be analyzed, (3) absorption chamber filled with KOH solution, (3a) mercury bubbler to prevent KOH from reaching reaction ampoule, (3b) gas buret, (3c) micrometer-driven syringe to measure volume of liberated oxygen (± 0.001 mL), (3d) bottle for adding or removing KOH solution, (3e,f,g) valves between chambers - constructed from ground joints with tips of plexiglass rods. (From Reference 27. With Permission.)

a mass loss of exactly 8% (1.6 mg) with an uncertainty of ± 0.02 mg. (Although mass can be measured more precisely than ± 0.02 mg, the mass of the sample continues to vary with temperature and the loss might not be more certain than ± 0.02 mg.) Using the value F = (1.6 ± 0.02)/20 in eq. 14 gives x = 0.18 ± 0.05. The formula of the superconductor is $YBa_2Cu_3O_{6.82\pm0.05}$.

4.0 OXYGEN EVOLUTION IN ACID

Reaction 1, in which O_2 evolves from superconductor when it is dissolved in acid, is the basis for another analytical procedure useful for $YBa_2Cu_3O_{7-x}$ when the oxygen content is greater than 6.5. Methods have been described involving volumetric (27)(28) or gas chromatographic measurement (29) of the evolved oxygen.

The most accurate analysis claimed in the literature uses the apparatus in Figure 2 (27) and gives a precision of ± 0.001 for x in the formula $YBa_2Cu_3O_{7-x}$. A 150-300 mg sample in a sealed, evacuated ampoule (2b) is broken with a magnetic hammer in a chamber (2) containing nitric acid solution. The evolved oxygen is swept by a stream of high purity CO_2 through a solution of KOH (3) that absorbs CO_2 but permits O_2 to rise into the thermostatically controlled gas buret (3b). A blank test in which CO_2 runs at 40 mL/min for 10 min produces 0.006 - 0.015 mL of gas in the buret. Superconductor samples subjected to the same conditions give 1.5 - 3 mL of gas. The equipment is calibrated by substituting the electrolysis chamber (2a) for the reaction chamber (2). Oxidation of water ($H_2O \rightarrow 1/2\, O_2 + 2H^+ + 2e^-$) to produce known quantities of oxygen demonstrated an accuracy of oxygen volume determination of ±0.009 mL, corresponding to a typical uncertainty in x of ±0.0008 in the formula $YBa_2Cu_3O_{7-x}$.

A gas chromatographic method for measuring oxygen evolution from superconductor dissolved in acid is based on the apparatus in Figure 3. A 25 to 30 mg solid sample is placed on the spoon in the reaction flask (G) that also contains 10 mL of 8 M HNO_3. After purging the system with He, the spoon is rotated 180° to drop the sample into the acid. The pulse pump (PP) circulates the gas through tube T containing solid NaOH and through the 1 mL volume loop L. After 20 to 30 min the system reaches equilibrium and the gas in the 1 mL loop is injected into a gas chromatograph equipped with

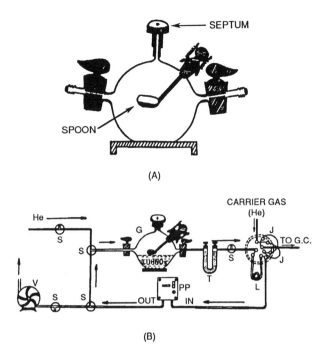

Figure 3: System for gas chromatographic measurement of O_2 evolution from superconductors dissolved in acid (29) (A) enlarged view of reaction flask, (B) circulating system including flask (G), NaOH U-tube (T), three-way valves (S), 1 mL loop (L), pulse pump (PP) and vacuum system (V), At the right is the gas chromatograph (G.C.). (From Reference 29. With Permission.)

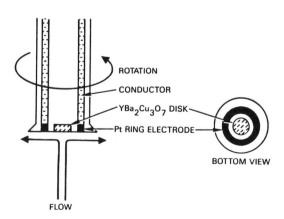

Figure 4: Rotating ring-disk electrode.

a molecular sieve column and thermal conductivity detector. The system is calibrated with injections of known volumes of oxygen through the septum at the top of flask (G). Five replicate injections gave a standard deviation of ±2% of measured gas volume.

5.0 ELECTROCHEMICAL INVESTIGATION OF SUPERCONDUCTOR OXIDATION STATE

A fascinating study of the chemistry of $YBa_2Cu_3O_{7-x}$ is based on the rotating ring-disk electrode in Figure 4 (30). A disk of superconductor and a Pt ring are embedded in the base of a rotating cylinder made of non-conductive material. The Pt ring is part of a circuit that includes a reference electrode and static auxiliary electrode immersed in the same solution as the rotating electrode. Rotation of the ring-disk electrode causes solution to flow up toward the electrode and then off to the side, as shown in Figure 4. The potential of the Pt ring can be adjusted to any desired value and the current flowing between the Pt ring and static auxiliary electrode as a result of oxidation-reduction chemistry can be measured. Disks made of Cu_2O, CuO, $YBa_2Cu_3O_{6.2}$ and $YBa_2Cu_3O_{6.99}$ were studied in solutions containing NaCl plus HCl, NaBr plus HBr or NaI plus HI.

Figure 5 shows a voltammogram for $YBa_2Cu_3O_{6.99}$ immersed in 14.4 mM HI plus 1.0 M NaI. The chemistry can be understood based on dissolution of the superconductor according to the stoichiometry

$$YBa_2Cu_3O_{7-x} + (14 - 2x)H^+ + (12 - 3x)I^- \rightarrow Y^{3+} + 2Ba^{2+} + 3CuI_2^- + (2 - x)I_3^- + H_2O \quad (15)$$

As the superconductor dissolves, CuI_2^- and I_3^- are carried within tens of milliseconds past the Pt ring electrode. In the potential range +0.2 to -0.5 V (vs. a saturated calomel reference electrode) a current near -0.1 mA flows due to reduction of the I_3^- at the Pt ring:

$$I_3^- + 2e^- \rightarrow 3I^- \quad (16)$$

At potentials more negative than -0.6 V, the current of -0.2 mA arises from reduction of CuI_2^-:

622 Chemistry of Superconductor Materials

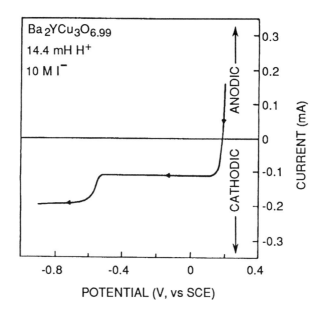

Figure 5: Current versus potential for rotating disk-ring electrode. (From Reference 30. With Permission.)

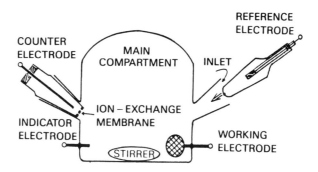

Figure 6: Apparatus used for coulometric titration of superconductor. (From Reference 32. With Permission.)

$$CuI_2^- + e^- \rightarrow Cu(0) + 2I^- \qquad (17)$$

Current from reaction 16 is proportional to the total Cu in the sample. Current from reaction 15 is proportional to (Cu^{2+} + 2Cu^{3+}), since Cu^{3+} oxidized twice as much iodide as Cu^{2+}. The relative currents from reactions 15 and 16 can be used to deduce the fraction of Cu in the Cu^{3+} state. A sample whose composition was $YBa_2Cu_3O_{6.99}$ by reductive thermogravimetric analysis appeared to be $YBa_2Cu_3O_{6.92}$ by the rotating ring-disk electrode.

The most interesting conclusion of the rotating ring-disk electrode experiment is the absence of a reduction wave due to any species other than CuI_2^- and I_3^-. The implication is that no free Cu^{3+} or any other reducible species with a lifetime beyond tens of milliseconds is produced when $YBa_2Cu_3O_7$ dissolves.

In another electrochemical study (31), thin films of $YBa_2Cu_3O_{7-x}$ sputtered onto a tungsten surface were anodically oxidized in 0.1 M KCl solution. The current flowing in successive oxidation processes was used to probe the oxidation state of the film.

The apparatus in Figure 6 was employed for a coulometric titration of $YBa_2Cu_3O_{7-x}$ (32). The main compartment into which the sample is added through the inlet contains a known excess amount of CuCl dissolved in HCl. (All solutions are handled in a glove box to avoid oxidation by air.) Oxygen evolved from the sample slowly reacts with Cu^+, oxidizing it to Cu^{2+}. After a period of time for the reaction to go to completion, a known, constant current of 1 - 5 mA is passed between the working electrode (Pt anode) and the counter electrode (cathode) that is separated from the main compartment by a cation-exchange membrane (Nafion). The working electrode oxidizes unreacted Cu^+ to Cu^{2+}. The potential, E, measured by the indicator (Pt) electrode depends on the relative amounts of Cu^+ and Cu^{2+} in the solution:

$$E = E°(Cu^{2+} | Cu^+) - \frac{RT}{F} \ln \frac{[Cu^+]}{[Cu^{2+}]} \qquad (18)$$

where R is the gas constant, T is temperature (K), F is the Faraday constant and $E°(Cu^{2+}| Cu^+)$ is the formal potential for the reaction $Cu^{2+} + e^- \rightarrow Cu^+$. When all of the Cu^+ has been oxidized to Cu^{2+} at the working electrode, the quotient $[Cu^+]/[Cu^{2+}]$ in eq. 18 changes rapidly and the voltage measured between the indicator and reference electrodes suddenly increases, signaling the end point. Knowing the current and time required to complete oxidation of the Cu^+, one can calculate how many electrons were required to oxidize the Cu^+, and therefore how much O_2 was released from the superconductor. This method is just another variety of redox titration, but reagent is generated electrochemically instead of being introduced from a buret. The end point is indicated by an electrode potential change instead of an indicator color change.

6.0 ASSESSMENT OF ANALYTICAL PROCEDURES

Among the procedures described in this chapter, reductive thermogravimetric analysis is the simplest, but not very accurate. For accurate analyses of slightly oxidized superconductors, the citrate-complexed copper titration is recommended. The difficult problem of assessing oxidation states of individual elements in Bi- and Tl-containing superconductors has not been addressed, and remains a significant challenge to analytical chemists.

7.0 REFERENCES

1. Müller, K.A., Bednorz, J.G., *Science* 237:1133 (1987).

2. Alp, E.E., Shenoy, G.K., Hinks, D.G., Capone, D.W.II, Soderholm, L., Schuttler, H.-B., Guo, J., Ellis, D.E., Montano, P.A., Ramanathan, M. *Phys. Rev. B* 35:7199 (1987).

3. (a) Straub, U., Krug, D., Ziegler, Ch., Schmeiber, Göpel, W., *Mat. Res. Bull.* 24:681 (1989), (b) Margerum, D.W., Chellappa, K.L., Bossu, F.P., Burce, G.L., *J. Am. Chem. Soc.* 97:6894 (1975), (c) Bossu, F.P., Chellappa, K.L., Margerum, D.W, *J. Am. Chem. Soc.* 99:2195 (1977), (d) Stevens, P., Waldeck, J.M., Strohl, J., Nakon, R., *J. Am. Chem. Soc.* 100:3632 (1978), (e) Kirsey, S.T. Jr., Neubecker, T.A., Margerum, D.W., *J. Am. Chem. Soc.* 101:1631 (1979).

4. Tokura, Y., Takagi, H., Uchida, S., *Nature* 337:345 (1989).

5. (a) Takagi, H., Uchida, S., Tokura, Y., *Phys. Rev. Lett.* 62:1197 (1989), (b) Huang, T.C., Moran, E., Nazzal, A.I., Torrance, J.B., *Physica C* 158:148 (1989),

6. Holland, G.F., Stacy, A.M., *Acc. Chem. Res.* 21:8 (1988).

7. (a) Santoro, A., Miraglia, S., Beech, F., *Mat. Res. Bull.* 22:1007 (1987), (b) Garbauskas, M.F., Green, R.W., Arendt, R.H., Kasper, J.S., *Inorg. Chem.* 27:871 (1988).

8. Shafer, M.W., Penney, T., Olson, B.L., Greene, R.L., Koch, R.H., *Phys. Rev. B* 39:2914 (1989).

9. Whangbo, M.-H., Evain, M., Beno, M.A., Geiser, U., Williams, J.M., *Inorg. Chem.* 27:467 (1988).

10. Harris, D.C., Hewston, T.A., *J. Sol. State Chem.* 69:182 (1987).

11. Barns, R.L., Laudise, R.A., *Appl. Phys. Lett.* 51:1373 (1987).

12. Shafer, M.W., de Groot, R.A., Plechaty, M.M., Scilla, G.J., *Physica C* 153-155:836 (1988).

13. Shafer, M.W., de Groot, R.A., Plechaty, M.M., Scilla, G.J., Olson, B.L., Cooper, E.I., *Mat. Res. Bull.* 24:687 (1989).

14. Salvador, P., Fernandez-Sanchez, E., Garcia Dominguez, J.A., Amador, J., Cascales, C., Rasines, I., *Sol. State Comm.* 70:71 (1989).

15. Harris, D.C., Vanderah, T.A., *Inorg. Chem.* 28:1198 (1989)

16. Rosamalia, J.M., Miller, B., *Anal. Chem.* 61:1497 (1989).

17. Harris, D.C., Hills, M.E., Vanderah, T.A., *J. Chem. Ed.* 64:847 (1987).

18. Nazzal, A.I., Lee, V.Y., Engler, E.M., Jacowitz, R.D., Tokura, Y., Torrance, J.B., *Physica C* 153-155:1367 (1988).

19. Kishio, K., Shimoyama, J., Hasegawa, T., Kitazawa, K., Fueki, K., *Jpn. J. Appl. Phys.* 26:L1228 (1987).

20. Appelman, E.H., Morss, L.R., Kini, A.M., Geiser, U., Umezawa, A., Crabtree, G.W., Carlson, K.D., *Inorg. Chem.* 26:3237 (1987).

21. Oku, M., Kimura, J., Hosoya, M., Takada, K., Hirokawa, K., *Fresenius Z. Anal. Chem.* 332:237 (1988).

22. Saito, Y, Noji, T., Hirokawa, K., Endo, A., Matsuzaki, N., Katsumata, M., Higuchi, N., *Jpn. J. Appl. Phys.* 26:L838 (1987).

23. Shafer, M.W., Penney, T., Olson, B.L., *Phys. Rev. B* 36:4047 (1987).

24. Fukushima, N., Yoshino, H., Niu, H., Hayashi, M., Sasaki, H., Yamada, Y., Murase, S., *Jpn. J. Appl. Phys.* 26:L719 (1987).

25. Gallagher, P.K., O'Bryan, H.M., Sunshine, S.A., Murphy, D.W., *Mat. Res. Bull.* 22:995 (1987).

26. Gallagher, P.K., O'Bryan, H.M., Cava, R.J., James, A.C.W.P., Murphy, D.W., Rhodes, W.W., Krajewski, J.J., Peck, W.F., Waszczak, J.V., *Chem. Mater.* 1:277 (1989).

27. Condor, K., Rusiecki, S., Kaldis, E., *Mat. Res. Bull.* 24:581 (1989).

28. Dou, S.X., Liu, H.K., Bourdillon, A.J., Savvides, N., Zhou, J.P., Sorrell, C.C., *Solid State Comm.* 68:221 (1988).

29. Parashar, D.C., Rai, J., Gupta, P.K., Sharma, R.C., Lal, K., *Jpn. J. Appl. Phys.* 27:L2304 (1988).

30. Rosamilia, J.M., Miller, B., *Anal. Chem.* 61:1497 (1989).

31. Salkalachen, S., Salkalachen, E., John, P.K., Froelich, H.R., *Appl. Phys. Lett.* 53:2707 (1988).

32. Kawamura, G., Hiratani, M., *J. Electrochem. Soc.* 134:3211 (1987).

33. Oku, M., Kimura, J., Omori, M., Hirokawa, K., *Fresenius Z. Anal. Chem.* 335:382 (1989).

17

Transport Phenomena in High Temperature Superconductors

Donald H. Liebenberg

1.0 INTRODUCTION

Transport in high temperature superconductors refers to both electron and phonon processes and will include in this section discussions of resistivity, critical currents and the onset of resistance in the intermediate state, Hall resistivity, magneto resistance, thermal conductivity, thermo-power, and tunneling transport. The procedures for measurements of these properties strongly interact with the analysis so that for each phenomenon the current practice and the limitations of the measurement will be described. These determinations are of importance for a fundamental understanding of the mechanisms of superconductivity and especially in the case of critical current measurements are essential for the technological development and practical application. As will be shown for the high temperature superconducting oxides, there is a significant variation of the transport properties dependent upon the synthesis/fabrication procedures.

2.0 RESISTIVITY MEASUREMENT

The first property measured to determine superconductivity has been the variation of resistance with temperature to determine the onset of 'zero' resistance. This temperature is denoted as $T_c(0)$, and lies below T_c, the transition temperature or the temperature at the midpoint of the drop from normal resistance to 'zero' (1). This property requires particular care in the measurement and analysis to provide a creditable datum. A typical schematic of the low temperature apparatus for this measurement is shown in Figure 1. Temperature measurement must provide for good thermal contact to the ceramic which is not an especially good thermal conductor (2). In some experiments the sample is located in a copper thermal shield as shown in the figure, thermally separated from the cryogenic bath and provided with an imbedded diode thermometer and heater coil to maintain temperatures above a cryogenic bath that is commonly either nitrogen or helium liquid (3). The use of a small refrigerator to maintain and control the sample volume temperature is an alternative technique. Thermometer lead wires as well as the electrical connections for resistivity need to be thermally isolated such as by heat sinking to a thermal reservoir at a similar temperature as the sample. Consideration must be taken of Joule heating produced in the course of the measurement. In the case of a thin film the heat conduction through the substrate is frequently the important thermal path to remove Joule heat so that the substrate must be thermally lagged for example by attachment to a copper block. Thermally conducting grease such as Dow Corning™ or Apiezon N™ are appropriate choices although Celvacene™ (CVC) is more easily removed after use.

Electrical connections for resistivity measurements usually include separate current and voltage leads and the resistance is measured by determining voltage changes with a constant current source across the sample. This technique avoids spurious voltages at the contact points of a two or three contact connection but introduces a consideration of current path separate from the voltage measurement that can lead to faulty measurements.

The application of the leads on ceramic samples can be a source of measurement difficulty, more so as will be discussed in the measurement of critical currents when larger current densities are required. With the $YBa_2Cu_3O_{7-x}$ (Y-Ba-Cu-O) material, where oxygen depletion at the surface can give a lower transition temperature or semiconducting material, the lead connection should penetrate the surface. This is frequently accomplished by using either silver or gold dust in a solvent, painting the contact pads on the ceramic and reannealing the material to drive off the solvent and reoxygenate the

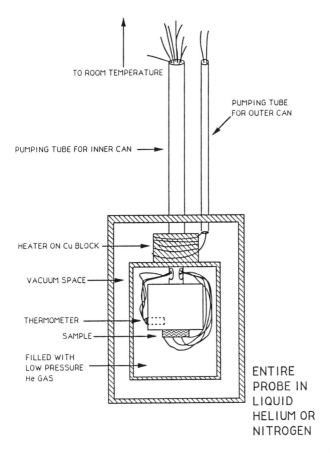

Figure 1: Typical apparatus for the measurement of resistivity after M. Osofsky, Naval Research Laboratory.

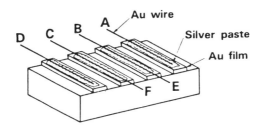

Figure 2: Schematic diagram of applied contact pads with gold film, silver paste and gold wire. (Ref. Sugimoto, I., Tajima, Y., Hikita, M., Low Resistance Ohmic Contact for Oxide Superconductor Eu-Ba-Cu-O, *Jpn. J. Appl. Phys.* 27:L864 (1988).

surface. Such pads with wires subsequently soldered with indium solder, sometimes using ultrasonic soldering techniques, have given reliable results. The reannealing process is carried out at temperatures of 400-500°C in air for about 4 hr. A cartoon of this process is shown in Figure 2. A recent recipe for this process used on Bi-Sr-Ca-Cu-O single crystals is described by Martin et al. (4). They attached 25 micron diameter gold wires using a silver paste and silver paint mixture which they cured by heating to 300°C in dry oxygen gas. They report resistance of 2-6 Ohm for an area of contact 10^{-4} cm^2. Alternative techniques have used an air drying silver paste to attach leads although these are less secure than with soldering, and in some cases the indium can be directly soldered to the ceramic, aided by the use of ultrasonic techniques. Since indium is soft the application of a small piece of indium pressed onto the sample has been effective. Then either the lead is soldered or after "tinning" with indium is pressed onto the pad. And Goldschmidt (5) has used an indium-gallium amalgam to wet the sample and then an indium solder. This technique has been used also with thin films; a Pb pencil applicator works well without scratching the film. The use of gold or silver pads cured with annealing has proven to be the most reliable although it is more time consuming. For the Bi-Sr-Ca-Cu-O material that in single crystal form is micaceous there are reports of resistivity measurements made with tungsten pressure contacts (6). When there is ready access to an evaporator the use of evaporated gold or silver pads to which leads can be soldered has provided low resistance connections.

For the usual dc measurement the constant dc current source should be capable of providing currents in the range 0.1-10 mA; for a typical bar of 1 mm square cross-section, 1 cm length, and a resistivity at 100 K of 50 μOhm-cm the voltage measured for a 1 mA current source would be 1 μV. Since even for a typical low value of the critical current density, 100 A/cm^2, the measurement current would be 1000 times less and thus have essentially no effect on the measurement. However, the measurement of 1 μV to a precision of 1% already requires care to assure that noise and thermal voltages are reduced well below this value. Currents of similar value are used for measurements in thin films.

Voltage measurements that require resolution of tens of nanovolts for signals of the order 1 μV (measured to 1%) can be made with commercially available instrumentation. To reduce noise in the

Transport Phenomena in High Temperature Superconductors 631

voltage measurement the wiring leads into the cryostat need to be matched in size and length, shielded, and soldered to vacuum pass-throughs to avoid thermal gradients at the connections. Heat sinking to equalize thermal gradients in the cryostat and to reduce heat leak to the sample and careful sample temperature control to reduce thermal contact voltages are necessary to obtain the accuracy and precision of the resistivity measurement.

Data analysis in the normal state of the superconductor is relatively straightforward, measurements of I, V, and T together with sample geometry are used to obtain resistivity vs temperature.

$$\rho = (V/I)(A/l) \qquad (1)$$

where A is the cross sectional area of the sample and l is the distance between the voltage leads. Sources of errors are possible in the cross sectional measurement for bulk or film materials that may not have uniform cross sections between the voltage leads. Errors may result from stray (such as thermally induced) voltages, poor electrical contacts to the sample, and a current path that is non-uniform. This latter error has been the source of an occasional claim to zero resistivity at much higher temperatures than could be verified. As will be discussed more fully in the section on critical current measurement the current is best introduced into the sample several cross sectional distances beyond the voltage leads. The possibility of surface effects may also inhibit an accurate measurement. Corrections to the temperature variation of the cross section and length are generally not made with the high temperature superconductors since for the thermal coefficient of expansion $\sim 10^{-4}$ K^{-1} over 200 K this effect would just begin to be significant at the 1% level. The effect of increasing measurement current is illustrated by Goldschmidt (5) for currents between 1 mA and 1 A; at the upper value a low temperature tail develops indicating that some fraction of the weak links in the bulk Y-Ba-Cu-O have gone normal with a current above the critical current.

In the superconducting state the observation of a near zero value on the voltage meter is frequently taken as the completion of the 'zero' resistance state. However, for the measurements in the above example a 10 nanovolt reading (the resolution of the voltmeter) would represent $\sim 10^{-7}$ Ohm-cm resistivity and this value is about 7 orders of magnitude greater than is needed to be assured of a superconducting state by a resistivity measurement (7). That is of course the reason for the skepticism expressed by Bednorz and Muller in their first paper until the Meissner effect could be determined. The field cooled flux exclusion is an effect unique to superconductivity, as compared with simply the zero resistivity that produces zero field cooled flux exclusion; see for example Poole, Datta, and Farach (8).

Measuring the magnetoresistance or the dependence of the resistivity on applied magnetic field requires the addition of a uniform magnetic field and is further complicated by the anisotropy of the high temperature superconductors. The magnetic field is oriented perpendicular to the larger surface area; this direction is usually perpendicular to the current direction also. Corrections for the geometry of the sample may be needed. Since the values of the lower critical field, H_{c1}, has a range 5-100 Gauss (although the exact value for any given sample is usually not well determined) some care must be given to shielding the standard resistivity measurements against small fields in a laboratory environment.

Resistivity measurements are also routinely made with an ac four probe technique. The wiring would follow according to Figure 1 and the measuring currents used would be in the range 0.1 - 10 mA with frequencies of ~100 Hz (9). For flux creep now known to modify susceptibility and critical current measurements care must taken with ac measurements of resistivity although for the low current densities involved the effect will not likely be observed except very close to T_c (10) or in a magnetic field.

An interesting alternative technique was discussed by Harris, et al. (11) who made a contactless measurement. A toroidal specimen is formed of the superconductor and is used as the tertiary winding of a small ferrite transformer. Signals from the secondary yield the resistivity in the normal state. Measurements were made at a 20 kHz frequency.

2.1 Survey of Results of Resistivity Measurements

The purpose of this section is not to review the more than 1000 papers that discuss resistivity measurements in the high temperature superconducting materials but to show some selected results, discuss in a limited way the interpretation, and to indicate some current avenues of research.

Single Crystal Resistivity: Single crystal samples of the high temperature superconductors provide information on the anisotropy of the resistivity and on the lower values of the normal state resistivity expected in the fully dense material. A Bi-Sr-Ca-Cu-O single crystal measurement was reported by Martin, et al. (4). In Figure 3 the resistivity as a function of temperature is shown in the case of a magnetic field perpendicular and parallel alignment with the a,b plane. These measurements on Bi-Sr-Ca-Cu-O which does not exhibit twinning also provided measurements in the a and b directions separately. The normal state resistivity is linear with temperature in the a,b directions but semiconductor-like in the c axis direction. Similar results have been reported for Y-Ba-Cu-O crystals (12). Recent measurements by Iwasaki et al. (13) on a Tl-Ba-Ca-Cu-O crystal (suggested to be 1212 composition) indicate broadening of the transition as the magnetic field is increased. When the field is perpendicular (perp) to the c axis, Figure 4a, this broadening is less, even at 21 Tesla, than when the field is parallel (para) to the c axis as in Figure 4b. In Figure 4b the broadening at 6 Tesla is large and there is also a knee that develops in the curve around 60 K. Also this shape of the curve is similar to results for Bi-Sr-Ca-Cu-O and the recent measurements for $YBa_2Cu_3O_7$ single crystals of Palstra et al. (12) The knee results from a transition from flux flow to flux creep that has been discussed recently by Malozemoff et al. (14).

Film Resistivity: Films of Y-Ba-Cu-O materials, Bi-Sr-Ca-Cu-O, and Tl-Ba-Ca-Cu-O have been extensively studied; an interesting recent report by Sun et al. (15) shows, in Figure 5, resistivities for Er-Ba-Cu-O (123 phase) oriented films with the magnetic field in both perpendicular Figure 5a and parallel Figure 5b directions to the c axis. The enhanced broadening for the field perpendicular to the a,b plane is similar to that for single crystals in Figure 4 a,b and there is also indication of a knee in the curve (14) suggesting a flux flow regime. This will be further discussed in relation to the critical current. For these films the normal resistivity just above the transition is about 100 μOhm-cm compared with values of 20 μOhm-cm for single crystals. Of interest is the ability of very thin films to obtain the superconducting state. T. Venkatesan, et al. (16) have shown that films as thin as 10 nm show a resistive transition to superconductivity while a 5 nm film did not show superconductivity to about 10 Kelvin. Other measurements show superconductivity

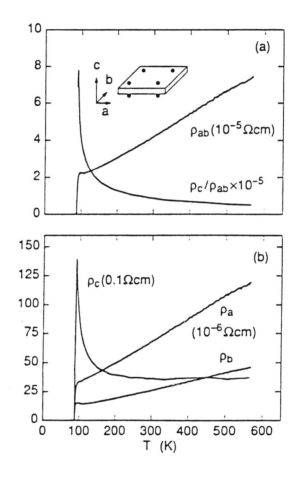

Figure 3: Resistivity of a single crystal Bi-Sr-Ca-Cu-O. (a) Averaged a,b plane resistivity and the ratio of c to a,b plane resistivity vs temperature. (b) Three components of the resistivity tensor. Ref. 4.

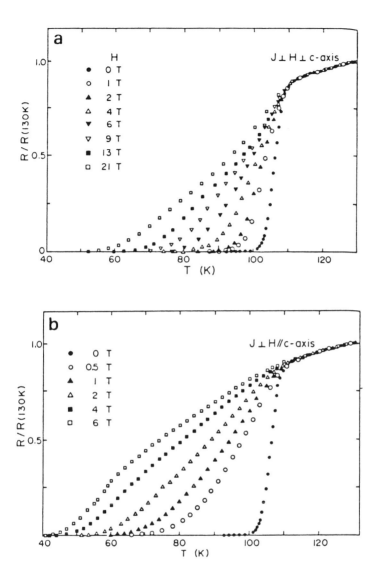

Figure 4: Temperature dependence of resistance of a single crystal Tl-Ba-Ca-Cu-O normalized by the value at 130 K at various magnetic fields for (a) H perpendicular to c-axis and (b) H parallel to c-axis. Ref. 13.

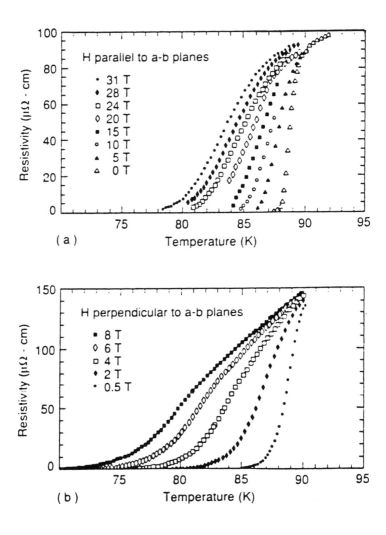

Figure 5: Resistivity of thin film $ErBa_2Cu_3O_{7-x}$. Top, H parallel to a,b planes; bottom, H perpendicular to a,b planes. Ref. 15.

for thicknesses down to 2 nm although the transition is broadened. These authors, Xi et al. (17), suggest that in all cases the system behaves as a 3D system and there is no crossover to 2D at their thinner films.

Bulk Ceramic Resistivity: There are many resistivity measurements on ceramic samples and an example is taken from D. Goldschmidt (5), in Figure 6, an expanded scale resistance vs temperature measurement where the normal state is 3 mOhm or with nominal sample dimensions about 40 μOhm-cm. There is very similar normal state resistance for good quality samples of single crystals, films, and bulk ceramics. Some additional information on the current dependence of the resistivity measurement is given by W. McGinnis, et al. (18). The silver painted leads had resistance of 50-100 mOhm although measurements were made with a pulsed current technique that will be discussed in the section on critical currents. The current densities were well below the critical current density of the sample and yet the depression of superconductivity is very large as shown in Figure 7 by McGinnis et al. (19).

2.2 Theoretical Notes

Theoretical discussion of the normal state remains incomplete. Models have been developed that depart from the usual interpretation of conduction by single type carriers in simple metallic bands. Two simple models were proposed by S. Bar-Ad, et al. (20) one involving a single wide band with nearly perfect electron-hole symmetry that is sensitive to dispersion in the c direction and to orthorhombic splitting. This model gives a better fit to the Hall carrier density. The split narrow band model with the upper part that consists of localized states and enters only into the chemical potential gives a better fit to the linear resistivity with temperature but the Hall carrier density and thermoelectric power saturate at values small compared with the bandwidth. The normal state resistivity remains a research problem; the solution is important to understanding the normal state and may be important in identifying the mechanisms leading to pairing in the superconducting state.

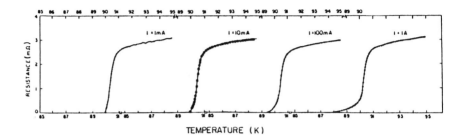

Figure 6: Resistivity of a bulk sample Y-Ba-Cu-O showing evolution of a low temperature tail as the current density increases. Ref. 5.

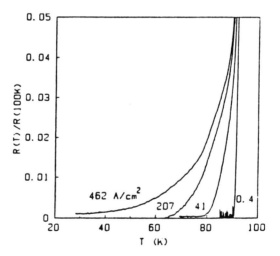

Figure 7: Resistivity of a bulk sample Y-Ba-Cu-O for a high normal resistance sample, evolution of the low temperature tail at current densities below critical. Ref. 19.

3.0 CRITICAL CURRENT DENSITY MEASUREMENTS

The determination of critical current density in the high temperature superconductors has importance for the fundamental understanding of the superconductivity and is a primary parameter in the technological development for applications. Both in bulk materials and thin film configurations the critical current density and the dependence on temperature and magnetic field limit the potential applications. There is ample evidence at present that the intrinsic critical current density is high enough to support numerous applications. Also clear is the extreme sensitivity of the critical current density to flux pinning mechanisms that can be related to materials chemistry, structure, and processing.

3.1 Measurement of Critical Current

Critical current measurements have been made with a variety of techniques. The indirect technique, that of obtaining the critical current from the magnetization response is discussed in Chapter 18. Direct transport measurements, using attached current and voltage leads, and indirect measurements requiring macroscopic current circulation will be discussed. Critical currents are desired as a function of both temperature and applied magnetic field since a variety of theories discuss the functional relationship. And applications may require either or both of these data.

The transport critical current is measured by using a four wire method for current and voltage leads often in a dc current mode although both ac and pulsed methods are in use. Techniques for placing the leads on the sample were described above (2.0 Resistivity Measurement) but critical current measurements frequently operate with large currents, 100 A or more, and thus a minimal contact resistance becomes important to avoid local sample heating at the contact interface. Several techniques have been reported. Low resistance contacts by Ekin et al. (21) are made with a thin gold film contact produced by sputter deposition of gold (or as an alternative, silver) with subsequent annealing at 500 - 600°C in an oxygen atmosphere. Their contact resistivity (contact resistance times the contact area) was in the range 10^{-10} Ohm-cm^2. More recently Suzuki et al. (22) discussed direct wire bonding to Y-Ba-Cu-O materials.

Silver wire was welded directly to Y-Ba-Cu-O bulk pellets with spot welding and ultrasonic bonding. The contact resistivities are reported at room temperature and 78 K; the latter values range from 2×10^{-3} to 3×10^{-8} Ohm-cm^2. Wires of 50 micron diameter were used and a contact area of about 0.03 mm^2 was measured optically. For currents of 100 A the heating would be of the order 10 mW so that larger wire size would be needed. For many samples this wire size would be adequate; however, for a 1 mm^2 bar with an expected critical current density $J_c = 10,000$ A/cm^2 the critical current would be 100 A and for a film 1 micron thick, 1 cm x 1 cm with $J_c = 10^6$ A/cm^2 the critical current would be 100 A. For a sample immersed in liquid helium the maximum heat flux without boiling is about 1 W/cm^2 and in liquid nitrogen is about 10 W/cm^2. In the above example if the heat at the contact area were released to a helium bath in no larger area than the contact resistance area the maximum convective heat transfer would be reached (23). Thus, low resistance contacts are essential as the critical current densities improve. The pressure contacts of tungsten as used in resistivity measurements (6) were noisy at lower temperatures when used for critical current measurements.

A technique for producing very low contact resistance is described by S. Jin, et al. (24). During the sintering of the material silver wire is embedded in the pellet. The pellet is sintered at 920°C for 40 hr in oxygen and slowly cooled to below 400°C. Wire stubs extending 2 cm beyond the pellet were then used in the connection. Two other variations were reported also by Jin (24); an embedded silver particle configuration and a silver clad patterned configuration. Contact resistivities of 10^{-12} Ohm-cm^2 were measured. A summary of electrical contact techniques to superconductors has been given by J. Talvacchio (25). The lower resistance techniques become necessary as the critical current increases.

The criterion for determining the critical current from measurements is not at all unique. The measurement proceeds from an experimental set-up similar to that shown in Figure 8 where the voltage is measured across the sample in this set-up for a pulse technique. For the more usual dc measurement a similar diagram with constant current source and dc voltmeters would replace the pulse generator, oscilloscope and differential amplifier/box-car averager. The advantages and disadvantages of these techniques will become more clear in the subsequent discussion. The typical results are shown

Transport Phenomena in High Temperature Superconductors 641

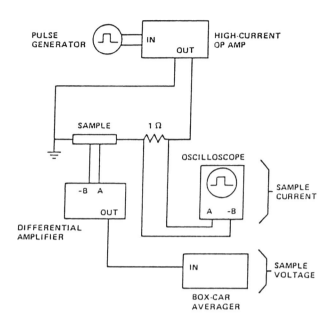

Figure 8: Pulsed current technique for the measurement of critical current. From Jones, T., McGinnis, W., Boss, R., Jacobs, E., Schindler, J., Rees, C., Tech. Doc. 1306 July 1988 Naval Ocean Systems Center, San Diego, CA.

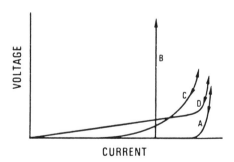

Figure 9: Voltage vs current and onset of resistance above the critical current for (A) "ideal case", (B) premature thermal runaway, (C) damaged sample, and (D) inadequate current transfer length between voltage and current taps as discussed for low temperature superconductors. Ref. 26.

in Figure 9 by curves A and C while curve B represent a premature thermal runaway (26). Curve D represents a sample that may show inadequate current transfer between current contact and voltage taps or lack of single phase material and percolation of superconductivity. These general results discussed by Ekin for low temperature superconductors are similar to results for the high temperature superconductors. At high currents the materials become normal conductors and will exhibit ohmic resistance. A voltage criterion, such as 0.1 μV/cm is considered to represent the value at which the current has become critical. Other criteria are used 1, 5, or 10 μV/cm since with short samples and the need for low currents a low voltage can not be determined due to system noise. As seen in the various curves of Figure 9 these criteria could give very different values of the critical current. This problem is exacerbated in most of the high temperature superconductors where a curve similar to C in Figure 9 is frequently observed and in addition the samples are often short and voltage leads only a few millimeters apart. Whether the curve C shape arises from intrinsic low current resistance or from impurities in the sample or from contact resistivity problems must be carefully determined in each measurement and in most of the published data has not been completely reported. Recognition of these problems has led some investigators to use pulsed measurements to avoid sample and contact heating in order to reach higher absolute currents in macroscopic samples (18). Pulsed measurements are more difficult to interpret when flux creep and flux flow are important issues. In addition response times of measuring instruments must be less than pulse durations.

The establishing of a criterion for critical current measurement led earlier investigators to formulate standards (27) and discuss for the composite (in the sense of low temperature superconducting wires that are fabricated with stranded superconductor in a copper or normal conductor matrix) superconductors problems such as current transfer voltage that can occur when the voltage tap is too close to the current contacts. The voltage (or more precisely the electric field) criterion is compared to a resistance criterion by R. Powell and A. Clark (28) and more recently by Ekin (29) and is shown in Figure 10. The electric field or voltage criterion is compared to the critical resistivity criterion and to the extrapolated offset J_c criterion for each curve at a constant magnetic field. The magnetic fields decrease from H_1, that

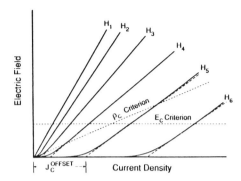

Figure 10: Critical current criteria as described in Ref. 29. The electric field vs. current density is shown as a function of magnetic field. The electric field criterion is set and the tangent at that intersection to the field vs current curve is extrapolated to obtain the offset J_c.

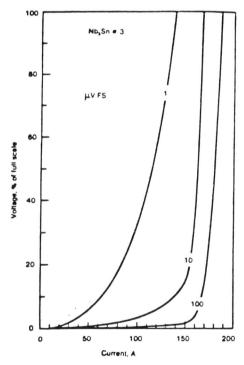

Figure 11: Voltage (in percent of full scale) vs current at 7 Tesla magnetic field for a commercial low temperature superconductor using a voltage criterion of different values to indicate the importance of specifying the criterion. Ref. 30.

is close to H_{c2}, to H_6. Some idea of the variation in critical current is obtained by comparing this value with the first appearance of voltage as indicated in Figure 10. The offset criterion is related to the critical-current distribution function and is an intrinsic criterion based on the concept of an average flux-flow J_c. Depending on the technological needs a non-zero but small resistance may be acceptable especially in high magnetic field so that the choice of criterion for I_c or J_c should be specified. Since most of the measurements to be discussed have been based on an electric field or voltage criterion, we will use that designation but understand that the recent result by Ekin may be a more useful indicator of J_c and has suggested advantage from a physics viewpoint. An example of the variation of the I-V curve at different voltage detection levels is shown in Figure 11 from Goodrich and Fickett (30) for the multifilamentary low temperature superconductor of Nb_3Sn at 7 Tesla.

The current transfer problem that had been identified with low temperature superconducting composites deserves additional mention for the high temperature superconductors, that in the bulk material are frequently not fully dense. Making the electrical connection in such a manner as to obtain uniform current distribution throughout the cross section of the material is difficult. The method described by Jin, et al. (24) with embedded wires or particles may provide for a significant improvement but the present techniques used to determine the critical current by a surface contact on the ceramic sample are subject to this problem. A discussion for the multifilamentary wire of Nb_3Sn is provided by Goodrich and Fickett (30) and this discussion is likely to be similar to the high temperature materials that are not fully dense.

Measurements of critical current are desired as a function of temperature and of magnetic field. The same care as discussed above (2.0 Resistivity Measurement) must be taken with temperature measurements; good thermal contact, shielding of the probe, and shielding and thermally anchoring the leads is required. In addition, the dependence of the temperature transducer upon magnetic field must be determined. Commercial transducers are available that have reduced sensitivity to applied magnetic fields including carbon glass, capacitive, and platinum transducers. The magnetoresistance of the transducers may need to be corrected but for these types is frequently small (31). Magnetic field measurements with rotating coils or Hall effect probes can be used to measure the field at the sample location. Many magnets in use currently have fairly uniform fields over volumes large compared to sample size, e.g., the usual superconducting solenoids.

3.2 Results of Critical Current Measurements

Thin films have shown the largest critical current densities, exceeding those of some conventional superconductors at temperatures of 4 Kelvin. The initial work in this field that provided a demonstration that critical currents were not intrinsically limited to a few hundred amps per square centimeter was published by Chaudhari et al. (32) where $J_c > 10^5$ A/cm^2 at 77 K and $> 10^6$ A/cm^2 at 4.2 K were established. Techniques for quality film fabrication are still being improved. At present the reactive sputtering techniques (33), frequently without post annealing, and the laser ablation techniques (16) have given films with T_c's of 90 K and J_c's of greater than 10^6 A/cm^2 at 77 K. Recent work by Hettinger et al. (31) has also shown that substantial current densities can be maintained in magnetic fields up to 15 Tesla. These improvements have occurred mostly during the last part of 1988 and further advances are expected including the ability to place these quality films on silicon substrates. An important feature of the reactively sputtered films of Buhrman (33) is the smooth (on the scale of 0.1 microns) upper surface. This smooth continuous upper surface is achieved now (except for an occasional "boulder") in laser ablated films and is essential to the fabrication of multilayer structures.

Film Critical Current Densities: Critical current densities of thin films have been reported by several hundreds of papers; a few representative but by no means inclusive are noted here in addition to those mentioned above. Desirable attributes of thin films for technology are: high transition temperature to zero resistance, high critical current, low substrate temperature during deposition, no high temperature post anneal, and atomically smooth surface without pinholes. A thermal coevaporation of yttrium, barium, and copper in an oxygen atmosphere have been deposited by Berberich (34) on substrates at 650°C with T_c's of 91 K on MgO and 89 K on SrTiO$_3$ without post anneal. Although critical currents of 10^6 A/cm^2 were obtained at 4 K, values of 10^4 A/cm^2 were found at 77 K. However, this technique combined with a short post anneal produced a film with T_c = 85 K on bare Si. Several contributions by the group at Bell Communications Research (35)(36)(37) discuss the laser ablation technique with $YBa_2Cu_3O_7$ films on $SrTiO_3$ and on Si with $BaTiO_3$/-$MgAl_2O_4$ buffer layers. Conventional optical lithography was used together with argon ion milling to produce micron-sized wires for four-probe measurements. Films of $YBa_2Cu_3O_7$ 200 nm thick with zero resistance T_c = 89 K have zero field critical current densities of 0.7 x 10^6 A/cm^2 at 77 K on $SrTiO_3$. No post anneal was used but the use of oxygen in the chamber during laser deposition was crucial. In a magnetic field the critical current density dropped significantly at 77 K (to about 100 A/cm^2 at 14 T (37)) although at temperatures of 60 K and a field of 14 T the value J_c was 10^5 A/cm^2. Their reported variation of J_c with temperature and magnetic field is shown in Figure 12a. Another result from Hettinger et al. (31) is shown in Figure 12b. For this Y-Ba-Cu-O epitaxial film critical currents greater than those in Nb_3Sn filaments can be carried above 20 K and at 15 T (compared to 4.2 K and 8 T). Plasma-assisted laser deposition techniques have been developed by Witanachchi et al. (38) to produce films on a silicon substrate at a temperature of 400°C. Films with a buffer layer of MgO showed higher values of T_c = 70 K and J_c ~ 10^3 A/cm^2 at 31 K. The lower substrate temperature is important to provide compatibility with semiconductor processing for potential hybrid electronics.

Other deposition techniques have been used (35) and a promising technique of chemical vapor deposition has been reported and substantial critical current densities obtained by Berry et al. (39) and by Watanabe et al. (40). Values of J_c at 77 K and at 2 T and 27 T were 4 x 10^5 and 6.5 x 10^4 A/cm^2 respectively (40). The surface of the film shows significant growth of c axis material in the plane as well as the desired normal to the plane growth. On the other hand this mixed phase growth may supply additional pinning centers for the improved critical current.

Films of Bi-Sr-Ca-Cu-O have been produced by a wide variety of techniques, a recent effort by Steinbeck et al. (6) shows results for a film with T_c = 90 K and J_c = 8 x 10^4 A/cm^2 at 77 K. This film was prepared on an MgO substrate and a packed micaceous-like structure is seen on the top surface of the film.

An oriented film of Tl-Ba-Ca-Cu-O (2212) on an MgO substrate was produced by Ichikawa et al. (41) by rf magnetron

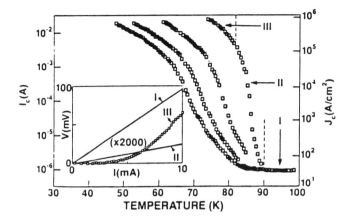

Figure 12a: Critical current of a Y-Ba-Cu-O film measured using 10 μV criterion as a function of temperature in magnetic fields of (from the right) 0.1, 5, 10, and 14 Tesla. The normal state region (I), fully superconducting state (III) and low resistance ohmic state (II) are indicated. The inset shows I-V characteristics of these three regions (III) region has been expanded by a factor 2000. Ref. 37.

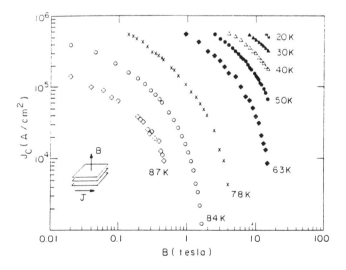

Figure 12b: Critical current density vs magnetic field applied normal to the film surface and current direction for a thin film microbridge of Y-Ba-Cu-O at various temperatures. Ref. 31.

sputtering and post-annealing. The highest value of zero resistance was found to give T_c = 102 K. At 77 K the value of critical current density was 1.2 x 10^5 A/cm^2. These films showed c axis orientation normal to the substrate.

Other substrates are being investigated for purposes such as obtaining reduced high frequency losses. The LaAlO$_3$ material has been used (42) for frequencies of 10 GHz. Reasonable critical current densities are found, J_c = 8 x 10^4 A/cm^2 at 77 K. A bilayer structure of the film was found; at the substrate surface the first layer had the c axis normal to the substrate and the second layer had the c axis in the plane.

Of interest for the future design of devices are techniques that can change the critical current after fabrication. The use of laser damage to produce a weak link for a Josephson junction was investigated by Buhrman (33). Ion irradiation has been used by White et al. (43) to reduce the critical currents in high quality films by orders of magnitude. Ions of Ne$^+$ at 1 MeV energy were used to ensure the range or penetration was greater than the film thickness of 200 nm. Values of J_c before irradiation were ~5 x 10^5 A/cm^2 and no sign of enhancement in J_c was obtained for fluences as low as 10^{11} Ne$^+$/cm^2. The decrease in J_c could be measured at five times this fluence and at ten times this fluence the authors estimate pinning centers introduced by atomic displacements every 5 nm. They argue that while some pinning centers are needed to maintain high critical current densities, when the pinning centers are overlapped on the scale of the coherence length the fluxoids can move freely between pinning centers, thus reducing the critical current density. Recent work by Roas et al. (44) finds an enhancement of J_c in epitaxial films of YBa$_2$Cu$_3$O$_7$ grown on SrTiO$_3$ and irradiated at 77 K with ^{16}O ions at 25 MeV energy. For fluences from 3.7 x 10^{13} to 2.1 x 10^{15} ions/cm^2 a steady increase in normal state resistivity and a nonlinear decrease of T_c is found. An enhancement in J_c in magnetic fields above 1 T is observed at 60 K and 77 K. The magnetic field was applied at an angle of 30 degrees to the c axis which was normal to the substrate plane. In Figure 13 these results are summarized. The very high J_c = 5.2 x 10^6 A/cm^2 at 77 K made measurements at 4 K difficult because of the contact resistance with evaporated silver contacts and bonded aluminum wires. Values of J_c >10^7 A/cm^2 at 4 K were found even at 6 T which is better than J_c values of Nb$_3$Sn commercial wires.

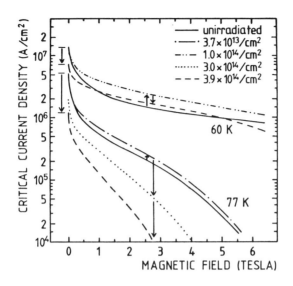

Figure 13: Critical current density vs. magnetic field as a function of increasing oxygen ion irradiation for Y-Ba-Cu-O at 60 and 77 K. Ref. 44.

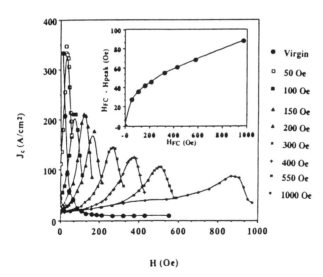

Figure 14: Critical current density vs applied magnetic field for zero field cooled case (virgin) and various field cooled values, H = 50, 100, 150, 200, 300, 400, 550, 1000 Oe. The enhancement peak value for the field cooled data vs magnetic field is shown in the inset. Ref. 48.

Single Crystal Critical Current Densities: Single crystals provide an opportunity to study the critical current anisotropy of the high temperature superconductors. One of the earlier measurements of critical current anisotropy in single crystal $YBa_2Cu_3O_7$ was inferred from magnetization techniques (45). [see Chapter 18] Recently a Bi-Sr-Ca-Cu-O (2212) single crystal was grown by Martin et al. (4). In addition to the resistivity measurements noted above the transport critical current densities were determined. An attempt was made to remove the 2D fluctuations that cause a rounding of the transition. A pulsed method with 5-10 microsec pulses, repetition at 50-100 Hz, and boxcar integrator averaging was used to avoid contact heating. They found different functional dependencies of J_c on T_c-T depending on whether the current was parallel or perpendicular to the ab plane. That is $J_c(perp) \sim (T_c-T)$ while $J_c(para) \sim (T_c-T)^{1/2}$. Neither of these results is in agreement with the expected functional dependence from the Ginzburg-Landau theory (46) $J_c \sim (T_c-T)^{3/2}$ or with the form from the Bardeen expression $J_c \sim (1-(T/T_c)^2)^{3/2}$.

Bulk Ceramic Critical Current Densities: The ceramic samples offer the most room for improvement of the critical current density and provide significant complications in interpreting and comparing the data. Only a sampling of the literature can be provided since the number of critical current measurements is large and to date the large values of J_c seen in films and single crystals have yet to be realized although as will be seen there is important progress. One of the earlier reports of transport critical current measurements by Cava et al. (47) finds a relatively low resistivity of 200-300 μOhm-cm and a $J_c > 1100$ A/cm^2 at 77 K. Many subsequent samples have exhibited values 1-500 A/cm^2.

The transport critical current of polycrystalline Y-Ba-Cu-O (123 phase) depends on previous magnetic field history (48) as shown in Figure 14. There is an enhancement in the field cooled critical current density compared to the zero field cooled measurements. McHenry et al. relate these results to the idea that intragranular diamagnetic current in zero field cooled field-increasing measurements are contributing substantially to the suppression of J_c. These hysteretic effects need further detailed understanding and must be reckoned with in technological applications.

The limitation on critical currents has been discussed by J. Clem (49) as due to weak link or Josephson coupling between grains.

This model which had been discussed much earlier in connection with granular classical superconducting films has been developed by many others (50). The nature of the weak link is still under much study, impurities at the grain boundaries were thought to be a problem but careful electron microscopy on well made samples indicates clean boundaries. The possibility of oxygen depletion and hence a lower temperature phase of the Y-Ba-Cu-O material has not been ruled out and in view of the difficulty in observing by photoemission the Fermi edge in single crystals without cleaving a fresh surface at low temperature this would seem a still likely possibility. Such an explanation is more difficult for Bi-Sr-Ca-Cu-O or Tl-Ba-Ca-Cu-O where the oxygen is known to be more stable. The experimental observations are mostly consistent with the weak link model that predicts greatly reduced critical currents in modest magnetic fields as the superconductivity in the weak links is quenched (51).

Measurements on ceramic samples of both $YBa_2Cu_3O_7$ and Bi-Sr-Ca-Cu-O have been made using a pulsed current technique by McGinnis et al. (18) and the temperature variation compared with the BCS value as obtained by Ambegaokar and Baratoff (52) $J_c \sim (T_c-T)^{3/2}$. The data have yielded different interpretations that may be dependent on sample preparation.

Recent studies of both $YBa_2Cu_3O_7$ and $DyBa_2Cu_3O_7$ by Aponte et al. (53) have shown that the best fit to the data is with the relation $J_c \sim (T_c-T)^v$ where $v=1$ is an average value. The actual fit is not exactly power law. A comparison with the previous crystal data would suggest that the current flow perpendicular to the a,b plane is controlling J_c in the ceramic polycrystalline material. Lobb (54) suggested that the departure from the expected $J_c \sim (T_c-T)^{3/2}$ value is due to the presence of strong fluctuations. Although Tinkham (10) suggests that flux creep can give the linear dependence near T_c the question of functional dependence between J_c and (T_c-T) is not fully answered at this time.

Oriented Grain Ceramic Critical Current Densities: In addition to the mixed orientation polycrystalline material various attempts have been made to align the c axis of the grains normal to the current flow direction in order to enhance the critical current density. Hampshire et al. (55) suspended the powders in a magnetic field of 0.5 T letting the alcohol solution evaporate and then compacting the aligned powder. While the J_c values were not large, 30-35 A/cm^2, they observed that at 4 K and magnetic fields of 22-27 T there was an abrupt drop of the critical current. Another texturing experiment by Jin et al. (56) obtained values of J_c ~10,000 A/cm^2. Recently Salama et al. (57) have reported texturing that gave 75,000 A/cm^2 in zero field and 30,000 A/cm^2 at fields of 0.6 T. At these current levels the question of contact resistance does become important and the actual measurements were reported to be likely lower limits. These experiments were repeated at the Naval Research Laboratory (66c) but with dc currents rather than pulsed currents and have shown that the high field value is above 3,000 A/cm^2 at 1 T. These results and others currently being announced indicate that substantial progress toward larger critical currents in bulk materials has been made in a relatively short time.

4.0 DISSIPATION IN THE INTERMEDIATE STATE

Dissipation or resistive behavior in a superconductor develops when the quantized vortices depin and cut across the current flow. Vortices are established in these superconductors either with an applied magnetic field or from the self field of a current. Visualization of the vortex structure in the high temperature superconductors has been studied with a Bitter decoration technique by Dolan et al. (58). The quantized nature of vortices has been studied in these materials by various techniques including microwave Josephson effect (59) and quantized currents in rings (60). These authors find that the quantum of flux is h/2e demonstrating that pairing occurs in these new materials as in the BCS theory and the classical superconductors.

An early discussion of dissipation is given by Anderson and Kim (61) and they discuss flux creep in the low dissipation regime. More recently Yeshurun and Malozemoff (62) have discussed the applicability of flux creep to the high temperature superconductors since these superconductors have relatively low pinning energy and the higher temperatures provide thermal activated flux motion. Tinkham (10) has presented a detailed theory. He applied the analytic theory of flux motion over a barrier developed by Ambegaokar and Halperin (63) and was able to describe the broadening to lower temperatures of the resistivity curves obtained in increasing magnetic fields. The barrier energy is

$$U_o = A(1 - T/T_c)^{3/2} / H \qquad (2)$$

and in the limit of low resistivity and driving force the resistivity is

$$\rho = \rho_n(U_o/kT)\exp(-U_o/kT) \qquad (3)$$

where ρ_n is the normal state resistivity. While this model works well in describing the low resistivity data Malozemoff et al. (14) have shown that the knee in the resistance curves developing at higher fields suggests a transition to a flux flow model that they develop. This model describes their new single crystal data and that by Palstra et al. (12). Recent data by Iwasaki et al. (13) for Tl-Ba-Ca-Cu-O shows similar knee structure at higher magnetic fields. Malozemoff also analyzes thin film measurements in terms of this model.

The situation in bulk material is more complex because the grain boundaries are shown to behave similarly to Josephson junctions and thus the measurements do not give directly the intrinsic transport currents. The magnetic field behavior as indicated in Figure 15 from Ekin (29) as well as in Figure 13 shows a rapid decrease of J_c with modest fields impressed on bulk material. This region represents the suppression of Josephson current in the distribution of junctions with various barrier heights. The flux flow region is in the non-linear regime and near the upper limit H_{c2} the I-V curves are more linear where the flux pinning is weak. Flux creep effects, when detectable (as in single crystals and thin films) result in a low resistivity linear I-V region extending up to the non-linear region.

The power law approximation of the voltage current characteristic for superconductors above I_c has been known for some time (64). Such studies have been made in Y-Ba-Cu-O (65) with results similar to those shown in Figure 16. The value of n in $V \sim I^n$ has been found to decrease as the magnetic field is increased, and of course becomes ohmic above H_{c2}. Another representation of the current voltage data is shown in Figure 17 from Enpuku et al. (66), log V vs 1/T for increasing currents (above critical). The expected near straight line arises from the flux creep model of Tinkham for $T/T_c \ll 1$.

Understanding flux motions and especially identifying the mechanism(s) for pinning at present remain a challenging opportunity. Measurements of a dissipation peak at near 45 K (well below T_c) in Bi-Sr-Ca-Cu-O samples oscillating in a magnetic field and interpreted

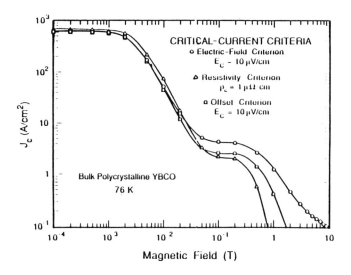

Figure 15: Critical current density vs. magnetic field for different criteria. Ref. 29.

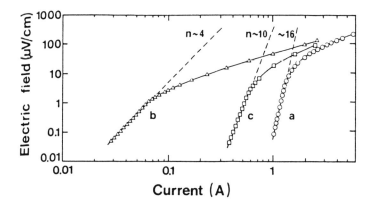

Figure 16: Logarithmic plot of electric field vs current for Y-Ba-Cu-O silver clad wire showing the value of n for different magnetic conditions. (a) for B = 0; (b) B = 27 mT; (c) as (b) after reduction of B to zero. Ref. 65.

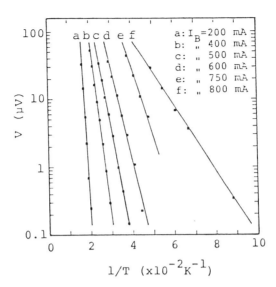

Figure 17: Temperature dependence of the voltage when the Y-Ba-Cu-O thin (800 nm) film is current biased. The linear dependence of log(V) on 1/T is observed. Ref. 66.

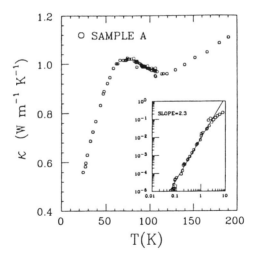

Figure 18: Temperature dependence of the thermal conductivity for sintered Bi-Sr-Ca-Cu-O (1112) with $T_c(0) = 90$ K. The inset shows data at very low temperatures (70 mK) and the line through the data points has a slope of 2.3. Ref. 69.

as flux lattice melting (66a) and measurements and interpretation of log(V) vs log(I) nonlinear curves as indicating a second order phase transition between the normal and superconducting states in a magnetic field (66b) are recent examples of ideas that are in the process of development and testing. And flux jumps, similar to those that have been seen in low temperature superconductors have been identified in recent transport critical current measurements (66c).

5.0 THERMAL CONDUCTIVITY

The transport of heat in metallic materials depends on both electronic transport and lattice vibrations, phonon transport. A decrease in thermal conductivity at the transition temperature is identified with the reduced number of charge carriers as the superconducting electrons do not carry thermal energy. The specific heat and thermal conductivity data are important to determine the contribution of charge carriers to the superconductivity. The interpretation of the linear dependence of the specific heat data on temperature in terms of defects of the material suggests care in interpreting the thermal conductivity results to be described.

Several techniques are available for thermal conductivity measurements, in the steady state technique a steady state thermal gradient is established with a known heat source and efficient heat sink. Since heat losses accompany this non-equilibrium measurement the thermal gradient is kept small and thus carefully calibrated thermometers and heat source must be used. A differential thermocouple technique and ac methods have been used. Wire connections to the sample can represent a perturbation to the measurement. Techniques with pulsed heat sources (including laser pulses) have been used; in these cases the dynamic response interpretation is more complicated.

Early results (67) for $YBa_2Cu_3O_7$ showed an increase in k (thermal conductivity) from lower temperatures to about 60 K, a slight decrease to near T_c and a nearly constant value between 100-150 K. The low temperature thermal conductivity in single crystal Bi-Sr-Ca-Cu-O (2212 phase) has been measured by Zhu et al. (68) and at temperatures less than 1 K they obtain a fit $k = 0.15\ T^2$ W/mK which they note is similar in temperature dependence to Y-Ba-Cu-O and also to a superconducting metallic glass, $Zr_{0.7}Pd_{0.3}$. A recent measurement in sintered samples by Peacor and Uher (69) determined the thermal conductivity of Bi-Sr-Ca-Cu-O in the (1112) and (2212) phases. Figure 18 shows their data for the (1112) phase sample that has $T_c(0) = 91$ K and these data are similar to other measurements in single crystals as well. Measurements to below 100 mK temperature did not show evidence for a crossover to a linear temperature dependence but rather a power law variation of k ~ T^2 is found (the exponent varied from 2 - 2.3) similar to the single crystal measurements. A peak in the conductivity curve is near 70 K with a value of 1 W/mK or about 1/25 the value of NaCl at this temperature. The peak is at a lower temperature than the transition temperature, 110 K. The lack of a linear temperature dependent term is correlated to the lack of a linear specific heat term. Thus, the linear terms are not intrinsic to the superconductor. The T^2 term might be interpreted as arising from phonon scattering on the tunneling states of a system similar to the situation in glasses. At temperatures above T_c the T^3 dependence of phonon boundary scattering is observed in contrast to the earlier results (67).

The effect of fast neutron fluence on thermal conductivity and thermopower has been determined by Uher and Huang (70). For fluences to 3×10^{18} n/cm^2 T_c decreases in Y-Ba-Cu-O to a temperature of 86 K, the thermal conductivity decreases and is without a peak above T_c and the thermopower starts from a negative value and approaches zero and becomes positive. As will be seen below the more usual value of thermopower is positive in the superconducting material but these authors note the variability dependent on sample preparation conditions.

6.0 THERMOPOWER

The thermopower or thermoelectric power is the electrostatic potential difference between the high and low temperature regions of a material with an impressed thermal gradient and zero electric current flow. The sign gives an indication of the sign of the charge carriers - positive for hole carriers.

Measurements in the La-Ba-Cu-O material were reported by Maeno et al. (71). At the optimum Ba concentration (0.15) the thermopower is positive and has a weak maximum of about 30 μV/K in the range 100-160 K as shown in Figure 19. For La-Cu-O the value is a factor 12 larger near 100 K and drops to low values beginning near 60 K. The data from Uchida et al. (72) show the decrease at lower temperatures for the pure nonsuperconducting La-Cu-O material. These authors do agree on the shape and absolute value for the stoichiometric Ba addition.

In single crystals of $YBa_2Cu_3O_7$ Ong et al. (73) reported out-of-plane thermopower linearly increasing with temperature above T_c from about 1 to 10 μV/K between 90 and 300 K. The in-plane thermopower exhibited a peak just above T_c, decreased to near 200 K and rose slightly to 300 K. The absolute value at maximum was 4 μV/K. This group has also determined the effect of selected impurities on the thermopower (74). Nickel and zinc were added and negative thermopower values measured as a function of concentration. The rapid change in thermopower implies that a very small change in band filling strongly affects the entropy transport. The thermopower of Tl-Ba-Ca-Cu-O (approximately the 2223 phase) bulk sample was determined by Mitra et al. (75) and showed a peak at about 170 K of S = 9 μV/K, decreasing to about 4 μV/K at 300K.

Theoretical interpretation is incomplete; early measurements on La-Sr-Cu-O by Hundley et al. (76) were suggested to indicate a phonon drag to explain some features although more recently a narrow band Hubbard model has been developed by Fisher (77).

7.0 HALL EFFECT

The Hall effect, an electric field perpendicular to both the impressed current flow and to the applied magnetic field, gives information about the mobility of the charge carriers as well as their sign. The Hall coefficient $R_H = E_y/J_xH_z$ is proportional to the reciprocal of the carrier density. The Hall coefficient is negative for electron charge carriers.

Early results of Hall coefficient measurements were presented in a review article by Tanaka (78). For the La-Ba-Cu-O material as a function of the increasing fraction of barium the Hall coefficient is positive, decreasing, and nearly temperature independent above T_c. These results are shown in Figure 20. For Y-Ba-Cu-O R_H increases

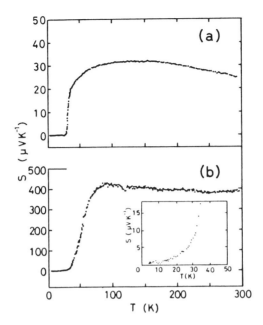

Figure 19: Temperature dependence of the thermoelectric power for the La-Ba-Cu-O material with the fraction of Ba either x = 0.15 (a) or x = 0 (b). Ref. 71.

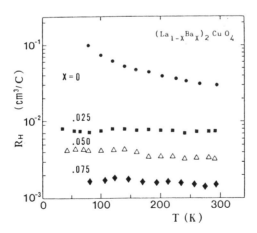

Figure 20: Temperature dependence of the Hall coefficient for $(La,Ba)_2CuO_4$. Ref. 78.

with decreasing temperature and with decreasing oxygen content. Early single crystal measurements were reported by Iye et al. (79). Their resistivity results show linear normal state resistance in the ab plane and semiconductor behavior in the c direction and may be taken as indication of quite good crystal quality although the resistivity is larger than for the current best crystals. The Hall coefficient for magnetic fields along the c axis shows that $1/R_H$ is linear in T in Figure 21. For a field applied in the ab plane the Hall coefficient is negative or electron like and the value is -2×10^{-4} cm^3/C.

For Bi-Sr-Ca-Cu-O (a mixture of 2212 and 2223 phases) the Hall coefficient is found (80) to be positive and decreasing from 5 to 3 ($\times 10^{-9}$ m^3/C) between 120 to 280 K. In the case of Tl-Ba-Ca-Cu-O (2223 phase) positive R_H decreasing from 5 to 3 ($\times 10^{-9}$ m^3/C) is measured by the Ong group. There are not yet single crystal measurements in these materials.

These results, Hall coefficient and thermopower, are fairly consistent in indicating that the carriers are holes in the ab plane of the Y-Ba-Cu-O and the Bi-Sr-Ca-Cu-O materials at temperatures just above T_c. Extrapolation to below T_c is assumed. The Hall coefficient in a single crystal suggests electron transport in the c direction for Y-Ba-Cu-O and for the newer materials of the T' class, $Nd_{2-x}Ce_xCuO_{4-y}$; the indication of electron pairing is also suggested by these measurements. Preliminary measurements including thermopower and resistivity for Nd-Ce-Cu-O are reported by Uji et al. (81) where a dependence on Ce concentration is given. The Hall coefficient in an applied field of 15 Tesla is measured by Wang et al. (82) for a single crystal Nd-Ce-Cu-O with T_c = 20 K as shown in Figure 22.

8.0 TUNNELING TRANSPORT

Electron pair tunneling through a thin layer of insulating material between two superconductors was predicted by Josephson and forms the basis for the Superconducting Quantum Interference Device (SQUID) with unique properties for superconducting electronic devices. Studies in low temperature superconductors have provided direct evidence for the superconducting energy gap and phonon spectra and extensive literature is available (83). Although measure-

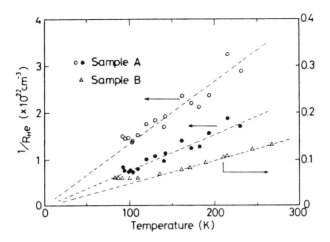

Figure 21: Temperature dependence of the reciprocal of the Hall coefficient for two samples of Y-Ba-Cu-O. Ref. 79.

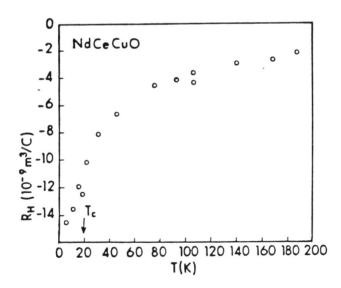

Figure 22: Temperature dependence of the Hall coefficient R_H in single crystal Nd-Ce-Cu-O with $T_c = 20$ K. The field of 15 Tesla is applied normal to the CuO_2 planes. Ref. 82.

ments at this time are tending toward similar values of the energy gap (and thus may be considered more reliable), the technique requires considerable care and neither the residual values of conductance at zero voltage nor the linearly increasing conductance above the gap value have been explained. In this section a brief discussion of tunneling and recent results will be given.

8.1 Josephson Effect

Electron pair tunneling across a gap in the absence of an applied electric or magnetic field gives a dc current of magnitude $I = I_o \sin(\phi)$ where I_o is the maximum or critical current and ϕ is the quantum phase factor across the gap. This current is only the dissipationless current, that is, the supercurrent. When the current exceeds the critical current of the junction a voltage V appears and the phase ϕ evolves with time, $d\phi/dt = 2eV/\hbar$. The supercurrent is modulated by the phase factor i.e., $\phi = \phi_o - 2eVt/\hbar$ and the frequency of the oscillation is $f = 2eV/\hbar$. Interference effects in addition to diffraction effects may be observed as macroscopic manifestations of quantum phase coherence of the superconducting electrons (84). At higher bias voltages there is a change in conductance above the energy gap and features of the phonon spectrum can be observed in the low temperature superconductors (83). A Josephson junction connected by a ring of superconducting material forms an rf SQUID (Superconducting Quantum Interference Device) in which a very small current flow can be detected, that current flow in response to magnetic flux through the ring. Detection is obtained with an inductive rf coupling to an external circuit. Two Josephson junctions inserted into a superconducting ring form a dc SQUID, called dc because a dc current can be used to bias the junctions to near their critical currents. A voltage output in response to a small input flux can be measured directly across the junctions although frequently an ac inductive coupling is needed. The description of SQUID operation involved the inductance, capacitance, and resistance of the ring and because of the small junction size and necessity of external coupling there can be interaction with other circuit elements and the introduction of noise from those elements. When the current across the junction is above critical the dissipation in the presence of a microwave field is marked by distinct Shapiro steps in the IV curves with a separation $\hbar/2e = 2.068$ microV/Ghz.

8.2 Tunneling Results

Josephson junctions can be made in a variety of configurations: a point contact between two superconductors can be adjusted to maintain a small gap, either a spatial gap, or a gap across an insulating barrier such as an oxide material; a film barrier of insulating or normal metal material can be used; a constriction in the superconductor also exhibits Josephson current behavior. A break junction technique has been used, most recently by Moreland et al. (85) in which a bulk of film material is fractured perhaps by bending the beam on which the material is mounted and then controlling the relaxation until a tunneling contact is obtained. The nature of the junction produced is of course quite uncertain but the method has been effectively used with the brittle ceramic superconductors (86). A thin oxide barrier such as produced by a controlled oxidation of aluminum has been used as a barrier and other film techniques using lithography have proved effective as illustrated for low temperature superconductors by Foglietti et al. (87). This IBM group has reported an eight level mask process with Nb-Pb alloy edge junction dc SQUID with a low value of extrinsic energy sensitivity. A constriction in a superconductor can act as a Josephson junction equivalent. Such a device has been made recently by the Raveau group (88) of $Tl_2Ba_2Ca_2Cu_3O_y$ material with a gradually engraved construction. The I-V characteristics are shown in Figure 23 and the Shapiro steps observed with the junction bathed in a microwave field at 9.169 GHz at 77 K is shown in Figure 24. These and earlier measurements have demonstrated that the steps are separated by values of 2.06 $\mu V/Ghz$, a value related to $h/2e$, demonstrating electron pair transport.

The conductance of a tunnel junction, dI/dV, measured against an increasing bias voltage displays evidence for the energy gap and phonon spectrum of a superconducting material (83). This technique has been applied to the high temperature superconductors with some success. So far it has not been possible to make a S-I-S tunnel junction of the high temperature superconducting materials using a prepared film insulating barrier. Thus, grain boundary barriers or break junction gaps have been used. Some success has been obtained with a S(HTS)-I-S(LTS) tunnel junction as shown by Lee et al. (89). They used a Y-Ba-Cu-O (123 phase) material with a native barrier and a Pb counter electrode. The results are shown in Figure 25. At

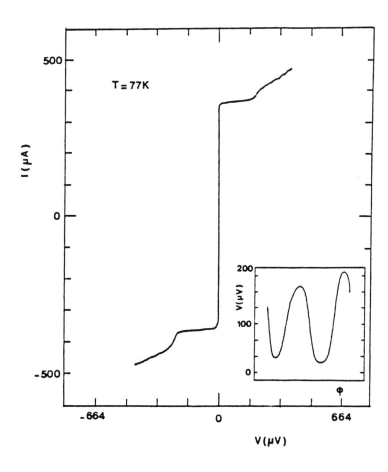

Figure 23: I-V characteristics of Tl-based ceramic with I_c = 350 µA and resistance for $I>I_c$ up to 30 Ohm. The inset shows the voltage vs applied magnetic flux. Ref. 88.

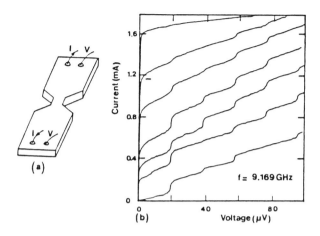

Figure 24: Schematic view of the sample with the constriction (a) and the Shapiro steps induced on the I-V characteristics by a microwave field at f = 9.169 GHz and T = 77 K. Ref. 88.

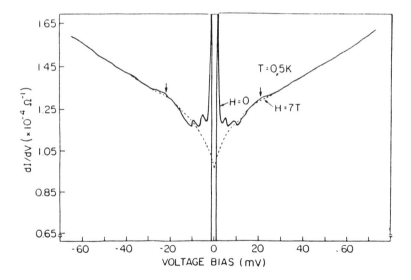

Figure 25: Differential conductance of a sandwich-type tunnel junction of Y-Ba-Cu-O(123) T_c = 60 K, formed with native oxide and a Pb counter electrode. Ref. 89.

a temperature 0.5 K, well below the transition temperature, in zero magnetic field the sharp edge of the Pb gap is seen and at increasing energy some evidence of the Pb phonon spectrum and then at near 20 mV a feature interpreted as the gap in Y-Ba-Cu-O is seen. This gap-like depression of the tunneling density of states is observed to go to zero at a temperature of 60K suggesting that the interface material next to the barrier is oxygen depleted Y-Ba-Cu-O with the lower T_c = 60 K. Other measurements are discussed in this reference with a conclusion that gap values $2\Delta/kT \sim 6$-7 are obtained from a variety of experiments for both the Y-Ba-Cu-O (123) and the Bi-Sr-Ca-Cu-O (2212). Questions of the anisotropy of the energy gap, the possible closure of this gap at some points or lines on the Fermi surface, and of the residual conductance that is found in many S(HTS)-I-S-(HTS) measurements remain to be definitively answered. The challenge to produce adequate artificial junctions remains important for the fundamental understanding and crucial for the realization of many devices.

ACKNOWLEDGEMENTS

It is a pleasure to acknowledge the many colleagues who have contributed information and provided illuminating discussions. Unfortunately this chapter could not contain all the information available and the choice remains with me. Especially to be thanked are Drs. R. Soulen, S. Wolf, and T. Francavilla who have read early manuscripts and contributed useful comments.

9.0 REFERENCES

1. Bednorz, J.G. and Muller, K.A., *Z phys.* B64:189 (1986). Also Wu, M., Ashburn, J., Torng, C., Hor, P., Meng, R., Gao, L., Huang, Z., Wang, Y. and Chu, C-W., *Phys. Rev. Lett.* 58:907 (1989).

2. See section: 4.0 Thermal Conductivity.

3. White, G. *Experimental Techniques in Low-Temperature Physics 3rd Edition*, Oxford University Press, London (1979).

4. Martin, S., Fiory, A., Fleming, R., Espinosa, G. and Cooper, A., *Appl. Phys. Lett.* 54:72 (1989).

5. Goldschmidt, D., *Phys. Rev.* B39:2372 (1989).

6. Steinbeck, J., Tsaur, B-Y., Anderson, A. and Strauss, A., *Appl. Phys. Lett.* 54, 466 (1989).

7. There are only a few measurements that have shown a 10^{-14} Ohm-cm resistivity; one method is by establishing a current flow in a ring of superconductor and monitoring the decay.

8. Poole, Jr., C., Datta, T. and Farach, H., *Copper Oxide Superconductors*, Wiley-Interscience, NY (1988).

9. Kwok, W.K., Crabtree, G.W., Hinks, D.G. and Capone, D.W., *Proc. 18th Int. Conf. on Low Temperature Physics.* Kyoto, 1987, *Japanese Journal of Applied Physics*, 26:1191 (1987).

10. Tinkham, M., *Phys. Rev. Lett.* 61:1658 (1988).

11. Harris, E.A., Bishop, J.E.L., Havill, R.L. and Ward, P.J., *J. Phys. C Solid State Phys.(U.K.)* 21:L673 (1988).

12. Palstra, T., Batlogg, B., van Dover, R., Schneemeyer, L. and Waszczak, J., *Appl. Phys. Lett.* 54:763 (1989).

13. Iwasaki, H., Kobayashi, N., Kikuchi, M., Kajitani, T., Syono, Y., Muto, Y. and Nakajima, S., *Physica* C159:301 (1989).

14. Malozemoff, A., Worthington, T., Yeh, N., Zeldov, E., McElfresh, M. and Holtzberg, F., Flux Creep and the Crossover to Flux Flow in the *Resistivity of High Temperature Superconductors Strong Correlations and Superconductivity*, Springer Series in Physics, ed. H. Fukuyama, S. Maekawa and A. Malozemoff, Springer-Verlag, Heidelberg (1989). Submitted.

15. Sun, J., Char, K., Hahn, M., Geballe, T. and Kapitulnik, A., *Appl. Phys. Lett.* 54:663 (1989).

16. Venkatesan, T., Wu, X.D., Dutta, B., Inam, A., Hegde, M., Hwang, D., Chang, C., Nazar, L. and Wilkens, B., *Appl. Phys. Lett.* 54:581 (1989).

17. Xi, X., Geerk, J., Linker, G., Li, Q. and Meyer, O., Preparation and Superconducting Properties of Ultrathin Y-Ba-Cu-O Films. *Appl. Phys. Lett.* 54:2367 (1989).

18. McGinnis, W., Jones, T., Jacobs, E., Boss, R. and Schindler, J., 1988 Applied Superconductivity Conference (*IEEE Transactions on Magnetics*).

19. McGinnis, W., Jones, T., Jacobs, E., Boss, R. and Schindler, J., Critical Current Densities for the High Temperature Ceramic Superconductors Y-Ba-Cu-O and Bi-Sr-Ca-Cu-O. 1988 Applied Superconductivity Conference, San Francisco, CA (1988).

20. Bar-Ad, S., Fisher, B., Ashkenazi, J. and Genossar, J., *Physica* C156:741 (1988).

21. Ekin, J.W., Larson, T., Bergren, N., Nelson, A., Swartzlander, A., Kazmerski, L., Panson, A. and Blankenship, B., *Appl. Phys. Lett.* 52:1819 (1988).

22. Suzuki, Y., Kusaka, T., Aoyama, T., Yotsuya, T. and Ogawa, S., *Appl. Phys. Lett.* 54:666 (1989).

23. For further discussion of heat transfer in cryogenic fluids, see R. J. Donnelly, Cryogenics, *Physics Vade Mecum*, ed. H.L. Anderson, AIP, NY (1981). For a recent discussion of contact resistance see also H. Jones, L. Cowey, and D. Dew-Hughes, *Cryogenics* 29:795 (1989).

24. Jin, S., Davis, M., Tiefel, T., van Dover, R., Sherwood, R., O'Bryan, H., Kammlott, G. and Fastnacht, R., *Appl. Phys. Lett.* 54:2605 (1989).

25. Talvacchio, J., *IEEE Trans.*, CHMT-12:21 (1989).

26. Ekin, J.W., *Materials at Low Temperatures*, ed. R.P. Reed and A.F. Clark, American Society for Metals, Metals Park, OH 44073 (1983). *Chap. 13 "Superconductors."*

27. Standard Test Method for D-C Critical Current Of Composite Superconductors B714-82. For example, see *1983 Annual Book of ASTM Standards*, Vol. 02.03, American Society for Testing and Materials, Philadelphia, PA 19103.

28. Powell, R. and Clark, A., *Cryogenics* 18:137 (1978).

29. Ekin, J., Offset Criterion for Determining Superconductor Critical Current, *Appl. Phys. Lett.* 55:905 (1989).

30. Goodrich, L. and Fickett, F., *Cryogenics* 22:225 (1979).

31. Hettinger, J., Swanson, A., Skocpol, W. Brooks, J. and Graybeal, J., *Phys. Rev. Lett.* 62:2044 (1989).

32. Chaudhari, P., Koch, R., Laibowitz, R., McGuire, T. and Gambino, R., *Phys. Rev. Lett.* 58:2683 (1987).

33. Lathrop, D., Russek, S., Tanabe, K., and Buhrman, R., Y-Ba-Cu-O Thin Films Grown by High Pressure Reactive Sputtering, *IEEE Trans. on Magn.* MAG-25 (in press).

34. Berberich, P., Tate, J., Dietsche, W. and Kinder, H., *Appl. Phys. Lett.* 53:925 (1988).

35. Inam, A., Hegde, M., Wu, X., Venkatesan, T., England P., Miceli, P., Chase, E., Chang, C., Tarascon, J. and Wachtman, J., *Appl Phys. Lett.* 53:908 (1988).

36. Wu, X., Inam, A., Hegde, M., Wilkens, B., Chang, C., Hwang, D., Nazar, L. and Venkatesan, T., *Appl. Phys. Lett.* 54:754 (1989).

37. England, P., Venkatesan, T., Wu, X., Inam, A., Hegde, M., Cheeks, T. and Craighead, H., *Appl. Phys. Lett.* 53:2336 (1988).

38. Witanachchi, S., Patel, S., Kwok, H. and Shaw, D., *Appl. Phys. Lett.* 54:578 (1989).

39. Berry, A., Gaskill, D., Holm, R., Cukauskas, E., Kaplan, K. and Henry, R., Formation of High Temperature Superconducting Films by Organometallic Chemical Vapor Deposition, *Appl. Phys. Lett.* 52:1743 (1988).

40. Watanabe, K., Yamane, H., Kurosawa, H., Hirai, T., Kobayashi, N., Iwasaki, H., Noto, K. and Muto Y., *Appl. Phys. Lett.* 54:575 (1989).

41. Ichikawa, Y., Adachi, H., Setsune, K., Hatta, S., Hirochi, K. and Wasa, K., Highly Oriented Superconducting Tl-Ca-Ba-Cu-Oxide Thin Films with 2-1-2-2 Phase, *Appl. Phys. Lett.* 53:919 (1988).

42. Mogro-Campero, A., Turner, L., Hall, E., Garbauskas, M. and Lewis, N., *Appl. Phys. Lett.* 54:2719 (1989).

43. White, A., Short, K., Dynes, R., Levi, A., Anzlowar, M., Baldwin, K., Polakos, R., Fulton, T. and Dunkleberger, L., *Appl. Phys. Lett.* 53:1010 (1988).

44. Roas, B., Hensel, B., Saemann-Ischenko, G. and Schultz, L., *Appl. Phys. Lett.* 54:1051 (1989).

45. Dinger, T., Worthington, T., Gallagher, W. and Sandstrom, R., *Phys. Rev. Lett.* 58:2687 (1987).

46. Tinkham, M., *Introduction to Superconductivity*, McGraw-Hill, NY (1980). See also Glover, R., Critical Currents in Thin Planar Films, *Rev. Mod. Phys.* 36:299 (1964).

47. Cava, R., van Dover, R., Murphy, D., Sunshine, S., Siegrist, T., Remeika, J., Rietman, E., Zahurak, S. and Epinosa, G., *Phys. Rev. Lett.* 58:1676 (1987).

48. McHenry, M., Maley, M. and Willis, J., *Phys. Rev.* B40:2667 (1989).

49. Clem, J., Dept. of Energy High Temperature Superconductivity Meeting, Argonne National Laboratory, January 20, 1987. More recently, see J. Clem, *Physica* C153-155:50 (1988).

50. Melikhov, E. and Gershanov, Y., Percolation Model of Ceramic High-Tc Superconductors. Critical Current and Current Voltage Characteristic, *Physica* C157:431 (1989).

51. Ekin, J., Braginski, A., Panson, A., Janocko, M., Capone II, D., Zaluzec, N., Flandermeyer, B., deLima, O., Hong, M., Kwo, J. and Liou, S., *J. Appl. Phys.* 62:4821 (1987).

52. Ambegaokar, V. and Baratoff, A., *Phys. Rev. Lett.* 10:486 (1963).

53. Aponte, J., Abache, H., Sa-Neto, A. and Octavio, M., Temperature Dependence of the Critical Current in High-T_c Superconductors, *Phys. Rev.* B39:2233 (1989).

54. Lobb, C., *Phys. Rev.* B36:3930 (1987).

55. Hampshire, D., Seutjens, J., Cooley, L. and Larbalestier, D., Anomalous Suppression of Transport Critical Current Below $B_{c2}(T)$ in Oriented Sintered Samples of $DyBa_2Cu_3O_7$, *Appl. Phys. Lett.* 53:814 (1988).

56. Jin, S., Tiefel, T., Sherwood, R., van Dover, R., Davis, M., Kammlott, G. and Fastnacht, R., Melt-textured Growth of Polycrystalline $YBa_2Cu_3O_7$ with High Transport J_c at 77 K, *Phys. Rev.* B37:7850 (1988).

57. Salama, K., Selvmanickam, L., Gao, L., and Sun, K., High Current Density in Bulk $YBa_2Cu_3O_x$ Superconductor, *Appl. Phys. Lett.* 54:2352 (1989).

58. Dolan, G., Chandrasekhar, G., Dinger, T., Feild, G. and Holtzberg, F., *Phys. Rev. Lett.* 62:827 (1989).

59. Olsson, H., McGrath, W., Claeson, T., Ericksson, S. and Johansson, L., *Proc. 18th Int. Conf. on Low Temperature Physics*, Kyoto, *Jpn. J. Appl. Phys.* 26:2113 (1987) Suppl.26-3.

60. Gough, C., Abel, J., Colclough, M., Formag, E., Keene, M., Mee, C., Muirhead, C., Rae, A., Sutton, S. and Thomas, N., *Proc. 18th Int. Conf. on Low Temperature Physics*, Kyoto, *Jpn. J. Appl. Physics.* 62:2117 (1987) Suppl. 26-3.

61. Anderson, P. and Kim, Y., *Rev. Mod. Phys.* 36;39 (1964).

62. Yeshurun, Y. and Malozemoff, A., *Phys. Rev. Lett.* 60:2202 (1988).

63. Ambegaokar, V. and Halperin, B., *Phys. Rev. Lett.* 22:1364 (1969).

64. Warnes, W. and Larbalestier, D., Determination of the Average Critical Current from Measurements of the Extended Resistive Transition. *IEEE Transactions on Magnetics*, MAG-23:1183 (1987) discusses this behavior for NbTi superconductor.

65. Evetts, J. and Glowacki, B., Relation of Critical Current Irreversibility to Trapped Flux and Microstructure in Polycrystalline $YBa_2Cu_3O_7$, *Cryogenics* 28:641 (1988).

66. Enpuku, K., Kisu, T., Sako, R., Yoshida, K., Takeo, M. and Yamafuji, K., Effect of Flux Creep on Current-Voltage Characteristics of Superconducting Y-Ba-Cu-O Thin Films, *Jpn. J. Appl. Phys.* 28:L991 (1989).

66a. Gammel, P. Schneemeyer, L., Waszczak, J., and Bishop, D., Evidence from Mechanical Measurements for Flux Lattice Melting in Single Crystal Y-Ba-Cu-O and Bi-Sr-Ca-Cu-O, *Phys. Rev. Lett.* 61:1666 (1989).

66b. Koch, R., Foglietti, V., Gallagher, W., Koren, G., Gupta, A., and Fisher, M., Experimental Evidence for Vortex-Glass Superconductivity in Y-Ba-Cu-O, *Phys. Rev. Lett.* 63:1511 (1989).

66c. Francavilla, T., Selvamanickam, V., Salama, K., and Liebenberg, D., The Effect of Magnetic Field on the Transport Critical Current of Grain-Oriented Y-Ba-Cu-O, International Conference on Critical Currents, Karlsruhe, FRG, 24-25 October (1989).

67. Morelli, D., Heremans, J., Swets, D., *Phys. Rev.* B36:3917 (1987).

68. Zhu, D-M., Anderson, A., Bukowski, E. and Ginsberg, D., *Phys. Rev.* B40:841 (1989).

69. Peacor, S. and Uher, C., *Phys. Rev.* B39:11559 (1989).

70. Uher, C. and Huang, W.-N, *Phys. Rev.* B40:2694 (1989).

71. Maeno, Y., Aoki, Y., Kamimura, H., Sakurai, J. and Toshizo, F., Transport Properties and Specific Heat of (La-Ba)-Cu-O, *Jpn. J. Appl. Phys.* 26:L402 (1987).

72. Uchida, S., Takagi, H., Ishii, H., Eisaki, H., Yabe, T., Tajima, S. and Tanaka, S., Transport Properties of (La-A)-Cu-O, *Jpn. J. Appl. Phys.* 26:L440 (1987).

73. Ong., N., Wang, Z., Hagen, S., Jing, T., Clayhold, J. and Horvath, J., Transport and Tunneling Studies on Single Crystals of $YBa_2Cu_3O_7$. *Proc. Int. Conf. on High Temperature Superconductors and Materials and Mechanisms of Superconductivity*, Interlaken (1988).

74. Ong, N., Jing, T., Wang, Z., Clayhold, J. and Hagen, S., Electronic Properties of $YBa_2Cu_3O_7$ in the Normal and Superconducting States. *Proc. 1st Asia Pacific Conf. on Condensed Matter Physics,* World Scientific, Singapore (1988).

75. Mitra, N., Trefny, J., Yarar, B., Pine, G., Sheng, Z., Hermann, A., *Phys. Rev.* B38:7064 (1988).

76. Hundley, M., Zettl, A., Stacy, A., Cohen, M., *Phys. Rev.* B35:8800 (1987).

77. Fisher, B., Grenossar, J., Lelong, I., Kessel, A. and Ashkenazi, J., - unpublished but see also reference 20.

78. Tanaka, S., High Temperature Superconductivity; Past, Present and Future. *Proc. 18th Int. Conf. on Low Temperature Physics, Jpn. J. Appl. Phys.* 26:2005 (1987) Supplement 26-3.

79. Iye, Y., Tamegai, T., Takeya, H., Takei, H., Transport Properties of Single Crystal YBa$_2$Cu$_3$O$_7$, *Jpn. J. Appl. Phys.*, Series I:46 (1988).

80. Clayhold, J., Ong, N., Hor, P. and Chu, C-W, The Hall Effect of the High-Tc Superconducting Oxides Bi-Ca-Sr-Cu-O and Tl-Ca-Ba-Cu-O, *Phys. Rev. B.* (1989). In press.

81. Uji, S., Aoki, H., and Matsumoto, T., Transport and Magnetic Properties of Nd-Ce-Cu-Oxides, *Jpn. J. Appl. Phys.*, 28:L563 (1989).

82. Wang, Z., Brawner, D., Chien, T., Ong, N., Tarascon, J., and Wang, E., Temperature Dependent Hall Effect in Nd$_{2-x}$Ce$_x$CuO$_4$ and "60 K" YBa$_2$Cu$_3$O$_y$ Single Crystals, International M^{2}S-HTSC Conference, Stanford, CA, 3-28 July 1989, to be published *Physica C*.

83. Josephson, B., *Superconductivity*, Ed. R. D. Park, Marcel-Dekker, NY, 1969.

84. Leggett, A., Macroscopic Quantum Tunnelling and Related Matters, *Proc. 18th Int. Conf. on Low Temperature Physics, Jpn. J. Appl. Phys.* 26-Suppl. 26-3:1986 (1987).

85. Moreland, J., Clark, A., Damento, M., and Gschneidner, K., *Physica* C153-155:1383 (1988).

86. Moreland, J., Ginley, D., Venturini, E., and Morosin, B., *Appl. Phys. Lett.* 55:1463 (1989).

87. Foglietti, V., Gallagher, W., Ketchen, M., Kleinsasser, A., Koch, R., and Sanstrom, R., *Appl. Phys. Lett.* 55:1451 (1989).

88. Robbes, D., Lam Chok Sing, M., Monfort, Y., Bloyet, D., Provost, J., and Raveau, B., *Appl. Phys. Lett.* 54:1172 (1989).

89. Lee, M., Kapitulnik, A., and Beasley, M., Tunneling and the Energy Gap in High-Temperature Superconductors, *Mechanisms of High Temperature Superconductivity*, Eds. H. Kamimura and A. Oshiyama, Springer Series in Materials Science vol. 11, Springer-Verlag, Berlin (1989).

18
Static Magnetic Properties of High-Temperature Superconductors

Eugene L. Venturini

1.0 INTRODUCTION

The two fundamental physical properties of a superconducting material are zero electrical resistance and the Meissner effect (1) which is the expulsion of an externally applied magnetic field. The Meissner effect arises because the lowest energy state for the superconductor in weak external fields occurs when the internal magnetic induction B is zero (2)(3). The existence of the Meissner effect can be shown to prove that superconductivity is a stable thermodynamic equilibrium state (4). The expulsion of a field leads to unique magnetic properties of superconductors (5-7) which are the topic of this chapter. The chapter is divided into four main parts: the static response of superconductors to low magnetic fields, the response in relatively high fields, the relaxation of magnetization (flux creep), and the distinct features of porous ceramic superconductors.

Magnetic measurements have played a major role in the discovery and understanding of high-temperature superconductors, and an overwhelming body of literature has appeared in less than three years. In general, the magnetic properties of high-temperature superconductors were first understood for "conventional" superconductors, and several classic books are particularly useful. Appropriate papers on high-temperature superconductivity are cited, but not exhaustively, and there is no claim that a particular reference is the first or even the "best" report concerning a given measurement.

There are four common instruments used to measure the magnetic response (moment) of a sample to an externally applied magnetic field, and each has strengths and weaknesses. Perhaps the most popular due to its sensitivity, ease of use and calibration, and accuracy is the **SQUID magnetometer** which employs an RF superconducting quantum interference device (SQUID) as the detector. The SQUID magnetometer measures the magnetic flux through a sense coil due to the sample magnetic moment. Also widely used is the **vibrating sample magnetometer** (VSM) which measures the changing dipolar field due to the magnetic moment of a vibrating sample. A third common instrument is the **Faraday balance** which determines the magnetic moment via the force exerted on a sample placed in a magnetic field gradient. The change in force (weight) is measured with a microbalance as a function of the average magnetic field strength. Finally, the **torque magnetometer** measures the torque on the sample when its magnetic moment is not aligned with an external magnetic field.

The VSM has the advantage of a fast response time and the ability to measure the sample moment in a continuously swept magnetic field. A minimum time constant of 1 second is typical for VSM studies of sample moments above 0.1 emu, and field sweeps are limited only by the inductance or power supply of the magnet. This fast response is particularly useful in studies of flux creep in superconductors (discussed below). In contrast, the SQUID magnetometer has the advantage of sensitivity: a moment of 10^{-6} emu is typically two orders of magnitude above the noise, allowing measurements on superconducting phases weighing a few micrograms distributed in a nonsuperconducting matrix weighing up to a few grams. However, the magnetic field must be stable to insure the sensitivity of a SQUID detector which limits the minimum response time to one or two minutes after a change in field strength. Hence, SQUID magnetometer data are obtained at discrete field values, although continuous temperature sweeps at a fixed field are possible. All data in this chapter were obtained with a commercial SQUID magnetometer (Biomagnetic Technologies, Inc., San Diego, CA, model VTS-905) with a temperature range of 2 to 400 K and a magnetic field range of approximately 1 Oe to 50 kOe (10^{-4} to 5 Tesla). Although the technical journals strongly prefer the SI system of units, most books and research scientists continue to rely on the more convenient and practical Gaussian cgs units such as emu, Gauss (G) and Oersted (Oe) for magnetic measurements; these units are used throughout this chapter.

2.0 LOW-FIELD MEASUREMENTS

2.1 Normal State Response

Consider a long thin cylinder of superconducting material placed in a weak uniform magnetic field H as shown in Figure 1. Figure 1(a) illustrates the magnetic response of the cylinder in the normal state, i.e., at a temperature T above the superconducting transition temperature T_c. The magnetic field lines penetrate the cylinder, inducing a small magnetic moment m parallel to the field. In contrast, the magnetic field does not penetrate the cylinder when $T < T_c$ as shown in Figure 1(b) and discussed in the next subsection. The magnetization M is just this moment divided by the volume V of the cylinder, M = m/V. The moment m is commonly measured in emu and the volume in cm^3, leading to a magnetization in emu/cm^3 or Gauss (G). For most superconductors in the normal state, M (or m) is proportional to the internal magnetic field H_i:

$$M = m/V = \chi_v \cdot H_i \qquad (T > T_c) \qquad (1)$$

where the proportionality constant χ_v is termed the volume magnetic susceptibility and is dimensionless in the cgs system. Actually, there is a negligible difference between the external magnetic field H_e and the internal field H_i for nonmagnetic and nonsuperconducting solids, but the distinction is important for superconductors as discussed below. The magnetic induction B inside the sample is defined by:

$$B = H_i + 4\pi M = (1 + 4\pi\chi_v) \cdot H_i \qquad (2)$$

678   Chemistry of Superconductor Materials

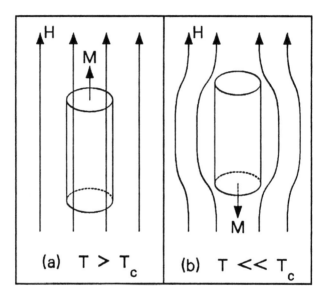

Figure 1: Magnetization M and magnetic field pattern for a long thin cylinder of material with a superconducting transition at temperature T_c placed in a small external field H: (a) at temperature $T > T_c$; (b) $T \ll T_c$.

Since it is much more convenient to weigh a sample than to measure its volume accurately, the magnetic response is often expressed as the moment per unit mass, m/W, where W is the weight of the cylinder. The proportionality between the internal field and the moment per unit mass is termed the mass magnetic susceptibility χ_g with units of cm^3/g. Obviously, χ_v and χ_g are simply related by the density of the cylinder $\rho = W/V$:

$$\chi_v = \rho \cdot \chi_g \qquad (3)$$

It is useful to consider typical values of different contributions to the magnetic susceptibility. One important component is the Curie susceptibility arising from isolated (i.e., noninteracting) paramagnetic ions with g-factor g and spin S which is given by:

$$\chi = N \cdot (g \cdot \mu_B)^2 \cdot S \cdot (S+1)/(3 \cdot k_B \cdot T) \qquad (4)$$

where N is the number of such ions, μ_B is the Bohr magneton, k_B is the Boltzmann constant, and T is the absolute temperature (8). The high-temperature superconductors contain Cu ions which are predominantly divalent. An isolated divalent Cu ion has a paramagnetic spin S = 1/2 and a g-factor g near 2, so Eq. 4 simplifies to:

$$\chi = (N/T) \cdot (\mu_B)^2/k_B = 6.230 \times 10^{-25} \cdot (N/T) \qquad (5)$$

The superconducting material YBa$_2$Cu$_3$O$_7$ weighs 666 g/mole or 222 g/mole Cu. If all the Cu ions were divalent and acted as isolated spins, N in Eq. 4 would be 2.7×10^{21} Cu^{+2} ions per gram. Hence, the mass susceptibility χ_g would be 5.6×10^{-6} cm^3/g at room temperature (295 K) and would increase inversely with decreasing temperature, tripling to 1.7×10^{-5} cm^3/g at 100 K.

Another additive term in the magnetic susceptibility arises from the temperature-independent core diamagnetism of all the ions in a solid. For YBa$_2$Cu$_3$O$_7$ the core diamagnetism is approximately -2×10^{-7} based on a calculation using Pascal's constants (9). This small negative contribution serves to reduce the total susceptibility. A third possible contribution arises from Van Vleck paramagnetism (10) caused by excited states in the atoms of the solid, while a fourth term is the band or Pauli paramagnetism minus the orbital diamagnetism (11) of the electrical carriers in the solid (holes in the case of $YBa_2Cu_3O_7$). Both the Van Vleck and Pauli contributions are positive and typically 10^{-5} to 10^{-6} cm^3/g, comparable to the Curie contribution above 100 K.

Finally, there is the question of interactions between the magnetic ions in the solid. There is a strong antiferromagnetic exchange coupling between the Cu ions in the CuO_2 sheets in high-temperature superconductors (actually superexchange coupling through the oxygen ions) (12). The insulating parent compounds such as La_2CuO_4 and $YBa_2Cu_3O_6$ exhibit long range antiferromagnetic ordering at temperatures near room temperature (12)(13). This ordering leads to a nearly temperature independent positive antiferromagnetic mass susceptibility below room temperature in the range of 10^{-7} to 10^{-6} cm^3/g from the Cu spins, in contrast to the inverse temperature dependence for a Curie susceptibility from Eq. 2. As carriers are added to the system, the correlation length for antiferromagnetic order decreases monotonically (as does the ordering temperature), and there is no long range antiferromagnetic order in the superconducting phase (14).

Typical normal state magnetic susceptibilities in the CuO-based high-temperature superconductors are nearly temperature independent above T_c and of order 10^{-6} cm^3/g. The values are not strongly affected by doping, remaining relatively constant from insulating antiferromagnetic La_2CuO_4 or $YBa_2Cu_3O_6$ with long range magnetic order to "metallic" $La_{1.85}Sr_{0.15}CuO_4$ or $YBa_2Cu_3O_7$ (15)(16). There is evidence from neutron scattering that the Cu ions continue to carry a spin in the metallic superconducting phase. The relative size of the various contributions to the magnetic susceptibility has not been firmly established for these systems, although estimates have been obtained for $La_{1-x}Sr_xCuO_4$ (17). However, it is clear that any strong positive <u>increase</u> in susceptibility with decreasing temperature reflects the presence of impurity phases with isolated paramagnetic Cu^{+2} ions, producing a Curie susceptibility. (Similarly, the detection of an electron paramagnetic resonance absorption near g = 2 indicates impurity phases containing isolated divalent Cu (18-20)).

2.2 Diamagnetic Shielding by a Superconductor

Diamagnetic shielding or flux exclusion is the response of a superconductor below T_c to a small applied magnetic field following zero-field cooling from above T_c. Figure 1(b) illustrates the situation for a superconducting cylinder in a small field H at a temperature well below T_c. The lowest energy state has the magnetic induction B equal to zero in the interior of the cylinder. Screening (shielding) supercurrents flow in the near surface region of the cylinder to exactly cancel the applied field in the interior, and the cylinder appears to have a large negative moment. The field relationship inside may be written as (21):

$$B = H_i + 4\pi M = 0 \qquad (6)$$

Hence $4\pi M = -H_i$, and the volume susceptibility for a long thin superconducting cylinder with its axis oriented parallel to the applied field is given by:

$$\chi_v = M/H_i = m/(V \cdot H_i) = -1/4\pi \qquad (7)$$

The shielding fraction is a comparison between the measured volume susceptibility and $-1/4\pi$, and it is given simply by $-4\pi\chi_v = -4\pi\rho\chi_g$.

There are several details which complicate this simple picture. All high temperature superconductors are type II superconductors, and they have a lower critical field strength H_{c1} above which the shielding is no longer complete. This is illustrated schematically in Figure 2 which shows the magnetization M versus increasing field H_i. The solid line indicates the response of a "perfect" type II superconductor. M increases linearly with H_i (slope = χ_v = $-1/4\pi$ from Eq. 7) when $H_i < H_{c1}$. There is a cusp at H_{c1}, above which the magnetic induction B is no longer zero in the lowest energy state, and magnetic flux enters the superconductor. When H_i reaches the upper critical field strength H_{c2}, the superconducting order parameter is zero, and the material acts as a normal metal for $H_i > H_{c2}$. For $H_{c1} < H_i < H_{c2}$ the flux in the superconductor is less than that in the normal state, and the material is said to be in the mixed state.

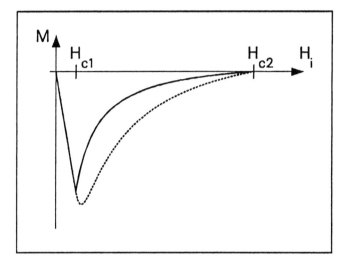

Figure 2: Shielding magnetization M versus increasing internal field H_i for a type II superconductor. The solid line shows the "ideal" response (i.e., with no flux pinning) with a cusp at the lower critical field H_{c1}; the dashed line shows the changes due to pinning. The material is superconducting for H_i below the upper critical field H_{c2} and a normal metal at higher fields.

There are two characteristic length scales in superconductors, the coherence length ξ and the penetration depth λ. The superconducting ground state is characterized by an order parameter which exhibits coherence over a spatial distance ξ, the coherence length for the paired carriers. The internal magnetic field H_i and supercurrent J_s vary over a characteristic distance λ, the penetration depth. For most type II superconductors including the high-T_c materials, $\lambda \gg \xi$. When a type II superconductor is in the mixed state, the flux appears as quantized units termed magnetic vortices, flux lines, or simply fluxoids. Each fluxoid contains one flux quantum $\phi_o = hc/2e = 2.07 \times 10^{-7}$ G·cm^2 where h is Planck's constant, c is the speed of light in vacuum, and e is the electron charge. A fluxoid consists of a cylindrical core where the superconducting order parameter rises from zero to its full value over a coherence length ξ. This core is surrounded by a cylinder of supercurrents flowing in a circular vortex pattern throughout a thickness equal to the penetration depth λ, generating a quantum of flux ϕ_o (22).

These flux lines are "pinned" at internal defects and by surface imperfections in all type II superconductors, leading to the magnetic response shown by the dashed line in Figure 2. Pinning has no effect when $H_i < H_{c1}$, since there is no flux entering the bulk of the superconductor (see below). However, the linear M-H curve has no cusp at H_{c1} when pinning is present; rather, there is a gradual deviation from linearity as flux lines overcome surface pinning and internal defect pinning to penetrate the superconductor. Finally, the magnetization reaches a minimum determined by the strength of the pinning and decreases for higher fields. Note that pinning has no effect on the value of H_{c2}, although the shape of the M-H curve in the mixed state is affected. Flux lines are not independent, and under appropriate conditions they can interact to form a flux lattice (Abrikosov lattice (22)) which exhibits collective pinning. The effects of this lattice are beyond the scope of this chapter.

When $H_i < H_{c1}$, B = 0 in the interior of the superconducting cylinder as shown in Figure 1(b), but magnetic flux enters the walls for a distance λ ($\approx$1400 Å for fields parallel to the a-b plane in $YBa_2Cu_3O_7$ at low temperatures (23)). Provided the sample is of macroscopic size, the penetration depth is negligible, but it becomes important for fine powders, fine-grained ceramics, and thin crystal plates. Two equations are useful for determining the correction for the penetration depth. First, the penetration depth is nearly temperature independent except when T approaches T_c (24):

$$\lambda(T) = \lambda(T=0) \cdot \{1 - (T/T_c)^4\}^{-1/2} \qquad (8)$$

For a field applied parallel to the broad face of a thin rectangular plate of thickness d, the penetration depth reduces the shielding susceptibility by (25):

$$\chi/\chi_o = 1 - \{(2\cdot\lambda)/d\}\cdot\tanh\{d/(2\cdot\lambda)\} \qquad (9)$$

where χ_o is the susceptibility when the penetration depth is negligible. Similar correction formulas for fine powder samples are available (26).

Another complication is the demagnetization correction due to the geometry of the specimen. Demagnetization (or the equivalent depolarization problem for dielectric bodies in an electric field) can only be solved analytically for an ellipsoid of revolution (27)(28). When H_e is applied parallel to one of the three axes of revolution, the magnetization is parallel to H_e, but the internal field H_i is given by (29):

$$H_i = H_e - 4\pi DM \qquad (10)$$

where D is termed the demagnetization factor. Substituting this relation for H_i in Eq. 7 and solving for M, we can define an experimental susceptibility χ_e relating M to the external magnetic field H_e:

$$\chi_e = M/H_e = \chi_v/(1 + 4\pi D \cdot \chi_v) = -1/\{4\pi\cdot(1 - D)\} \qquad (11)$$

Since D is between 0 and 1 in all cases and $\chi_v < 0$, demagnetization produces an enhanced experimental susceptibility compared to the true volume susceptibility.

Some common shapes have the following demagnetization factors (30). D = 0 for fields applied parallel to an infinite sheet and D = 1 for fields normal to the sheet (i.e., χ_e becomes infinite). D = 1/2 for fields applied normal to the axis of an infinite cylinder and D = 0 for fields parallel to the axis (hence the choice of a "long thin" cylinder parallel to H in Figure 1). Finally, D = 1/3 for fields in any direction if the sample is a sphere. Measurements with the applied field normal to a thin crystal or a thin film must be corrected for demagnetization which can be very large. For example, the measured low-field susceptibility was $-4750/4\pi$ with the field normal to a 0.3 μm-thick $Tl_2Ca_2Ba_2Cu_3O_x$ film with a 3x3 mm^2 cross-section (31).

An example of a diamagnetic shielding or flux exclusion measurement is shown in Figure 3 where we plot the magnetization versus external field strength at 5 K for a thin crystal of $YBa_2Cu_3O_{7-\delta}$. This crystal has dimensions 1.10 x 0.62 x 0.040 mm^3, and the data were taken with the field applied parallel to the longest dimension. The demagnetization factor D is approximately 0.02 for this geometry, so complete diamagnetic shielding would be reflected in an experimental susceptibility $\chi_e = -1.02/4\pi$ from Eq. 11 compared to the measured value of $-0.96/4\pi$; hence the shielding fraction is 0.94. There is a slight correction for the penetration depth λ which is 1400 Å for fields parallel to the a-b plane of $YBa_2Cu_3O_{7-\delta}$: using the thickness d = 40 μm in Eq. 9, we obtain $\chi/\chi_o = 0.993$. Hence, the finite penetration depth reduces the susceptibility for complete diamagnetic shielding by nearly 1%, effectively raising the measured shielding fraction to 0.95. Clearly, as T approaches T_c and λ increases dramatically according to Eq. 8, the measured shielding susceptibility will fall rapidly. This decrease in χ with increasing temperature provides a direct measurement (23) of $\lambda(T)$ using Eq. 9, provided H_i remains below H_{c1} at all temperatures.

The data in Figure 3 also provide a determination of the lower critical field H_{c1}. The shielding response will remain linear in the applied field until flux lines start to enter the crystal. Assuming a negligible surface barrier to flux penetration, H_{c1} is determined by the point where the M-H curve deviates from linearity. In Figure 3 this occurs between 280 and 300 Oe, in good agreement with a recent estimate of 250±50 Oe from magnetic relaxation (32), but somewhat higher than estimates of 120(10) Oe obtained by the same method of deviation from

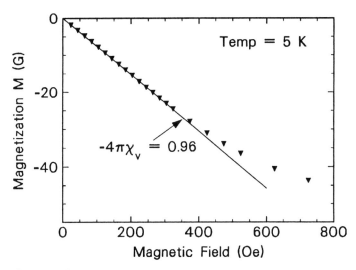

Figure 3: Shielding magnetization at 5 K versus increasing external field for a superconducting $YBa_2Cu_3O_{7-\delta}$ single crystal.

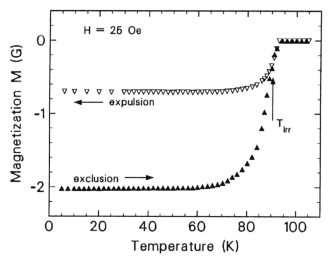

Figure 4: Flux exclusion (shielding) versus increasing temperature (solid triangles) and flux expulsion (Meissner effect) versus decreasing temperature (open triangles) in a 25 Oe external field for a superconducting $YBa_2Cu_3O_{7-\delta}$ single crystal. Exclusion and expulsion are equal for temperatures above the irreversibility point T_{irr} ($\approx$90.5 K at 25 Oe); T_c is 92 K.

linearity in the M-H curve (33) and 180(20) Oe obtained from flux exclusion versus temperature (34).

2.3 Magnetic Flux Exclusion and Expulsion

Diamagnetic shielding is measured by applying a small magnetic field after cooling the superconductor below T_c in nearly zero field. This effect is also termed flux exclusion since the superconductor supports screening currents near the surface to exclude the magnetic flux associated with the applied field. In contrast, one can apply the external field at a temperature T above T_c and measure the magnetization as the sample cools. In sufficiently small fields ($H_i < H_{c1}(T)$), the lowest energy state of the superconductor has B = 0, and the sample achieves this state by expelling the magnetic flux associated with the external field. This flux **expulsion** is termed the Meissner effect after its discovery by Meissner and Ochsenfeld in 1933 (1) (it should be called the Meissner-Ochsenfeld effect).

Figure 4 compares the flux exclusion and flux expulsion versus temperature measured in a field of 25 Oe applied parallel to the long axis of the same single crystal of $YBa_2Cu_3O_{7-\delta}$ discussed above. For temperatures above T_c = 92 K the magnetization is slightly positive ($\approx 6 \times 10^{-5}$ G), but becomes large and negative at lower temperatures for both expulsion and exclusion measurements. This is one of the simplest tests for superconductivity: an abrupt decrease in the measured susceptibility. Typically, a SQUID magnetometer can easily detect a superconducting volume fraction of 0.001% since χ_v is $-1/4\pi$ compared to a normal state χ_v of 10^{-6} or less. The rapid decrease in exclusion with increasing temperature beginning near 75 K in Figure 4 occurs because H_i exceeds $H_{c1}(T)$. Flux lines enter the superconducting crystal, and the magnetization decreases. This offers another method to determine $H_{c1}(T)$ by measuring exclusion versus temperature as a function of magnetic field (34).

Note that the expulsion and exclusion responses are equal for T close to T_c. This "reversible" region extends from T_c down to an "irreversibility" temperature T_{irr} which is a function of the applied field, varying as $(H_e)^{2/3}$ (35). The lack of reversibility at temperatures below T_{irr} indicates pinning of the magnetic flux: the

Meissner state requires that the sample expel the flux as it cools, and this cannot occur if the flux lines are pinned by impurities, defects or perhaps an intrinsic pinning mechanism (36). If there were no flux pinning, the exclusion and expulsion data would overlap at all temperatures. Hence, T_{irr} is that temperature below which the flux expulsion does not reach a true equilibrium Meissner state. If the field is removed following field cooling to a temperature $T < T_{irr}$, the sample exhibits a "remanent moment" opposite in sign to the Meissner signal. In fact, this moment equals the difference between the exclusion and expulsion curves at temperature T (see Figure 4). This equality demonstrates the importance of flux pinning in understanding magnetization data of high-temperature superconductors (37).

In the previous section we defined the shielding fraction as the ratio of the measured volume susceptibility following zero field cooling to the complete shielding value of $-1/4\pi$ (neglecting the finite penetration depth and including a correction for demagnetization). In a similar fashion, we can define a Meissner fraction as the ratio of the measured volume susceptibility to $-1/4\pi$ following field cooling. The shielding fraction will always be greater than the Meissner fraction for temperatures below T_{irr} where flux pinning occurs. Why is this important? Consider the situation where the superconducting portion of the sample is a relatively thick "shell" on a nonsuperconducting interior (38)(39). Provided that the shell thickness is large compared to the penetration depth λ, the flux will be excluded from the entire body by screening currents in the shell. Hence the shielding fraction will approach unity, indistinguishable from the situation where the entire sample is superconducting. In contrast, the nonsuperconducting interior will not expel the flux in a Meissner measurement, and only the superconducting shell will produce a Meissner signal. Hence, the Meissner fraction provides a reliable lower estimate on the "true" superconducting fraction (with the same caveats about penetration depth and demagnetization correction as in the shielding fraction).

If the entire body is superconducting, the Meissner fraction will be reduced due to flux pinning (40). How can one determine whether a low Meissner fraction is due to flux pinning and/or due to a sample problem like superconducting "shells"? One approach is to measure both the shielding and Meissner fractions as a function of applied field. If the two fractions agree at low temperature, the entire sample must be superconducting. There is always a reversible region sufficiently close to T_c where a true equilibrium Meissner state is achieved (i.e., where exclusion and expulsion are equivalent). If $H_{c1}(T)$ exceeds H_i during cooling in the reversible region, the sample will achieve a true Meissner state, and the exclusion and expulsion will remain in complete agreement at all lower temperatures. The experimental difficulty is choosing a sufficiently small measurement field to satisfy this criterion.

Our approach has been to measure both exclusion and expulsion at a low temperature (typically 10 K for $YBa_2Cu_3O_{7-\delta}$) and plot the ratio versus applied field. Figure 5 shows exclusion and expulsion measurements at 10 K versus applied fields from 2.5 to 700 Oe. The Meissner data are nearly independent of field above approximately 60 Oe, suggesting that the trapped flux is strongly pinned at higher fields when T decreases through T_{irr}. This also means that the Meissner fraction will be anomalously low if the measurement field is "too large". The shielding data are linear in field to approximately 300 Oe as previously shown in Figure 3 (a different crystal was used for the data in Figure 5). However, the shielding magnetization saturates near 600 Oe in Figure 5, while it is still increasing at 725 Oe in Figure 3. This implies that the crystal used for Figure 5 has less flux pinning at 10 K than that used for Figure 3. The onset of nonlinear behavior in the M-H exclusion curve indicates that the internal field H_i has reached H_{c1} at that temperature. Stronger fields place the sample in the mixed state, and pinning will prevent the flux from entering the sample freely. More pinning will result in a higher shielding magnetization at a given $H_i > H_{c1}$ due to less complete flux penetration.

Figure 6 shows the exclusion/expulsion ratio versus applied field using the data from Figure 5. The inset to Figure 6 emphasizes the behavior at low fields: as the measurement field is reduced below 10 Oe, the exclusion/expulsion ratio decreases smoothly to unity. This confirms that the entire crystal is superconducting, since the shielding fraction for this crystal is approximately 0.97 from the linear portion of the shielding M-H curve. There is a decided advantage to plotting the exclu-

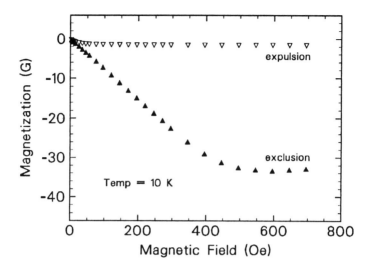

Figure 5: Flux exclusion and flux expulsion at 10 K versus external magnetic field for a superconducting $YBa_2Cu_3O_{7-\delta}$ single crystal.

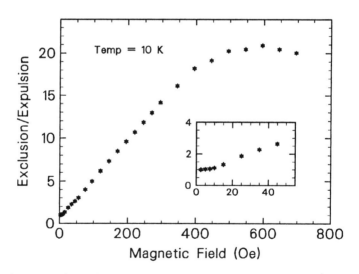

Figure 6: Ratio of flux exclusion to flux expulsion at 10 K versus external magnetic field for a superconducting $YBa_2Cu_3O_{7-\delta}$ single crystal. The inset shows that this ratio approaches unity at very low fields.

sion/expulsion ratio as in Figure 6: the complications due to demagnetization and penetration depth are removed since they enter both exclusion and expulsion equally in the limit of no pinning. Hence a ratio of unity implies complete superconductivity, provided that the shielding fraction is unity.

3.0 HIGH-FIELD MEASUREMENTS: HYSTERESIS LOOPS AND CRITICAL CURRENT DENSITY

Magnetization measurements at high magnetic fields $H_e \gg H_{c1}$ can be used to study flux pinning and to extract a magnetization critical current density J_{cm}. Figure 7 illustrates schematically a complete isothermal hysteresis loop for a typical type II superconductor with strong flux pinning (41). The dashed line shows the initial shielding response versus increasing field following cooling to the measurement temperature in zero field. At the highest field to the right of Figure 7 the sample is well into the mixed state, and the shielding current is uniform and equal to the critical current density throughout the sample. If the field sweep is then reversed, there is a rapid reversal of the shielding current near the surface, while the flux which entered during the increasing field cycle remains pinned within the sample. This results in a positive external magnetization (i.e., parallel to the external field) and a substantial hysteresis (difference) between M for increasing and decreasing fields. As the decreasing field sweep crosses zero and approaches a maximum in the opposite direction, the shielding currents again become uniform at the critical current density, but in the reverse sense. Hence, as the sweep direction is reversed again at the far left in Figure 7, the magnetization reverses sign and hysteresis is observed. Note that after the initial increasing field sweep (dashed line), the magnetization is reasonably symmetric about zero field, and the curve shown as a solid line can be retraced indefinitely.

For H_e much larger than H_{c1} the magnetization in Figure 7 is nearly the same on the initial (dashed) and repetitive (solid) curves for a given increasing field value. Hence one can accurately determine the hysteresis ΔM for "large" fields using only a partial loop where the field is increased from zero to a maximum

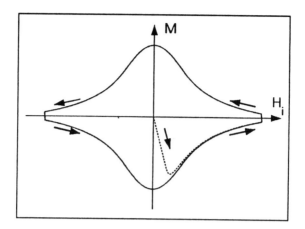

Figure 7: Sketch of a complete magnetization hysteresis loop versus internal magnetic field H_i. The dashed line is the initial response following zero field cooling, while the solid line is the response for repeated cycles between maximum and minimum field strengths. The arrows indicate the direction of the field sweep, and the hysteresis is the difference in magnetization at a given field during increasing and decreasing field sweeps.

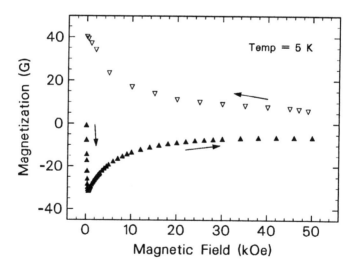

Figure 8: Partial magnetization hysteresis loop at 5 K for a superconducting $YBa_2Cu_3O_{7-\delta}$ single crystal. The solid triangles represent the response to increasing fields while the open triangles are for decreasing fields.

Static Magnetic Properties of High-Temperature Superconductors 693

value and then decreased back to zero. Figure 8 shows such a partial hysteresis loop measured at 5 K with the magnetic field applied along the long axis of the same $YBa_2Cu_3O_{7-\delta}$ single crystal used for Figure 3. The arrows indicate the field change direction beginning near zero field; increasing field data are shown as solid triangles, while open triangles represent decreasing fields. The large negative magnetization due to the shielding supercurrents reaches a minimum near 800 Oe as flux lines penetrate the crystal. As the field strength increases to 50 kOe, the magnetization decreases as more flux enters the crystal and the critical current density decreases in the stronger field. As the field is decreased below 50 kOe, the magnetization reverses direction rapidly to a positive value nearly equal in magnitude to the negative magnetization recorded at the highest increasing field.

This behavior reflects strong pinning of the flux lines in the crystal at 5 K and is explained by the Bean critical state model for hard (strongly pinned) type II superconductors (42-47). In fact, the hysteresis in magnetization $\Delta M = M(H_e^-) - M(H_e^+)$, where H_e^- denotes decreasing field and H_e^+ denotes increasing field, is proportional to the critical current density J_{cm} flowing in the sample (the subscript "m" is used to distinguish this result from the critical current density measured by direct transport and denoted J_{ct}). Two common geometries for magnetization field-loop (hysteresis) measurements are with the field applied parallel to the axis of a cylinder (or thin disk) of diameter d or normal to one face of a rectangular slab with cross-section axb. In the case of the cylinder:

$$J_{cm} = 30 \cdot \Delta M / d, \tag{12}$$

while for the rectangular slab with a > b:

$$J_{cm} = (20 \cdot \Delta M / b) / (1 - b/3a). \tag{13}$$

We obtain $\Delta M(H_e)$ from the data in Figure 8 and calculate $J_{cm}(H_e)$ using the rectangular slab Eq. 13 where a = 0.62 mm and b = 0.040 mm. The result is shown in Figure 9 where the magnetization critical current density J_{cm} falls monotonically with increasing field strength. Actually, Eq. 13 must be modified for

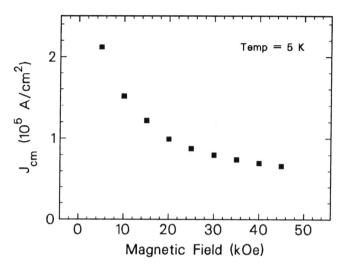

Figure 9: Magnetization critical current density J_{cm} at 5 K versus external magnetic field calculated from the hysteresis loop data in Figure 8 using the Bean critical state model [42].

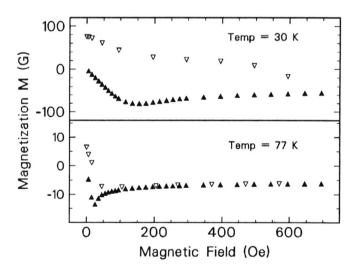

Figure 10: Comparison of partial magnetization hysteresis loops at 30 and 77 K for a superconducting $Tl_2Ca_1Ba_2Cu_2O_8$ single crystal (T_c is 105 K). The lack of hysteresis at 77 K is due to weak flux pinning.

the case of anisotropic critical current observed in all the high-temperature superconductors (48). J_{cm} in Figure 9 reflects critical current flowing along the c axis of the $YBa_2Cu_3O_{7-\delta}$ crystal and is in good agreement with similar measurements on other crystals (48-50).

Certain conditions must be satisfied for Eqs. 12 and 13 to be valid: (1) the applied field H_e must be large compared to H_{c1}, (2) the field variation across the sample must be small, and (3) the maximum field in the loop must be large enough to insure that critical currents are induced in the same sense throughout the sample (51). Condition (3) can be expressed as a minimum field strength H_p for complete penetration of critical current to the center of the sample (41,52). The situation for H_e comparable to or less than H_{c1} is considerably more complicated (53). Condition (2) is not satisfied when the flux lines are weakly pinned. Weak pinning combined with high temperatures in the copper-oxide based superconductors leads to "giant flux creep" (54) which is the subject of the next section.

Figure 10 shows the dramatic effects of weak pinning by comparing isothermal hysteresis loops at 30 and 77K in a $Tl_2Ca_1Ba_2Cu_2O_8$ single crystal. These data were recorded with the field applied along the crystal c axis (normal to the large surface), and the solid triangles represent increasing field strength while the open triangles show decreasing fields. The hysteresis loop at 30 K is relatively open, suggesting reasonably strong flux pinning (the hysteresis persists to 10 kOe). Conditions (1) to (3) above are satisfied, and we can use Eq. 13 to estimate $J_{cm} \approx 5.3 \times 10^4$ A/cm^2 at 30 K in 400 Oe. In contrast, there is virtually no hysteresis above 200 Oe at 77 K for this same crystal (note the change in vertical scale). The negative magnetization and lack of hysteresis suggests reversible flux concentrations, the opposite of strong pinning. The thermal energy at 77 K results in easy flux motion (see below), and condition (2) above is not satisfied. Hence the Bean model is not applicable, and Eqs. 12 or 13 cannot be used to obtain J_{cm}. Small hysteresis has also been reported for $YBa_2Cu_3O_{7-\delta}$ (50,55-57), $Bi_2Sr_2CaCu_2O_8$ (58-60) and $Tl_2Ca_2Ba_2Cu_3O_{10}$ (61)(62).

The comparison of critical current density obtained from direct transport measurements (termed J_{ct}) with that inferred from magnetization hysteresis data J_{cm} is controversial when the flux pinning is weak, i.e., at high temperatures in $YBa_2Cu_3O_{7-\delta}$ or moderate temperatures in the Bi-Sr-Ca-Cu-O and Tl-Ca-Ba-Cu-O superconductors (58)(61)(63-68). J_{ct} from a pulsed measurement can exceed J_{cm} from hysteresis by two orders of magnitude when flux motion is rapid (58)(61), but J_{cm} from the critical state model (Eq. 12) may not be valid when pinning is weak. The agreement is considerably better at low temperatures where flux motion is limited (58)(61)(68). A further complication in this comparison is caused by the different time scales of the two experiments: magnetization data in a SQUID magnetometer are recorded over minutes (isothermal hysteresis loops with 50 points typically require 8 hours), while transport data are frequently obtained using submillisecond pulses to minimize sample heating at the current contacts (58)(61)(63). In fact, the amount of hysteresis can depend on the rate of field sweep (64), and vibrating sample magnetometers have a decided advantage over SQUID magnetometers in this situation. Since flux creep is a dynamic process, J_{ct} and J_{cm} must be measured on the same time scale for a realistic comparison.

4.0 MAGNETIZATION RELAXATION OR GIANT FLUX CREEP

Thermally activated flux motion or creep has been studied extensively in conventional type II metallic superconductors (69-74). This motion is governed by the thermal energy relative to the average flux pinning energy. The thermal energy is higher in high-T_c superconductors because of the higher possible measurement temperatures, while the pinning energy is considerably smaller due to the short coherence lengths (54). Hence, flux moves easily in high-T_c materials (75-77) compared to conventional superconductors, leading to the phenomenon termed "giant flux creep" (54) where the magnetization changes by a large fraction on a time scale of minutes. An alternate interpretation of magnetization relaxation in terms of a superconducting "glass" state has been proposed (75)(78).

The basic measurement of flux creep is magnetization versus time (relaxation) in a constant field at fixed temperature (73). The

Static Magnetic Properties of High-Temperature Superconductors 697

magnetization changes as flux lines move into or out of a superconductor. Inward flux creep in the mixed state is studied by stabilizing the sample at a fixed temperature at one field strength (typically zero field) and then applying a stronger field $H_e > H_{c1}$. Flux motion into the sample is reflected by the decrease in magnetization with time as it relaxes toward the equilibrium Meissner state for that field and temperature. Outward flux creep is analogous: the sample is stabilized at the measurement temperature in a high field $H_e > H_{c1}$, the field is changed to a lower value (typically zero field), and the magnetization relaxes with time toward the equilibrium Meissner state (which is zero magnetization for zero field).

Typical magnetization relaxation data (65) at 10 K are shown in Figure 11 with the external magnetic field applied normal to a 0.7 μm-thick $Tl_2Ca_2Ba_2Cu_3O_{10}$ (Tl-2223) film on $SrTiO_3$. The solid triangles reflect flux motion out of the film following a step change from 10 kOe to nearly zero field. The open triangles show flux creep into the film after a step change from nearly zero field to 500 Oe. These situations are quite different: for flux creep out of the film, the measured signal reflects critical current flow at nearly zero field arising from pinned flux remaining after a field sweep from the mixed state to zero field (top data in Figure 11). An important point is the history of the mixed state preparation: we have found that consistent results are obtained if the large field (10 kOe for this case) is applied at an elevated temperature (> 50 K for Tl-2223) where flux motion is rapid and a near equilibrium state is achieved quickly (as shown by the absence of hysteresis). Then the sample can be cooled to the measurement temperature prior to lowering the field strength. For flux creep into the film, the shielding response reflects critical current flow to exclude the applied field (bottom data in Figure 11). Clearly, the flux motion is similar in these two physically distinct measurements.

The semi-log plot in Figure 11 emphasizes the logarithmic time dependence of the magnetization signal for times between 200 and 4000 seconds. These data are consistent with thermally activated flux motion (54)(63)(65)(69-74)(79-86). The data for flux motion out of the film were extended to 7500 seconds with no observed change in behavior. Similar relaxation data (76) for $La_{1.8}Sr_{0.2}CuO_4$ ceramic showed logarithmic time dependence over

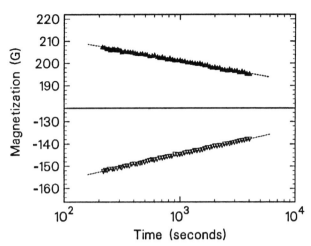

Figure 11: Magnetization relaxation versus log(time) at 10 K in fields applied normal to a 0.7 μm-thick $Tl_2Ca_2Ba_2Cu_3O_{10}$ superconducting film (T_c is 108 K). The solid triangles show relaxation due to flux creep out of the film in zero applied field after cooling in a 10 kOe field and removal of the field. The open triangles represent flux creep into the film in a 500 Oe field applied following zero field cooling. The dashed lines show the fit to Eq. (13).

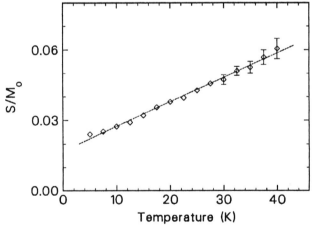

Figure 12: Ratio of flux creep rate S(T) to the magnetization at 1 second $M_o(T)$ versus temperature. S and M_o are obtained by fitting data such as those in Figure 11 using Eq. (13). The slope of the dashed line corresponds to an effective flux pinning potential U = 83 meV according to a simple thermally activated flux creep model which yields Eq. (14).

four decades from 10 seconds to 24 hours. The zero of time for the magnetization data in Figure 11 is defined as the time when the field sweep in our SQUID susceptometer reaches the measurement field. Due to the design of this instrument, there is a dead time of approximately 200 seconds after the field sweep stops before useful data can be recorded.

The dashed lines in Figure 11 represent a simple two-parameter linear fit to the logarithmic time dependence:

$$M(t,T) - M_{eq}(T) = M_o(T) + S(T) \cdot \ln(t) \qquad (14)$$

where $M(t,T)$ is the magnetization at time t and temperature T, $M_{eq}(T)$ is the equilibrium magnetization for the Meissner state at the measurement field and temperature (M_{eq} is zero for flux creep out of the sample when the measurement field is zero), $M_o(T)$ is the extrapolated magnetization at t = 1 second, and $S(T) = d(M(t,T)-M_{eq}(T))/d\ln(t)$ is the creep rate (63)(73)(83-85). Measurements at temperatures between 5 and 40 K for flux creep into the film in 500 Oe show a similar logarithmic time dependence. The simple time dependence in Eq. 14 is valid over a limited range of field, temperature and time in an activated flux creep model (63)(73)(83-85), and we have frequently observed strong deviations experimentally at both short and long times, particularly for single crystals.

The two fitting parameters in Eq. 14 can be used to extract an average pinning potential U by fitting the magnetization time dependence for data taken over a range of temperatures (83-85):

$$S(T)/M_o(T) = k_B T/U \qquad (15)$$

where k_B is the Boltzmann constant. Figure 12 plots this ratio $S(T)/M_o(T)$ for flux creep out of the film versus temperature, demonstrating the linear relation given by Eq. 15. The dashed line corresponds to U = 83 meV (65), which is a rather small pinning potential. This weak flux pinning with activated motion explains the hysteresis collapse at 77 K shown in Figure 10. Considerably more detailed treatments of magnetization relaxation than Eq. 15 have been published (63)(81)(84)(85).

Eq. 15 predicts no flux creep at T=0 (no thermal energy),

while the data in Figure 12 do not extrapolate to zero at T=0. This suggests that the creep in these films may be driven by mechanisms more complex than simple thermal activation including a distribution of activation energies (63). Other $Tl_2Ca_2Ba_2Cu_3O_{10}$ films with higher transport critical current densities show somewhat lower pinning potentials between 50 and 60 meV. In contrast, magnetization relaxation measurements on a $Tl_2Ca_2Ba_2Cu_3O_{10}$ single crystal with the field applied along the c axis (normal to the broad face) show a pinning potential of only 15 meV, consistent with the weak pinning in $Bi_2Sr_2CaCu_2O_8$ (60)(80). It is clear that grain boundary pinning has an important influence on flux motion and transport critical currents in granular thin films or bulk ceramics, and the pinning strength will differ at grain boundaries and within individual grains. Further, the presence of impurity phases can also dramatically alter the magnetization hysteresis and flux pinning.

5.0 PROBLEMS WITH POROUS AND WEAK-LINKED CERAMICS

A final topic in magnetization studies of high-temperature superconductors is the problem of ceramic samples where porosity and the presence of both **intergranular** and **intragranular** supercurrents complicate the analysis. Soon after the discovery (87)(88) of copper-oxide-based high-temperature superconductors, transport measurements (89) on ceramic samples showed a dramatic drop in critical current in the presence of very modest magnetic fields. This behavior was attributed to weak links acting as Josephson junctions between grains, leading to a two order of magnitude drop in J_{ct} between zero field and 100 Oe. A similar decrease occurs in the intergranular magnetization shielding supercurrents (90)(91) which has been termed "grain decoupling" (90) at low magnetic fields. The effects of granularity on critical current density and magnetization have been treated theoretically (92).

Typical shielding data at 5 K versus low field are shown in Figure 13 for a sintered, oxygen-annealed $YBa_2Cu_3O_{7-\delta}$ bulk ceramic plate cut to a rectangular plate of dimensions 8 x 6 x 0.7mm^3. Preparation conditions are given elsewhere (90). The plate was oriented with the long direction parallel to the applied field to minimize demagnetization effects. The shielding response is linear in field below about 20 Oe (dashed line) with a slope $\chi_e = -1.02/4\pi$, exhibits a nonlinear field dependence between 20 and 60 Oe, and becomes linear again between 60 and 200 Oe (solid line) with a slope $\chi_e = -0.69/4\pi$.

Consider first the linear response at very low field. The demagnetization factor with the field along the longest dimension is $D \approx 0.05$ (27), so from Eq. 11 we obtain a volume susceptibility $\chi_v = (1-D)\cdot\chi_e = -0.97/4\pi$. Hence the shielding fraction is 97%, corresponding to nearly complete screening from the entire volume of the sample. However, the measured density obtained from the approximate dimensions and weight is 4.45 g/cm^3 or 70% of the X-ray density of 6.36 g/cm^3 for stoichiometric YBa$_2$Cu$_3$O$_7$ using published lattice constants. Hence, the maximum superconducting volume is 70% of the plate due to the porosity of the sintered ceramic, while the shielding response suggests 97% is superconducting. This discrepancy is explained by the ability of the near surface shielding supercurrents to screen the interior of the plate. Low-field shielding data do not provide an accurate estimate of the superconducting fraction due to this screening effect which hides the interior porosity and any nonsuperconducting impurity phases as well. This fact dictates our preference for the Meissner fraction as indicated in section 2.3 despite the complications due to flux pinning.

Next, consider the nonlinear shielding response between 20 and 60 Oe in Figure 13. This is the field range where the transport critical current decreases by two orders of magnitude in typical ceramic YBa$_2$Cu$_3$O$_{7-\delta}$ (89-91), and the magnetization is reflecting grain decoupling in the plate. Although J_{ct} is small above 100 Oe in weakly linked material, the shielding can remain substantial due to intragranular supercurrents. The nonlinear region is the crossover from intergranular to intragranular screening. The volume susceptibility above 60 Oe in Figure 13 is $\chi_v = \chi_e = -0.69/4\pi$, where we have not used a demagnetization correction since the appropriate sample "shape" or effective dimensions are not determined in the decoupled grain regime. This problem has been solved for the ideal case of a spherical sample consisting of isotropic spherical decoupled grains close-packed on a cubic

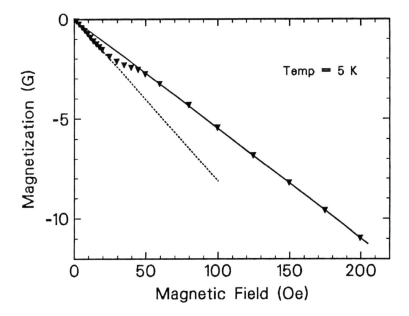

Figure 13: Shielding magnetization at 5 K versus external magnetic field for a thin plate cut from sintered superconducting $YBa_2Cu_3O_{7-\delta}$ ceramic. The dashed line indicates shielding of the entire volume by near surface supercurrents below 20 Oe. **Intergranular** supercurrents decrease rapidly above 20 Oe, resulting in essentially complete "grain decoupling" above 60 Oe. The solid line indicates **intragranular** shielding which dominates the magnetization at higher fields.

array (93). One way to test for grain decoupling is to measure the diamagnetic shielding for different orientations and look for demagnetization enhancements based on the shape of the entire sample. Full demagnetization corresponds to coupled grains, while little demagnetization implies decoupled grains (90).

Note that there is excellent agreement between the shielding fraction $-4\pi\chi_v = 0.69$ and the measured density of 70% of the x-ray density. We have found empirically that the mass susceptibility shielding fraction defined as $-4\pi\rho\chi_g$, where ρ is the x-ray density and χ_g is the mass susceptibility from Eq. 3, is nearly equal to one in a fully superconducting ceramic for fields above the grain decoupling field but below H_{c1}, independent of the porosity of the sample. The indication that grain decoupling has occurred is a magnetization response similar to that in Figure 13 where there are two well-defined linear regions. These two linear regions are not observed when the sample is ground to a powder such that the grains are always decoupled, and there is no obvious crossover from coupled to decoupled (intergranular to intragranular) shielding when the sample is almost completely dense such as hot-pressed ceramic. Mixed phase ceramics with grains which are decoupled at all fields show a nonlinear response even at the lowest fields, and the magnetization can be used to obtain an effective grain size if the penetration depth is known (26). Similar grain decoupling and low J_{ct} values in modest magnetic fields have been reported in ceramic samples of the newer Bi-Sr-Ca-Cu-O and Tl-Ca-Ba-Cu-O superconducting systems (62)(94).

Finally, consider the effect of grain decoupling on high-field magnetization hysteresis loops. The measured response is solely the intragranular mixed state behavior including shielding supercurrents and flux pinning. The critical current can be estimated using a modified Eq. 12 to accommodate the individual grain morphology and alignment, provided that "d" is the grain diameter rather than the bulk sample dimension. Note that an increase in hysteresis reflects better intragranular flux pinning, but it does **not** imply an improvement in the macroscopic intergranular critical current density. For example, neutron irradiation (95) or melting and quenching (96) have produced dramatic increases in hysteresis and thus J_{cm} for $YBa_2Cu_3O_{7-\delta}$ ceramic, but these data alone do not imply any improvement in J_{ct}.

A combination of DC magnetization, AC susceptibility, inductive and microstructural data can be used to determine an intrinsic J_{cm} value in ceramic material (39)(62)(90-92)(94)(97-99). An alternate approach is to disperse isolated grains from a ground superconducting powder in epoxy and induce grain alignment by applying a strong magnetic field while the epoxy hardens. If the grain morphology and relative concentration in the epoxy are known, magnetization hysteresis determines the anisotropic, intragranular J_{cm} (100)(101). These techniques yield values for intragranular J_{cm} in the 10^6-10^7 A/cm^2 range for high-temperature superconductors, and J_{ct} measurements on thin films confirm that these are realistic values.

One final point concerns the detection of impurity phases in ceramic samples. A substantial fraction of the common impurity phases contains isolated divalent paramagnetic Cu ions which have a Curie susceptibility given by Eqs. 4 and 5. Hence this contribution becomes strongest at low temperatures. Figure 4 shows that the low-field exclusion and expulsion in the superconducting phase(s) are temperature-independent at low temperatures, so the presence of a paramagnetic impurity phase(s) is detected by a diamagnetic exclusion or expulsion which becomes smaller in magnitude (less negative) as the temperature decreases. In particular, the Meissner signal can only increase as the temperature decreases (including flux creep, penetration depth, and H_{c1} considerations), so a decreasing Meissner signal at low temperature indicates that paramagnetic impurity phases are present. Shielding data are less reliable since surface screening currents can hide impurity phases in the interior of the sample. Impurity phases will also appear in high-field magnetization data: the superconducting diamagnetic response remains negative and decreases toward zero with increasing field in the mixed state (Figures 2 and 10), while a paramagnetic impurity phase will have a positive contribution that increases linearly with increasing field. Hence at sufficiently high field (but below H_{c2}), the impurity phase will become apparent as a linear field response in the magnetization data.

6.0 CONCLUDING REMARKS

Magnetization is a versatile and fundamental probe for superconducting materials, including the high-temperature copper-oxide-based superconductors. The normal state magnetic susceptibility contains information about both the electrical carriers and the copper spin system in the superconductor plus the behavior of any impurity phases present. The low-field magnetization in the superconducting state includes flux exclusion (diamagnetic shielding) and flux expulsion (Meissner effect) responses which differ due to flux pinning in single crystals and due to pinning, porosity, grain decoupling, and impurity phases in typical sintered ceramic material. Low-field data can be used to determine the fraction of superconducting material and the lower critical field H_{c1}. High-field data are used to study flux pinning and magnetization hysteresis which can be related to the critical current density versus field and temperature. Weak pinning and the associated "giant flux creep" are major concerns in high-temperature superconductors, and magnetization relaxation directly addresses this phenomenon. Two substantial challenges for application of these materials are to increase flux pinning and to improve intergranular supercurrents in ceramic samples. Magnetization measurements are essential in assessing progress on these challenges.

ACKNOWLEDGEMENTS

This work was performed at Sandia National Laboratories, supported by the U. S. Department of Energy under Contract No. DE-AC04-76DP00789. Several conspirators in high-temperature superconductivity have provided materials and ideas including T. L. Aselage, R. J. Baughman, D. S. Ginley, J. F. Kwak, B. Morosin, J. E. Schirber, and C. P. Tigges.

7.0 REFERENCES

1. W. Meissner and R. Ochsenfeld, *Naturwissen*, 21:787 (1933).

2. D. Shoenberg, *Superconductivity*, Cambridge Univ. Press, London, Ch. 6 (1952).

3. P.G. De Gennes, *Superconductivity of Metals and Alloys*, translated by P.A. Pincus, W.A. Benjamin, New York, Ch. 5 (1966).

4. D. Saint-James, G. Sarma and E.J. Thomas, *Type II Superconductivity*, Pergamon, Oxford, p. 4 (1969).

5. D. Saint-James, G. Sarma and E.J. Thomas, *ibid.*, Ch. 1, 3, 8.

6. M. Tinkham, *Introduction to Superconductivity*, McGraw-Hill, New York, Ch. 1, 3, 5.(1975).

7. A.C. Rose-Innes and E. H. Roderick, *Introduction to Superconductivity*, 2nd ed., Pergamon, Oxford, Ch. 2, 4, 6, 12, (1978).

8. J.H. Van Vleck, *The Theory of Electric and Magnetic Susceptibilities*, Oxford, London, p. 233, (1932).

9. Landolt-Bornstein, New Series, Group II: Atomic and Molecular Physics, Vol. 2, *Magnetic Properties of Coordination and Organo-Metallic Compounds*, ed. by E. Konig, Springer-Verlag, Berlin, pp. 1-1 to 1-25, (1966).

10. H.J. Zeiger and G.W. Pratt, *Magnetic Interactions in Solids* Oxford, London, pp. 405-408, (1973).

11. A.A. Abrikosov, *Fundamentals of the Theory of Metals*, North-Holland, New York, p. 173-178, (1988).

12. G. Shirane, Y. Endoh, R.J. Birgeneau, M.A. Kastner, Y. Hidaak, M. Oda, M. Suzuki, T. Murakami, *Phys. Rev. Lett.* 59:1613 (1987).

13. J.M. Tranquada, D.E. Cox, W. Kunnmann, H. Moudden, G. Shirane, M. Suenaga, P. Zolliker, D. Vaknin, S.K. Sinha, M.S. Alvarez, A.J. Jacobson, D.C. Johnston, *Phys. Rev. Lett.* 60:156 (1988).

14. R.J. Birgeneau, D.R. Gabbe, H.P. Jenssen, M.A. Kastner, P.J. Picone, T.R. Thurston, G. Shirane, Y. Endoh, M. Sato, K. Yamada, Y. Hidaka, M. Oda, Y. Enomoto, M. Suzuki, T. Murakami, *Phys. Rev.* B 38:6614 (1988).

15. H. Takagi, T. Ido, S. Ishibashi, M. Uota, S. Uchida, Y. Tokura, *Phys. Rev.* B 40:2254 (1989) and references therein.

16. Y. Yamaguchi, M. Tokumoto, S. Waki, Y. Nakagawa, Y. Kimura, *J. Phys. Soc. Japan* 58:2256 (1989) and references therein.

17. D.C. Johnston, *Phys. Rev. Lett.* 62:957 (1989).

18. J.A.O. de Aguiar, A.A. Menovsky, J. van der Berg, H.B. Brom, *J. Phys.* C 21:L237 (1988).

19. S. Tyagi, M. Barsoum, K.V. Rao, *Phys. Lett.* A 128:225 (1988).

20. F. Mehran, S.E. Barnes, T.R. McGuire, T.R. Dinger, D. Kaiser, F. Holtzberg, *Solid State Commun.* 66:299 (1988).

21. E.A. Lynton, *Superconductivity*, Metheun, London, p. 23 (1969). Actually, all three fields (B, H_i, and M) are zero in the interior, but it is convenient to treat the superconductor as if it were a magnetic body where M is the equivalent magnetization produced by the screening supercurrents.

22. A.M. Campbell and J.E. Evetts, *Critical Currents in Superconductors*, Taylor and Francis, London, p. 3, (1972).

23. L. Krusin-Elbaum, R.L. Greene, F. Holtzberg, A.P. Malozemoff, Y. Yeshurun, *Phys. Rev. Lett.* 62:217 (1989).

24. D. Shoenberg, *op. cit.*, p. 143 and 197.

25. D. Shoenberg, *ibid.*, p. 165.

26. D.E. Farrell, M.R. DeGuire, B.S. Chandrasekhar, S.A. Alterovitz, P.R. Aron, R.L. Fagaly, *Phys. Rev.* B 35:8797 (1987) and references therein.

27. J.A. Osborn, *Phys. Rev.* 67:351 (1945).

28. E.C. Stoner, *Phil. Mag.* 36:803 (1945).

29. A.H. Morrish, *The Physical Principles of Magnetism*, pp. 8-11, Wiley, New York, (1965).
30. S. Chikazumi and S.H. Charap, *Physics of Magnetism*, Wiley, New York, pp. 19-24 (1964).
31. E.L. Venturini, J.F. Kwak, D.S. Ginley, R.J. Baughman, B. Morosin, *Physica* C 162-164: 673 (1989).
32. Y. Yeshurun, A.P. Malozemoff, F. Holtzberg, T.R. Dinger, *Phys. Rev.* B 38:11828 (1988).
33. A. Umezawa, G.W. Crabtree, J.Z. Liu, T.J. Moran, S.K. Malik, L.H. Nunez, W.L. Kwok, C.H. Sowers, *Phys. Rev.* B 38:2843 (1988).
34. L. Krusin-Elbaum, A.P. Malozemoff, Y. Yeshurun, D.C. Cronemeyer, F. Holtzberg, *Phys. Rev.* B 39:2936 (1989).
35. A.P. Malozemoff, "Macroscopic Magnetic Properties of High Temperature Superconductors", in *Physical Properties of High Temperature Superconductors I*, ed. by D. M. Ginsberg, World Scientific, Singapore, Ch. 3, and references therein (1989).
36. L. Schimmele, H. Kronmuller, H. Teichler, *Phys. Stat. Solid* (b) 147:361 (1988).
37. A.P. Malozemoff, L. Krusin-Elbaum, D.C. Cronemeyer, Y. Yeshurun, F. Holtzberg, *Phys. Rev.* B 38:6490 (1988).
38. D.S. Ginley, E.L. Venturini, J.F. Kwak, R.J. Baughman, B. Morosin, J.E. Schirber, *Phys. Rev.* B 36:829 (1987).
39. R.L. Peterson, *Phys. Rev.* B 40:2678 (1989).
40. L. Krusin-Elbaum, A.P. Malozemoff, Y. Yeshurun, D.C. Cronemeyer, F. Holtzberg, *Physica* C 153-155:1469 (1988).
41. W.J. Carr, Jr., *AC Loss and Macroscopic Theory of Superconductors*, Gordon and Breach, New York, Ch. 4, (1983).
42. C.P. Bean, *Phys. Rev. Lett.* 8:250 (1962).
43. H. London, *Phys. Lett.* 6:162 (1963).
44. Y.B. Kim, C.F. Hempstead, A.R. Strnad, *Phys. Rev.* 129:528 (1963).
45. C.P. Bean, *Rev. Mod. Phys.* 36:31 (1964).
46. W.A. Fietz and W.W. Webb, *Phys. Rev.* 178:657 (1969).

47. A.M. Campbell and J.E. Evetts, *op. cit.*, pp. 45-81.
48. E.M. Gyorgy, R.B. van Dover, K.A. Jackson, L.F. Schneemeyer, J.V. Waszczak, *Appl. Phys. Lett.* 55:283 (1989).
49. T.R. Dinger, T.K. Worthington, W.J. Gallagher, R.L. Sandstrom, *Phys. Rev. Lett.* 58:2687 (1987).
50. G.W. Crabtree, J.Z. Liu, A. Umezawa, W.K. Kwok, C.H. Sowers, S.K. Malik, B.W. Veal, D.J. Lam, M.B. Brodsky, J.W. Downey, *Phys. Rev.* B 36:4021 (1987).
51. R.L. Fleischer, H.R. Hart, Jr., K.W. Lay, F.E. Luborsky, *Phys. Rev.* B 40:2163 (1989).
52. A.P. Malozemoff, T.K. Worthington, Y. Yeshurun, F. Holtzberg, P.H. Kes, *Phys. Rev.* B 38:7203 (1988).
53. P. Chaddah, K.V. Bhagwat, G. Ravikumar, *Physica* C 159:570 (1989).
54. Y. Yeshurun and A.P. Malozemoff, *Phys. Rev. Lett.* 60:2202 (1988).
55. D.E. Farrell, B.S. Chandrasekhar, M.R. DeGuire, M.M. Fang, V.G. Kogan, J.R. Clem, D.K. Finnemore, *Phys. Rev.* B 36:4025 (1987).
56. R.N. Shelton, R.W. McCallum, M.A. Damento, K. Gschneider, Jr., *Intern. J. Mod. Phys.* B 1:401 (1988).
57. G.M. Stollman, B. Dam, J.H.P.M. Emmen, J. Pankert, *Physica* C 159:854 (1989).
58. R.B. van Dover, L.F. Schneemeyer, E.M. Gyorgy, J.V. Waszczak, *Appl. Phys. Lett.* 52:1910 (1988).
59. J. van den Berg, C.J. van der Beek, P.H. Kes, J.A. Mydosh, M.J.V. Menken, A.A. Menovsky, *Supercond. Sci. Techn.* 1:249 (1989).
60. H. Kumakura, K. Togano, E. Yanagisawa, H. Maeda, *Appl. Phys. Lett.* 55:185 (1989).
61. J.F. Kwak, E.L. Venturini, R.J. Baughman, B. Morosin, D.S. Ginley, *Cyrogenics* 29:291 (1989).
62. J.R. Thompson, J. Brynestad, D.M. Kroeger, Y.C. Kim, S.T. Sekula, D.K. Christen, E.D. Specht, *Phys. Rev.* B 39:6652 (1989).
63. A.P. Malozemoff, T.K. Worthington, R.M. Yandrofski, Y. Yeshurun, *Intern. J. Mod. Phys.* B 1:1293 (1988).

64. E.M. Gyorgy, R.B. van Dover, S. Jin, R.C. Sherwood, L.F. Schneemeyer, T.H. Tiefel, J.V. Waszczak, *Appl. Phys. Lett.* 53:2223 (1988).

65. E.L. Venturini, J.F. Kwak, D.S. Ginley, B. Morosin, R.J. Baughman in *Science and Technology of Thin Film Superconductors*, ed. by R.D. McConnell and S.A. Wolf, Plenum, New York, p. 395, (1989).

66. B. Placais, P. Mathieu, Y. Simon, *Solid State Commun.* 71:177 (1989).

67. U. Dai, N. Hess, L.R. Tessler, G. Deutscher, G. Vetter, F. Queyroux, N. Bontemps, R. Mahoum, M. Lagues, P. Mocaer, *Appl. Phys. Lett.* 55:1135 (1989).

68. T.R. McGuire, D. Dimos, R.H. Koch, R.B. Laibowitz, J. Mannhart, *IEEE Trans. Magnetics* 25:3218 (1989).

69. Y.B. Kim, C.F. Hempstead, A.R. Strnad, *Phys. Rev. Lett.* 9:306 (1962).

70. P.W. Anderson, *Phys. Rev. Lett.* 9:309 (1962).

71. Y.B. Kim, C.F. Hempstead, A.R. Strnad, *Phys. Rev.* 131:2486 (1963).

72. P.W. Anderson and Y.B. Kim, *Rev. Mod. Phys.* 36:39 (1964).

73. M.R. Beasley, R. Labusch, W.W. Webb, *Phys. Rev.* 181:682 (1969).

74. Y.B. Kim and M.J. Stephen in *Superconductivity*, Vol. 2, ed. by R.D. Parks, Marcel Dekker, New York, Ch. 19, (1969).

75. K.A. Muller, M. Takashige, J.G. Bednorz, *Phys. Rev. Lett.* 58:1143 (1987).

76. A.C. Mota, A. Pollini, P. Visani, K.A. Muller, J.G. Bednorz, *Phys. Rev.* B 36:4011 (1987).

77. C. Giovannella, G. Collin, P. Rouault and I.A. Campbell, *Europhys. Lett.* 4:109 (1987).

78. I. Morgenstern, K.A. Muller, J.G. Bednorz, *Z. Phys.* B 69:33 (1987).

79. M. Tinkham, *op. cit.*, Ch. 5.

80. Y. Yeshurun, A.P. Malozemoff, F. Holtzberg, *J. Appl. Phys.* 64:5797 (1988).

81. Y. Yeshurun, A.P. Malozemoff, T.K. Worthington, R.M. Yandrofski, L. Krusin-Elbaum, F.H. Holtzberg, T.R. Dinger, G.V. Chandrashekhar, *Cryogenics* 29:258 (1989).

82. P.H. Kes, J. Aarts, J. van den Berg, C.J. van der Beek, J.A. Mydosh, *Supercond. Sci. Techn.* 1:242 (1989).

83. M.E. McHenry, M.P. Maley, E.L. Venturini and D.S. Ginley, *Phys. Rev.* B 39:4784 (1989).

84. C.W. Hagen, R.P. Griessen, E. Salomons, *Physica* C 157:199 (1989).

85. K. Yamafuji, T. Fujiyoshi, K. Toko, T. Matsushita, *Physica* C 159:743 (1989).

86. H.S. Lessure, S. Simizu, S.G. Sankar, *Phys. Rev.* B 40:5165 (1989).

87. J.G. Bednorz and K.A. Muller, *Z. Phys.* B 64:189 (1986).

88. M.K Wu, J.R. Ashburn, C.J. Torng, P.H. Hor, R.L. Meng, L. Gao, Z.J. Huang, Y.Q. Wang, C.W. Chu, *Phys. Rev. Lett.* 58:908 (1987).

89. J.W. Ekin, A.I. Braginski, A.J. Panson, M.A. Janocko, D.W. Capone II, N.J. Zaluzec, B. Flandermeyer, O.F. de Lima, M. Hong, J. Kwo, S.H. Liou, *J. Appl. Phys.* 62:4821 (1987) and references therein.

90. J.F. Kwak, E.L. Venturini, D.S. Ginley, W. Fu, in *Novel Superconductivity*, ed. by S.A. Wolf and V.Z. Kresin Plenum, New York, p. 983, (1987).

91. D.C. Larbalestier, M. Daeumling, X. Cai, J. Seuntjens, J. McKinnell, D. Hampshire, P. Lee, C. Meingast, T. Willis, H. Muller, R.D. Ray, R.G. Dillenburg, E.E. Hellstrom, R. Joynt, *J. Appl. Phys.* 62:3308 (1987).

92. J.R. Clem, *Physica* C 153-155:50 (1988).

93. J.R. Clem and V.G. Kogan, *Japan J. Appl. Phys.* 26: Suppl. 26-3 (Proc. 18th Intern. Conf. Low Temp. Phys.), p. 1161 (1987).

94. H. Kupfer, S.M. Green, C. Jiang, Yu Mei, H.L. Luo, R. Meier-Hirmer, C. Politis, *Z. Phys.* B 71:63 (1988).

95. J.R. Cost, J.O. Willis, J.D. Thompson, D.E. Peterson, *Phys. Rev.* B 37:1563 (1988).

96. M. Murakami, M. Morita, N. Koyama, *Japan J. Appl. Phys.* 28:L1125 (1989).

97. H. Kupfer, I. Apfelstedt, W. Schauer, R. Flukiger, R. Meier-Hirmer, H. Wuhl, *Z. Phys.* B 69:159 (1987).
98. D.C. Larbalestier, S.E. Babcock, X. Cai, M. Daeumling, D.P. Hampshire, T.F. Kelly, L.A. Lavanier, P.J. Lee, J. Seuntjens, *Physica* C 153-155:1580 (1988).
99. E. Babic, D. Drobac, J. Horvat, Z. Marohnic, M. Prester, *Supercond. Sci. Technol.* 2:164 (1989).
100. M.M. Fang, V.G. Kogan, D.K. Finnemore, J.R. Clem, L.S. Chumbley, D.E. Farrell, *Phys. Rev.* B 37:2334 (1988).
101. M.M. Fang, D.K. Finnemore, D.E. Farrell, N.R. Bansal, *Cryogenics* 29:347 (1989).

Part IV
Structure-Property Considerations

19

Electronic Structure and Valency in Oxide Superconductors

Arthur W. Sleight

1.0 INTRODUCTION

In view of the other chapters in this book, very little introduction of this chapter is required. Both oxide superconductors based on copper, e.g., $YBa_2Cu_3O_7$, and bismuth, e.g., $(Ba,K,Rb)(Bi,Pb)O_3$, will be considered as high T_c superconductors. The upper T_c for cuprates has not increased above 122 K in the last year. However, the T_c in the $(Ba,K)BiO_3$ system has been pushed up to 34K (1), and a T_c of 37K has been reported in the $(Ba,K,Rb)BiO_3$ system (2). Thus, we now have cubic, noncuprate superconductors with Tc's as high as that originally reported by Bednorz and Müller for the cuprate superconductors. We will be exploring the common features of the bismuth-based and copper-based high T_c oxides, and we will be relating these common features to electronic structure and to possible mechanisms for high Tc. Consideration will be given to mixed valency, high covalency, band structure, magnetic properties, charge fluctuations, polarizibility, defects and stability. A more complete discussion of valency in inorganic solids generally has been recently given (3).

2.0 MIXED VALENCY AND THE PARTIALLY FILLED BAND

A feature common to both bismuth and copper oxides is the availability of three consecutive oxidation states: Cu^I, Cu^{II}, Cu^{III} and

Bi^{III}, Bi^{IV}, Bi^{V}. All of these oxidation states are well known to chemists, with the exception of Bi^{IV}. We know that s^1 cations such as Bi^{IV} are unstable when concentrated. Thus, we do not normally find discrete compounds where bismuth is entirely in the tetravalent state. However, we do find Bi^{IV} in dilute systems, both insulating and metallic systems (3).

The lowest oxidation states for copper, $Cu^I(3d^{10})$, and for bismuth, $Bi^{III}(6s^2)$, are diamagnetic. The middle valency, $Cu^{II}(3d^9)$ or $Bi^{VI}(6s^1)$, is always a one electron state. The highest oxidation state is always diamagnetic in the case of bismuth, $Bi^V(6s^0)$. However, the $3d^8$ state for Cu^{III} is paramagnetic for cubic (octahedral or tetrahedral) copper but is diamagnetic for square planar copper. It would seem significant that all copper oxide superconductors have a square planar arrangement of oxygens around copper. There may be a fifth, or a fifth and sixth oxygen, ligand; however, these are always at much longer distances. Thus, the mixed valency analogy between bismuth and copper is complete for the structures where superconductivity is observed. Both the lower and upper oxidation states are diamagnetic and the middle oxidation state is spin one half.

The mixed valency in any given superconductor may be viewed as involving just two oxidation states: $Cu^{II,III}$ in $La_{2-x}Sr_xCuO_4$, $Cu^{I,II}$ in $Nd_{2-x}Ce_xCuO_4$ and $Bi^{IV,V}$ in $Ba_{1-x}K_xBiO_3$. It is essential for metallic properties and for superconductivity that the two oxidation states be present on one crystallographic site. Thus, $YBa_2Cu_3O_6$ is neither metallic nor superconducting because Cu^I and Cu^{II} occupy distinct crystallographic sites and there is no mixed valency on either site. On the other hand, $YBa_2Cu_3O_7$ is metallic and superconducting, and this is related to $Cu^{II,III}$ mixed valency in the CuO_2 sheets.

The mixed valency situation may be equally well described in terms of bands. The band to be considered in the case of bismuth is the $Bi6s$-$O2p$ band. The band in the case of copper is the $Cu3d_{x^2-y^2}$-$O2p$ band. Both of these are σ^* bands. For the lowest oxidation state (Cu^I or Bi^{III}), these bands are filled; for the highest oxidation state (Cu^{III} or Bi^V), these bands are empty. According to simple band structure considerations, we would expect metallic properties for any partial filling of these σ^* bands in bismuth and copper oxides. While metallic properties are indeed observed for most of the intermediate band fillings, insulating properties are always found for the half-filled band in both cases, *i.e.*, the situation for pure Bi^{IV} or pure Cu^{II}.

There are two competing mechanisms to produce an insulating state for the half-filled σ^* band. One is through valency disproportionation, and this is the case for $BaBiO_3$, *i.e.*, $2Bi^{IV} \rightarrow Bi^{III} + Bi^{V}$. If Bi^{IV} had not disproportionated in $BaBiO_3$, we would expect this compound to be metallic because it would have a half-filled band. Many have been confused by this disproportionation description because of confusion about the meaning of valent states as opposed to real charges. We will come back to the subject later.

Disproportionation of Bi^{IV} in $BaBiO_3$ leads to two distinct sites for Bi^{III} and Bi^{V}. Thus, we formulate this compound as $Ba_2Bi^{III}Bi^{V}O_6$. The Bi-O distances are those expected of these oxidation states (3). The real charges on Bi^{III} and Bi^{V} are very much lower than the oxidation states due to covalency effects. Thus, the real charge difference between Bi^{III} and Bi^{V} is much less than two. An equivalent description of the disproportionation situation in $BaBiO_3$ is the term charge density wave (CDW).

There is another way to achieve an insulating state for the half-filled band, and that is through antiferromagnetic ordering. This happens for all of the copper oxides which contain copper only as Cu^{II}. Thus, spin pairing of electrons on Cu or Bi sites is always related to the insulating state of the Cu^{II} and Bi^{IV} oxides. For the copper oxides, it is spin pairing between adjacent Cu^{II} sites. In the case of Bi^{IV}, spin pairing occurs on alternate Bi sites creating Bi^{III} and Bi^{V}. The magnetic situation in the Cu^{II} oxides may also be described as a spin density wave (SDW).

We will refer to the two competing mechanisms for localization at the half-filled band as disproportionation and magnetic. We could say that if $BaBiO_3$ did not become insulating through disproportionation, it would have become a $BaBi^{IV}O_3$ antiferromagnetic insulator. However, muon spin rotation experiments have completely ruled out that unlikely possibility (4). We could also say that if Cu^{II} oxides did not become antiferromagnetic insulators, disproportionation into Cu^{I} and Cu^{III} would have occurred. This is, of course, exactly what usually happens for Ag^{II} and Au^{II}. For AgO, the oxidation state of silver is clearly not two. This compound is instead $Ag^{I}Ag^{III}O_2$.

Schematic band structures for the copper and bismuth oxides just discussed are given in Figure 1. The hypothetical half-filled band

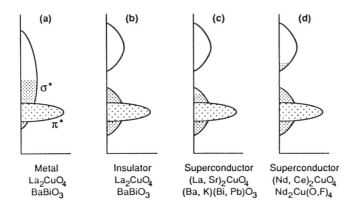

Figure 1: Schematic band structures for pure and doped $BaBiO_3$ and La_2CuO_4. The metallic compounds (a) are hypothetical materials without disproportionation or magnetic interactions which cause band splitting (b) when the σ^* band is half filled. Both p-type (c) and n-type (d) conductors are shown.

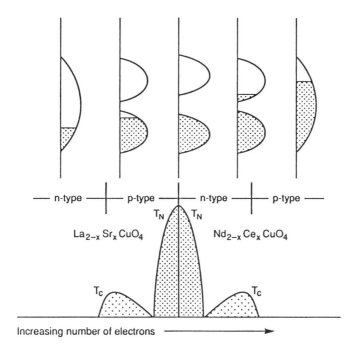

Figure 2: Schematic phase diagram (lower) and schematic band structure (upper) for A_2CuO_4 phases, both n- and p-types.

metals do not exist due to electron-electron correlations which cause band splitting. The π^* band is placed so that it is always below the Fermi level. This is consistent with all band structure calculations. However, band structure calculations are not reliable on this issue because they ignore the electron-electron correlations which split the σ^* band and which could cause the filled portion of the σ^* band to drop below the top of the π^* band. The primary assurance that the π^* band always lies below the Fermi level is that an obvious symmetry exists at the Fermi level, i.e., doping either p- or n-type produces essentially the same result (Figure 2). This would not be expected if the π^* band overlapped the Fermi level. The consensus from nuclear resonance studies is also that there are carriers in only the σ^* band (5).

3.0 VALENT STATES VS. REAL CHARGES

An understanding of the chemistry of the copper and bismuth superconductors is hardly possible without a good understanding of oxidation states or valent states. These states are basically unrelated to the magnitude of the real charges on cations and anions (3). The term valent state originated in organic chemistry where carbon is tetravalent, nitrogen is trivalent, oxygen is divalent, and hydrogen is univalent. Carbon is tetravalent even in diamond and in graphite where its real charge is obviously zero.

The term oxidation state is an outgrowth of valent states; now oxygen is the standard. Oxygen is by definition in an oxidation state of minus two unless it bonds to itself. Thus, oxidation states are entirely unrelated to the actual charge on oxygen or on the cations in oxides. Oxygen is divalent in O_2 but has an oxidation state of zero in this case. Knowing the real charges on cations and anions may sometimes be useful, but real charges are not nearly as useful as oxidation states or valent states. One only has to look at one example: SiO_2. Any estimate of the real charges in SiO_2 leads to the conclusion that oxygen is better represented as O^{1-} than as O_{2-}. Thus, we might describe SiO_2 as $Si^{2+}O_2^{1-}$. We then must account for the lack of paramagnetism from Si^{2+} and O^{1-}. Of course, these unpaired electrons on Si^{2+} and O^{1-} ions are paired in the bonds, but this is clearly a very clumsy way to describe the bonding electrons in SiO_2. There is no such problem with the oxidation state approach where SiO_2 is $Si^{IV}O_2^{-II}$. Bonding in SiO_2 is highly covalent. Thus, we expect real charges to be much lower than the oxidation state assignments. The real charges may be estimated using the electronegativity values of silicon and oxygen.

The systems of valent states and oxidation states introduced by chemists are not *merely* electron accounting systems. They are *the* systems which allow us to understand and predict which ratios of elements will form compounds and also suggests what are the likely structures and properties for these compounds (3). In the case of highly covalent compounds, the actual occupancy of the parent orbitals may seem to be very different than that implied from oxidation states if ionicity were high. Nonetheless, even some physicists have recognized the fundamental validity and usefulness of the chemist's oxidation state approach where the orbitals may now be described as symmetry or Wannier orbitals (6).

For significantly covalent compounds, such as the copper and bismuth oxides, we will not assume that valence electrons should be assigned to either the cations or to oxygen. Rather these electrons are in the bonds between the cations and anions. The terms cation and anion are only used to differentiate between atoms that take on small positive charges from those that take on a small negative charge. The most foolish question which has been commonly asked of the oxidized copper oxides is whether the holes produced are on oxygen or copper. The covalency in these systems is too high to make that a meaningful question. The atoms in the CuO_2 sheets each contribute nearly equally to the states near the Fermi level (7). The greater preponderance of oxygen states relative to copper states in this energy region is due essentially entirely to the fact that there are twice as many oxygen as copper atoms in these all important sheets. In fact, we know that the mobile holes must spend significant time on both copper and oxygen in order to explain the delocalization of the holes.

The oxidation state situation in the oxidized copper oxides is O^{-II} and mixed Cu^{II}-Cu^{III} by definition. There is no implication in this designation that the charge on copper has greatly increased beyond that of a pure Cu^{II} oxide. The electron was not removed from copper, but rather an antibonding electron was removed from the Cu-O bond.

4.0 STABILIZATION OF O^{-II} AND HIGH OXIDATION STATES

We know that the O^{-2} ion is not stable in the gas phase. Such a species would spontaneously lose an electron and become O^{-1}. Nonetheless, we correctly conclude that oxygen is a divalent in nearly all of its compounds, including O_2 and O_2^{-2}. This is not surprising in view of the differences in real charges, oxidation states and valent states just discussed. However, there are two very different ways to describe the stabilization of divalent oxygen. One way has a covalent approach, the other an ionic approach. In highly covalent compounds such as O_2, CO_2, SO_3, and OsO_4, we would refer to oxygen as divalent even though the real charge on oxygen is zero or just slightly negative. Thus in this case, the problem of stabilizing the O^{-2} ion never occurs because the real charge on oxygen never approaches two.

In oxides of high ionicity, it is useful to employ another approach to rationalize the existence of divalent oxygen. This alternative approach has particular relevance to the copper and bismuth oxides despite the covalency of the Cu-O and Bi-O bonds. Although the O^{-1} ion is not stable in the gas phase, it can be stabilized as a fully doubly negatively charged O^{-2} ion in a solid through Madelung forces. The cations surrounding oxygen in a solid produce an electrostatic field intense enough to capture the second electron. This means that we have so stabilized the oxygen $2p$ levels that the second electron may be added despite the strong on-site Columbic forces which repel this second electron.

One of the common features of the copper and bismuth oxides superconductors is the presence of highly electropositive cations such as Ba^{II}. These cations may have several important roles such as enabling certain framework structures and increasing the covalency of the Cu-O or Bi-O bonds (8). One very crucial role of the electropositive cations is the stabilization of oxygen $2p$ levels to an extent which allows for unusually high oxidation states for copper and bismuth. We know that in the binary Cu/O and Bi/O systems neither $Cu_2^{III}O_3$ nor $Bi_2^{V}O_5$ can be produced even at high oxygen pressure. However, introduction of highly electropositive cations stabilize oxygen $2p$ states to such an extent that Cu^{III} and Bi^{V} can exist. Lacking the electropositive cations, Cu^{III} or Bi^{V} would readily oxidize O^{II} to O_2^{-II} or O_2. Good examples of Cu^{III} and Bi^{V} are in $KCuO_2$ and $KBiO_3$ where potassium serves the role of stabilization of oxygen $2p$ states sufficiently to allow Cu^{III} or Bi^V to exist.

The disproportionation of Bi^{IV} and Cu^{II} is shown schematically in Figure 3. This disproportionation can only be understood if one takes into account the high covalency and the stabilization of Bi^V and Cu^{III} by highly electropositive cations such as Ba^{II}. The main point is that there is a redistribution of electrons in the Bi-O and Cu-O bonds. Although a charge density wave is produced, the magnitude of the charges is much less than would be implied in an ionic model. The situation is in fact very much analogous to what occurs in $BaTiO_3$ on going through its ferroelectric transition (Figure 3). The main difference is that disproportionation is an antiferroelectric, rather than ferroelectric, process.

5.0 POLARIZIBILITY

Another common feature of copper and bismuth oxides is that we must consider more than one electronic shell of the bismuth and copper atoms when attempting to understand their oxidation states and bonding. For copper we must consider both the *3d* and *4s* orbitals. For bismuth, we must consider both the *6s* and *6p* orbitals. A number of consequences follow, especially for the lowest oxidation state: Cu^I or Bi^{III}. We normally describe Bi^{III} as a $6s^2$ cation, ignoring the covalency with oxygen. However, we know that the bonding of Bi^{III} in oxides generally shows strong evidence of *6s-6p* hybridization. A pure $6s^2$ cation would be expected to take on highly symmetric environments. Instead, we typically find short bonds on one side of bismuth and much longer bonds on the opposite side. We may understand this on the basis that a filled *6s* core on bismuth interferes with Bi-O bond formation using the empty *6p* orbitals on bismuth. In other words, the radius of the filled *6s* core is such that the σ bonds based on Bi*6p* (formally empty) and O*2p* (formally filled) overlap becomes significantly antibonding due to the overlap of the Bi *6s* (formally filled) with the O*2p* (formally filled). This problem can be overcome through *6s-6p* hybridization on bismuth. The filled core of Bi^{III} is no longer centered on the nucleus (Figure 4). Strong covalent bonds may now form on one side of bismuth due to this polarization of the core. However, good covalent bonds have been further destabilized on the other side of the bismuth. Very little *6s-6p* hy-

Disproportionation of Bi^{IV} and Cu^{II}

$$-\overset{IV}{Bi}-O-\overset{IV}{Bi}-O-\overset{IV}{Bi}-O-\overset{IV}{Bi}-O-\overset{IV}{Bi}-O-\overset{IV}{Bi}-$$

$$-O-\overset{V}{Bi}-O-\overset{III}{\underset{\uparrow}{Bi}}-\overset{}{\underset{\downarrow}{}}-O-\overset{V}{Bi}-O-\overset{III}{\underset{\uparrow}{Bi}}-\overset{}{\underset{\downarrow}{}}-O-\overset{V}{Bi}-O-\overset{III}{\underset{\uparrow}{Bi}}-$$

$$-O-\overset{III}{Cu}-O-\overset{I}{\underset{\uparrow}{Cu}}-\overset{}{\underset{\downarrow}{}}-O-\overset{III}{Cu}-O-\overset{I}{\underset{\uparrow}{Cu}}-\overset{}{\underset{\downarrow}{}}-O-\overset{III}{Cu}-O-\overset{I}{\underset{\uparrow}{Cu}}-$$

Bond Disproportionation in $BaTiO_3$ (d^0)

$$-\overset{-}{O}-:.-\overset{+}{Ti}-:.-\overset{-}{O}-:.-\overset{+}{Ti}-:.-\overset{-}{O}-:.-\overset{+}{Ti}-:.-\overset{-}{O}-:.-\overset{+}{Ti}-:.-\overset{-}{O}-:.-\overset{+}{Ti}-:.-\overset{-}{O}-:.-\overset{+}{Ti}-:.$$

$$-\overset{-}{O}::\overset{+}{Ti}-:-\overset{-}{O}::\overset{+}{Ti}-:-\overset{-}{O}::\overset{+}{Ti}-:-\overset{-}{O}::\overset{+}{Ti}-:-\overset{-}{O}::\overset{+}{Ti}-:-\overset{-}{O}::\overset{+}{Ti}-$$

Figure 3: Disproportionation and distortions in the perovskite structure. Arrows represent the spin of antibonding electrons in the Bi^{III}-O^{-II} and Cu^{I}-O^{-II} bonds. For $BaTiO_3$, the dots represent bonding electrons, σ and π.

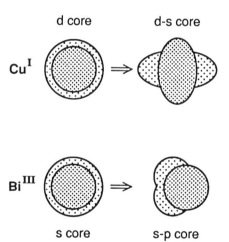

Figure 4: Polarization of the cores of Bi^{III} and Cu^{I}. Oxygen ligands are unable to form strong bonds using the empty Cu 4s or Bi 6p orbitals due to interference of the 3d or 6s cores. Core polarization, d-s, for copper solves this problem and gives two-fold linear coordination for Cu^{I} and other d^{10} cations such as Ag^{I} and Hg^{II}. Core polarization, s-p, for bismuth results in strong bonds on one side of Bi^{III} and weak bonds on the opposite side.

bridization is actually required to have this very dramatic effect on the bonding.

The situation for Cu^I is very much analogous to that of Bi^{III}, yet the effect on bonding is very different. In the case of Cu^I, it is the filled $3d$ core which interferes with bonding based of the formally empty $4s$ and $4p$ orbitals of copper and the filled $2p$ and $2s$ orbitals of oxygen. The antibonding problem which results from a spherical $3d^{10}$ core is alleviated through $3d$-$4s$ hybridization (Figure 4). This normally gives a disk shaped core. Thus, two strong bonds form along the unique axis of this disk. Bonds are now even less likely perpendicular to this unique axis of the disk shaped core. Thus, it is easy to understand the common two-fold linear coordination of Cu^I to O^{-II}.

We might well have asked why $3d$-$4s$ hybridization on Cu^I did not lead to an egg shaped core instead of a disk shaped core. To a first approximation, these two situations would indeed to be degenerate. With hybridization to an egg shaped core, the occurrence of Cu^I in three-fold of four-fold planar coordination should be no surprise. This, of course, actually happens in the n-type copper oxides such as $Nd_{2-x}Ce_xCuO_4$.

High polarizibility of atoms or ions requires easy hybridization and/or large size. We see that for both copper and bismuth we have ample opportunity for hybridization that could lead to high polarizibility. For copper, it is $3d$-$4s$ hybridization; for bismuth, it is $6s$-$6p$ hybridization. We expect this polarizibility to be more pronounced for the lower oxidation states of bismuth and copper. Oxygen polarizibility ($2s$-$2p$) may also be significant.

Another type of polarizibility results from the near degeneracy of the metal levels and the oxygen $2p$ levels. This is directly related to the high covalency in these systems; thus, this type of polarizibility will be greater for the higher oxidation states of copper and bismuth. Both of these polarizibility contributions are likely very important for theories of superconductivity based on charge fluctuations.

6.0 DEFECTS AND INHOMOGENEITIES

Most of our high T_c superconductors are highly inhomogeneous on a scale that is likely important for both normal state and superconducting properties. Some have taken this as evidence for phase separation. However, most of these electronically inhomogeneous materials appear to be single phase by our usual thermodynamic and structural definitions.

We will take the $La_{2-x}Sr_xCuO_4$ system as an example to discuss the relationship between point defects and carrier concentration. The point defect purposely introduced is the substitution of Sr for La. For the sake of this discussion, we will ignore other known point defects in this system such as oxygen interstitials and oxygen vacancies. The Sr substitution initially creates regions of electron delocalization in an insulating matrix of La_2CuO_4. The electronic properties of this single phase material are only perturbed in the vicinity of the dopant, Sr. This is a complicated system to model because as x becomes larger, the Sr rich regions change character, become larger, and become more numerous. Eventually, the regions strongly influenced by Sr connect and the materials appear homogeneous by some of our probes. However, the $La_{2-x}Sr_xCuO_4$ system is not really homogeneous for any value of x except zero. A consequence of this inhomogeneity is that some probes may be more sensitive to the Sr rich regions and others may be more sensitive to the Sr poor regions. Thus, reconciling the results of two studies may be impossible, even when conducted on the same sample, because each experiment probed two distinctly different regions of an electronically inhomogeneous material that is truly a single phase. The situation becomes even more complex on consideration of the oxygen stoichiometry and the presence of extended defects.

The lack of homogeneity pervades most of the systems where high T_c superconductivity has been observed. The n-type copper oxides, e.g., $Nd_{2-x}Ce_xCuO_4$ and $Nd_2CuO_{4-x}F_x$ are particularly plagued with this problem because superconductivity exists over such a small range of x. The inevitable variations of x on a microscopic level will inevitably lead to materials which are not electronically homogeneous.

The $YBa_2Cu_3O_{6+x}$ system can in principle give strictly homogeneous compounds at x = 0.0 and x = 1.0. Thus, $YBa_2Cu_3O_7$ could be a good example of a highly homogeneous high Tc superconductor. However, it tends to be plagued with extended defects such as the 110 twinning. It is also difficult to obtain an x value of exactly 1.00. Furthermore, some oxygen tends to be in the wrong chain position, and oxygen in the correct chain position is off the axis so that the chain Cu-O-Cu bond angles are not 180°.

Other high T_c superconductors which might be regarded as relatively homogeneous are $YBa_2Cu_4O_8$ and $Y_2Ba_4Cu_7O_{15}$, compounds with structures closely related to that of $YBa_2Cu_3O_7$. However, a recent structure determination of $Y_2Ba_4Cu_7O_{15}$ has shown (9) the presence of interstitial oxygen, again indicating inhomogeneity. A complication to the understanding of all of these $YBa_2Cu_3O_7$ related phases is that there are two crystallographic sites for copper. This tends to lead to an uncertainty in the carrier concentration in the CuO_2 sheets. Also, the carriers outside the CuO_2 sheets are less mobile than those in the sheets. Thus to the carriers in the CuO_2 sheets, the lattice appears to be inhomogeneous even for ideal compositions because of the random nature of carrier placement in the chains.

In the $YBa_2Cu_3O_{6+x}$ system, it has become common to speak of the 90 K phase and the 60 K phase. This is highly misleading because structural and thermodynamic studies do not support such a two phase description. Furthermore, we know inhomogeneity of electronic properties does not necessarily suggest phase separation. The common observation in the $YBa_2Cu_3O_{6+x}$ system is that T_c is close to 60 K from x = 0.6 to 0.7 and is close to 90 K from x = 0.9 to x = 1.0. Between x = 0.7 and x = 0.9, there is a rapid rise in Tc. However, structural and thermodynamic studies indicate that this is a single-phase solid-solution range (10). We nonetheless expect this range to be electronically inhomogeneous. A solid solution cannot be microscopically homogeneous. In compositional ranges where T_c changes rapidly with x, materials will naturally appear electronically inhomogeneous due to the local variations in x. On the other hand, we expect electronically homogeneous materials in the situation where T_c changes only slowly with x. Thus, there is no reason to invoke phase separation to understand the regions of apparent electronic homogeneity and inhomogeneity in the $YBa_2Cu_3O_{6+x}$ systems.

If one plots T_c vs. carrier concentration for $YBa_2Cu_3O_{6+x}$ or $Y_{1-x}Pr_xBa_2Cu_3O_7$ phases, a relatively smooth plot is obtained without plateaus. The conclusion is then that the carrier concentration in the CuO_2 sheets of $YBa_2Cu_3O_{6+x}$ phases is a strongly nonlinear function of x. The Cu-O distances along the c axis appear to be a measure of the relative carrier concentration in the chains and sheets. These distances also vary in a nonlinear fashion with x.

There is evidence that some phase separation does occur in the $YBa_2Cu_3O_{6+x}$ system at certain values of x. This evidence is primarily from electron diffraction studies where many different superstructures have been observed. Unfortunately, it has not been generally possible to assign definite composition or structures to these various superstructures. Furthermore, it appears that all but one of the superstructures results from metastable structures (10). The stable superstructure has a doubled b axis which is presumed due to the absence of every other chain. Thus, the ideal composition of this phase is $YBa_2Cu_3O_{6.5}$. The so-called 60 K phase actually exists at oxygen contents somewhat richer than this.

Point defects are also highly prominent in the Tl,Pb,Bi/Ba,Sr,-Ca/Cu/O superconductors. Cation vacancies frequently occur. Some Tl is found on Ca sites, and there is evidence for Ca on the Sr/Ba site. Some Bi is found on both Sr and Ca sites. Both oxygen interstitials and vacancies apparently can occur. Present evidence suggests that compounds with the ideal structures and compositions would not be metallic or superconducting. There are also strong indications that these materials at their ideal compositions are in fact too unstable to be prepared.

Inhomogeneities have also plagued the $Ba(Bi,Pb)O_3$ and $(Ba,K)BiO_3$ superconductors. These are other examples of solid solution systems, and all solid solutions are necessarily inhomogeneous.

It is highly doubtful that a good understanding of the high T_c superconductors will be obtained unless we learn to model the inhomogeneities that abound. If we limit our attention to the possible homogeneous compounds $YBa_2Cu_3O_7$, $YBa_2Cu_4O_8$ and $Y_2Ba_4Cu_7O_{15}$, we have complex structures and no trends of T_N and T_c with carrier concentrations. Thus by focussing only on these three phases, we would be ignoring the bulk of the interesting data that have been obtained. Modeling the inhomogeneities is essential and highly challenging.

7.0 STABILITY

Classical BCS theory dictated that superconductors with the highest T_c's would not be thermodynamically stable phases. Softening the phonons would raise T_c but would ultimately lead to structural instabilities. Increasing the density of states at the Fermi level would raise T_c but would eventually lead to an electronic instability.

Although we remain uncertain of the mechanism of superconductivity in the high T_c materials, the suggestion of metastability remains.

Trivial examples of metastability are solid solutions. Because these are inherently defect systems, they cannot be thermodynamically stable at low temperatures. Most of our high T_c superconductors need to be regarded as solid solutions which are then necessarily metastable phases. We could dismiss this as an irrelevant observation on the basis that solid solutions are merely required in order to adjust the carrier concentration to appropriate levels. However, we seem unable to generally make stable high T_c superconductors. One could even suggest that there is a correlation between T_c and metastability: the higher the Tc, the more unstable.

We do not need to regard $YBa_2Cu_3O_7$ as a solid solution, but it is not a thermodynamically stable phase at any temperature/pressure condition (11). Thermodynamic stability exists in the $YBa_2Cu_3O_{6+x}$ system in a certain region of temperature and oxygen pressure, but only for values of x between zero and about 0.6. Even these compositions appear not to be thermodynamically stable at room temperature and below.

Compositions in the system $YBa_2Cu_3O_{6+x}$ system with high x values and high T_c are made by oxidation of $YBa_2Cu_3O_{6+x}$ phases with lower value of x. Thus, the highest T_c phases in this system are made through an intercalation reaction. We know that essentially all phases made by intercalation are not actually thermodynamically stable even though we frequently speak loosely about equilibrating the degree of intercalation.

Our most blatant examples of thermodynamically unstable phases are just those phases where we have the highest T_c's, the $(A'O)_m A_2 Ca_{n-1} Cu_n O_{2+2n}$ family where A' is Tl or Bi and may be partially Pb, A is ideally Ba or Sr, m is one or two, and n may range up to about five. Many structure determinations have now been carried out on these phases using both single crystal x-ray diffraction data and powder neutron diffraction data. Virtually all these studies, covering many different compositions, have shown two features which could not be present in compounds thermodynamically stable to low temperature. Firstly, the positions of some atoms are not well ordered in three dimensions. Secondly, there are always high levels of point defects such as cation vacancies. These factors combined have made it exceedingly difficult to describe the precise structure and composition of these phases. In view of the compositional ranges, these phases must be regarded as solid solutions. Current evidence suggests that the solid solution ranges in general do not extend to the ideal composition. For example, the solid solution range for the "$Bi_2Sr_2CaCu_2O_8$" phase does not include the composition $Bi_2Sr_2CaCu_2O_8$ (12). Only the end members of solid solutions can be thermodynamically stable at low temperature. Therefore, none of the compositions in the observed range of the "$Bi_2Sr_2CaCu_2O_8$" phase can be stable at low temperatures. Thus, the evidence that the $(A'O)_m A_2 Ca_{n-1} Cu_n O_{2n+2}$ phases are entropy stabilized is overwhelming. This conclusion is further bolstered by the fact that exotherms of compound formation are not observed when these materials are synthesized from the component binary oxides. Furthermore, the only successful syntheses are at high temperatures close to the melting temperatures in these systems.

We thus conclude that our oxide superconductors with the highest T_c's, the $(A'O)_m A_2 Ca_{n-1} Cu_n O_{2n+2}$ family, are endothermic compounds or entropy stabilized compounds. They only exist because of stabilization by disorder and defects. Thus, they are thermodynamically stable only at high temperatures. Entropy stabilized oxides are rare but well established in the case of $CuFe_2O_4$, Fe_2TiO_3 and $Al_6Si_2O_{13}$ (mullite). In all of the known cases of entropy stabilized oxides, the important entropy term has been identified as point defects (13). We tentatively conclude therefore that point defects are also an essential component for the high T_c superconductors in the $(A'O)_m A_2 Ca_{n-1} Cu_n O_{2n+2}$ family.

We must of course ask the question of why the structures for $(A'O)_m A_2 Ca_{n-1} Cu_n O_{2n+2}$ family are not stable structures at low temperatures. The answer has seemed obvious since the first structures were solved. These structures will simply not give reasonable interatomic distances for all atoms regardless of the crystallographic parameters used. Thus, the CuO_2 sheets and the region close to them have the expected interatomic distances, and these rigid bonds fix the size of the a and b cell edges. Given this constraint, there is no possibility of reasonable interatomic distances for the Tl-O or Bi-O bonds. Thus, the structure found at high temperatures cannot be stable at lower temperatures. At high temperatures, disorder and entropy allow these structures to form in a very narrow stability range. Once formed, these phases are kinetically highly stable and do not tend to decompose into the phases which are thermodynamically stable at the lower temperatures.

There are two possible exceptions to the rule that high T_c superconductors are metastable. Both $YBa_2Cu_4O_8$ and $Y_2Ba_4Cu_7O_{15}$ appear to be stable under the conditions where they form, and they are superconductors without further oxidation. The remaining question to be answered is whether or not they are thermodynamically stable at room temperature and below. This is a difficult question to answer. Calorimetry can compare the heats of formation of these compounds with other compounds known in the Y/Ba/Cu/O system. However, it is always possible that some of the most stable phases in the Y/Ba/Cu/O system have not yet been prepared because they are kinetically inaccessible.

Structural refinements of the high temperature superconductors have nearly always indicated a poorly determined part of the structure. For $(A'O)_mA_2Ca_{n-1}Cu_nO_{2n+2}$ family, it is the A'O sheet region. For $YBa_2Cu_3O_7$ and $Y_2Ba_4Cu_7O_{15}$, it is the oxygen of the Cu-O chains. For $La_{2+x}Sr_xCuO_4$ superconductors, it is the oxygen above and below the CuO_2 sheets. In all cases, this lack of structure definition is showing the instability of the structure and pointing to the region of the structure responsible for the instability.

8.0 T_c CORRELATIONS

For the p-type copper oxides that superconduct, both carrier concentration and Cu-O distance correlate well with T_c. If one plots either carrier concentration or Cu-O distance vs. T_c, a maximum in T_c is found. The trend is shown schematically in Figure 2.

A problem with the two correlations is that they are highly correlated with each other. Changing carrier concentration is changing the number of electrons in the $Cu3d_{x^2-y^2}$-$O2p$ σ^* band. Thus, we expect that the Cu-O distance in the CuO_2 sheets will increase as the number of electrons increases regardless of whether the copper oxide is n-type or p-type.

In the $La_{2-x}A_xCuO_4$ series where A may be Ba, Sr, Ca, or Na, T_c initially increases with increasing x but the Cu-O distance is also decreasing. Which should we really be correlating with T_c, the carrier concentration or the Cu-O distance? The answer seems to be both. Application of pressure for a fixed composition causes an increase in

T_c, and this then suggests that the correlation between T_c and Cu-O distance is meaningful. On the other hand, the case for a correlation between carrier concentration and T_c is even more convincing. In the p-type materials of the $La_{2-x-y}Sr_xR_yCuO_{4-z}$ series, there are many examples of T_c actually decreasing as the Cu-O distance decreases. However, in all of these cases, an excellent correlation between carrier concentration and T_c remains (14).

We can conclude then that our best correlation with T_c is the carrier concentration of electrons or holes. The correlation of T_c with Cu-O distance appears to be real, but is easily overwhelmed by the correlation with carrier concentration. In cases where the Cu-O distance is known but the carrier concentration is basically unknown, we can suspect that the strong correlation between T_c and Cu-O distance may be basically related to the correlation between carrier concentration and Cu-O distance.

Knowing that T_c is highly correlated with carrier concentration does not necessarily make life simple for those synthesizing superconductors. Frequently, we are unable to obtain the ranges of carrier concentration we would like to investigate. Even more troubling is that many of the oxide superconductors are electronically inhomogeneous at certain carrier concentration ranges. As previously discussed, this does not necessarily suggest phase segregation. Actually, there may be no composition in some systems where lack of homogeneity is not very troubling. These problems with homogeneity in no way invalidate the general trends we see with carrier concentration and T_c. However, it does mean that the actual values of carrier concentration may not relate well to correct theory because measured properties are from samples which have variations of carrier concentration on a microscopic level.

It currently appears that the only way to systematically vary the carrier concentration is through point defects. Band structure calculations had suggested that the carrier concentration in the $Cu3d_{x^2-y^2}$-$O2p$ σ^* band could be altered by overlap with the $Bi6p$ band in the $Bi_2Sr_2Ca_{n-1}Cu_nO_{2n+4}$ superconductors. However, this conclusion is unreasonable from a chemistry point of view because it is akin to the oxidation of Cu^{II} by Bi^{III} or the even more implausible the oxidation of O^{-II} by Bi^{III}. We now know that the band structure calculations gave erroneous results because they were based on the ideal structures, which deviate greatly from the real structures.

Recent band structure calculations have confirmed that the Bi$6p$ band moves away from the Cu$3d_{x^2-y^2}$-O$2p$ σ^* band as one approaches the real structure of the Bi$_2$Sr$_2$Ca$_{1-n}$Cu$_n$O$_{2n+4}$ phases (15). The question of whether or not the Tl$6s$ band overlaps the Fermi level is still unsettled. However, the T$_c$ in (TlO)$_m$Ba$_2$Ca$_{n-1}$Cu$_n$O$_{2n+2}$ superconductors is so readily altered by doping on Ba and Ca sites that it appears that Tl$6s$ band overlap is not a significant factor in determining the carrier concentration of the Cu$3d_{x^2-y^2}$-O$2p$ σ^* band.

Something other than carrier concentration must be strongly influencing T$_c$ in the copper oxide based superconductors. It does not seem likely that the carrier concentration in the (TlO)$_m$Sr$_2$Ca$_{n-1}$Cu$_n$O$_{2n+n}$ phases greatly exceeds that in La$_{2-x}$Sr$_x$CuO$_4$ systems. Hall data, in fact, indicate an even lower carrier concentration in the (TlO)$_m$Sr$_2$Ca$_{n-1}$Cu$_n$O$_{2n+2}$ superconductors than in the La$_{2-x}$Sr$_x$CuO$_4$ superconductors. On the other hand, muon spin rotation studies show a universal correlation between T$_c$ and the relaxation rate (16). The relaxation rate is directly proportional to the carrier concentration divided by the effective mass of the carriers. There is currently no good way to separate these two terms which impact the relaxation rate. However, it is tempting to conclude that the T$_c$ correlation with relaxation rate means that the effective mass of the carriers is decreasing on going from La$_{2-x}$A$_x$CuO$_4$ to YBa$_2$Cu$_3$O$_7$ to (BiO)$_m$Sr$_2$Ca$_{n-1}$Cu$_n$O$_{2n+2}$ to (TlO)$_m$Ba$_2$Ca$_{n-1}$Cu$_n$O$_{2n+2}$ superconductors. The higher Tc's of the Bi and Tl containing phases would then be associated with mixing of Bi$6p$ and Tl$6s$ states into the Cu$3d_{x^2-y^2}$-O$2p$ σ^* band. This hybridization adds neither carriers nor total number of states to the σ^* band, but it could change the band width and the effective mass of charge carriers in this band.

9.0 MECHANISM FOR HIGH T$_c$

There is still no consensus for a mechanism for the high T$_c$ in the cuprate superconductors. Nonetheless, we have learned much about the electronic structure of such materials. Some proposed theories may now be discarded. The discovery of the n-type cuprate superconductors was the clinching evidence needed to discard theories based on some unique feature of an oxygen $2p$ band, a π^* band, or overlapping bands. The central question now for the cuprate superconductors is the fate of spin correlations as the Cu^{II} based insulators are doped. Certainly, these spin correlations are diminished with either n-type or p-type doping. Long range magnetic order disappears, and the materials become metallic. One line of thinking is that it is a vestige of this magnetic state which is indirectly responsible for superconductivity. A more traditional view is that magnetism and superconductivity tend to be mutually exclusive phenomena. The first role of doping is then to weaken the spin correlations sufficiently so that charge fluctuations can then dominate and produce superconductivity. Only this latter view is compatible with both the Bi-based and the Cu-based superconductors, and only this view will be further discussed here.

Many of the features common to both the copper-based and bismuth-based superconductors have been discussed earlier in this chapter. One might say that these two types of superconductor differ in their isotope effects in a way that suggests a different mechanism, but this is not really the case. It is now generally agreed that the oxygen isotope effect in $YBa_2Cu_3O_7$ is finite but quite small. However, it is also known that this fact alone cannot be used to discount a possible critical importance of the electron-phonon interaction in such superconductors. The oxygen isotope effect is pronounced in both $La_{2-x}A_xCuO_4$ and $Ba_{1-x}K_xBiO_3$ superconductors (17)(18). Thus, we can be rather certain that the electron-phonon interaction is important for all these superconductors, including the cuprate superconductors with T_c's above 100 K.

I continue to believe that the mechanism for high T_c lies in the tendency for disproportionation of the middle valence state, i.e., Cu^{II} or Bi^{IV} (Figure 3). Such a tendency is well known in chemistry, as previously discussed in this chapter. A problem has been, and remains, that this tendency to disproportionate is not well understood in terms of physics. Nonetheless, this is a real phenomenum which inherently gives the electron-electron attraction required for superconductivity. Such an approach to forming Cooper pairs necessarily means that there will be an electron-phonon interaction. Still, there may be important charge fluctuations that are not so strongly coupled to phonons, e.g., on-site polarizibility. This may also make a significant contribution to increasing T_c. A serious problem in developing this view is that charge fluctuations in ferroelectrics such as $BaTiO_3$ (Figure 3) have never been well understood from a physics perspective.

10.0 REFERENCES

1. Jones, N.L., Parise, J.B., Flippen, R.B., and Sleight, A.W., Superconductivity at 34 K in the K/Ba/Bi/O system, *J. Solid State Chem.* 78: 319 (1989).

2. Tseng, D., and Ruckenstein, E., Some improvement in the T_c of bismuth-based superconducting compounds, *Materials Letters* 8: 69 (1989).

3. Sleight, A.W., Valency Considerations in Oxide Superconductors, *Proc. Welch Conf. on Chem. Res.*, XXXII. Valency, p. 123, (1989).

4. Uemura,Y.J., Sternlieb,B.J., Cox, D.E., Brewer, J.H., Kadono, R. Kempton, J.R., Kiefl, R.F., Kreitzman, S.R., Luke, G.M., Riseman, T., Williams, D.L., Kossler, W.J., Yu, X.H., Stronach, C.E., Subramanian, M.A., Gopalakrishnanan, J., and Sleight, A.W., Absence of magnetic order in $(BaK)BiO_3$, *Nature* 335: 151 (1988).

5. Walstedt, R.E. and Warren, W.W., *Science* 248:1082 (1990).

6. Anderson, P.W., Principles of Electron Function in Compounds and Crystals, *Proc. Welch Conf. on Chem. Res.* p. 11, XXXII. Valency, (1989).

7. Arko, A.J., *Proceedings of the International Conference of Superconductivity*, High-Temperature Superconductivity, Stanford Univ. (1989).

8. Sleight, A.W., High-Temperature Superconductivity in Oxides, in *Chemistry of High-Temperature Superconductors*, (Nelson, Whittingham, George, eds.) p. 2, Amer. Chem. Soc., Wash. D.C. (1987).

9. Bordet, P., Chaillout, C., Chenavas, J., Hodeau, J.L., Marezio, M., Karpinski, J. and Kaldis, E., Structure determination of the new high-temperature superconductor $Y_2Ba_4Cu_7O_{14+x}$, *Nature* 334: 596 (1988).

10. Beyers, R., Ahn, B.T., Gorman, G., Lee, V.Y., Parkin, S.S.P., Ramirez, M.L., Roche, K.P., Vazquez, J.E., Gür, and Huggins, R.A., Oxygen ordering, phase separation and the 60 K and 90 K plateaus in $YBa_2Cu_3O_x$, *Nature* 340:619 (1989).

11. Sleight, A.W., Chemistry of High-Temperature Superconductors, *Science* 242:1519 (1988).

12. Bloomer, T.E., Golden, S.J., Lange, F.F., and Vaidya, K.J., Processing and characterization of single phase superconducting thin films in the Bi-Sr-Ca-Cu-O system from chemical precursors, *Materials Research Society Fall (1989) Meeting*, abstract M7.122.

13. Navrotsky, A., in *Solid State Chemistry*, Edited by Cheetham and Day, Clarendon Press, Oxford, (1987), p.382.

14. Subramanian, M.A., Gopalakrishnan, J. and Sleight, A.W., Depression of Tc with Rare Earth Substitution for La in $La_{2-x}Sr_xCuO_4$ Superconductors, *J. Solid. State Chem* 84:413 (1990).

15. Whangbo, M.-H., private communication.

16. Uemura, Y.J., Luke, G.M., Sternlieb, B.J., Brewer, J.H., Carolan, J.F., Hardy, W.N., Kadono, R., Kempton, J.R., Kiefl, R.F., Kreitzman, S.R. Mulhern, P., Riseman, T.M., Williams, D.L., Yang, B.X., Uchida, S., Takagi, H., Gopalakrishnan, J., Sleight, A.W., Subramanian, M.A., Chien, C.L., Cieplak, M.Z., Xiao, G., Lee, V.Y., Statt, B.W., Stronach, C.E., Kossler, W.J., and Yu, X.H., Universal Correlation between T_c and ns/m* (Carrier Density/Effective Mass) in High-T_c Cuprate Superconductors, *Phys. Rev. Lett.* 62: 2317 (1989).

17. Crawford, M.K., Kunchur, M.N., Farneth, W.E., McCarron, E.M., and Poon, S.J., Variable ^{18}O isotope effect in $La_{2-x}Sr_xCuO_4$, *Phys. Rev. B*. 41:282 (1990).

18. Hinks, D.G., Richards, D.R., Dabrowski, B., Marx, D.T., and Mitchell, A.W., *Nature* 335: 419 (1988); Batlogg, B. Cava, R.J., Rupp, L.W., Mujsce, A.W., Krajewski, J.J., Remeika, J.P., Peck, W.F., Cooper, A.S., and Espinosa, G.P. *Phys. Rev. B* 61:1670 (1988).

20
Electron-Electron Interactions and the Electronic Structure of Copper Oxide-Based Superconductors

Jeremy K. Burdett

1.0 INTRODUCTION

This chapter will describe ways to understand the electronic properties of solids with special emphasis on the structures of the high-T_c copper oxide based superconductors which form the basis for this book. We shall not attempt to discuss present theories of superconductivity, but will be concerned with the development of a theoretical framework to describe the electronic structure of these materials upon which a physical mechanism must be based. The approach is based on the coalescence of two viewpoints, both simplifications of 'the quantum mechanical truth' which enable us to gain a feel for the important electronic effects which control structure and the arrangement of the electrons in the energy states open to them. Although the focus of physical theories of superconductivity has lain with 'exotic' and non-conventional ideas, we will see that many of the geometrical features of these materials may be understood in quite conventional chemical terms. Given that structure controls function in all areas of chemistry, the importance of such a viewpoint will become apparent. We can too, glean important information concerning the electronic structure of these materials by comparing the predictions different electronic models make for some specific geometrical observations.

2.0 ONE- AND TWO-ELECTRON TERMS IN THE ENERGY

Chemists are familiar with the Tanabe-Sugano diagrams used as a basis for understanding the electronic spectra of transition metal complexes (1)(2). A typical picture is shown in Figure 1 for the specific case of the octahedral d^8 configuration. At the far left-hand side of the diagram the energetics of the system are described in terms of the free ion. The energies of the levels are usually given in terms of the Racah parameters A, B and C, which describe various combinations of the Condon-Shortley parameters used to define the different types of electron-electron interactions present. These two-electron terms in the energy are of the Coulomb and Exchange type, and both are coulombic in origin, containing an operator of the form r_{ij}^{-1}, the inverse of the distance between electrons i and j. The coefficient describing the contribution of the Racah A parameter to the energy depends only upon the number of electrons, and thus the energy differences between the atomic terms are controlled by the relevant B and C parameter coefficients. B and C are often found to be related by a factor of 4, so one parameter is sufficient to describe these energy differences. At the far right-hand side of the diagram the energetics in the strong-field limit are heavily influenced by the one-electron terms, Δ, the t_{2g}/e_g splitting in this particular example. In order to calculate the energies of the electronic states here, we use the wavefunctions constructed by consideration of the relevant occupation of the one-electron orbitals set by the crystal field splitting Δ, and then calculate the contribution from the two-electron terms mandated by this electronic description. At the left hand-side of the diagram, the weak field limit, the energies of the electronic states are obtained by taking the wavefunctions set by the two-electron operator r_{ij}^{-1}, and then using the one-electron parameter Δ as a perturbation. (There is an assumption that A, B and C represent the same physical property at both left and right-hand sides of the picture.) There are some differences between the predictions of the two approaches. For example, the Crystal Field Stabilization Energy for the electronic ground states of high-spin d^2 and d^7 systems depend upon which régime is in force. We shall see below the reason for this. Both approaches are ways to include the effects of both one- and two-electron energy terms. Recall that the one-electron terms are those energetic contributions arising from that part of the Hamiltonian

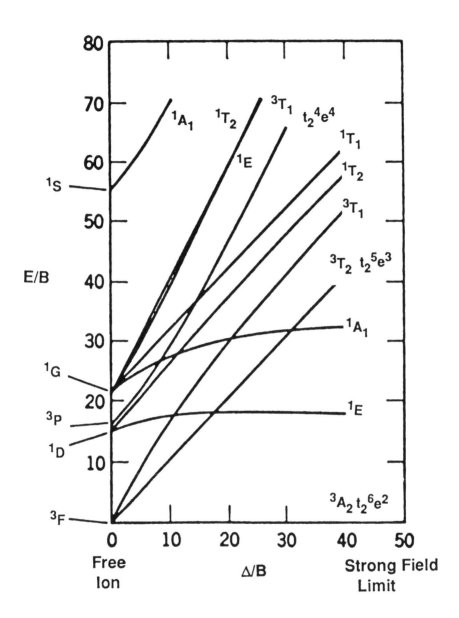

Figure 1: Tanabe-Sugano diagram for the d^8 configuration.

which contains r_i^{-1}, where r_i is the distance between electron i and the nucleus. Examples are the Hückel α and β integrals. The 'overlap forces' of the chemist are thus one-electron terms. The two-electron terms are those containing r_{ij}^{-1}, where r_{ij} is the distance between electrons i and j. They include the Coulomb repulsion and Exchange terms.

When viewing the structures of molecules and solids, chemists often focus on the viewpoint of the right-hand side of the diagram by considering the one-electron energy as dominant in a first approximation, and then considering the effects of the two-electron terms in more detail (the strong-field limit). Many of the simple orbital theories used in molecular chemistry ignore the latter completely in a formal sense, but close examination often shows that such many-body effects are subsumed into the parameters of the one-electron model. As an example consider the Hückel β parameter, used extensively to understand the properties of conjugated organic molecules. The model may be used to study, amongst other things, the π-delocalization energy, the absorption maxima of these hydrocarbons and their redox potentials. However the numerical values of β which need to be used to get good agreement with experiment vary. In these three examples the values are -0.69, -2.71 and -2.37eV respectively (3). Thus we should realize that the use of such simple (but often very successful) models often hides some deeper considerations.

There are several examples we could use to illustrate the interplay of the various factors which control the structures of molecules, but consider the case (2) of complexes of Ni^{+2}. At the octahedral geometry with two electrons in an e_g orbital it is easy to generate via the symmetric and antisymmetric direct products of e_g, three electronic states $^3A_{2g}$, $^1A_{1g}$ and ^1E_g. Their energetic ordering is given at the right-hand side of Figure 1, obtained in the manner just described. Let us focus on the two lowest energy states of this example. The energetics associated with a D_{4h} distortion can be readily understood by consideration of the energy changes of the e_g *orbitals* on distortion (Figure 2a) usually associated with changes in the one-electron part of the energy, namely via changes in orbital overlap. The ^1E_g state is Jahn-Teller unstable at this geometry, and one component of it is rapidly stabilized during a distortion which lengthens two trans ligands to such an extent that they are lost from the complex altogether, leaving a low-spin (diamagnetic) square-

Electronic Structure of Copper Oxide-Based Superconductors 739

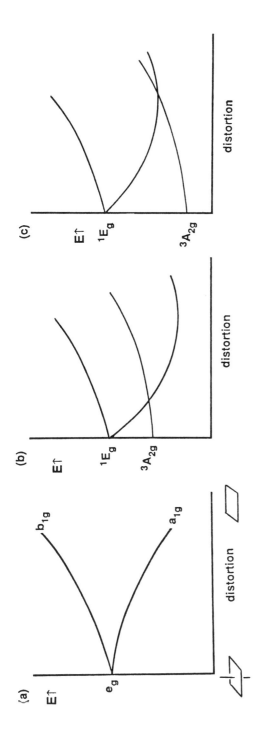

Figure 2: (a) Splitting apart in energy of the e_g levels of the octahedral complex as two trans ligands are removed and four in-plane ligands shorten their metal-ligand distances. (b), (c) Two possibilities for the relative stability of singlet and triplet states on distortion.

planar molecule (Figures 2b,c). The $^3A_{2g}$ state is destabilized during such a distortion as may readily be seen from the orbital picture. One electron has to reside in each component of e_g. From these figures we can thus say that if a given Ni^{+2} complex is paramagnetic (high-spin) then it will be octahedral, but if it is diamagnetic (low-spin) then it will be square planar. What is less quantifiable is the answer to the following question. Given a Ni^{+2} ion and a collection of ligands, will the system be high or low spin? I.e., which diagram will be appropriate, 2b or 2c. It is the balance between the change in the one- and two-electron terms in the energy on distortion which is very difficult to accurately measure or estimate. If the geometry itself does not change dramatically on moving from high-spin to low-spin then the situation is a little simpler. Use of Jørgensen's empirical rules (4) which allow estimation of Δ and B for a given collection of metal and ligands is often useful in making a prediction in this case. In general, if we define a parameter U, the energy cost associated with placing two electrons in the same orbital then the condition for stability of the singlet relative to the triplet, for the problem of two electrons and two orbitals separated in energy by Δ, is roughly $\Delta > U$. In fact to be more correct we need to include in addition to U the change in exchange energy associated with the spin change. For the d^4 configuration, for example, the coulomb part of the 'pairing energy' is 2C, and the exchange part 6B + 3C in terms of the Racah parameters. In our discussions below of the copper oxide problem, we will however, face the more complex problem associated with the additional change in geometry.

There are some generalizations though which are useful. Electron-electron repulsions (A, B, C) in second and third row transition metal ions are smaller than those in their first row analogs, a result often attributed to the larger size of the heavier atom, with its commensurately larger electron-electron separations. Contrarily Δ is larger for the heavier analogs. Thus Δ/B lies further to the right of Figure 1 for the heavier transition elements than for the first row ones. The chemical effects of these trends are of textbook importance. Second and third row complexes are invariably of low spin and are colorless, compared with their high-spin and frequently highly colored, first row analogs.

There is another important effect which influences the ordering of the electronic states. Square cyclobutadiene is an organic molecule with strong electronic similarities to the Ni^{+2} problem. It has two electrons in an e_g orbital. On distortion this doubly degenerate level splits apart in energy in just the same way as the e_g orbital does in the Ni^{+2} example (Figure 3a). At the square geometry this configuration leads to a triplet state $^3A_{2g}$ and three singlets. (No degenerate electronic state results from this electronic configuration in this point group and the molecule is pseudo-Jahn-Teller unstable via a second order mixing of $^1B_{1g}$ and $^1A_{1g}$ states.) Our discussion above suggests that it is the triplet which will be the lowest energy state, and from calculations close to the Hartree-Fock limit it is indeed found (5) that the triplet state lies lower in energy than any singlet state. In Hartree-Fock theory the mean-field approximation is used which requires each electron to move in the mean field of all of the others. This is not true of course. Electrons being of the same charge will tend to avoid each other and become correlated. Configuration interaction partly takes care of this electron correlation, by mixing excited states into the ground states via the r_{ij}^{-1} operator. Inclusion of configuration interaction for the cyclobutadiene problem depresses the lowest singlet so that it is calculated (6) to lie beneath the triplet at this geometry (Figure 3b). This effect is expected to be most important for singlet states, since in the triplet state the electrons are already correlated to some degree. Another example which shows some of the quantitative aspects of the problem is shown in Figure 4 for the lowest energy states of atomic magnesium. It has long been known that of the two states arising from the (3s3d) configuration, it is the singlet which lies lowest in energy, rather than the triplet which simpler theories would suggest. Strong configuration interaction with the ^1D state arising from the $(3p)^2$ configuration depresses the ^1D state arising from the (3s3d) configuration. The effect is quite large, and a numerical calculation (1)(7) which reproduces the experimentally observed energies leads to a heavy mixing of the two states such that now the '(3s3d)' state contains some $(3p)^2$ character. Thus as a result of the mixing we would write $\Psi'(3s3d) = a\Psi(3s3d) + b\Psi(3p)^2$. In this particular case the mixing is heavy; the lower energy state contains 25% $(3p)^2$ character. It is configuration mixing of this type which is behind the differences in CFSE mentioned earlier for the d^2 and d^7 systems in the strong and weak field limits.

There is another way to gain some understanding of the concept of configuration interaction, and how the idea may be used to understand

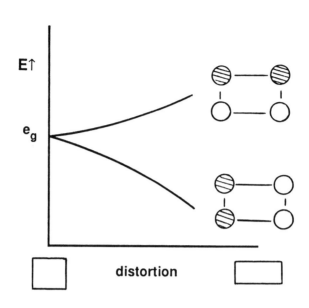

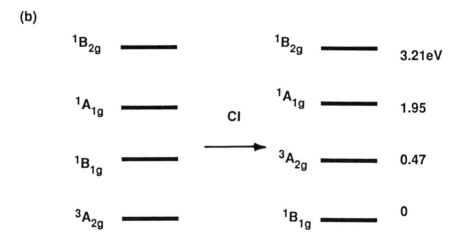

Figure 3: (a) Splitting apart in energy of the e_g levels of the square cyclobutadiene molecule as two opposite C-C bonds are shortened and the other two lengthened. (b) The effect of configuration interaction on the relative energies of the electronic states at the square geometry.

electronic structure, and that is to see how the electronic description of a molecule, such as that of H_2 varies with internuclear distance. Recalling the traditional molecular orbital approach (Mulliken-Hund = MH) to the chemical bonding problem we would write a singlet wavefunction of the type (ignoring normalization and antisymmetrization for the time being)

$$^1\Psi_{MH} = (\phi_1 + \phi_2)(1)(\phi_1 + \phi_2)(2)$$

$$= \phi_1(1)\phi_1(2) + \phi_2(1)\phi_2(2) + \phi_1(1)\phi_2(2) + \phi_2(1)\phi_1(2) \qquad (1)$$

to describe the state. Two electrons reside in $\phi_1 + \phi_2$, the lower energy bonding orbital formed by overlap of the two H 1s atomic orbitals, ϕ_1, ϕ_2 on the two centers. We can readily evaluate the one-electron energy of this problem by setting $<\phi_1 | H | \phi_1> = <\phi_2 | H | \phi_2> = \alpha$ and $<\phi_1 | H | \phi_2> = \beta$. This is the simple Hückel model and the total one-electron energy (for the two electrons) is $2(\alpha + \beta)$ where $\alpha, \beta < 0$. A natural outcome of the formulation of eqn. 1 is that the two electrons are allowed to simultaneously reside on one of the two atoms via the presence of the terms $\phi_1(1)\phi_1(2) + \phi_2(1)\phi_2(2)$. If the electrostatic interaction energy of two electrons located on the same atom is U, then the crudest way to estimate the two-electron part of the energy is to multiply U by the probability of finding the two electrons on the same atom. This is 1/2, so that the total energy is simply

$$E = 2(\alpha + \beta) + U/2 \qquad (2)$$

Such a simple treatment ignores the exchange energy which we would get by using the antisymmetrized form of eqn 1. As the H-H distance increases the chance of two electrons residing on the same atom becomes less likely and the molecular orbital model less appropriate. Imagine the extreme case at infinite separation, where each neutral H atom holds just one-electron and never two, in the lowest energy arrangement. Now we would write a localized (Heitler-London = HL) wavefunction to describe this state of affairs of the form.

$$^1\Psi_{HL} = \phi_1(1)\phi_2(2) + \phi_2(1)\phi_1(2) \qquad (3)$$

In the molecule there is another singlet electronic state, one of higher energy than $^1\Psi_{MH}$ obtained by locating two electrons in the antibonding orbital. (The energy separation between bonding and antibonding levels is simply equal to 2β.) This is given by

$$^1\Psi'_{MH} = (\phi_1 - \phi_2)(1)(\phi_1 - \phi_2)(2)$$

$$= \phi_1(1)\phi_1(2) + \phi_2(1)\phi_2(2) - \phi_1(1)\phi_2(2) - \phi_2(1)\phi_1(2) \qquad (4)$$

and its energy is readily obtained as

$$E' = 2(\alpha - \beta) + U/2 \qquad (5)$$

One way to describe the electronic situation at an intermediate geometry can be obtained by starting off with Ψ_{MH} and mixing in a contribution from this higher energy electronic state as a configuration interaction, viz.

$$^1\Psi_{CI} = a\,^1\Psi_{MH} + b\,^1\Psi'_{MH}$$

$$= \lambda(\phi_1(1)\phi_1(2) + \phi_2(1)\phi_2(2)) + \phi_1(1)\phi_2(2) + \phi_2(1)\phi_1(2) \qquad (6)$$

where λ varies from 0 (H-L) to 1(M-H). This allows a gradual change from the M-H to H-L description. The operator used in the process is just the on-site interelectron repulsion r_{ij}^{-1}. The new energy is simply $E_{CI} = 2(\alpha + \beta) + (\lambda^2/2)U$ and indicates a decrease in electron-electron interaction energy which we naturally associate with the improved electron correlation as described earlier.

There is a triplet state for this molecule which is simply written as

$$^3\Psi = (\phi_1 + \phi_2)(1)(\phi_1 - \phi_2)(2) + (\phi_1 + \phi_2)(2)(\phi_1 - \phi_2)(1) \qquad (7)$$

with an energy of 2α. Thus the condition for stability of the ground state singlet relative to the triplet is $2(\alpha+\beta) + (\lambda^2/2)U < 2\alpha$, i.e., $|4\beta| > \lambda^2 U$, or $U/|4\beta| < 1/\lambda^2$. This is a result similar to that of the balance between one- and two-electron terms in the energy which determine the geometries of high spin and low spin complexes. Notice that

Electronic Structure of Copper Oxide-Based Superconductors 745

configuration interaction provides an extra stabilization of the singlet over the triplet, a result analogous to that found in cyclobutadiene.

Finally there is another singlet state, with the same orbital occupation as $^3\Psi$. Its wavefunction is simply written as

$$^1\Psi'' = (\phi_1 + \phi_2)(1)(\phi_1 - \phi_2)(2) - (\phi_1 + \phi_2)(2)(\phi_1 - \phi_2)(1) \qquad (8)$$

It is less stable than the triplet by the exchange interaction between the electrons.

An alternate route to the electronic problem starts off at the HL limit and adds in the effect of the overlap interactions between the atomic orbitals ϕ_1 and $\phi_2 (= \beta)$. Our treatment is adapted from that in reference 8. The complete set of singlet states of the problem are

$$\Psi_0 = 2^{-1/2}[\phi_1(1)\phi_2(2) + \phi_2(1)\phi_1(2)]$$

$$\Psi_1 = \phi_1(1)\phi_1(2)$$

$$\Psi_2 = \phi_2(1)\phi_2(2) \qquad (9)$$

Ψ_0 is just a normalized version of $^1\Psi_{HL}$ used earlier. The energies of these states are easy to calculate. Ψ_0 does not allow the two electrons to reside in the same orbital (energy = 2α) but this is not the case in Ψ_1 and Ψ_2 (energy = $2\alpha + U$). The matrix element coupling Ψ_0 to Ψ_1 and Ψ_2 is easily seen to be $2^{-1/2}\beta$, which leads to a Hamiltonian matrix of the form

$$\begin{pmatrix} 2\alpha & 2^{-1/2}\beta & 2^{-1/2}\beta \\ 2^{-1/2}\beta & 2\alpha + U & 0 \\ 2^{-1/2}\beta & 0 & 2\alpha + U \end{pmatrix} \qquad (10)$$

to give a ground state energy of

$$E = 2\alpha + (1/2)U - (4\beta^2 + (1/4)U^2)^{1/2}$$

$$\sim 2(\alpha + \beta) + (1/2)U - \beta\kappa^2 \qquad (11)$$

where $\kappa = U/4\beta$. The new wavefunction, correct to a normalization constant, is

$$\Psi'_0 = \Psi_0 + 2^{-1/2}((1+\kappa^2)^{1/2} - \kappa)(\Psi_1 + \Psi_2) \qquad (12)$$

Notice the appearance both here and above of the vital parameter $U/4\beta$ which is the measure of the importance of the two-electron terms to the one-electron ones. This discussion is exactly analogous to the Hubbard treatment of solids which we will study below.

In describing the electronic states of molecules there are therefore two important effects to be considered associated with electron-electron interactions. The first, approachable using Hartree--Fock theory, is the repulsion associated with electron pairs occupying the same region of space. This is how such interactions are accommodated in the strong-field limit of the transition metal complex of Figure 1 and in the H_2 molecule via equation 2. These are of two types, coulomb and exchange. The second effect is associated with electron correlation and is dramatically demonstrated numerically by the figures for atomic magnesium in Figure 4. For the H_2 molecule we showed analytically how a reduction in electron-electron repulsion occurs by using either configuration interaction with the M-H states (equation 6) or by switching on the effect of the one-electron interactions with the H-L states. The results show that the delocalized model will be most appropriate for short internuclear distances and a localized model for longer ones. A useful parameter with which to describe the transition between the two is the parameter κ. The one-electron terms increase in magnitude as the overlap increases, i.e., as the interatomic distance decreases. The structures of solids, really large molecules, can be approached using a similar philosophy. Systems which lie to the left-hand side of Figure 5 can usefully be described by a molecular orbital or band model where the importance of interorbital interactions, via the resonance or interaction integral of the chemist or the hopping or transfer integral of the physicst, are of paramount importance. In this case the electrons are delocalized throughout the material. Systems corresponding to the right-hand side of Figure 5 are described by localized models where the electrons are non-itinerant. It is not uncommon to find a system where one part of the orbital problem is descibed by localized, and the other by delocalized interactions (9). Transition metal oxides are a case in point.

Electronic Structure of Copper Oxide-Based Superconductors 747

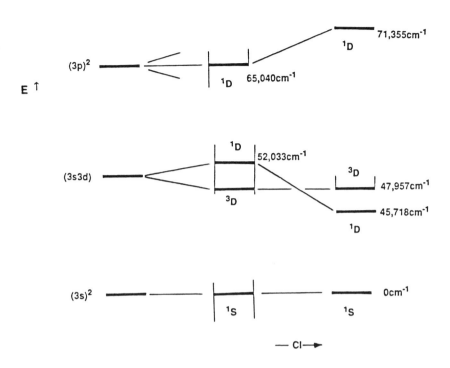

Figure 4: Effect of configuration interaction on some of the states of atomic magnesium.

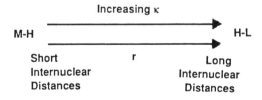

Figure 5: Areas of validity of Mulliken-Hund (M-H) and Heitler London (H-L) viewpoints. κ is the ratio $U/4\beta$, the ratio of one of the two-electron terms to a multiple of one of the one-electron terms.

In the hybrid situations, the middle ground of Figure 5, although we recognize the importance of these many-body terms, we shall seek orbital or one-electron arguments to probe the nuances of geometrical structure. This is just as true when looking at the interatomic separations of transition metal compounds (traditionally treated by crystal field theory, but now studied using molecular orbital ideas) as it will be in these superconductors.

3.0 ENERGY BANDS OF SOLIDS

To develop the energy bands of solids it is useful to recall the standard method of generating the energy levels of molecules. Our approach will follow the detailed discussion of reference 10. Let us take a molecule which is described by the point group G and has a collection of symmetry elements $g \in G$, and a set of basis orbitals $\{\phi_i\}$. For reasons that will become clear choose G to be a cyclic group which would, for example, describe the π orbitals of the benzene molecule, or the equilateral triangular structure of cyclopropenium, $C_3H_3^+$. The symmetry elements g simply take one orbital to another one in the ring. For example a C_3^n operation takes ϕ_1 to ϕ_{1+n}. There are the same number of symmetry elements in this group as π (for example) orbitals in the ring so each symmetry element takes the starting orbital to a different orbital of the ring. We may construct symmetry adapted linear combinations of the basis orbitals in the standard group theoretical way. For the kth irreducible representation of G, the resulting molecular orbital, ψ_k, is

$$\psi(k) = \sum_g \chi(k,g) g \phi_1 \tag{13}$$

where $\chi(k,g)$ is the character of the kth irreducible representation corresponding to the operation g. If there is only one possible $\psi(k)$ as in the benzene case then the energy of the molecular orbital is found by substitution into the wave equation

$$E_k = <\psi(k)|H|\psi(k)>/<\psi(k)|\psi(k)> \tag{14}$$

For the case where there are two orbitals of the same symmetry, for example the π orbitals of the $B_3N_3H_6$ molecule then they will interact with each other and solution of a secular determinant is needed to get their energies.

Now let us move to solids. Here instead of a point group, the system is described by a space group. The simplest way to appreciate how the electronic structure evolves is to envisage an infinite chain of atoms (orbitals) described by a translation group T with a collection of symmetry elements t ϵ T, and a set of basis orbitals $\{\phi_i\}$. We usefully point out the analogies to the molecular case just described. Just as in the cyclic molecule where there was a collection of symmetry operations which took the starting orbital to each other orbital in the ring, so the translation group contains a collection of symmetry elements which take the starting orbital to every other orbital of the chain. Using this new group we can generate the symmetry adapted linear combination of orbitals in an exactly analogous way to that of equation 13.

$$\psi(k) = \sum_t \chi(k,t) t \phi_1 \qquad (15)$$

Just as for the cyclic group each irreducible representation is represented once in the collection of 'molecular' (now called crystal) orbitals, and there are now a very large number of them corresponding to the number of atoms in the crystal. For the translation group however, the characters take on a particularly simple form.

$$\chi(k,t) = \exp(ikr) \qquad (16)$$

where k is defined a little differently to the molecular case to take into account the infinite nature of the solid. If a is the repeat distance along the chain, then r is the distance associated with a particular translation t, k is called the wavevector and takes values from 0 to π/a. Thus the discrete set of levels found for the molecule is replaced by a continuous set in the solid, the energy band of orbitals. Figure 6 shows the situation for the solid-state analog (polyacetylene) of the benzene molecule. At the bottom of the band where k = 0, there are no nodes between the orbitals. At the top of the band where k = π/a there are nodes between each adjacent orbital pair. The similarity to the molecular picture is thus a clear one. k may be regarded as a node counter. Notice that the width of the band is given on this simple model by 4β, where β is the interaction (transfer or hopping) integral

750 Chemistry of Superconductor Materials

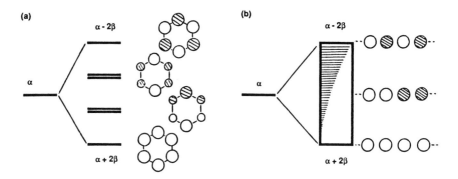

Figure 6: (a) Energy levels and orbitals of the π-system of benzene, and how they are related to the energy levels of polyacetylene $(CH)_n$. (b) Notice how the top and bottom of the energy band lie at exactly the same energy as the most antibonding and most bonding orbitals respectively of benzene. The shading in the box represents the infinite collection of energy levels which compose the energy band.

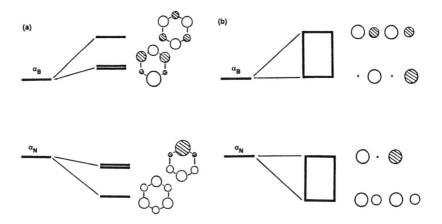

Figure 7: A similar diagram to that of Figure 6, except that the heteroatomic molecule $B_3N_3H_6$, and the hypothetical polymer (BNH_2) is used.

between the atomic orbitals concerned.

The extension of the picture to the case of a chain containing different atoms is straightforward. Now we have to solve a secular determinant and the result is two energy bands split apart in energy. The quantitative similarity to the case of the $B_3N_3H_6$ molecule is shown in Figure 7 where we use the (unknown) borazine analog of polyacetylene. It is a relatively simple matter to draw out the form of the wavefunctions for various k. If $|\alpha_N| > |\alpha_B|$ (as it is) then we would describe the deeper lying (bonding) band as largely N in character and the higher lying (antibonding) band as largely B in character, just as we would to describe the molecular orbitals. For the case of the copper oxides the difference in the α values is small and the mixing between the two is large. This mixing is sometimes described as 'covalency' of the Cu-O bond. Notice that the antibonding character, not only of the molecular orbitals, but of the crystal orbitals comprising the energy band, increases as the energy increases.

The electronic situation we have described is directly applicable to two chemical situations which we will discuss in this chapter. The first is the series of systems with empirical formula PtL_4X (L=amine, X=halogen) shown in Figure 8a. Here the system is one-dimensional and the important orbital for our purposes is the z^2 orbital of the PtL_4 fragment which lies parallel to the chain direction. The energy bands are developed in Figure 9a, a picture very similar to that of Figure 7. The second situation is found in the two-dimensional sheets of stoichiometry CuO_2 found in the high T_c superconductors. Here the relevant copper orbital is x^2-y^2 and the approach we have used applies to the two (x,y) directions defined by the lobes of this orbital. The amount of metal or non-metal character in the highest energy band will clearly depend upon the relative energies of the atomic orbitals which go into the model. For copper and oxygen the relevant ionization energies are close and so the two will be heavily mixed as a result of their interaction. In both the platinum chain compounds and in the copper oxides these bands are half occupied, or close to it. Before concluding this section we should note two points. First the orbital model will only be a good description in those cases where the ratio of the one to two-electron energy terms is large. Second, even if these conditions are satisfied we need to add on the two-electron terms to the simple model, in both of the ways described for molecules in Section 2.0.

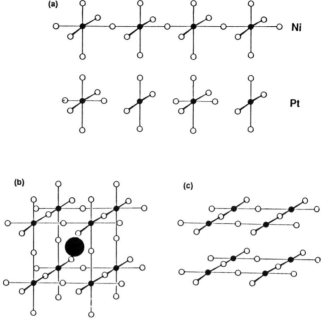

Figure 8: (a) The one dimensional platinum chain compound PtL_4X^{+2} chains (X = halide, L = NH_3, $NEtH_2$ etc.) and its undistorted nickel analog (11). (b) The perovskite structure. (c) The structure of the two dimensional CuO_2 sheets found in all copper-containing high-T_c superconductors.

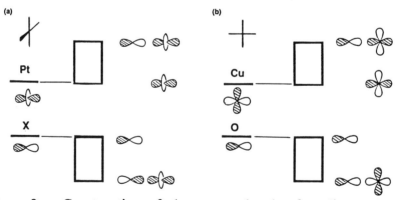

Figure 9: Construction of the energy bands of undistorted one dimensional chains and two dimensional sheets of (a) the platinum amine halide system where the z^2 orbital is used, and (b) the CuO_2 sheet where the $x^2 - y^2$ orbital is used. Notice the strong similarities in orbital character, and compare with Figure 7b.

4.0 THE PEIERLS DISTORTION

Chemists are very familiar with the Jahn-Teller distortion in molecules which lifts the degeneracies of electronic states. In Figure 2 we showed how the geometrical change associated with a Jahn-Teller distortion led to a stabilization of one component of the degenerate ^1E_g state of an octahedral Ni^{+2} complex. This resulted from the half filling of the e_g pair of orbitals. In solids exactly analogous distortions are associated with half full energy bands. (In fact distortions can and do occur at other fillings but we will not consider them here.) Electron counting in the platinum chains and in the 2-1-4 superconductor (La_2CuO_4) quickly shows that the highest energy orbital of the building block of the solid is half full. Let us assume for the present that all of the electrons are paired up so that all the levels in the lower half of the energy band are full of paired electrons. A typical diamagnetic metal should result. In the platinum example a d^7 configuration results which leads to half-filling of the z^2 orbital of the square planar PtL_4 unit, and in the d^9 2-1-4 compound by half-filling of the x^2-y^2 orbital of a CuO_4 unit. Notice that these d^n labels allow us to effectively count electrons. We do not mean to imply that they reside in orbitals containing 100% d character. As we have already noted heavy admixture of central atom and ligand orbitals often occurs. The result is then a half-filled band in the solid.

Figure 10a shows how a distortion of the type shown cleanly splits the band into two, converting what would be a metal into a semiconductor or insulator. The splitting pattern should be compared with the molecular one of Figure 2a. Figure 10b shows another view of the distortion where we plot the energy of the orbitals against k, the wavevector. This emphasizes that it is the levels around the Fermi level (the highest filled level of the solid for the cases here) which are most strongly affected by the geometrical change. Thus removal of electron density from this band by doping the material with electron acceptors will tend to decrease the electronic driving force for such a distortion. Analogously, addition of electron density will do the same. Figure 11a also shows a description of the change in the wave functions which describe the energy bands. Notice that the lower energy (and full) band of the two is now largely associated with the square planar Pt atom and the higher energy (and empty band) is associated with the six coordinate Pt atom. Thus we would describe

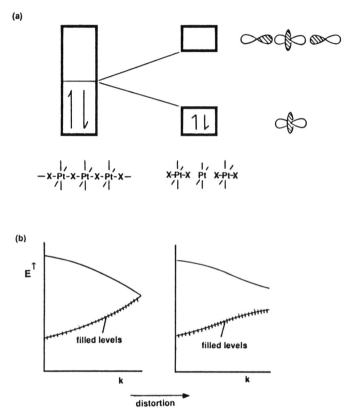

Figure 10: (a) Splitting apart of the half-filled band on the asymmetric distortion of Figure 8a. This is the higher of the two bands of Figure 9(a). Notice the change in orbital character as a result of a Peierls distortion for the platinum chain. (b) Plot of orbital energy vs k, showing the orbitals most strongly affected.

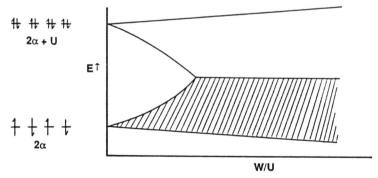

Figure 11: The Hubbard picture for the half-filled band. Energy vs the parameter W/U.

the distorted structure as a mixed valence one containing Pt^{II} and Pt^{IV} centers. In the analogous copper oxide case $Cu^{II} \to Cu^{III} + Cu^{I}$. Clearly from Figure 10a there is a one-electron stabilization for the process, but the distortion puts two electrons onto the same atom ($Pt^{III} \to Pt^{II}$) during the localization. A large Coulombic repulsion (the two-electron energy term U) should result. Whether the distortion proceeds or not is a balance between the two.

Before proceeding further we should think a little more about the meaning of this Coulomb U which we have just associated with two electrons being localized on the same atom (Pt^{II}). In fact of course the electrons are paired in the diamagnetic metallic state too, so where does this U actually come from? A large contribution to its magnitude comes from the changes in the ionization potentials of the atom with oxidation state. In general there seems to be a roughly quadratic dependence of ionization potential on ion charge, the more highly ionized the atom, the more tightly bound the remaining valence electrons and the higher the next ionization potential. With such a quadratic dependence of ionization energy, the process $M^{+n} \to M^{+(n-1)} + M^{+(n+1)}$ will not be favored. Thus the ingredients that go into this two-electron energy term may well be rather complex.

Since electron-electron interactions appear to be less energetically important for heavier transition metals it is then not too much of a surprise to discover that whereas for these platinum halides a distorted, alternating structure is found, the analogous structure is not found for the nickel analog of the one dimensional chain (11). In terms of higher dimensional problems a distorted structure of this type is found for the (three-dimensional) perovskite-like structure (Figure 8b) of $CsAuCl_3$, but not for the (two dimensional) perovskite-like copper oxide case. The observation is related to the molecular one described earlier. Pt^{II} and Pd^{II} are never octahedral, but square planar (i.e., 'distorted' in the present language) and diamagnetic. Ni^{II} complexes are often octahedral (i.e., undistorted) and high spin. Similar considerations lead us to expect that the copper oxide system will not distort in a similar fashion. Indeed it does not. There are however, two mechanisms for stabilizing the undistorted structure. One we study in the next section, the other is the generation of a ferromagnetic insulator.

Up until now, in this section, we have placed all of the electrons, paired, in the lower half of the energy band. There is, of course, another way to arrange them. This is the ferromagnetic state, where each level of the band contains a single unpaired electron. Since there are no free states open to the electrons with this configuration the system is an insulator. This arrangement is clearly the solid-state analog of the molecular triplet state, and similar energetic condiderations should apply to those discussed above concerning their relative stabilities. If the energy band is centered at an energy α, and has a width W (analogous to Δ in the transition metal case) then the one-electron energy of the half-full band is $\alpha - W/\pi$ for the diamagnetic metal, (W > 0) and simply α for the ferromagnetic insulator. Just as in the molecular case the two-electron contribution to the energy is zero for the ferromagnet but is equal to U/4 for the metal. Thus the condition for stability of the metal is just $W/\pi > U/4$ or $W/U > \pi/4$, a result very similar to that derived for the molecular cases, both those of the transition metal complex and for the hydrogen molecule. As we noted specifically for the transition metal example, the model contains many simplifications. We should include in the solid, not only the on-site repulsions, U, but the inter-site repulsions, V, and also the inter-site exchange, J. Inclusion of these terms leads (12) to the more complex result $W/\pi > U/4 - V + 3J/2$. One of the very difficult problems in this area, which should be apparent already, is the numerical estimation of terms such as these.

5.0 ANTIFERROMAGNETIC INSULATORS

So far we have concentrated on an electronic description of these systems in which we have used a simple prescription of one- and two-electron energy terms, much akin to the discussion of the high and low spin complexes of Ni^{+2}. One important topic yet to be discussed is the importance of electron correlation. We can use in a straightforward manner the ideas developed around this topic for the hydrogen molecule. Imagine a solid composed of weakly interacting atoms such as that described toward the right-hand side of Figure 5. Here we know that the material will be dominated by on-site electron-electron repulsion U so that movement of an electron from one center to another will energetically be unfavorable. Clearly an insulator will result from this state of affairs. For the case where U is small then the diamagnetic conducting state is a possibility.

Hubbard (13) elucidated a mathematical description of the change from one situation to another for the simplest case of a half-filled s band of a solid. His result is shown in Figure 11. For ratios of W/U greater than the critical value of $2/\sqrt{3}$ then a Fermi surface should be found and the system can be a metal. This critical point is associated with the Mott transition from metal to insulator. At smaller values than this parameter, then, a correlation, or Hubbard, gap exists and the system is an antiferromagnetic insulator. Both the undoped 2-1-4 compound and the nickel analog of the one dimensional platinum chain are systems of this type. At the far left-hand side of Figure 11 we show pictorially the orbital occupancy of the upper and lower Hubbard bands.

The antiferromagnetic state described by the occupation of the lower Hubbard band is stabilized by inclusion of such electron correlation, but the ferromagnetic analog is not. This is a result exactly analogous to the stabilization of the lowest singlet state in cyclobutadiene below the triplet. For the simple density of states used by Hubbard in his treatment he showed in fact that the condition for ferromagnetism was

$$U/\sqrt{((2U + W)^2 - 4nWU)} > 1 \qquad (17)$$

an impossibility since the parameter n <1. This is not a general result since we know indeed that ferromagnets do exist. In addition the treatment does not include the inter-site coulomb and exchange terms described in the previous section.

There is another way of generating an antiferromagnetic insulator, and that is within the framework of Hartree-Fock theory (i.e., without correlation). Slater suggested (14) that if unrestricted Hartree-Fock theory were used then an electronic situation of this type could arise too. In orbital terms his formulation also involves the mixing in of higher energy orbitals of the band into the ground state. In fact the state which gets mixed in is the solid state equivalent of the singlet H_2 function given by Equation 8. Figure 12 shows how the metallic state slowly is transformed to the antiferromagnetic insulating state as a result of such mixing. We now have to use separate boxes for spin up and spin down electrons. The gap between occupied and unoccupied levels arises purely as a result of electron-electron repulsions. This behavior is sometimes called a Spin Density Wave. At the

758 Chemistry of Superconductor Materials

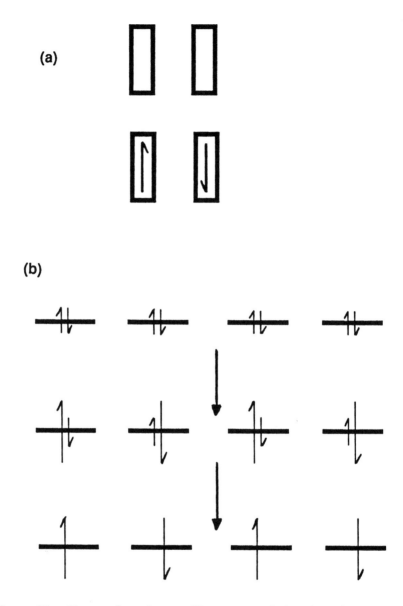

Figure 12: Generation of an antiferromagnetic insulator in the Slater approach. (a) The description of the insulator in terms of occupation of spin up and spin down bands and (b) how the up and down spins on adjacent atoms change as the on-site repulsion is gradually increased. When the on-site repulsion is infinite two electrons are not located on the same atom. Compare this diagram with the bottom left-hand picture of Figure 11.

'infinite U' limit then each electron must be localized on separate atoms (the very bottom of Figure 12b) but for small U then the electron is not so localized. The two approaches to generation of the antiferromagnetic state have similar, but not identical origins. It is not clear at the time of writing this review whether the antiferromagnetic insulator found for these high-T_c oxides arises via a spin density wave or via correlation. In Table 1 we summarize the electronic requirements for these half-filled band situations in terms of four parameters, the change in the one-electron energy, and the contributions from U and V, the on-site and inter-site Coulomb repulsions respectively, and J, the intersite exchange repulsions (U > V > J). These are expressions evaluated by Whangbo (12) for the simplest case of a half-filled one-dimensional band. Relative to the metallic state the ferromagnetic insulator loses all of the one-electron stabilization afforded the diamagnetic half-filled band (W/π) but gains in electron-electron repulsion. The Peierls distortion gains in one-electron energy, Δ_{dist} (< 0) but loses in terms of electron-electron repulsion. (δ is a system-dependent weight and takes the value of 0 for a metal and 1/2 for an antiferromagnetic insulator.) The antiferromagnetic insulator in the Slater scheme loses one-electron energy (Δ_{he}) but makes large gains in relief of Coulomb U. Relief of coulomb repulsion is a feature of the Hubbard approach too. The details of the sizes of the parameters involved will determine which state lies lowest in energy. Examples of each are known, and we refer the reader to an interesting article (15). Electron correlation is important in other ways since it influences the details of the Fermi surface.

6.0 ELECTRONIC STRUCTURE OF COPPER OXIDE SUPERCONDUCTORS

In what follows we should bear in mind that the generation of a diamagnetic metallic state (irrespective of whether it is a superconductor or not) will not be favored by a half-filled band of electrons. Either a Peierls distortion or the generation of an antiferromagnetic insulating state will result, with a ferromagnet being less likely for the reasons discussed. Superconductivity in these materials is in fact only observed if electrons are removed, or (less commonly to date) added to the half-filled band. Considerable effort is underway to theoreti-

Table 1. Energetics of Half-Filled Band Possibilities (12)

System	Stability	Stability relative to Metallic State
Metallic	$2\alpha - W/\pi + (U+4V-2J)/4$	0
Ferromagnetic	$2\alpha + 2(V-J)$	$W/\pi - (U-4V+6J)/4$
Peierls Distortion	$2\alpha - W/\pi + (U+4V-2J)/4 + \Delta_{dist} + \delta^2(U-4V-2J)/4$	$\Delta_{dist} + \delta^2(U-4V+2J)$
Antiferromagnetic	$2\alpha - W/\pi + (U+4V-2J)/4 - \Delta_{he} - \delta^2(U-2J)$	$-\Delta_{he} - \delta^2(U-2J)$

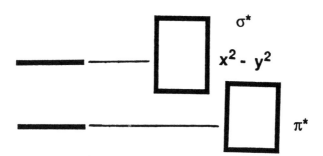

Figure 13: Energy bands for a CuO_2 plane showing the important σ^* and π^* bands. The band derived from z^2 varies in energy depending upon the location of the fifth and sixth oxygen atoms around copper, and is not shown.

cally describe this very complex electronic situation, some using variants of the ideas described here and others quite different. We shall focus here on how the structure of the solid is controlled by the electron occupancy of the energy bands in terms of the crystal orbitals themselves, but bearing in mind the complications to this picture provided by these many-body terms.

We have already described the generation of the x^2-y^2 band of the CuO_2 sheet, of universal occurrence in all high T_c copper oxides. Figure 13 shows the derivation of the other metal bands (16) of the system. The local CuO_6 (d^9) unit is Jahn-Teller unstable and leads to a local geometry containing four short (in-plane) and two long (out-of-plane) Cu-O distances. The d orbitals of such a local structure split apart in energy in a very characteristic way. We use the angular overlap method (3) to list the energies of several fragments in Figure 14. Recall that the angular overlap parameter e_σ increases with decreasing metal-ligand distance since this represents an increase in metal-ligand interaction. Thus systems with short Cu-O distances will have energy bands which are wider than those with longer ones. The drop in energy of the z^2 orbital as ligands are removed is quite apparent, and in all geometries where an ML_4 plane of atoms is preserved the highest orbital is the $x^2 - y^2$ one. Since the structures of these copper-based superconductors are derived from the perovskite structure, which contains linear chains of atoms ...O-Cu-O-Cu-O... etc, linked into a three dimensional network, a reasonable approximation is to study the interactions of the metal atom orbitals with the ligands in just the same way as in a molecular complex. Thus, because of the geometry the 't_{2g}' set of metal orbitals may only be involved in π bonding with the oxygen atom, and the $x^2 - y^2$ and z^2 orbitals are exclusively involved in σ-bonding. This simplifies our discussion immensely. We shall show later an important role of the π orbitals in these systems. The large lobes of the z^2 orbital point along the z direction where, depending upon the structure, there are either no atoms or one or two atoms with long Cu-O distances. The highest occupied orbital for Cu^{II} is therefore one which is almost completely $x^2 - y^2$ strongly mixed with some oxygen character. It is of interest to ask at this point how configuration interaction, demonstrably of importance at x = 0 in the generation of an antiferromagnetic insulator, changes this picture. Notice that the bottom of the '$x^2 - y^2$' band (Figure 9b) is purely d in character by symmetry, but increas-

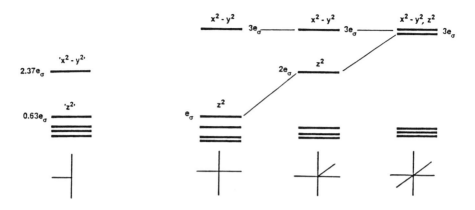

Figure 14: Energies of σ metal d-orbitals as a function of geometry using the parameters of the angular overlap model.

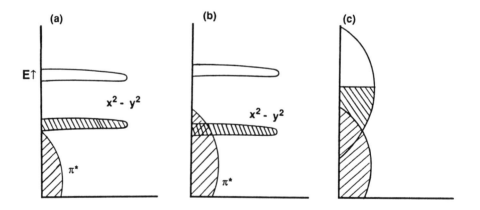

Figure 15: (a) One-band model where electrons are removed from the lower Hubbard band on doping. (b) One-band model where electrons are removed from the π^* band of the material on doping. (c) Similar to (b) but where the system is a normal metal described by $(W/U) > (W/U)_{crit}$ of Figure 11.

ingly heavy mixing occurs as the energy increases. Thus not only does the size of the antibonding interaction increase, but also the amount of oxygen character in the wavefunctions. Configuration interaction, which mixes in states derived from occupation of these higher energy orbitals will thus be expected to increase the oxygen character of the ground state, a result confirmed by numerical calculation. Compare this result with our earlier one for the effect of configuration interaction in the magnesium atom. An analogous picture would result if the antiferromagnetic state was of the Slater type.

From Figure 13 it is clear that the next lowest energy band of the solid will be one which contains the metal-oxygen π type orbitals. π-type interactions in these materials can be large and produce a band which lies lower in energy than σ, as confirmed (17) by good band structure calculations on these materials. Is it possible to end up with a situation where electrons are removed from this penultimate energy band rather than from the $x^2 - y^2$ band which has received most of our attention? This is not a trivial question and in studying this possibility reveals some of the competing elements in the electronic structure problem. Figure 15 shows three ways to put these ideas together for these compounds where we have drawn a level filling appropriate to a half-filled band. In Figure 15a is shown the result for the case where the correlation splitting of the $x^2 - y^2$ band is large enough to produce an insulator, but not large enough so that the lower Hubbard band is depressed below the level of the π band. This model then suggests that when removal of electron density from the system occurs, it takes place from $x^2 - y^2$. In Figure 15b is shown the result for the case where the lower Hubbard band is indeed depressed below the level of the π band. This model would suggest that when removal of electron density from the system occurs, it takes place from the π-type orbitals. (We could easily envisage a third case where density was removed from both bands - a two-band model.) Which is the correct state of affairs is, once again, a balance between the one- and two-electron terms in the energy. Figure 15c shows the case where the ratio of W/U is such that a diamagnetic metal is formed. We will in fact adopt the model of Figure 15a or 15c, and provide below some geometrical evidence for its correctness. In addition, experiments in the 2-1-4 series where electrons have been *added* to the d^9 situation, show that these species often behave in a similar way to those where electrons have been removed. We might expect rather different behavior if in one case electrons were removed from a delocalized π-band but added to a partially localized σ-band.

The picture which we will use for the 2-1-4 compound is thus a simple one (18). The x^2-y^2 band (antibonding between Cu and O) is approximately half full. Doping either with a 2-electron or 4-electron donor (Sr^{+2} or Ce^{+4}) results in the generation of a band less than and greater than half full respectively. The Cu-O distance changes on doping in agreement with our model which predicts less overall bonding the more electrons in the antibonding x^2-y^2 band. The bond lengths increase in the order Sr^{+2} doping (less than half-full) < no doping (half full) < Ce^{+4} doping (greater than half full). The results are basically the same as we would expect from a system with and without the inclusion of configuration interaction, since this process involves the admixture of Cu-O antibonding orbitals too. Note that in qualitative terms the bond lengthening will be the same whether it is the σ^* or π^* band which is variably occupied here.

The 1-2-3 compound ($YBa_2Cu_3O_{7-\delta}$) with $\delta = 0$ contains puckered CuO_2 sheets which sandwich CuO_3 chains of atoms. Thus the metal atoms of the sheets are in square pyramidal coordination with a long axial bond (a memory of the Jahn-Teller instability of the octahedral structure) and the metal atoms of the chains in square planar coordination. Because the Cu-O distances in the planes are significantly shorter than in the sheets, as Figure 16a shows, the x^2-y^2 bands of the two units (actually the z^2-y^2 band of the copper atom in the chain because of its geometrical orientation) lie at different energies. Recall that the width of the band (4β on our simple model above) is controlled by the size of the interaction integrals along the chain direction, but the location of the band will be fixed by the sum total of all four interactions of the coordinated oxygen atoms with the z^2-y^2 orbital. In fact the two bands do overlap considerably such that there is electron transfer at $\delta = 0$ from planes to chains. This leads to an increase in positive charge on the Cu^{II} plane atoms and a decrease in positive charge on the Cu^{III} chain atoms. The electron transfer here is very important since it is an internal doping mechanism to move the electron filling away from half full.

The electronic properties should depend crucially on the details of the local geometry on this model, via the relative location of the two bands concerned. We see this in the dependence of T_c on δ, which measures the oxygen stoichiometry in the 1-2-3 compound, shown in

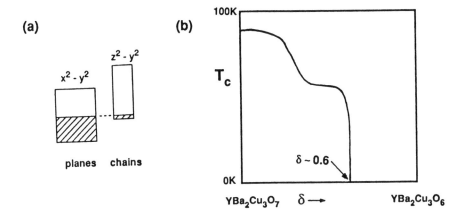

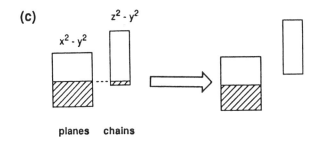

Figure 16: (a) Location of the $x^2 - y^2$ (planes) and $z^2 - y^2$ (chains) bands in the 1-2-3 superconductor showing plane-chain electron transfer and the removal of the half-filled $x^2 - y^2$ band. (b) Variation in T_c with oxygen atom stoichiometry showing the dramatic plunge at around $\delta = 0.6$. (c) Location of the $x^2 - y^2$ (planes) and $z^2 - y^2$ (chains) bands in the 1-2-3 system for $\delta > 0.6$ showing how plane-chain electron transfer is switched off and the half-filled $x^2 - y^2$ band is generated.

Figure 16b. Here we will only discuss one aspect of the interesting variation in structure with this parameter and that is one which may be directly related to the precipitous drop in T_c at $\delta \sim 0.6$. Loss of oxygen from the $\delta=0$ structure occurs from the chains and generates T-shape and dumbbell geometries in place of the square planar units. (The electronic effect of this is discussed elsewhere (19).) As seen in Figure 14 the highest levels of such units lie below those of the square planar levels. By the time $\delta = 1$ all of the empty (z^2-y^2) square planar levels have disappeared, the x^2-y^2 band is half full and the material should not be metallic. Indeed $YBa_2Cu_3O_6$ is a semiconductor. However superconductivity has disappeared by $\delta \sim 0.6$ when there will still be some x^2-y^2 levels available for this internal doping. It turns out that at around this stoichiometry the short Cu-O distances perpendicular to the chain (1.85Å) become even shorter (1.80Å) (20). This structural change has been called (21) 'the c-axis anomaly'. As shown in Figure 16c the result is to push the z^2-y^2 band to higher energy and thus to switch off the charge transfer. Now the half-filling of the x^2-y^2 band is restored and superconductivity has disappeared. The material becomes an antiferromagnetic insulator. Thus avoidance of the half-filled band is essential for the generation of a metal, and we can understand the present observations using some quite conventional chemical arguments. Consensus as to why such systems become high-T_c superconductors has not yet been achieved. In this article we will not even speculate.

7.0 THE ORTHORHOMBIC-TETRAGONAL TRANSITION IN 2-1-4

The focus of the previous section was on the interesting control of the electronic properties by the geometry. By concentrating on the orbital characteristics of the system, but bearing in mind the important effects of the two-electron energy terms, we were able to produce a working model of the system. Here we look at the factors controlling the energetics of the tetragonal and orthorhombic forms of the 2-1-4 compound ($La_{2-x}Sr_xCuO_4$) which has been the subject of much discussion. Is it a Peierls distortion or is there something unusual about it? We shall tie the results to some details concerning the local geometry in the 1-2-3 compound ($YBa_2Cu_3O_{7-\delta}$). Here there are puckered CuO_2 sheets containing a five coordinate, square pyramidal copper with an axial-basal angle (ξ) greater than 90°.

The structures of the two forms of the 2-1-4 compound are shown in Figure 17a,b. The sheets are flat in the tetragonal and rumpled in the orthorhombic variant. It is now well established that as x increases from zero the energy difference between the tetragonal form and the more stable orthorhombic form decreases (5). The material ceases to become a superconductor once x is larger than ~0.2, when the tetragonal form is now the only arrangement found, but we will not be concerned with this (important) observation. Increasing x in $La_{2-x}Sr_xCuO_4$ leads to a depletion of the highest energy band of the system, which we have described as x^2-y^2. A calculated energy difference curve (22) between the two structures using band theory is shown in Figure 18. It is compatible with experiment, in that the tetragonal structure is increasingly favored with decreasing d count. Notice the crossover in relative stability around d^9 where the x^2-y^2 band is half-full. Elsewhere (23) we discuss the shape of curves of this type and how they can be indicators of distortions of the Peierls type. The shape of this curve tells us definitely that this distortion is not of the Peierls type where a gap is opened at the Fermi level on distortion. Studies (24) by Whangbo and coworkers have also shown that this is not a Peierls distortion not only on symmetry grounds, but also by calculation. No gap is opened.

Very interestingly a similar curve is found for the variation with d-count of the magnitude of the puckering distortion of the five-coordinate arrangement of the $YBa_2Cu_3O_7$ compound in Figure 17c. For d^6 systems $\xi = 90°$, but for d^9 systems $\xi > 90°$. Thus increasing d-count leads to a more pyramidal geometry. There is no experimental evidence to test this out since systematically doped systems have not been studied, but the calculated plot is exactly what would be expected from the well-established variation in molecular geometry with d-count for molecular five-coordinate transition metal systems (25). In all three cases, one molecular and two from the solid-state, it is the stabilization associated with the x^2-y^2 band or orbital (Figure 19), during the distortion which shifts the energy minimum of the structure away from those geometries with 90° and 180° O-Cu-O angles. The driving force is the relief of strong antibonding interactions between metal x^2-y^2 and ligand σ orbitals. The destabilization of the occupied levels favors the tetragonal

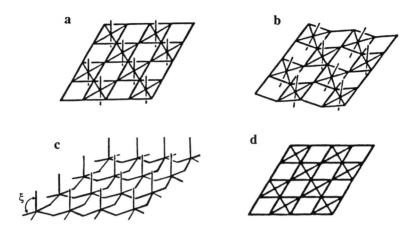

Figure 17: Schematic structures of (a) tetragonal and (b) orthorhombic forms of the 2-1-4 compound, (c) the 1-2-3 compound, and (d) a single CuO_2 plane.

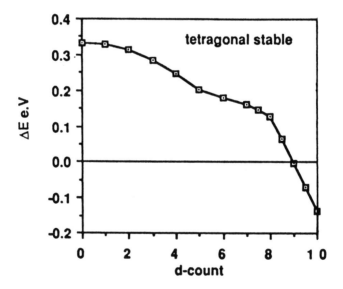

Figure 18: Calculated energy difference curve between the tetragonal and orthorhombic forms of the 2-1-4 compound as a function of d-count. Most of the plot is chemically meaningless, but it is the part around the d^9 count which is of most interest.

Electronic Structure of Copper Oxide-Based Superconductors   769

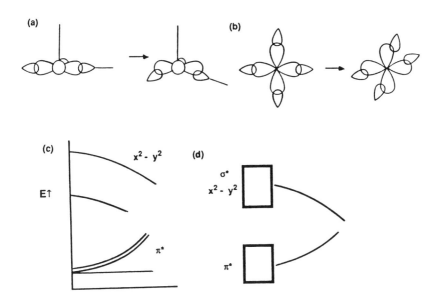

Figure 19: Stabilization of $x^2 - y^2$ levels by loss of overlap in (a) puckering of 1-2-3, (b) the tetragonal to orthorhombic distortion in 2-1-4. The geometrical change has been exaggerated for effect. (c) Change in energy associated with increase of ξ in the square pyramid. (d) Change in energy of the σ^* and π^* bands during a distortion away from those structures with O-Cu-O angles of 90° and 180°.

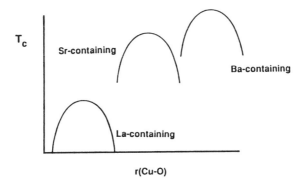

Figure 20: Schematic variation in T_c with copper-oxygen distance in three series of superconductors. (Adapted from reference 28).

structure as shown in Figure 19. The occupation of the x^2-y^2 band is reduced, as x increases from zero in the 2-1-4 compound, and the contribution to the stabilization energy on distortion from this source decreases. It leads eventually to the tetragonal structure becoming the lowest energy alternative. Notice that if it was depletion of the π band which was occurring here (one possible model noted above) then the geometrical distortion would have gone in the opposite direction with d-count.

We believe then that the puckering distortion of the 1-2-3 compound and the orthorhombic distortion of the 2-1-4 compound arise via the same mechanism, controlled by the population of the x^2-y^2 band. The difference in the copper coordination number distinguishes the two. A puckering distortion is less appropriate in the six-coordinate 2-1-4 system than for the five-coordinate 1-2-3 one, simply because there are oxygen atoms on both sides of the CuO_2 plane in the six-coordinate case. Thus a more complex distortion is found for the 2-1-4 compound. Electronic arguments are probably not the whole story here however. The orthorhombic to tetragonal transition in the 2-1-4 compound, and structural details in other high-T_c systems, have been shown (26)(27) to be strongly influenced by interactions, either left out of the tight-binding calculation completely or not well modeled by it. These include ionic interactions between the copper oxide lattice and the cations, and the mutual repulsions of the oxygen atoms. As in almost all chemical problems of this type the overall picture is a mix of the two types of chemical forces.

8.0 SUPERCONDUCTIVITY AND SUDDEN ELECTRON TRANSFER

At this point it is pertinent to point out some general features of the electronic structure of these superconductors which may have a crucial bearing on their electronic properties. Figure 20 shows schematically (28) how T_c varies with Cu-O distance in three different series of superconductors. A common feature is a rise to a maximum and then a fall. For the 2-1-4 series, since increasing x is associated with a shortening of the Cu-O distance, a similar plot is seen for T_c versus x. Now as the Cu-O distance decreases, we have noted that the band width W increases. Certainly too as this distance decreases, the size of the copper-oxygen interaction increases so that the electrons may be more equally shared between copper and oxygen. In molecular orbital language the metal and oxygen coefficients will become more nearly equal. A natural result of this is a reduction in the effective value of U. Thus shortening the Cu-O distance leads to a larger value of W/U and the system moves from left to right across the diagram of Figure 11. Is there a critical point on this picture where the electronic description rapidly changes? We can make a molecular analogy at this stage with the plot of Figure 21 which shows the electronic situation for the formation of a NaCl molecule from its atomic constituents. The molecule itself is best described by writing it as Na^+Cl^-, whereas the isolated atoms form a lower energy state than the separated ions. In the language of the present chapter the large Coulomb U associated with the last two electrons in Cl^-, relative to Cl·, destabilize the atom pair. There is thus a critical region where there is an avoided crossing of the two curves. At this point Mulliken described (29) the electronic situation of a 'sudden electron transfer' as the system moves through this region. Here the electron is rapidly transferred from sodium to chlorine. A similar description should apply to the solid, and a vital question (and one which we will not try to answer) is whether superconductivity is directly coupled to this electronic situation by a vibration which sweeps the system through this region. It would not necessarily be a soft vibration of the type prevalent in the lower temperature superconductors. The Cu-O distance is fixed at a given distance by the x^2-y^2 electron density and the sizes of the electropositive cations as described above. The Cu and O character in the orbitals is controlled by the Cu-O distance and also perhaps by the electronegativity of the other atoms in the structure. A qualitative argument would suggest that the more electronegative these atoms, the larger their involvement in the Cu/O framework and the smaller the electron density on the copper atoms, a result directly controlling the value of U.

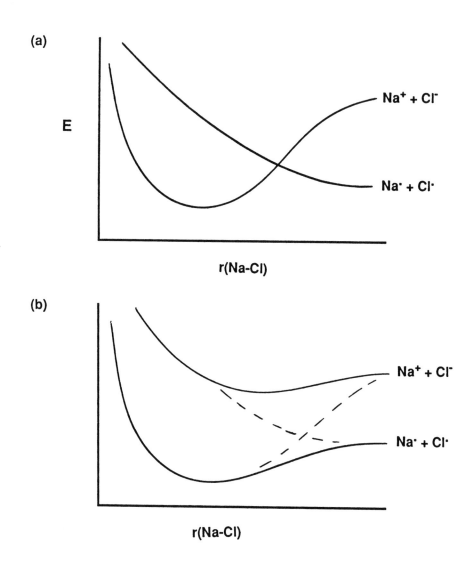

Figure 21: Sudden electron transfer in sodium chloride. Curves before (a) and after (b) the avoided crossing.

9.0 SOME CONCLUSIONS

A conventional electronic picture has allowed us to understand some fundamental structural aspects of these materials. By recognizing the importance of the balance between the one- and two-electron terms in the energy we have been able to make analogies between these solid-state systems and ones from other areas of chemistry to provide insights into their behavior. The philosophy has been a clear one. Although the many-body terms in the energy are clearly important, and crucially control the electronic behavior, we have been able to understand much of what is going on by looking at the one-electron terms. We have, for example, shown how the bond lengths in the 2-1-4 compound respond to a change in the x^2-y^2 electron density in accord with the Cu-O antibonding nature of this band. The orthorhombic-tetragonal phase transformation in this compound is driven by a local geometric distortion which we have described not only in terms of the Cu-O antibonding nature of the band, provided structural evidence that the important band is of σ^* and not π^* symmetry. We predict similar control of the puckering distortion of the 1-2-3 compound but there is no experimental data to test the hypothesis. Most strikingly perhaps the structural change at $\delta \sim 0.6$ gives an explanation of the loss of superconductivity using a simple orbital argument. Thus although unconventional models have been suggested for the mechanism of superconductivity several structural features of these materials are explicable using conventional orbital ideas. The approach presented here has been used to make predictions concerning other systems. In terms of the similarities between the platinum chain compounds and the chemistry described here, might it be possible to make a superconductor by doping the recently made (antiferromagnetically insulating) nickel analog (11) of these systems?

The discussion of the stability of various electronic situations in these molecules and solids has focussed on the balance of one- and two-electron forces. One of the big problems of course is to actually calculate the magnitude of these, and then to decide how the structure of these materials determines the details of the overall electronic structure. In addition there is the other problem, which we mentioned in connection with the molecular Hückel model, of the frequent ambiguity in what the parameters of these models actually mean. The factors behind the Coulomb U associated with the Peierls distortion for example are a complex blend of electron-electron and electron-nuclear forces. Finally we stress once again how the structural details are of crucial importance in controlling the electronic properties.

ACKNOWLEDGEMENTS

This research has been supported by The University of Chicago and by the National Science Foundation in a grant to The Science and Technology Center for High-Temperature Superconductivity. I would like to thank John Goodenough, Raj Kulkarni, Kathryn Levin and John Mitchell for many useful conversations. M-H. Whangbo read the manuscript.

10.0 REFERENCES

1. Griffith, J.S., *Theory of Transition Metal Ions*, Cambridge University Press, London (1961).
2. Orgel, L.E., *An Introduction to Transition Metal Chemistry: Ligand Field Theory*, Wiley, NY (1960).
3. Burdett, J.K., *Molecular Shapes*, Wiley, NY (1980).
4. Jørgensen, C.K., *Absorption Spectra and Bonding in Complexes*, Pergamon, London (1961).
5. Borden, W.T., Davidson, E.R., *Ann. Rev. Phys. Chem.* 30:125 ((1979).
6. Jafri, J.A., Newton, M.D., *J. Amer. Chem. Soc.*, 100:5012 (1978).
7. Balcher, A., *Phys. Rev.*, 56:385 (1939).
8. Ashcroft, N.W., Mermin, N.D., *Solid State Physics*, Saunders, Philadelphia (1976).
9. Goodenough, J.B., *Magnetism and the Chemical Bond*, Wiley, NY (1963).
10. Burdett, J.K., *Progress in Solid State Chemistry*, 15:173 (1984).
11. Toriumi, T, Wada, Y, Mitani, T., Bandow, S., *J. Amer. Chem. Soc.*, 111:2341 (1989).

12. Whangbo, M-H., *J. Chem. Phys.*, 73:3854 (1980), 75:4983 (1985).
13. Hubbard, J., *Proc. Roy. Soc.*, A276:238 (1963), A277:237 (1964), A281:401 (1964).
14. Slater, J.C., *Phys. Rev.*, 82:538 (1951).
15. Adler, D., *Rev. Mod. Phys.*, 40:714 (1968).
16. Burdett, J.K., Kulkarni, G.V., in *Inorganic Compounds With Unusual Properties*, ed. M.K. Johnson, R.B. King, P.M. Kurtz, C. Kutal, M.L. Norton and R.A. Scott, American Chemical Society (1989).
17. Freeman, A.J., Massidda, S., Yu, J., in *Chemistry of High-Temperature Superconductors II*, eds Nelson, D.L., George, T.F. ACS Syposium Series # 377, (1988).
18. Burdett, J.K., Kulkarni, G.V., *Phys. Rev.* (B) 40:8908 (1989).
19. Burdett, J.K., Kulkarni, G.V., Levin, K., *Inorg. Chem.* 26:3650 (1987).
20. Renault, A., Burdett, J.K., Pouget, J-P., *J. Solid State Chem.* 71:587 (1987).
21. Cava, R.J., Batlogg, B., Rabe, K.M., Reitman, E.A., Gallagher, P.K., Rupp, L.W., *Physica*, C156:523 (1988).
22. Burdett, J.K., Kulkarni, G.V., *Chem Phys. Lett.* 160:350 (1989).
23. Burdett, J.K., Accts. *Chem. Res.*, 21:189 (1988).
24. Whangbo, M-H., Evain, M., Beno, M.A., Williams, J.M., *Inorg. Chem.*, 26:1829 (1987).
25. Rossi, A.R., Hoffmann, R., *Inorg. Chem.* 14:365 (1975).
26. Baetzold, R.C., *Phys. Rev.*, B38:304 (1988).
27. Whangbo, M-H., Evain, M., Beno, M.A., Geiser, U., Williams, J.M., *Inorg. Chem.*, 27:467 (1988).
28. Whangbo, M-H., Kang, D.B., Torardi, C.C., *Physica* C158:371 (1989).
29. Mulliken, R.S., *J. Chim. Phys. Phys.-Chim.Biol*, 61:20 (1964).

Appendix A
Guide to Synthetic Procedures

The reader may consult this compilation to quickly find procedures, or references thereto, for preparation of the various high-temperature superconducting oxide systems. The information provided here is intended to serve as a "starting point" for those seeking to prepare these fascinating compounds. Those seeking modifications of these procedures and/or more up-to-date protocols will find it useful to cite the references given here in the *Science Citation Index* and/or to consult *Chemical Abstracts*.

This guide is organized according to the **ideal** chemical formula of the desired phase. The reader will be directed to the section(s) of this text containing pertinent synthetic information. In many cases, additional references to the literature that contain synthetic details have been included. Most of the information provided pertains to the synthesis of bulk polycrystalline and single-crystal samples; the preparation of high-quality superconducting thin films has been comprehensively treated in other books and only incidental information is included here.

Those unfamiliar with preparative chemistry will find Chapter 5 to be especially useful as a general review of basic techniques in solid state synthesis and crystal growth. Other general information is found in Chapter 6, (section 1.0) and Chapter 8, (section 2.0). Appendix B lists several texts dealing with synthetic solid state chemistry, and a short review of superconductor sample preparation is given in Chapter 5 of the book by Poole, Datta, and Farach.

Several special issues of the *Journal of Crystal Growth* have appeared that focus on high-T_c superconductors and contain a wealth of procedural information:

Crystal Growth 1989, Proceedings of the Ninth International Conference on Crystal Growth (Sendai, Japan, 22-25 August 1989), J. Chikawa, J.B. Mullin, and J. Woods., Eds., *J. Crystal Growth* 99(1-4):915-975 (1990).

High-T_C Superconductors: Materials Aspects II, D. Elwell, M. Schieber, and L. Schneemeyer, Eds., *J. Crystal Growth* Special Issue 91(3):249-469 (1988).

High-T_C Superconductors: Materials Aspects, D. Elwell, M. Schieber, and L. Schneemeyer, Eds., *J. Crystal Growth* Special Issue 85(4):563-665 (1987).

The procedures that follow are for bulk polycrystalline samples unless indicated otherwise.

$La_{2-x}Sr_xCuO_4$

1. Crystals: Chapter 5, section 3.3 ("$La_{2-x}AE_xCuO_4$...")

2. R.J. Cava, R.B. van Dover, B. Batlogg, and E.A. Rietman, Bulk Superconductivity at 36 K in $La_{1.8}Sr_{0.2}CuO_4$. *Phys. Rev. Lett.* 58(4):408 (1987).

3. J.M. Tarascon, L.H. Greene, W.R. McKinnon, G.W. Hull, and T.H. Geballe, Superconductivity at 40 K in the Oxygen-Defect Perovskites $La_{2-x}Sr_xCuO_4$. *Science* 235:1373 (1987).

4. T.C. Huang, J.B. Torrance, A.I. Nazzal, and Y. Tokura, A Study of the Superconducting $La_{2-x}Sr_xCuO_4$ System by X-ray Powder Diffraction. *Powder Diffraction* 4(3):152 (1989).

$YBa_2Cu_3O_7$

1. Chapter 5, references 15-17

2. Chapter 7, section 2.0

3. Crystals: Chapter 5, section 3.2 ("Overview of Flux Growth"), section 3.3 ("$Ba_2YCu_3O_7$")

4. Thick films: Chapter 7, section 2.4 ("Films")

5. R.J. Cava, B. Batlogg, R.B. van Dover, D.W. Murphy, S. Sunshine, T. Siegrist, J.P. Remeika, E.A. Rietman, S. Zahurak, and G.P. Espinosa, Bulk Superconductivity at 91 K in Single-Phase Oxygen Deficient Perovskite $Ba_2YCu_3O_{9-x}$. *Phys. Rev. Lett.* 58:1676 (1987).

6. W. Sadowski and H.J. Scheel, Reproducible Growth of Large Free Crystals of $YBa_2Cu_3O_{7-x}$. *J. Less-Common Metals* 150:219 (1989).

7. D.W. Murphy, D.W. Johnson, Jr., S. Jin, and R.E. Howard, Processing techniques for the 93 K superconductor $Ba_2YCu_3O_7$. *Science* 241:922 (1988).

8. G.F. Holland and A.M. Stacy, Physical properties of the quaternary oxide superconductor $YBa_2Cu_3O_7$. *Acc. Chem. Res.* 21:8 (1988).

9. $RBa_2Cu_3O_7$, R = rare earth: J.M. Tarascon, W.R. McKinnon, L.H. Greene, G.W. Hull, and E.M. Vogel, Oxygen and Rare-Earth Doping of the 90-K Superconducting Perovskite $YBa_2Cu_3O_{7-x}$. *Phys. Rev.* B 36(1):226 (1987); L.F. Schneemeyer, J.V. Waszczak, S.M. Zahorak, R.B. van Dover, and T. Siegrist, Superconductivity in Rare Earth Cuprate Perovskites. *Mater. Res. Bull.* 22:1467 (1987).

10. R.L. Meng, C. Kinalidis, Y.Y. Sun, L. Gao, Y.K. Tao, P.H. Hor, and C.W. Chu, Manufacture of Bulk Superconducting $YBa_2Cu_3O_{7-x}$ by a Continuous Process. *Nature* 345:326 (1990).

Appendix A: Guide to Synthetic Procedures 779

11. F. Holtzberg and C. Field, Crystal Growth and Characterization of the High Temperature Superconductor $YBa_2Cu_3O_{7-x}$. *Eur. J. Solid State Inorg. Chem.* 27(1/2):107 (1990).

$YBa_2Cu_4O_8$; $Y_2Ba_4Cu_7O_{15}$

1. J. Karpinski, E. Kaldis, E. Jilek, S. Rusiecki, and B. Bucher, Bulk Synthesis of the 81-K Superconductor $YBa_2Cu_4O_8$ at High Oxygen Pressure. *Nature* 336:660 (1988).

2. D.E. Morris, J.H. Nickel, J.Y.T. Wei, N.G. Asmar, J.S. Scott, U.M. Scheven, C.T. Hultgren, and A.G. Markelz, Eight New High-Temperature Superconductors with the 1:2:4 Structure. *Phys. Rev. B* 39(10):7347 (1989).

3. D.E. Morris, N.G. Asmar, J.Y.T. Wei, J.H. Nickel, R.L. Sid. and J.S. Scott, Synthesis and Properties of the 2:4:7 Superconductors $R_2Ba_4Cu_7O_{15-x}$ (R = Y, Eu, Gd, Ho, Er). *Phys. Rev. B* 40(16):11406 (1989).

4. D.M. Pooke, R.G. Buckley, M.R. Presland, and J.L. Tallon, Bulk Superconducting $Y_2Ba_4Cu_7O_{15-x}$ and $YBa_2Cu_4O_8$ Prepared in Oxygen at 1 Atm. *Phys. Rev. B* 41(10):6616 (1990).

5. U. Balachandran, M.E. Biznek, G.W. Thomas, B.W. Veal, and R.B. Poeppel, Synthesis of 80 K Superconducting $YBa_2Cu_4O_8$ Via a Novel Route. *Physica C* 165:335 (1990).

6. S. Jin, H.M. O'Bryan, P.K. Gallagher, T.H. Tiefel, R.J. Cava, R.A. Fastnacht, and G.W. Kammlott, Synthesis and Properties of the $YBa_2Cu_4O_8$ Superconductor. *Physica C* 165:415 (1990).

7. R.G. Buckley, J.L. Tallon, D.M. Pooke, and M.R. Presland, Calcium-Substituted Superconducting $RBa_2Cu_4O_8$ with T_c=90 K Prepared at One Atmosphere. *Physica C* 165:391 (1990).

8. E. Kaldis and J. Karpinski, Superconductors in the $Y_2Ba_4Cu_{6+n}O_{14+n}$ Family. *Eur. J. Solid State Inorg. Chem.* 27(1/2):143 (1990).

$Bi_2Sr_2Ca_{n-1}Cu_nO_{2n+4}$

1. In this formula, n denotes the number of CuO_2 layers in each defect-perovskite slab of the structure. The different phases are often referred to by numerical acronyms that are given in Chapter 6, section 1.0, Table 1.

2. General synthesis information is given in Chapter 6, section 2.0. Synthesis of the individual phases, including Pb-substituted compounds, is discussed and referenced thoroughly in this chapter in sections 3.1-3.3.

3. Crystals, n = 2: Chapter 5, section 3.1 ("Float Zone")

4. Crystals, n = 1,2,3: Chapter 5, section 3.3 ("Superconductors in the Bi-Sr-Ca-Cu-O System")

5. Crystals and powder: J.M. Tarascon, Y. Le Page, and W.R. McKinnon, Synthesis and Structural Aspects of Doped and Undoped High T_c Bismuth Cuprates. *Eur. J. Solid State Inorg. Chem.* 27(1/2):81 (1990).

6. Crystals and powder: C.C. Torardi, M.A. Subramanian, J. Gopalakrishnan, E.M. McCarron, J.C. Calabrese, K.J. Morrissey, T.R. Askew, R.B. Flippen, U. Chowdhry, and A.W. Sleight, Synthesis, Structure, and Properties of $A_2B_2Ca_{n-1}Cu_nO_{2n+4}$ Superconductors (A/B = Bi/Sr or Tl/Ba, and n = 1,2,3). In *High Temperature Superconductivity: The First Two Years*, R.M. Metzger, Ed., Gordon & Breach, NY (1989).

$Tl_2Ba_2Ca_{n-1}Cu_nO_{2n+4}$ and $TlA_2Ca_{n-1}Cu_nO_{2n+3}$; A = Ba, Sr

1. In these formulas, n denotes the number of CuO_2 layers in each defect-perovskite slab of the structure. The different phases are often referred to by numerical acronyms that are given in Chapter 6, section 1.0, Table 1.

Appendix A: Guide to Synthetic Procedures   781

2. General synthesis information is given in Chapter 6, section 2.0. Synthesis of the individual phases, including Pb-substituted compounds, is discussed and referenced thoroughly in this chapter in sections 4.1-4.3.

3. 2223 phase: Chapter 5, section 2.2 ("Precursor Techniques")

4. Crystals, general comments: Chapter 5, section 3.3 ("Superconductors in the Tl-Ba-Ca-Cu-O System")

5. Review article: M. Greenblatt, S. Li, L.E.H. McMills, and K.V. Ramanujachary, Chemistry and Superconductivity of Thallium-Based Cuprates, in *Studies of High-Temperature Superconductors*, A.V. Narliker, Ed., Nova Science Publishers, NY (1990).

6. Crystals and powder: C.C. Torardi, M.A. Subramanian, J. Gopalakrishnan, E.M. McCarron, J.C. Calabrese, K.J. Morrissey, T.R. Askew, R.B. Flippen, U. Chowdhry, and A.W. Sleight, Synthesis, Structure, and Properties of $A_2B_2Ca_{n-1}Cu_nO_{2n+4}$ Superconductors (A/B = Bi/Sr or Tl/Ba, and n = 1,2,3). In *High Temperature Superconductivity: The First Two Years*, R.M. Metzger, Ed., Gordon & Breach, NY (1989).

$Pb_2Sr_2Ln_{1-x}Ca_xCu_3O_8$ (Ln = Y, Rare Earth)

1. Crystals: Chapter 5, section 3.3 ("$Pb_2Sr_2M_{1-x}Ca_xCu_3O_8$")

2. Crystals and powder: R.J. Cava, B. Batlogg, J.J. Krajewski, L.W. Rupp, L.F. Schneemeyer, T. Siegrist, R.B. van Dover, P. Marsh, W.F. Peck, Jr., P.K. Gallagher, S.H. Glarum, J.H. Marshall, R.C. Farrow, J.V. Waszczak, R. Hull, and P. Trevor, Superconductivity Near 70 K in a New Family of Layered Copper Oxides. *Nature* 336:211 (1988).

3. L.F. Schneemeyer, R.J. Cava, A.C.W.P. James, P. Marsh, T. Siegrist, J.V. Waszczak, J.J. Krajewski, W.P. Peck, Jr., R.L. Opila, S.H. Glarum, J.H. Marshall, R. Hull, and J. M. Bonar, Crystal Growth and Substitutional Chemistry of $Pb_2Sr_2MCu_3O_8$. *Chem. Materials* 1:548 (1989).

4. M. Masuzawa, T. Noji, Y. Koike, and Y. Saito, Preparation of the High-T_c Superconductor $Pb_2Sr_2Y_{0.5}Ca_{0.5}Cu_3O_8$ with Zero Resistance at 75 K. *Japan. J. Appl. Phys.* 28(9):L1524 (1989).

5. J.S. Xue, M. Reedyk, Y.P. Lin, C.V. Stager, and J.E. Greedan, Synthesis, Characterization and Superconducting Properties of $Pb_2Sr_2(Y/Ca)Cu_3O_{9-x}$ Single Crystals. *Physica C* 166:29 (1990).

$Ba_{1-x}K_xBiO_3$

1. Chapter 5, section 2.2 ("Control of Oxygen Partial Pressure to Alter Constituent Volatility")

2. Crystals: Chapter 5, section 3.3 ("$Ba_{1-x}K_xBiO_3$")

3. Powder, Crystals, and Thin Films: Chapter 9, sections 4.1, 4.2, and 4.3, respectively

4. Crystals, Powder: Chapter 10, section 6.0

$BaPb_{1-x}Bi_xO_3$

1. Powder, Crystals, and Thin Films: Chapter 9, sections 4.1, 4.2, and 4.3, respectively

2. Crystals, Powder: Chapter 10, section 4.0

$Nd_{2-x}Ce_xCuO_4$

1. Chapter 11, section 4.0 (includes preparations of $Nd_{2-x}Th_xCuO_4$ and $Nd_2CuO_{4-x}F_x$)

2. Crystals: Chapter 5, section 3.3 ("$Nd_{2-x}Ce_xCuO_4$"); Chapter 11, section 5.0

3. Thin films: Chapter 11, section 5.0

4. H. Takagi, S. Uchida, and Y. Tokura, Superconductivity Produced by Electron Doping in CuO_2-Layered Compounds. *Phys. Rev. Lett.* 62(10):1197 (1989).

5. Y. Tokura, H. Takagi, H. Watabe, H. Matsubara, S. Uchida, K. Hiraga, T. Oku, T. Mochiku, and H. Asano, New Family of Layered Copper Oxide Compounds with Ordered Cations: Prospective High-Temperature Superconductors. *Phys. Rev. B* 40(4):2568 (1989).

6. M.F. Hundley, J.D. Thompson, S.-W. Cheong, Z. Fisk, R.B. Schwarz, and J.E. Schirber, Bulk Superconductivity above 30 K in T*-phase Compounds. *Phys. Rev. B* 40(7):5251 (1989).

$La_{2-x}Sr_xCaCu_2O_6$

1. R.J. Cava, B. Batlogg, R.B. van Dover, J.J. Krajewski, J.V. Waszczak, R.M. Fleming, W.F. Peck Jr., L.W. Rupp Jr., P. Marsh, A.C.W.P. James, and L.F. Schneemeyer, Superconductivity at 60 K in $La_{2-x}Sr_xCaCu_2O_6$: the Simplest Double-Layer Cuprate. *Nature* 345:602 (1990).

Appendix B
Further Reading in Superconductivity and Solid State Chemistry

Review articles that address chemical aspects of oxide superconcductors:

Solid State Chemistry of High-T_c Superconductors: Materials and Characterization, E. Kaldis, Ed., Special issue of the *Eur. J. Solid State Inorg. Chem.* 27(1/2) (1990).

R.J. Cava, Superconductors Beyond 1-2-3. *Scientific American* August:42 (1990).

R.J. Cava, Structural Chemistry and the Local Charge Picture of Copper Oxide Superconductors. *Science* 247:656 (1990).

R.M. Hazen, Crystal Structures of High-Temperature Superconductors. In *Physical Properties of High-Temperature Superconductors*, Vol. II, D.M. Ginsberg, Ed., World Scientific, Teaneck, NJ (1990).

Properties of High T_c Superconductors, M.B. Maple, Ed., Special Issue of the *MRS Bulletin* XV(6) (1990).

M. Greenblatt, S. Li, L.E.H. McMills, and K.V. Ramanujachary, Chemistry and Superconductivity of Thallium-Based Cuprates. In *Studies of High-Temperature Superconductors*, A.V. Narliker, Ed., Nova Science Publishers, NY (1990).

D.P. Matheis and R.L. Snyder, The Crystal Structures and Powder Diffraction Patterns of the Bismuth and Thallium Ruddlesden-Popper Copper Oxide Superconductors. *J. Powder Diffraction* 5(1):8 (1990).

H. Mueller-Buschbaum, The Crystal Chemistry of High-Temperature Oxide Superconductors and Materials with Related Structures. *Angew. Chem. Int. Ed. Engl.* 28:1472 (1989).

H. Mueller-Buschbaum, Oxometallates with Planar Coordination. *Angew. Chem. Int. Ed. Engl.* 16:674 (1977).

C.N.R. Rao and B. Raveau, Structural Aspects of High-Temperature Cuprate Superconductors. *Acc. Chem. Res.* 22(3):106 (1989).

A. Reller and T. Williams, Perovskites - Chemical Chameleons. *Chem. Britain* December:1227 (1989).

High T_C Superconductors, M.B. Maple, Ed., Special Issue of the *MRS Bulletin* XIV(1) (1989).

K. Yvon and M. Francois, Crystal Structures of High-T_C Oxides. *Z. Phys. B - Condensed Matter* 76:413 (1989).

G. Nardin, L. Randaccio, and E. Zangrando, A Unified Description of the Crystal Structures of Copper-Containing Ceramic Superconductors. *Acta Crystallogr.* B45: 521 (1989).

A.W. Sleight, Chemistry of High-Temperature Superconductors. *Science* 242:1519 (1988).

B. Raveau, C. Michel, M. Hervieu, and J. Provost, Structure and Non Stoichiometry, Two Important Factors for Superconductivity in Mixed-Valence Copper Oxides. *Rev. Solid State Sci.* 2(2,3):115 (1988).

D.W. Murphy, D.W. Johnson, Jr., S. Jin, and R.E. Howard, Processing Techniques for the 93 K Superconductor $Ba_2YCu_3O_7$. *Science* 241:922 (1988).

G.F. Holland and A.M. Stacy, Physical Properties of the Quaternary Oxide Superconductor $YBa_2Cu_3O_7$. *Acc. Chem. Res.* 21:8 (1988).

J.M. Williams, M.A. Beno, K.D. Carlson, U. Geiser, H.C. Ivy Kako, A.M. Kini, L.C. Porter, A.J. Schultz, R.J. Thorn, H.H. Wang, M.-H. Whangbo, and M. Evain, High Transition Temperature Inorganic Oxide Superconductors: Synthesis, Crystal Structure, Electrical Properties, and Electronic Structure. *Acc. Chem. Res.* 21:1 (1988).

J.B. Torrance, Y. Tokura, and A. Nazzal, Overview of the Phases Formed by Rare Earth Alkaline Earth Copper Oxides. *Chemtronics* 2(9):120 (1987).

Books on oxide superconductors that address chemical aspects:

B. Raveau, C. Michel, M. Hervieu, and D. Groult, *Crystal Chemistry of High T_c Superconducting Copper Oxides*, Springer-Verlag, NY (in preparation).

Studies of High-Temperature Superconductors, A. Narlikar, Ed., Nova Science Publishers, NY (1990).

High Temperature Superconductivity, J.W. Lynn, Ed., Springer-Verlag, NY (1990).

F.S. Galasso, *Perovskites and High-T_c Superconductors*, Gordon & Breach, NY (1990).

High Temperature Superconductors: Relationships Between Properties, Structure, and Solid-State Chemistry; J.D. Jorgensen, K. Kitazawa, J.M. Tarascon, M.S. Thompson, and J.B. Torrance, Eds., *Mat. Res. Soc. Symp. Proc.* Vol 156, Materials Research Society, Pittsburgh, PA (1989).

High Temperature Superconductivity: The First Two Years, R.M. Metzger, Ed., Gordon & Breach, NY (1989).

C.P. Poole, Jr., T. Datta, and H.A. Farach, *Copper Oxide Superconductors*, John Wiley & Sons, NY (1988).

Chemistry of Oxide Superconductors, C.N.R. Rao, Ed., Blackwell Scientific Publications, Oxford (1988).

Chemistry of High-Temperature Superconductors II, D.L. Nelson and T.F. George, Eds., American Chemical Society, Washington, D.C. (1988).

Chemistry of High-Temperature Superconductors, D.L. Nelson, M.S. Whittingham, and T.F. George, Eds., American Chemical Society, Washington, D.C. (1987).

Books relating to solid state chemistry: Structure, Synthesis, and Structure-Property Relations

A.F. Wells, *Structural Inorganic Chemistry*, 5th ed., Clarendon Press, Oxford (1984).

O. Muller and R. Roy, *The Major Ternary Structural Families*, Springer-Verlag, NY (1974).

H.D. Megaw, *Crystal Structures: A Working Approach*, W.B. Saunders, Philadelphia (1973).

B.G. Hyde and S. Andersson, *Inorganic Crystal Structures*, John Wiley & Sons, NY (1989).

D.M. Adams, *Inorganic Solids - An Introduction to Concepts in Solid-State and Structural Chemistry*, John Wiley & Sons, NY (1974).
G.M. Clark, *The Structures of Non-Molecular Solids*, Applied Science Publishers, London (1972).

M.F.C. Ladd, *Structure and Bonding in Solid State Chemistry*, John Wiley & Sons, NY (1979)

F. Galasso, *Structure and Properties of Inorganic Solids*, Pergamon Press, NY (1970).

R.C. Evans, *An Introduction to Crystal Chemistry*, 2nd ed., Cambridge University Press, London (1964).

Structure and Bonding in Crystals, M. O'Keefe and A. Navrotsky, Eds., Vols. I & II, Academic Press, NY (1981).

M. Krebs, *Fundamentals of Inorganic Crystal Chemistry*, McGraw-Hill, NY (1968).

R.J.D. Tilley, *Defect Crystal Chemistry and its Applications*, Chapman & Hall, NY (1987).

Nonstoichiometric Oxides, O.T. Sorensen, Ed., Academic Press, NY (1981).

F.D. Bloss, *Crystallography and Crystal Chemistry - An Introduction*, Holt, Rinehart, and Winston, NY (1989).

G. Burns and A.M. Glazer, *Space Groups for Solid State Scientists*, 2nd ed., Academic Press, NY (1990).

A.R. West, *Solid State Chemistry and its Applications*, John Wiley & Sons, NY (1984).

A.K. Cheetham and P. Day, *Solid State Chemistry Techniques*, Clarendon Press, Oxford (1987).

C.N.R. Rao and J. Gopalakrishnan, *New Directions in Solid State Chemistry*, Cambridge Univ. Press, Cambridge (1986).

H. Schmalzried, *Solid State Reactions*, 2nd ed., Verlag Chemie, Weinheim (1981).

B.R. Pamplin, Ed., *Crystal Growth*, Pergamon Press, NY (1975).

D. Elwell and H.J. Scheel, *Crystal Growth from High-Temperature Solutions*, Academic Press, NY (1975).

W.D. Kingery, H.K. Bowen, and D.R. Uhlmann, *Introduction to Ceramics*, 2nd ed., John Wiley & Sons, NY (1976).

M.F. Berard and D.R. Wilder, *Fundamentals of Phase Equilibria in Ceramic Systems*, R.A.N. Publishers, Marietta, OH (1990).

R.E. Newnham, *Structure-Property Relations*, Springer-Verlag, NY (1975).

J.B. Goodenough, *Magnetism and the Chemical Bond*, John Wiley & Sons, NY (1963).

R. Hoffmann, *Solids and Surfaces: A Chemist's View of Bonding in Extended Structures*, VCH Publishers, NY (1988).

J.F. Nye, *Physical Properties of Crystals*, Clarendon Press, Oxford (1985).

P.A. Cox, *Electronic Structure and Chemistry of Solids*, Oxford University Press, Oxford (1987).

C.N.R. Rao and K.J. Rao, *Phase Transitions in Solids*, McGraw-Hill, NY (1978).

Formula Index

(See also *Appendix A* for synthetic procedures)

$ACuO_2$ (A = alkali metal), 55, 720

$(ACuO_{3-x})_m(AO)_n$ series, 107ff
 crystal-chemical description, 191-200, 561
 description as layered structures, 191-200
 schematic drawings, 108

AgO, 54
 disproportionation, 716

$BaBiO_3$, 382-92
 band structure schematic, 717
 CDW, 351
 charge configuration, 351, 384, 389-91, 716
 crystallographic data, 387
 doping, 348-51
 ferroelectricity, 367
 oxygen content, 405
 polymorphs, 390-1
 structure, 348-52, 385-9
  bond lengths, 386
  parent, 383
  schematic, 388
 synthesis, 391-2

$BaCuO_2$, 67-70

$Ba_{1-x}K_xBiO_3$, 47-9, 347-69, 380-422, 410-9
 applications, 367-9
 band structure schematic, 717

Formula Index   791

charge density wave, 413
crystal growth, 233, 235, 244, 248, 354-8, 412-3
discovery, 47, 381, 411
effect of K-substitution, 352
infrared properties, 361-2
isotope effects, 364, 732
magnetic properties, 361
magnetic susceptibility, 48
mechanism, 365-7
metal-insulator transition, 413-4
mixed valency, 715
optical properties, 361-2
oxygen content, 413
pressure effects, 363-4
specific heat, 362
structure, 348ff, 414-9
   evolution from $BaBiO_3$, 414
   orthorhombic, crystallographic data, 418
   -property correlation, 419-422
superconducting composition, optimal, 413-4
synthesis, 355-9, 411-3
   annealing, 413
   polycrystalline, 355-6
   thin films, 358-9
T_c-composition diagram, 421
thermal stability, 355-6
transport properties, 359-61
variations with x-value
   charge density wave, 417
   structure, 354, 414-9
      crystallographic data, 416
   superconductivity, 355, 414-9

$BaLa_4Cu_5O_{13.4}$, 73-5

$BaPb_{1-x}Bi_xO_3$, 46-9, 76, 347-69, 380-422, 396-406
applications, 367-9
charge density wave, 403
crystal growth, 233, 356-8, 405-6
dielectric anomalies, 403
discovery, 46-7, 381
disproportionation, 349, 403
effect of Pb-substitution, 352
infrared properties, 361-2
isotope effects, 364
magnetic properties, 361
mechanism, 365-7
metal to semiconductor transition, 397
optical properties, 361-2
oxygen content, 405
phase transitions, 402
pressure effects, 363-4
resistivity, 45
specific heat, 362
structure, 201, 348ff, 403-5
   crystallographic data, 404
   neutron powder diffraction data, 203
   -property correlation, 419-422
superconducting composition, 397
   crystallographic data, 404
   symmetry of, 402
synthesis, 355-9, 405-6
   polycrystalline, 355-6
   thin films, 358-9

T_c-composition diagram, 421
thermal stability, 355-6
transport properties, 359-61
variations with x-value
  properties, 396-7
  structure, 354, 397-403
    crystallographic data, 399, 401
      sensitivity to preparative conditions, 402
    superconductivity, 354, 400

$BaPbO_3$, 392-6
  charge configuration, 392
  crystal growth, 395
  crystallographic data, 394
  oxygen content, 405
  properties, 395-6
  structure, 392-5
  synthesis, 395

$BaPb_{1-x}Sb_xO_3$, 226, 382, 406-10
  comparison with Bi-system, 407
  crystal growth, 233
  ionization potentials, 408
  structure, 409
    -property correlation, 419-422
  synthesis, 410
  T_c-composition diagram, 421
  valence of Sb, 409

$Bi_2Sr_2Ca_{n-1}Cu_nO_{2n+4}$, 190
  atomic displacements in rock salt layers, 499
  acronyms, 260
  band structure calculations, 498, 500
  cation substitution studies, 328-41
    by other 3d metals, 332-7
      crystallographic data, 334
      magnetic rare earths, 331
  chemical analysis by electron microscopy, 545-57, 580-2
  crystal growth, 233, 240-2
  crystal-chemical description, 198-9, 217, 491-3
  discovery, 257
  electron microscopy studies, 578-589
  grain boundaries, 582, 585
  Hall effect, 660
  hole doping, 493
  incommensurate structures, 133ff, 330-2
  intergrowth defects, 129-133
  interlayer bonding, 498, 580
  lead-substituted, 557
  lone pair effects, 133ff
  modulated structures, 332ff, 336, 578-86
    schematic, 584
    twinning of, 582
  $n = 1$
    crystal growth, 235
    melting behavior, 230
    positional parameters, 506-7
    structure, 262, 312, 492
    synthesis, 265-6
    X-ray powder diffraction pattern (calc), 523

Formula Index   793

n = 2
   cation deficiency, 546
   composition by AEM, 551-7, 581
   crystal growth, 232, 235, 240
   magnetization hysteresis, 695
   nonstoichiometry, 545, 547, 580-2, 728
   positional parameters, 506-7
   single-phase region, 269
   structure, 312, 492, 581
   synthesis, 266-70
   X-ray powder diffraction pattern (calc), 524
n = 3
   composition diagram for 110 K phase
   flux creep, 697-8, 700
   positional parameters, 506-7
   single-phase region, 274
   structure, 312
   synthesis, 270-3
   X-ray powder diffraction pattern (calc), 526
nonstoichiometry, 273, 493
oxygen interstitials, 586-9
structural description, 491-3, 580
structural distortions, 496-500
structures, 106-114, 258, 312
   of Fe congener, 336, 491
   positional parameters, 506-7
synthesis, 263-4
T_c values, 260
thermal conductivity, 655-7

$Bi_2Sr_2(Ln_{1-x}Ce_x)_2Cu_2O_{10-y}$, 260

$Ca_{0.86}Sr_{0.14}CuO_2$, 201
   single-crystal data, 204

$CaTiO_3$, 485
   structure, 35, 486
      compared to $YBa_2Cu_3O_7$, 486

CuO, 54
   crystal growth, 232

Cu_2O, 54

Fe_3O_4, 49

K_2NiF_4, 190
   structure, 89, 195, 197, 490

$La_{2-x}Ba_xCuO_4$, 64-9, 71-2, 76-9, 314ff
   crystal growth, 236
   Hall effect, 658-9
   neutron powder diffraction data, 212
   thermopower, 657-9
   variation of T_c with Cu-Cu distance, 316

$La_{2-x}Ca_xCuO_4$, 64-9, 76, 314ff
    crystal growth, 236
    variation of T_c with Cu-Cu
        distance, 316

La_2CoO_4, 62

La_2CuO_4, 62-7, 70-2, 190
    antiferromagnetism, 680
    band structure schematic, 717
    atomic displacements in
        orthorhombic form, 499
    chemical substitutions in, 84ff, 314-322
    crystal growth, 232, 235
    hole doping, 491
    interrelationships of T, T', T*
        structures, 320
    structure, 108, 113, 195, 197, 312, 487, 567
        changes with La substitution, 317
        rock-salt/perovskite blocks, 487
        positional parameters, 505
    substitution of La, 314-8
        by other rare earths, 317
    substitutional effects, 566ff
    thermopower, 658
    X-ray powder diffraction pattern (calc), 521

$La_{1.8}Na_{0.2}CuO_4$
    positional parameters, 505

La_2NiO_4, 62

$La_{2-x}Sr_xCaCu_2O_6$, 71-2
    crystal-chemical description, 198, 213
    structural parameters, 215
    structure, 108

$La_{2-x}Sr_xCuO_4$, 64-9, 71-2, 76-8, 314ff
    band structure schematic, 717
    comparison with Nd_2CuO_4 systems, 320
    crystal growth, 233, 236-8
    dislocation structures, 568-70
    inhomogeneity, defects, 724
    magnetism, normal state, 680
    Meissner fraction as f(x), 237
    orthorhombic-tetragonal transition, 766-70
    oxidation states, 609, 715
    oxygen isotope effect, 732
    stability, 729
    phase diagram schematic, 717
    structure (see also La_2CuO_4), 312, 487, 490-1, 567
        rock-salt/perovskite blocks, 487
        positional parameters, 505
    substitution of Cu, 320-2
        resistivity curves, 321
        variation of T_c, 338
    substitution of La
        by other rare earths, 315-16
        T_c correlations, 730

substitutional effects, 566ff
T_c correlations, 729
thermopower, 658
variation of T_c with Cu-Cu distance, 316
variation with x-value
  lattice parameters, 314-5
  magnetism, 314
  resistivity, T_c, 314, 316, 567
  transport properties, 314-317
X-ray powder diffraction pattern (calc), 522

$La_{8-x}Sr_xCu_8O_{20-y}$, 59

$Li_{1+x}Ti_{2-x}O_{4-y}$, 50-3
  crystal growth, 233

Ln_2CuO_4, 64, 67

MO (M = Ti,V,Nb,Mn,Fe,Co, Ni,Cu), 30-4
  crystal growth, 233

$MgAl_2O_4$
  structure, 48

Na_xMoO_3, 44

Na_xReO_3, 44

Na_xWO_3, 38-44
  crystal growth, 233

$Nd_{2-x}Ce_xCuO_4$, 112, 427-45
  band structure schematic, 717
  cerium oxidation state, 431
  comparison with La_2CuO_4 systems, 320
  composition range, 437
  copper oxidation state, 427, 442
  crystal chemistry, 428-31
  crystal growth, 244, 436-7, 442-4
  discovery, 318
  electron diffraction, 438
  Hall effect, 660-1
  inhomogeneity, defects 724
  iodometric analysis, 438
  Meissner fraction, 437
  mixed valency, 715
  normal-state resistivity, 432
  oxidation states, 609
  oxygen stoichiometry, 438, 442
  phase diagram schematic, 717
  rare earth analogs, 432-5
  related systems, 318, 320
    interrelationship of T, T', T^* structures, 320
    other rare earth analogs, 432-5
  resistivity curves, 319
  single-crystal studies, 318-9, 436
  structural parameters, 443

structure (see also Nd_2CuO_4), 312, 428-31
substitution of Nd by other rare earths, 318
  by La, 320
  $Eu_{1.85-x}La_xCe_{0.15}CuO_4$, 320
  $Nd_{1.85-x}La_xTh_{0.15}CuO_4$, 320
  $Nd_{1.85-x}Y_xCe_{0.15}CuO_4$, 320
  variation of T_c with Cu-Cu distance, 316
superlattice modulation, 436-8, 442
synthesis, 437-42
thermogravimetric analysis, 438
thin films, 444
thorium analog, 427, 431-2, 435, 439
two-phase nature, 436
variation with x-value
  lattice parameters, 315, 318, 431, 433
  magnetic ordering temperature, 432
  resistivity, 318, 432
  T_c, 432-5
X-ray powder diffraction pattern, 443

Nd_2CuO_4, 70, 112
  band structure schematic, 717
  electron doping, 431ff
  interrelationships of T, T', T* structures, 320
  schematics, 429
  stability fields, 429
  properties, 432
  structure, 69, 113, 312, 428-9
  substitution of Nd by La, 315

$Nd_2CuO_{4-x}F_x$, 244, 427, 431ff, 435
  band structure schematic, 717
  chemical analysis, 441
  inhomogeneity, defects 724
  resistivity curves, 434
  stability vs. pO_2, 440
  superlattice, 441
  synthesis, 439-41

$Pb_2Sr_2Ln_{1-x}Ca_xCu_3O_8$ (Ln = Y, rare earth)
  crystal growth, 233, 242-4
  crystal-chemical description, 216-7, 495-6
  neutron powder diffraction data, 215
  structure, 109-12, 134ff, 245, 497
    positional parameters, 514-15
  thermogravimetric curves, 246
  X-ray powder diffraction pattern (calc), 539

PdO, 54

Formula Index 797

$RBa_2Cu_3O_{7-x}$ (R = rare earth), 322
  resistivity curves, 323
  structures, 111

$SrLaCuO_4$, 55

$SrTiO_{3-x}$, 36-8, 73
  crystal growth, 233

$Sr_{n+1}Ti_nO_{3n+1}$, 191

TiO_{1-x}, 31

$Tl_{1+x}A_{2-y}Ln_2Cu_2O_9$ (A = Sr,Ba), 114, 139
  structure, 115

$TlBa_2Ca_{n-1}Cu_nO_{2n+3}$
  acronyms, 260
  crystal growth, 273-80
  crystal-chemical description, 198-9, 217, 259
  electron microscopy studies, 589-604
  hole doping, 500
  incommensurability, 137
  intergrowth defects, 129ff
  lone pair effects, 134-7
  n = 1
    positional parameters, 510-11
    structure, 494
    X-ray powder diffraction pattern (calc), 532
  n = 2
    positional parameters, 510-11
    structure, 108, 494
    X-ray powder diffraction pattern (calc), 533
  n = 3
    composition by AEM, 557
    positional parameters, 218, 510-11
    structure, 110, 494
    X-ray powder diffraction pattern (calc), 534
  n = 4
    positional parameters, 510-11
    structure, 110
    X-ray powder diffraction pattern (calc), 535
  n = 5, 590
    disorder, defects, 594-7
  structures, 106-14, 258, 494
    positional parameters, 510-11
  synthesis, 273-80
  T_c values, 260
  thallium oxidation state, 137

$TlSr_2Ca_{n-1}Cu_nO_{2n+3}$
  crystal growth, 273-80
  crystal-chemical description, 198-9, 259
  electron microscopy studies, 589-604
  lead-substituted series, 138, 260, 279-80, 495

chemical composition by
AEM, 601
electron microscopy studies,
597-604
flux exclusion behavior, 599
oxidation states of metals,
600
positional parameters, 512-13
site disorder, 600
stoichiometry, 600
X-ray powder diffraction
patterns (calc), 536-9
synthesis, 273-80
structures, 106-114

$Tl_2Ba_2Ca_{n-1}Cu_nO_{2n+4}$, 190
acronyms, 260
band structure calculations, 498,
500
composition by AEM, 557
crystal growth, 233, 242, 273-80
crystal-chemical description,
198-9, 217, 493-5
discovery, 257
electron microscopy studies,
589-604
incommensurability, 137
intergrowth defects, 129ff
lone pair effects, 134-7
$n = 1$
composition by AEM, 557
displacements in rock salt
layers, 499
positional parameters, 508-9
structure, 108, 262, 494

substitution with Cd, 602-3
X-ray powder diffraction
pattern (calc), 528
$n = 2$
composition by AEM, 557
magnetization hysteresis loop,
694
positional parameters, 508-9
refined structural parameters,
219
structure, 108, 494
X-ray powder diffraction
pattern (calc), 529
$n = 3$
composition by AEM, 557
flux creep, 697-8, 700
Josephson junction, 663-5
magnetization hysteresis, 695
positional parameters, 508-9
refined structural parameters,
219
structure, 110, 494
X-ray powder diffraction
pattern (calc), 530
$n = 4$
composition by AEM, 590
disorder, defects, 591-4
grain boundary interface, 595
positional parameters, 508-9
structure, 110
X-ray powder diffraction
pattern (calc), 531
structural description, 493-5
structural distortions, 496-500
structures, 106-14, 258, 494
positional parameters, 508-9
synthesis, 227-8, 264-5, 273-80

Formula Index   799

T_c values, 260
thallium oxidation state, 137
thermopower, 658

$YBa(Fe,Cu)_2O_5$
  neutron powder diffraction data, 204
  structural description, 205

$YBa_2Cu_3O_6$
  bond distance data, 153
  structure, 89, 112, 115
    atomic position data, 152
    compared with fully oxygenated, 152, 195-6
    compared with perovskite, 195-6
    positional parameters, 503-4
  substitution with Ca, 576-9
    variation of unit cell with, 579
  X-ray powder diffraction pattern (calc), 516

$YBa_2Cu_3O_{7-x}$
  analysis by iodometric titration, 610ff
  antiferromagnetism, 680
  bond distance data, 153
  copper oxidation state, 122
  critical current density, 645-54
  crystal growth, 232-3, 238ff
  crystallographic shear, 205-13
  defects, 124-9, 724
  and carrier concentration, 725
  detwinning, 240
  discovery, 79ff, 146, 476-83
  electron microscopy studies, 116-41
  electronic structure, 764-6
  films, critical current density, 646ff
  films, thick, 300-2
  flux flow, creep, 633
  flux pinning centers, 571
  grain boundaries, 574-6
  grain size problems, 291
  Hall effect, 658-61
  inhomogeneity, electronic, 725
  Josephson junction, 663-6
  magnetic properties
    critical current density, 694-5
    flux exclusion vs expulsion, 686-91
    flux pinning, 693
    hysteresis loop, 692-3
    magnetization vs H; T, 685-6
    normal state, 680
  magnetic susceptibility calculation, 679
  microstructure, 571ff
  mixed valency, 715
  orthorhombic-tetragonal transition, 150-4, 239, 571
  oxidation state analysis, 609-24
  oxidation states, 610
  oxygen elimination mechanisms, 169-74
  oxygen isotope effect, 732
  oxygen ordering, 121ff
  oxygen vacancy ordering, 161-9

oxygen variability, 88, 114ff, 240
  as a function of pO_2 and temperature, 239-40
  bond distances as a function of, 153
  properties as a function of, 151
  T_c as a function of, 241
penetration depth, 683
phase diagram study to identify, 476-83
phase separation, 725
stability, 226, 727
structural description, 488-90
structure, 80, 89, 109, 111, 146-50, 312, 486, 489, 571-2
  changes in with oxygen content, 150-4
  compared with oxygen-deficient, 195-6
  compared with perovskite, 149, 195-6, 486
  neutron powder diffraction data, 148
  positional parameters, 503-4
substitution of Ba, 324
  by Sr, 322
substitution of Cu, 90-1, 174-85, 324-8
  changes in bond lengths, 328
  changes in properties, 326ff
    effect of sample history, 326
  O-T transition, 324
  oxygen uptake, 326
  role of chains vs. planes, 324-9
  site occupation studies, 326
  structural models/mechanisms for, 178-85
  variations in lattice parameters, 324-5
  variations in T_c, 327-9
substitution of Y, 92, 174-5
  by other rare earths, 322-4
  resistivity curves, 323
  substitution of Ba, 324
superstructures, 118ff, 123
synthesis
  low-temperature compared with high, 290-1
  precursors, 293-8, 302-5
  solution processes, 292-306
tetragonal, structure, 117, 150-4
thermal conductivity, 656-7
thermopower, 655, 658
thick film lithography, 297
twinning, 124-9, 571-3
  structural models for, 154-61
weak link effects, 700-3
X-ray powder diffraction pattern (calc), 517

$YBa_2Cu_4O_8$, 112
stability, 226, 729
structural description, 206, 208, 210, 490
structure, 113, 489
  atomic positional coordinates, 209, 503-4
substituted with Ca, 576

X-ray powder diffraction pattern
 (calc), 518

Y_2BaCuO_5, 58

$Y_2Ba_4Cu_7O_{15}$, 112
 stability, 729
 structural description, 206, 210, 490
 structure, 112
  refined parameters, 211
  positional parameters, 503-4
 X-ray powder diffraction pattern
  (calc), 519

Subject Index

(See also *Formula Index*)

A-15 structure, 12-14, 20, 21
ablation
　laser, 444
absorption
　zero field microwave, 249
activation energies, 350
additive proportions, 453
aliovalent substitutions, 427
Alkemade lines, 464
alloys, superconducting, 8-13
analytical electron microscopy (AEM), 545-58, 562ff
analytical procedures, 247, 609-24, 545-58
anisotropic materials, 249
anisotropy, 249
　shape, 249
　resistivity, 249
annealing, 240

methods, 357
antibonding, 500, 578, 719, 721, 751, 764
antiferromagnetic insulator, 380, 756ff
antiferromagnetism, 326, 335, 381, 680, 716
　superconductivity relationship, 381
applications, 367
atomic displacements, 493ff
atomic jumps
　sequence of, 174
atomic magnesium, 741
atoms
　coordination of, 193
Aurivillius phases, 190, 259, 580

B-1 structure, 15, 30
band structure, 714, 716
　Cu-O sheets, 761

band structure calculations, 495, 498, 597
band theory, 366
bands, 715
Bardeen, Cooper and Schrieffer, 353
basis vectors, 466
BCS theory, 320, 353, 362, 364, 726
Bednorz and Mueller, 70, 73, 74, 75, 82, 83
behavior
  invariant, 455
  univariant, 455
bipolarons, 367
bismuth cuprates (see also *Formula Index*), 347
  bilayers, 109
  comparison with thallium cuprates, 261-2
  composition by AEM, 545-58, 564ff
  crystal chemistry, 491-3
  crystal morphology, 261
  discovery, 257
  electron microscopy studies, 561-604
  lead-substituted, 272-4
  lone pair effects, 261
  stability, 727-9
  structural distortions, 496-500
  structural formulas, 107ff, 260
  structural principles, 106-10, 261
  structures, 258, 262
  synthesis, 265-73, 263
  T_c values, 260

  X-ray diffraction data, 491ff
bismuthates, non-copper, 347-69, 380-422
  comparison with cuprates, 381
bivariant, 455, 457
bond
  Cu-O, 340
bond length
  Cu-O4, 328
  optimum Cu-O, 435
bond valence, 501
borides, 15-16, 23ff
boundary,
  insulator-superconducting, 587
  metal-insulator, 587
boundary curves, 463
Bragg angle, 466
Bragg equation, 466, 562
Bravais lattices, 466-7
Bridgeman-Stockbarger growth, 231
bulk
  oriented grain critical current, 651
  resistivity, 637

c-axis anomaly, 766
capped square antiprisms, 214
carbides, 15-16
carrier
  concentration, 724
  density, 366
  reservoir, 328
  type, 365
cation substitution, 310-41, 586

Ce-doped neodymium materials, 339
cell doubling, 240
"ceramic", 77
ceramic processing, 174
chains, 328, 335
  corner-sharing near-squares, 205
  Cu-O, 313, 322, 324, 328
  double, 205
  edge-sharing near-squares, 205
  effects of substitution in, 338ff
  vs. planes, 324-9, 335, 338
characterization
  magnetic properties, 675-705
  oxidation state analysis, 609-24
  single crystals, 247-9
  transport properties, 627-66
  X-ray diffraction, 450-83
charge
  carriers, 365
  density wave (CDW), 351-2, 381, 716
  disproportionation, 350
  fluctuations, 723, 732
  imbalance, 324
  reservoir, 380
  transfer, 338, 339, 340
  transfer compounds, 28
charge density wave (CDW), 352, 361, 415, 716
  stabilized energy gap, 361
chemical analysis
  by electron microscopy, 545-58, 564ff
  by titration methods, 609-624
chemical composition, 562
  by electron microscopy, 545-58, 564ff
chemical substitutions
  cationic, 310-41
  crystal chemistry, 84
  electron diffraction studies, 566ff
  in lanthanum cuprate, 84
  in perovskites, 84-6
  in yttrium barium copper oxide, 90
chemical system
  components in, 452
chemistry
  role of, 311
  substitutional, 244, 310-41
Chevrel phases, 23-4
chronology of superconductivity, 3ff, 22, 81
Chu, C.W., 77, 79
cleavage, 580
coherence length, 166, 313, 571, 683
commensurate modulation, 330, 333, 493, 586
complexes of Ni^{3+}, 738
components, 453
composition triangle, 460-4
compositional information, 247
compounds
  intermediate, 459
  line, 459
  Mn-doped Bi, 335
  non-copper containing, 244
  nonstoichiometric, 348
  thallium containing, 242
compressive stress, 430
Condon-Shortley parameters, 736

configuration interaction, 741, 745
congruent composition, 230
congruently melting phases, 230
contact resistance, 639
contacts, 628-30, 640
containers, 226, 236-7
Cooper pairs, 362, 732
cooperative displacements, 602
copper
  coordination, 311ff
  coordination geometries, 56-60, 122
  ionic radii, 56
  linearly-coordinated, 244
  oxidation states, 52-5, 88, 122
    determination, 611ff
  -oxygen bond distances, 55-61, 500, 764
  -oxygen interaction, 771
  substitution of by impurity metal atoms, 175
  three-fold coordinated, 169
  trivalent compounds, 55
  valence, optimum, 324, 338
copper oxide superconductors (see also *Formula Index*), 380ff
  chemical features, 52ff, 88, 380
  common features, 380, 488, 561
  discovery, 52, 70-83
  electronic structure, 759-66
  families, 84, 85
  microstructural/microchemical variations, 561-604
  structural principles, 106-14, 200-2
copper-oxygen sheets, band structure, 761

correlation gap, 757
correlations
  Cu-O distances with doping, 500, 764ff
  hole density with T_c, 602, 725, 729
  T_c with cation substitution, 338-41
  T_c with Cu-Cu distance, 316
  T_c with formal copper valence, 324, 332
  T_c with in-plane Cu-O distance, 500ff, 730
  T_c with number of layers, 129, 261, 597
  with T_c, 729-31
covalency, 55-6, 714, 720, 751
covalency effects, 366
covalent compounds, 719
covalent-ionic bond character, 55
criteria for establishing superconductivity, 6
critical current, 2, 360, 571
  contacts, 639-40
  criterion, 642-44
  cryogen, 640
  current transfer, 644
  density, by magnetization, 691-6
  from transport vs. magnetic measurements, 695-6
  magnetic field dependence, 642-6
  oxygenation effects, 650-2
  results, 645-6, 651
  temperature dependence, 649-53
  theoretical limitation, 650-1

transport measurements, 639, 644-5
critical magnetic field, 2, 361, 681ff
critical parameters, 2, 6
critical transition temperature, 2, 81
  see also T_c
crucible materials, 236-7
crucibles, 226, 236
  cleaning, 226
  ceramic, 236-7
crystal chemistry, 190
  of layered structures, 191-200
crystal field stabilization energy, 736
crystal growth methods, 229-36
crystal lattice, 466
crystal orbitals, 749
crystal structure
  T, 339
  determination by X-ray diffraction, 465ff
  of various cuprate superconductors, 485ff
crystal systems, 466
crystalline substance, 465-6
crystallographic databases, 464ff
crystallographic data for cuprate superconductors, 485
crystallographic shear, 205
crystals, 357
Cu1, 326, 328
Cu2, 326, 328
cuboctahedron, 193
cuprates (see also copper oxide)
  structural formulas, principles, 52-61, 106-14
  substituted, 338
Curie temperature, 330
cyclobutadiene, 740

d orbitals, energies, 762
d values, 468
  for cuprate superconductors, 485ff
Debye temperature, 353
defect structures, 562
defective layers, 195
defects, 124-41, 493ff, 564, 576, 594, 723ff
degrees of freedom, 452, 455
delocalization, 352, 719
demagnetizing factor, 249
dense ceramics, 356
detwinning, 240
diamagnetic shielding, 681ff
  calculation of fraction, 681
diffraction
  atom type effect, 468
  atom position effect, 468
  by X-rays, to determine structure, 465ff
diffuse diffraction peaks, 240
diffuse scattering, 162
diffuse streaks, 162
directional solidification, 230
discovery of superconductivity, 4
dislocations, 564, 576, 594
displacements from ideal positions, 498

Subject Index

disproportionation, 122, 361, 716, 721, 732
dissipation
  flux creep, 633, 652-3
  flux flow, 652-3
  flux pinning, 652-3
  power law approximation, 652-3
  temperature variation, 652-3
  weak links, 631-2, 651
distance
  critical, 341
  Cu-Cu, 320, 341
  in-plane Cu-Cu, 317, 340
distortions, structural, 496ff, 738
domain structures, 355
doping, 335, 348, 718, 732, 764
  compression-tension relationship, 430, 435
  electron, 430ff, 444
  hole, 430
  hole vs. electron, 428
  in T, T', T* structures, 430
  symmetry, 340-1
doping mechanism, 330, 335

E21 structure, 30
electrochemical deposition, 358
  analysis, 621
electron
  -electron attractive force, 353
  -electron interaction, 353, 718, 735-74
  localization mechanisms, 367, 716
  phonon coupling, 351, 361

phonon interaction, 351-2, 362, 366, 732
pocket, 500
transfer, 350, 770ff
electron-doped superconductors, 427-45
electron-doping, systematics, 431-7
  lattice tension, 435
electron microscopy, 561-604
  chemical analysis, 545-58, 564ff
  microanalysis, 564
  microdiffraction, 564
  phase identification, 564
  studies, 116-41
  techniques, 562-6
electronegativity, 771
electronic
  coupling, 328
  instability, 726
  models, 735
  specific heat, 362
  structure, 714-32, 735-74
elemental analysis
  by electron microscopy, 247, 545-58, 564ff
elements, superconducting, 8-11, 36
energy
  barrier, 454
  minimum, 454
energy bands of solids, 748
entropy stabilized compounds, 728
epitaxial thin films, 359
equations
  Hall coefficient, 658
  Josephson coefficient, 662

Meissner effect
  resistivity, 627-8, 652-3
  voltage/current power law, 653
  vortex barrier energy, 652-3
equilateral triangles, 462
equilibrium, 450
  state, 454
  univariant, 455
  phase, 477
eutectic
  binary, 463
  composition, 459
  point, 459
  reaction, 459
  ternary, 463
eutectoid, 459
exitons, 367
extended defects, 124-41, 564

Faraday balance, 676
Fermi level, 500, 718, 726, 731
ferroelectric transition, 721-2, 732
ferromagnet
  classic, 335
ferromagnetic insulator, 755
ferromagnetism, 335
films, 633, 639, 645-8
  thick, preparation, 297ff
float zone technique, 229, 232
fluorine substitution, 341, 435
fluorination, 431
fluorite layers, 114, 139, 260
fluorite type structure, 114
flux, 232, 356, 652-3
  choosing, 234
  table of standard, 235

flux creep, 696-700
flux exclusion vs. expulsion, 685-91
flux growth method, 232-6
flux pinning, 570-1, 574, 602, 653, 685-91
formal oxidation state, 718
  analysis, 609-624
  copper, 122, 310, 317, 324, 330, 332, 338-40, 501, 561, 566, 715
  in contrast to real charges, 718ff
  lead, 600
  optimum value for copper, 338, 501, 602
  thallium, 600
four-fold planar coordination, 147
future bismuthate research, 365

gap, 361
  superconducting energy, 362 498
gels, 314
Gibbs phase rule, 451ff
grain boundaries, 232, 368, 562, 571, 591
grain size, 291
granular effects, magnetic, 700-5
granular superconductivity, 292
graphite intercalation compounds, 26
growth
  crystal, 224-49, 356
  Czochralski, 230
  flux, 229
  grain, 230

habit, 249
low temperature crystal, 354
melt, 229
parameters, 234
sintering, grain, 229
solution, 232
top-seeded solution, 236, 244

H11 structure, 30
half-filled band, energetics, 759
Hall coefficient
  positive, 339
Hall effect, 658-660
Hall mobility, 367
Hanawalt search method, 472
Hartree-Fock, 741
Hc2, 361
Heitler-London wavefunction, 743
heterogeneity, 562, 726
high pressure studies, 28, 47, 50, 79
history of superconductivity, 2ff
hole-doping, 488, 500, 561, 578, 602
  crystal chemistry of, 430-1
  vs. electron-doping, 430
hole density, 501
holes, 107, 141, 488, 491, 493ff
homogeneity, 578
  electronic, 725
homologous series, 589
HREM, 562ff
Hubbard treatment of solids, 746, 757
Hueckel parameter, 738
hybridization, 723
between the Cu 3d and O 2p levels, 310, 338
hydrogen
  metallic, 367
hydrogen reduction, 616ff
hydrothermal technique, 357
hysteresis loops, 691-6

ideal structures, deviations from, 496-500
impurities, 225
  magnetic, 339
impurity phases, 228
  identification by XRD, 470ff
  magnetic effects of, 704
incommensurate superstructure, 133ff, 330, 491
incommensurability effects, 106ff
incongruently melting phases, 232
indexing X-ray diffraction data, 472ff
inhomogeneities, 578, 723-6
  "electronic", 724
instability, 726-9
insulator-metal boundary, 587
insulator-superconducting boundary, 587
intercalation, 727
intercalation compounds, 26-8
interfaces, 562, 594
intergrowth defects, 129ff
intergrowths, 242, 428, 430, 485, 490, 561, 563, 570, 591, 594
interlayer bonding, 580
interplanar spacing, 468
interstitial oxygen, 493, 586, 589, 725

inverse spinel, 49
iodometric titration methods, 609-624
ionic mobility of oxygen, 571
IR, 362
irradiation effects, 648, 657
"irreproducible superconductors", 26
isobar, 455
isopleth, 462
isotherm, 455
isotope effects, 364, 732

Jahn-Teller distortion, 54, 56, 59-62, 64, 738, 753, 761
Jorgensen's empirical rules, 740
Josephson effect, 662
  device, 18
  junctions, 359, 368
  tunnel junction arrays, 368

kinetically stable, 454, 728
Kitazawa, K., 83

lanthanum cuprate, 61-8
lattice defects, 360
lattice mismatch, 430, 496ff
Laves phases, 11
layered structures, 191
layers
  CuO_2, 330, 340
  double Bi-O, 335
lead containing cuprates
  structural data, 495ff

"lever principle", 457, 462, 479
limitations of the phase rule, 454
liquidus line, 457
lithography, 297
local structural disorder, 339
lone pair effects, 133ff, 139, 141, 261-3

magnetic field effects, 632-3, 645-6
magnetic ion effects, 320ff
magnetic moment per RE, 330
magnetic pair breaking, 320, 339
magnetic properties, 675-705
  instrumentation, 676
magnetic susceptibility
  calculations, 679
magnetization, 678
  relaxation, 696-700
magnetometers, 676
magnetoresistance, 632
materials
  n-type, 340, 341
  preparation, 355ff
  p-type, 340, 341
Matthias, B.T., 73
maximum transition temperature, 19
mechanism for high T_c, 311, 361, 714, 731-3, 773, 771
  sudden electron transfer, 771
Meissner effect, 6, 7, 74, 291, 675ff, 685-91
Meissner fraction, 77, 249, 685-91
melt growth, 230
metal-insulator boundary, 587

Subject Index 811

metallic behavior, 314
metastable, 454, 726-9
microchemical variations, 561ff
microstructural characterization, 562, 566ff
microstructural inhomogeneities, 166, 561
microstructure-chemistry relationships, 562
microtwinning, 125, 240
microwave absorption, 249
microwave switch, 368
Miller indices, 468
minimum value of the free energy, 454
mismatch, 430, 496ff
mixed valency, 106-7, 714, 715
mixed-valent copper oxides, 71
Michel and Raveau, 71, 73
modulation, 121, 341
Moessbauer studies, 324, 363, 366
molecular beam epitaxy, 359
monotectic, 459
Montgomery method, 249
mosaic spreads, 355
Mott transition, 757
Mueller and Bednorz, 70, 73, 74, 82, 83
Mulliken-Hund wavefunction, 743
muon spin resonance, 432, 587, 716, 731
mutual solubility, 457

n, 365
n-type, 340
 vs. p-type, 340

n-type superconductors, 427-45
near-zero field microwave absorption, 249
niobium oxides, 20
nitrides, 15-16
nonstoichiometric melt, 232
nonstoichiometry, 545-58, 562
non-transition metal systems, 25
normal spinel, 49

Onnes, Heike Kamerlingh, 4, 5
optical devices, 368
optical microscopy, 481
orbital
 antibonding band, 340
 overlap, 338
orbital theories, 738
order-disorder transition, 166
ordered structures, 240
ordering
 antiferromagnetic, 314
 schemes, 165
organic superconductors, 28-30
organometallic precursors, 226
orthorhombic-tetragonal transformation, 324, 571, 766ff, 770
oxidation state (see formal oxidation state)
oxidation state analysis, 609-624
oxidation state, stabilization of high, 720ff
oxide precursors, 225
oxide synthesis, 225
oxides, superconducting, 17-22
oxygen
 analysis, 609-624

812 Chemistry of Superconductor Materials

defects/interstitial, 438
diffusion of, 155
evolution, 610, 619
inhomogeneity, 116ff
local ordering, 121
nonstoichiometry, 114ff
partial pressure, control of, 228
stoichiometry, 150, 240
superstructure, 121
vacancies, 116, 359
vacancy ordering, 161, 162

p type, 340, 366
pairing mechanism, 380
particle size vs. magnetization, 292
Pauli paramagnetism, 680
Peierls distortion, 753ff, 759, 766, 774
penetration length/depth, 291, 683ff
peritectic, 459
  reaction, 459
peritectoid, 459
perovskite, 347
perovskite structure, 21, 30, 34-7, 40, 89, 191-2, 213, 348, 383, 485, 561
  blocks, 107, 195, 200-2, 487
  cation deficient, 38
  chemical substitution in, 84ff
  compounds with, 202
  "doubled", 87
  layers, 259, 580, 589
  metal-oxygen sublattice, 349
  ordered systems, 40, 45, 86, 87

oxygen deficient, 36, 37
  -rocksalt structures, 213-20
  "tripled", 87-9
perovskite-rocksalt structures, 213-20
peroxide, 610
phase
  3d metal substituted, 335
  Bi-based cuprate, 313
  CDW, 352
  definition of, 452
  disproportionation, 355
  identification by XRD, 465ff
  non-superconducting fully RE-doped Bi, 332
  RE-doped 123, 322, 330
  rule, 451
  separation, 725
  T, 311, 317, 320
  T', 311, 317, 320
  T*, 318, 320
  transformation, 366
  transition, 155
phase diagram, 451, 454-64
  BaO-CuO-YO$_{1.5}$, 238-9, 478
  BaO-PbO$_2$-Bi$_2$O$_3$, 406
  Bi-Sr-Cu-O, 241, 267
    lead-substituted, 274
  binary eutectic, 233
  construction, 464
  identification of YBa$_2$Cu$_3$O$_7$, 476-83
  of a binary system, 455
  resource literature, 464
  studies, 464-5
  ternary, 463
  unary, 454

X-ray diffraction and, 476ff
phase relations in solid state
    systems, 464
phase rule, 451
    limitations, 454
phase separation, 725
phenomenological trends, 341
phonons, 351
pinning mechanisms, 360
planar faults, 576
planes
    CuO_2, 311, 313, 320, 322, 324, 328, 333, 335, 338, 380, 576, 761
    compression vs. tension, 430
    energy bands, 760
    hole-doping of, 488
    role compared to chains, 324-9, 335, 338
plasma
    edge, 361
    frequency, 361
plastic deformation, 564
platinum chain compounds, 751
point defects, 726
points
    invariant, 455
    triple, 455
polarizability, 721-3
porosity effects, magnetic, 700-5
positional parameters, 501ff
powder X-ray diffraction, 465ff
    data tables for cuprate superconductors, 485ff
precipitation methods, 293-7
precipitation phenomenon, 358
precursor synthetic methods, 227-8, 293-306
prereacted precursor, 227
pressure effects, 363
properties
    electron transport, 359
    infrared, 361ff
    magnetic, 361
    optical, 361ff
    physical, 359ff

Racah parameters, 736
Raman, 362
random spinel, 49, 50
Raveau and Michel, 71, 73
reflectivity, 361
refractory materials, 234
repeat unit, 466
resistivity, 627-37
    correction in magnetic field, 642-4
    magnetic field dependence, 633
    normal state theory, 633
    measurements, 628-32, 640
    results, 633, 637
resonance
    muon spin, 339
rf-sputtered films, 358
Rietveld analysis, 146
rocksalt structure, 30-4, 192-5, 195, 213, 485, 491, 580
    layers, 107-9, 200-2, 259, 487, 496ff, 589
        distortions/electronic properties, 496ff

rocksalt-perovskite structures, 213-20
Ruddlesden-Popper type phases, 191, 259, 384
Rutherford Backscattering Spectroscopy (RBS), 247

samples
  homogeneous, 227
  homogeneity, 229
  identification by XRD, 465-83
Scanning Electron Microscope, 247
scattering
  differential anomalous X-ray, 326
Seebeck coefficient, 657-8
Shapiro steps, 662-3
shear defects, 570
shielding fraction, 688ff
short range order, 366
silicides, 23
single crystals, 224
  critical current density, 650
  Hall effect, 658ff
  structure determination, 465ff
  thermopower, 657
single phase region, 459
sintering, 230
site disorder, 591ff
sodium chloride structure, 21, 30-4, 192-5, 485
sol-gel process, 288, 313
solid solution, 317, 576, 727
solid state reaction technique, 224-7, 290, 314

solidus line, 457
solution processes, 292ff
Sommerfield parameter, 362
specific heat, 362
spectrometer
  energy dispersive X-ray, 247
spin correlations, 732
spin density wave (SDW), 716, 757
spin polarized neutron diffraction, 432
spinel structure, 21, 30, 48-51
square-bipyramidal, 216
SQUID, 660ff, 676
stability, 726-9
  conferred by entropy, 728
  electronic, 726
  kinetic, 728
stacking faults, 330
starting material, 225-6
states
  global equilibrium, 454
  local equilibrium, 454
  metastable, 454
static
  disorder, 154
  CDW, 351
steric factor, 501
stoichiometric nonuniformity, 360
structural data for cuprate superconductors, 485
structural distortions, 496ff
structural formulas, 107-14
structural modulation, 332, 578-587
structural principles, 106ff

structure determination by X-ray diffraction, 465ff
structure-property relationships, 604
substitution
  3d metal, 332
  anionic, 244
  at the Cu site, 322
  at the rare earth site, 322
  cationic, 310-41
  defects, 493
  fluorine, 341
  for Cu by a 3d metal, 326
  magnetic ion, 320
  rare-earth, 435
substrates, 291, 645-50
sulfides, 23ff
superconducting properties
  as a function of the oxygen content, 151
superconductivity
  alloy systems exhibiting, 8, 11-14
  borides exhibiting, 15, 16, 24
  carbides exhibiting, 15
  ceramic materials exhibiting, 15ff
  chronology, 3ff, 9, 12, 22, 81
  criteria for establishing, 6, 10
  definition, 4
  discovery, 4, 5, 9, 12
  discovery of "high-temperature", 70-83, 310, 476
  elements exhibiting, 8, 10
  history, 2ff,
  intercalation compounds, 28
  intermetallics exhibiting, 11ff
  mechanism, 361, 714, 731-2, 773
    sudden electron transfer, 771
  nitrides exhibiting, 15
  Nobel prizes in physics, 6
  organic compounds exhibiting, 28ff
  oxides exhibiting, 17ff, 21ff
  phenomenon, 4
  phosphides exhibiting, 15
  selenides exhibiting, 15
  sulfides exhibiting, 15, 23ff
  ternary compounds exhibiting, 23ff,
superconductor
  conventional BCS, 320, 339
  crystal-chemical descriptions, 200-2
  detection as minor phase, 249
  electron-doped, 244, 428
  electron-type, 341
  fluorine-doped, 439
  n-type, 339
  p-type, 339
superfluidity, 4
superlattice, 436, 491ff
superlattice structure, 366, 491ff
superstructure, 114-24, 491ff, 580, 726
syntactic intergrowths, 591
syntectic, 459
synthesis (see *Formula Index* and *Appendix A*), 313-4
  low temperature, 226
  of samples, 224
  powder, 356
  solid state, 224-49, 290

solution techniques, 287-306, 314
technique, 114ff, 313
systems
  binary, 453, 455
  condensed, 457
  electron-doped, 432
  isomorphous, 457
  metathetical, 453
  multicomponent, 454
  phases in, 453
  reciprocal, 453
  ternary, 453, 459
  unary, 453

T structure (see K_2NiF_4; La_2CuO_4), 311, 317, 320, 340-1, 428-31
  schematic, 429
  stability field, 429
T' structure (see also Nd_2CuO_4), 311, 317-8, 320, 339-41, 428-31, 436
  electron-doping, 431ff
  schematic, 429
  stability field, 429
T* structure, 318-20, 428-31
  schematic, 429
  stability field, 429
tables
  chronology of discoveries, 9
  copper-oxygen bond distances, 56
  crystallographic information, 501ff
  cuprate superconductors, 85
  d-spacings, 501ff

high T_c systems with A-15 structure, 13
intercalation compounds, 28
milestones in superconducting oxides, 22
non-transition metal systems, 25
organic systems, 29
powder patterns, 501ff
superconducting borides and sulfides, 24
superconducting carbides, nitrides, borides, 16
superconducting rock-salt-type systems, 33
T_c values for elements, alloys, 8
transition temperatures in lithium titanates, 53
tungsten bronze type superconductors, 41
X-ray diffraction patterns/data, 501ff
Tanabe-Sugano diagrams, 736
Tanaka, S., 76, 83
tantalum oxides, 20
T_c (see *correlations*)
  correlations, 729-31
  dependence on magnetic ions, 320
  depression of, 339
  maximum, 340
  mechanism, 731-3
  variation with Cu-Cu distance, 316
techniques
  precursor, 227
  standard ceramic, 227

Subject Index

TEM, 562ff
tenorite, 54, 232
ternary compounds,
  superconducting, 23ff
tetragonal-orthorhombic transformation, 324, 571, 766, 770
tetravalent Pr, 322
thallium cuprates, 129, 134ff
  bands, 495
  bilayers, 109
  comparison with bismuth cuprates, 261-2
  crystal chemistry, 493-5, 589
  crystal morphology, 261
  difficulty of synthesis, 263
  discovery, 257
  electron microscopy studies, 589-604
  stability, 727-9
  structural data, 493ff
  structural distortions, 496-500
  structural formulas, 107ff, 259-60, 589
  structural principles, 106-10, 261
  structures, 258, 262
  synthesis, 263-5, 273-80
  T_c values, 260
  tetragonal distortion, 54
thallium oxidation state, 500, 600
thallium oxide, 227
  toxicity, 227, 264-5
  volatility, 264-5
thallium-oxygen bond
  requirements, 497-8
thermal
  conductivity, 628, 656-7

parameters, 501ff
stability, 226
treatment, 227
thermodynamically stable, 454, 727
thermogravimetric analysis, 226, 326, 438, 442, 616ff
thermo-mechanical techniques, 240
thermopower, 657-8
thin film preparation, 358
tie-line, 456-7, 459, 462
tight-binding calculation, 770
titration
  iodometric, 438, 442
  methods, 609-624
top-seeded solution growth, 234-6
toxicities, 227
transition
  ferroelectric phase, 367
  insulator to metal, 352
  orthorhomic-tetragonal structural, 324
  O-T, 324, 326
reentrant superconducting
  resistive, 360
transition temperature determination, 3
transport phenomena, 627-66
triangular prism, 460
trivariant, 457
TTF.TCNQ, 28
tubes
  sealed, 228
  sealed gold, 240
tungsten, 20
tungsten bronzes, 18, 26, 38-44
tunneling spectroscopy, 362

tunneling transport, 657, 660-6
twin boundaries, 125, 154
 density of, 161
twin laws, 155
twin obliquity, 155
twin walls/boundaries, 571
 structure of, 155
 movement of, 172
twinned individuals, 155
twinning, 154, 571, 582
 by pseudo-merohedry, 155
 domains, 126
 incipient, 169
two dimensional
 analogs, 369
 character, 366

unit cells, 466

vacancy ordering, 165
valence
 Cu, 322, 324
valence electrons vs. critical temperature, 12
Van Vleck paramagnetism, 679
vapor phase transport, 229-30
vibrating sample magnetometer, 676
volatile constituents, 227, 234
vortices, 652

Wannier orbitals, 719
wavevector, 749
weak links, 360, 651, 700-5

XPS studies, 437
X-ray diffraction, 228, 247, 451, 465-83
 analytical method, 450-83
 data for cuprate superconductors, 485ff
 databases, 464-5, 470
 Hanawalt method, 474-5
 identification strategy, 471-6
 indexing powder patterns, 481
 JCPDS-PDF, 470-1
 Miller indices, 468
 physical basis of, 468
 powder, 228, 470
 powder patterns for cuprate superconductors, 485ff
 reference data, 471
 single-crystal method, 470
 standard patterns, 471-2
 use to identify 92 K superconductor, 476-83
X-ray fluorescence, 474

yttria-stabilized zirconia cell, 439

zero-field cooling, 681